CAMBRIDGE LIBRARY COLLECTION

Books of enduring scholarly value

Life Sciences

Until the nineteenth century, the various subjects now known as the life sciences were regarded either as arcane studies which had little impact on ordinary daily life, or as a genteel hobby for the leisured classes. The increasing academic rigour and systematisation brought to the study of botany, zoology and other disciplines, and their adoption in university curricula, are reflected in the books reissued in this series.

The Life, Letters and Labours of Francis Galton

A controversial figure, Sir Francis Galton (1822–1911), biostatistician, human geneticist, eugenicist, and first cousin of Charles Darwin, is famed as the father of eugenics. Believing that selective breeding was the only hope for the human race, Galton undertook many investigations of human abilities and devoted the last few years of his life to promoting eugenics. Although he intended his studies to work positively, for eradicating hereditary diseases, his research had a hugely negative impact on the world which subsequently bestowed on Galton a rather sinister reputation. Written by Galton's colleague, eugenicist and statistician Karl Pearson (1857–1936), this four-volume biography pieces together a fascinating life. First published in 1930, the first part of Volume 3 focuses on Galton's later research on correlation, personal identification, and eugenics. Pearson himself was later appointed the first Galton professor of eugenics at University College London.

Cambridge University Press has long been a pioneer in the reissuing of out-of-print titles from its own backlist, producing digital reprints of books that are still sought after by scholars and students but could not be reprinted economically using traditional technology. The Cambridge Library Collection extends this activity to a wider range of books which are still of importance to researchers and professionals, either for the source material they contain, or as landmarks in the history of their academic discipline.

Drawing from the world-renowned collections in the Cambridge University Library, and guided by the advice of experts in each subject area, Cambridge University Press is using state-of-the-art scanning machines in its own Printing House to capture the content of each book selected for inclusion. The files are processed to give a consistently clear, crisp image, and the books finished to the high quality standard for which the Press is recognised around the world. The latest print-on-demand technology ensures that the books will remain available indefinitely, and that orders for single or multiple copies can quickly be supplied.

The Cambridge Library Collection will bring back to life books of enduring scholarly value (including out-of-copyright works originally issued by other publishers) across a wide range of disciplines in the humanities and social sciences and in science and technology.

The Life, Letters and Labours of Francis Galton

VOLUME 3 – PART A:
CORRELATION, PERSONAL
IDENTIFICATION AND EUGENICS

KARL PEARSON

CAMBRIDGE
UNIVERSITY PRESS

CAMBRIDGE UNIVERSITY PRESS

Cambridge, New York, Melbourne, Madrid, Cape Town,
Singapore, São Paolo, Delhi, Tokyo, Mexico City

Published in the United States of America by Cambridge University Press, New York

www.cambridge.org
Information on this title: www.cambridge.org/9781108072427

This edition first published 1930
This digitally printed version 2011

ISBN 978-1-108-07242-7 Paperback

THE

LIFE, LETTERS AND LABOURS

OF

FRANCIS GALTON

Printing Arts Co. Ltd. sc.

Sincerely yours
Francis Galton

Francis Galton, aged 66, from the copperplate prepared for *Biometrika*, Vol. II.

THE
LIFE, LETTERS AND LABOURS
OF
FRANCIS GALTON

BY
KARL PEARSON
GALTON PROFESSOR, UNIVERSITY OF LONDON

VOLUME III A

CORRELATION, PERSONAL IDENTIFICATION
AND EUGENICS

CAMBRIDGE
AT THE UNIVERSITY PRESS
1930

Bookplate of Samuel Galton.

PRINTED IN GREAT BRITAIN

The farther aspect of the matter lies in the opinion I have formed of what Galton's influence will be upon the future. Even since his death I see what strides in public acceptance the doctrine he preached has made. The dominant race of the future, the leading nation of civilisation, will not be the one with the greatest material resources, nay, not even the one with the greatest wealth of tradition; it will be the one which can claim to have the finest breed of men and women, physically and mentally. Civilisation has gained nothing from rivalry in destructive warfare; it can gain enormously from the rivalry of nations in rearing their future generations from the most efficient of their citizens. Galton was the first to realise this great truth, to preach it as a moral code, and to lay the foundations of the new science which it demands of man. In the centuries to come, when the principles of Eugenics shall be commonplaces of social conduct and of politics, men, whatever their race, will desire to know all that is knowable about one of the greatest, perhaps the greatest scientist of the nineteenth century. I have endeavoured to put together many things of which the knowledge in another fifty years will have perished, or not improbably the documents on which that knowledge could be based will be distributed in many directions. I have to the extent of my judgment and powers given an account of Galton's scientific work and of his social ideas, so that all that is essential to an appreciation of his labour and thought will be found in these volumes without the need for continual reference to widely scattered papers, and in the future to still more widely scattered letters.

With regard to Francis Galton's letters a word must be said here. I owe a deep debt of thanks to his relatives and friends for the immense mass of correspondence which has been placed at my disposal. Galton's own letters cover a period of at least eighty-five years, and the family letters stretch over a century. During that time profound changes have taken place in the manner of thought and in the habits of the dwellers in this country, and nothing can illustrate these changes better than the letters interchanged between the members, old and young, of a large family. We learn from such a century of letters much of the social history of our own country. We pass from an age when people travelled on horseback or in coaches to an epoch of aeroplanes and motor-boats; we note that it was once an open question whether it was wiser to invest in canal or railway shares, and we trace the changes from private to joint-stock banks. We see brought forcibly before us the passage from sail to steam; and—as the chief interest—we grasp how this evolution influenced the minds of those who were spectators of it. This century of Galton family letters would in the future be of high value to the social historian of our country, and it is with grief that I think of its dispersion. In a biography like the present there is small excuse for publishing letters which do not directly bear on the characterisation of its subject, but in picking out for publication letters from the many placed at my disposal my delight in social history may have occasionally led me to err in choosing letters which depict Galton's family environment even more significantly than they illustrate his keen affection for four generations of his kinsfolk.

PREFACE

AGAIN after a long interval the third and final volume of this *Life* appears. The delay is traceable to the same difficulties as arose in the case of the second volume, namely the high cost of producing nowadays a work of this character. As it was the generous help of Mr Lewis Haslam which enabled the second volume to be printed, so I have to record my gratitude to two friends who have assisted me to obtain the funds requisite on the present occasion. In the first place Professor Henry A. Ruger of Columbia University, New York, a former postgraduate worker in the Galton Laboratory, interested Miss Dorothy Chase Rowell in Galton's writings, and in the second place Dr F. A. Freeth reported my need to Mr Henry Mond. I wish to place on record here my deep gratitude to Miss Rowell and Mr Mond, whose gifts so far supplemented the proceeds of the sales of the first two volumes that I ventured to send the third to press.

It may be said that a shorter and less elaborate work would have supplied all that was needful. I do not think so, and there are two aspects of the matter to which I should like to refer. The writer of biographies usually belongs to the literary world, and is too often a minor light of that world. I have no claim to literary distinction of any order. I have written my account because I loved my friend and had sufficient knowledge to understand his aims and the meaning of his life for the science of the future. I have had to give up much of my time during the past twenty years to labour which lay outside my proper field, and that very fact induced me from the start to say, that if I spend my heritage in writing a biography it shall be done to satisfy myself and without regard to traditional standards, to the needs of publishers or to the tastes of the reading public. I will paint my portrait of a size and colouring to please myself, and disregard at each stage circulation, sale or profit. Biography is thankless work, but at least one can get delight in writing it, if one writes exactly as one chooses and without regard to the outside world! In the process one will learn to know—as intimately as any human being can know another—a personality not one's own; that is the joy of spending years over a biography where there is a wealth of material touching the mental output, the character and even the physical appearance of the subject.

If a work is to be printed, even twenty years after a man is dead some things, some strong opinions and some names, must still be omitted. Our lives are too closely entwined with those of others not to call for some reticence even after two decades have elapsed. Still I think the reader will find in these volumes a portrait of Galton which represents without undue repression, and without uncritical adulation, the man as I knew him, and as I have learnt from his writings and letters to interpret him.

While the circumstances detailed in the preface to my second volume led to a great extension of the original plan of this work, I felt the exclusion of many of these charming family letters was not justified by the introduction of so much scientific detail, and thus I have added them as an additional chapter to this volume. To Galton's niece, Mrs Lethbridge, I owe the privilege of publishing the selection from letters which, after the death of his sister Emma in 1904, her Uncle wrote to her almost weekly. They give the most perfect characterisation of Galton in his relationship to his family.

One apology I must make if the reader feels that in the chapter on the last decade of Galton's life the biographer has introduced too much of himself. To me that last decade was essentially bound up with our joint work for a subject we both had closely at heart; and I believe that for Galton himself our common aim—the establishment of Eugenics as an accepted branch of science—was a leading, if not the principal, purpose of those years. My own enthusiasm may possibly have deceived me, but I believe Galton during that decade lived more in the struggles and difficulties of our infant Laboratory than in any other phase of his wide interests. The sympathy and help he always so readily tendered to his friends may again have misled me, but I think the history of the Laboratory he founded and finally endowed was also the essential history of his own life in those last years. At any rate such is the aspect of Galton's many-sided nature that I then saw most closely, and it is accordingly that which I am best fitted to render account of. To me his final crusade for eugenic principles was the crowning phase of a life whose labours in medicine, evolution, anthropology, psychology, heredity and statistics directly fitted him to be the teacher and prophet of the new faith.

I have to express my gratitude to various societies and editors of journals for permission to reproduce the illustrations that accompanied Francis Galton's letters and papers. In particular, to the Royal Institution for permission to use the figures illustrating Galton's lectures of 1877, to the Royal Anthropological Institute for permission to use the diagrams of Galton's memoir of 1885; and to the Editor of *Nature* for permission to use Galton's diagrams or other figures from that journal. The permission of the Royal Society to reproduce illustrations to Galton's memoirs was granted when my second volume was published. The copyright in Galton's books belongs to the University of London. The copyright in most of the letters and photographs belongs to those members of the Galton and Darwin families who provided me with them, and permission to reproduce them again must be obtained from those members, as well as from myself (if the second reproduction be made from this volume).

While I must again renew my thanks to many who have aided me in this as in the earlier volumes, I am under deep obligations to my colleagues Professor C. J. Sisson and Miss Ethel M. Elderton for assistance in the toil of proof-reading; if in a few instances I have not followed their obviously better judgment, I trust they will not despise me for being of a perverse heart. To Dr Julia Bell I owe the expenditure of too many of her free hours for several years in the preparation of the ample index to this work; while to my Wife,

Margaret V. Pearson, I am indebted for the heavy task of aiding in selecting and of afterwards transcribing the numerous letters and papers, which has very greatly lightened my own labours. I cannot conclude without a word of thanks for the care which my printers, the Cambridge University Press, have devoted to the preparation of this work and the endeavours they have always made to meet the very varied requirements of its illustration.

KARL PEARSON.

THE GALTON LABORATORY,
 UNIVERSITY OF LONDON.
  March 22, 1930.

> Yff Ony thyng Amysse be
> blame connyng, and nat me:
> I desyr þe redar to be my frynd,
> yff þer be ony amysse, þat to amend.
>> (Mary Mavdleyn, *Digby Mystery.*)

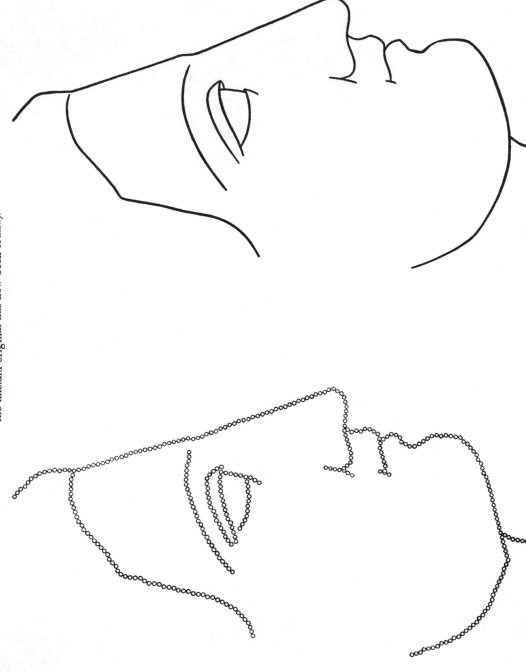

EXTRA PLATE

The Greek Girl of the "Just Perceptible Difference" Lecture of 1893 (see Vol. II, p. 309 and footnote; the mislaid original has now been found).

The reader should place this plate some ten feet from his eye, and gradually approach it, noting the distance at which the 342 small circles become distinguishable.

CONTENTS OF VOLUME III^A

ILLUSTRATIONS TO VOLUME III^A

In the Pocket at the end of this volume:

(a) Supplementary Pedigree of Distinguished Ancestors of Francis Galton and Charles Darwin

(β) Galton's Types of Finger-Print Patterns reduced from the framed enlargements, once in Galton's Anthropometric Laboratory, now in the Anthropometric Laboratory at University College, London

Galton.

Bookplate of Tertius Galton.

ERRATA TO VOLUME I

p. 53. Lines 11 and 19–20, *for* "John Hubert Barclay Galton" *read* "Hubert John Barclay Galton.

p. 150. Plate LI. The long horizontal object above the mantel-mirror is an oriental pipe not a lance.

p. 161. On p. 160 we see that Tertius Galton was proposing a visit to the English Lakes, and it would appear from Emma Galton's diary that this actually took place. It is not clear whether Tertius Galton's serious illness occurred at Keswick in the English Lakes, or at "Keswick" the home of the Gurneys near Norwich on the homeward journey. In the letter on p. 162 Galton is speaking of Keswick in the Lakes, but it is not always easy in the diaries of Emma and Francis to distinguish between visits to Lakeland and to the Gurneys' home.

p. 168. Line 9. The mysterious "Missourian" of Galton's letter to his Father is very probably Galton's misspelling for "Mesosaurian." Not only in his boyhood and his college days, but even to the last decade of his life, Galton's spellings could be erratic. In one of his letters to me he excuses his spelling by the darkness in which he is writing. It is probable therefore that he judged the spelling of words by seeing them, and he may only have *heard* this fossil lizard spoken of, and not seen the name written.

Pedigree Plate A. *Immediate Ancestry and Collaterals of Sir Francis Galton* in pocket at end of Vol. I. Last line but one, seventh column of names, *for* "F. M. Cormford" *read* "F. M. Cornford."

To

M. S. P. and M. V. P.

whose unstinted sympathy and aid
have enabled me to complete my task

CHAPTER XIV

CORRELATION AND THE APPLICATION OF STATISTICS TO THE PROBLEMS OF HEREDITY

"It is full of interest of its own. It familiarises us with the measurement of variability, and with curious laws of chance that apply to a vast diversity of social subjects. This part of the inquiry may be said to run along a road on a high level, that affords wide views in unexpected directions, and from which easy descents may be made to totally different goals to those we have now to reach. I have a great subject to write upon, but feel keenly my literary incapacity to make it easily intelligible without sacrificing accuracy and thoroughness."

Natural Inheritance, p. 3.

A. *Introductory.* Thus wrote Francis Galton in 1889 when the significance of correlation and its measurement had impressed themselves upon him. Up to 1889 men of science had thought only in terms of causation, in future they were to admit another working category, that of correlation, and thus open to quantitative analysis wide fields of medical, psychological and sociological research. Turning to the writings of Turgot and Condorcet, who felt convinced that mathematics were applicable to social phenomena*, we realise to-day how little progress in that direction was possible because they lacked the key—correlation—to the treasure chamber. Condorcet often and Laplace† occasionally failed because this idea of correlation was not in their minds. Much of Quetelet's work and of that of the earlier (and many of the modern) anthropologists is sterile for like reasons.

Galton turning over two different problems in his mind reached the conception of correlation : *A* is not the sole cause of *B*, but it contributes to the production of *B*; there may be other, many or few, causes at work, some of which we do not know and may never know. Are we then to exclude from mathematical analysis all such cases of incomplete causation? Galton's answer was: "No, we must endeavour to find a quantitative measure of this degree of partial causation." This measure of partial causation was the germ of the broad category—that of correlation, which was to replace not only in the minds of many of us the old category of causation, but deeply to influence our outlook on the universe. The conception of causation—unlimitedly profitable to the physicist—began to crumble to pieces. In no case was *B*

---

* "Un grand homme [Turgot], dont je regretterai tousjours les leçons, les exemples, & sur-tout l'amitié, étoit persuadé que les vérités des Sciences morales & politiques, sont susceptibles de la même certitude que celles qui forment le système des Sciences physiques, & même que les branches de ces Sciences qui, comme l'Astronomie, paroissent approcher de la certitude mathématique." *Discours préliminaire, Essai sur l'application de l'analyse à la Probabilité des Décisions*, p. i, Paris, 1785.

† See for example Laplace's memoir in *Mémoires de l'Académie des Sciences* for 1783, pp. 693–702, where he entirely overlooks the correlation between the size of the population and the number of births in evaluating what is really the probable error of the birth-rate.

simply and wholly caused by A, nor indeed by C, D, E and F as well! It was really possible to go on increasing the number of contributory causes, until they might involve all the factors of the universe. The physicist was clearly picking out a few of the more important causes of A, and wisely concentrating on those. But no two physical experiments would—even if our instruments of measurement, men and machines, were perfect—ever lead to absolutely the same numerical result, because we could not include all the vast range of minor contributory causes. The physicist's method of describing phenomena was seen to be only fitting when a high degree of correlation existed. In other words he was assuming for his physical needs a purely theoretical limit—that of perfect correlation. Henceforward the philosophical view of the universe was to be that of a correlated system of variates, approaching but by no means reaching perfect correlation, i.e. absolute causality, even in the group of phenomena termed physical. Biological phenomena in their numerous phases, economic and social, were seen to be only differentiated from the physical by the intensity of their correlations. The idea Galton placed before himself was to represent by a single numerical quantity the degree of relationship, or of partial causality, between the different variables of our ever-changing universe. How far he was successful forms the subject-matter of this chapter.

I have said that Galton came to this fundamental conception from two aspects. The first problem was that of inheritance. To take an illustration: A character in the Father does not determine absolutely the like character in the Son; it is only one out of many contributory factors. The character is only a partial expression of the Father's germ-plasm; so it is with the Son's character—it is not at all a full expression of his germ-plasm. Again, the Son is not a product only of his Father's germ-plasm, but of his Mother's also, and those of both parents in their turn are products of innumerable ancestral stirps leading us back through long eons of evolution. Nor is the somatic or bodily character of the Son a product only of heredity, it is the integration of a number of factors acting throughout his prenatal and postnatal growths. From the physicist's standpoint of causation there was no way at all to attack this problem, the causes were too indefinite and elusive to be individually grasped and measured. They could only be dealt with one at a time—the measure of the resemblance of offspring to parent, a partial causation, led Galton to the idea of correlation.

The second problem which impressed itself on Galton's mind was that of correlation in the narrow biological sense. The word itself appears to have originated with Cuvier who denoted by it an association between two organs or characters of a family—thus the occurrence of a split hoof with a particular form of tooth, so that from the discovery of one organ a prediction could be made as to the nature of others. It has been said that Cuvier's conception did not involve causation*. I do not know that any correlationist of to-day would assert that the knowledge of the length of the femur, which would enable him to closely predict the length of the humerus, is an assertion of

---

* See C. Herbst, *Handwörterbuch der Naturwissenschaften*, Bd. III, S. 621, Jena, 1913.

causation in a sense different from that of Cuvier; he would merely think in terms of associations with differing grades of intensity. Be this as it may, Galton's second idea of measuring the degree of relationship arose from the fact that he had recognised that two characters measured on a human being are not independent, they vary with each other. The femur of man has its characters associated with those of the humerus.

Galton did not realise immediately that his two problems admitted of the same solution. His first actual attempts at solution of the inheritance problem were based on the weight of the seeds of mother and daughter plants. In the first place he used, about 1875, some seed like that of cress (see Vol. II, p. 392), and he started by endeavouring to correlate grades or ranks. This could not be very successful because the regression curve and the "isograms" (see Vol. II, p. 391) are not linear, but extremely complicated curves. Later in 1875 (*ibid.* p. 187) we find him experimenting with Darwin's assistance on the weight and diameter of sweet-pea seeds, and here he reached his first "regression line." I reproduce (p. 4) from Galton's data in a note-book the first "regression line" which I suppose ever to have been computed. I have recalculated the constants and redrawn the line. It is for sweet-pea diameters in mother and daughter plants. The correlation coefficient is ·33, almost exactly 1/3. Two points must here be noticed. First the parental mean is considerably higher than the offspring mean. If the offspring mean denotes that of the general population, this would indicate that Galton's parental population was not a random sample of the original general population. Secondly the means of the diameters of the daughter plant peas for each size of mother plant pea, give a series of points of rather irregular distribution, which conforms as well to a sloping straight line as to any other form of curve. Here we have the origin of Galton's "regression straight line." We see that as size of mother pea increases, so does size of daughter pea, but whether in excess or defect of mean the daughter pea does not reach the deviation of the mother's diameter from the mean value, the offspring is less a giant or a dwarf than the mother pea. This is Galton's phenomenon of regression. In this case the variabilities of mother and daughter peas were approximately equal, and Galton reached the idea that the slope of the regression line would measure the intensity of resemblance between mother and daughter. If there were no slope the diameter of daughter pea would be the same for all diameters of mother pea. If it sloped at 45°, i.e. a slope of unity, the daughter pea's diameter would be exactly that of the mother pea's, supposing their means were the same; if they were not, the deviations from their respective means would still be equal.

It is strange that both Galton and Mendel should have started from peas, the former from sweet and the latter from edible peas. Galton tells us distinctly why he chose the former, namely because he would not be troubled to the same extent by variation in size of peas within the same pod. We must leave it to the future to judge whether the correlational calculus, which has sprung from Galton's peas, is or is not likely to be of equal service with

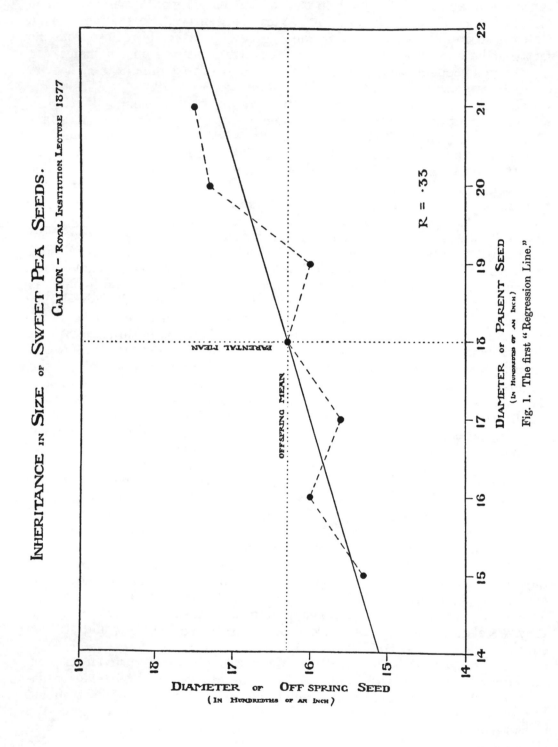

INHERITANCE IN SIZE OF SWEET PEA SEEDS.
GALTON – ROYAL INSTITUTION LECTURE 1877

R = ·33

DIAMETER OF PARENT SEED
(IN HUNDREDTHS OF AN INCH)

DIAMETER OF OFFSPRING SEED
(IN HUNDREDTHS OF AN INCH)

Fig. 1. The first "Regression Line."

the vast system of factorial genetics which has arisen from Mendel's peas—and this even in the theory of heredity. We see now what Galton might have done, he might have provided us with data to check Johansen's later bean-weight experiments, he might have thrown light on the "pure line." He might possibly have reached the correlation coefficient instead of the regression slope in his first attempt to get a measure of correlation. Whatever he might have done, he reached the idea of regression before he reached that of the coefficient of correlation. As long as he was dealing with heredity in the same sex, the approximate equality of variabilities in the two generations preserved him from any great error.

Galton was driven to his second problem by Bertillon's system for the identification of criminals. Bertillon claimed, as I remember Dr Garson did at a much later date, that the measurements chosen were practically independent. Galton needed a criterion to show whether such measurements as head length, foot length, stature, etc. were or were not associated. He saw that the problem closely resembled that of heredity, but he was troubled by the fact that the slope of his regression line depended on the units in which its two component variables were measured. It was not till more than 13 years* after his first attack on the subject that Galton realised, namely in 1889 during a walk in Naworth Park, that the two problems were identical, provided each character were measured in its own variability as unit (see our Vol. II, p. 393). With that provision the slope of the regression line becomes what we now term the coefficient of correlation. It is needful to realise this history of Galton's progress: namely that he reached regression and even the constancy of the array variabilities 12 to 14 years before he formulated his coefficient of correlation, in order to understand fully the sequence of his memoirs on this topic.

One further fact it is necessary to bear in mind in order to measure his achievements. He started like Quetelet from the normal curve as describing the deviations of a population or of any selected population, e.g. that of an array of offspring from a parent of given character. He did not start with a general definition of correlation and see whither that would lead him. His justification was that he was dealing with anthropometric characters or measurements on living forms whose deviations from type approximately followed this special law of distribution. Thus he naturally reached a straight regression line, and the constant variability for all arrays of one character for a given value of a second†. It was, perhaps, best for the progress of the correlational calculus that this simple special case should be promulgated first; it is so easily grasped by the beginner. But it has had the disadvantage that certain branches of science, as psychology for example, have rarely got further, and, without taking the trouble to apply tests, adopt linear

---

* In his *Natural Inheritance*, 1889, p. 79, Galton says his sweet-pea data were collected more than 10 years previously. His lecture at the Royal Institution, Feb. 1877, shows that he was then already in possession of sweet-pea data, and the first measurements seem to have been made in 1875.

† What we now term "homoscedasticity."

regression and homoscedasticity where it is quite inappropriate. It is interesting to note how the history of the spread of knowledge follows with halting steps the history of its discovery.

Again, if the reader anticipates that Galton was a faultless genius, who solved his problems straightaway without slip or doubtful procedure, he is bound to be disappointed. Some few creative minds may have done that, or appear to have done it, because, the building erected, they left no signs of the scaffolding; but the majority of able men stumble and grope in the twilight like their smaller brethren, only they have the persistency and insight which carries them on to the dawn.

B. *The First Idea of "Regression."* I think these conceptions will be well illustrated if we consider Galton's first paper dealing with the subject of regression, namely the lecture entitled: *Typical Laws of Heredity*, which he gave on February 9, 1877 at the Royal Institution. It is the next forward step he took after the memoir of 1875, in which he had propounded for the first time the continuity of the germ-plasm. See our Vol. II, pp. 184–8. The paper itself embraces three fundamental sections, which I will take in logical sequence if not that of the paper itself.

First: an account of the experimental data on sweet-peas. Galton assumes here that sweet-peas are invariably self-fertilised, a result which from my own observation I consider only partially true. There is also a further difficulty here: he does not take the *average* seed of the mother plant as representing the maternal character. He takes seeds of equal weight which may have been the ordinary produce of large-seeded plants, or the exceptional produce of small-seeded plants, and treats these as representing the parental character. This very fact would in itself involve regression in the offspring seeds, and leaves unsettled two important questions: (i) whether in the average result from all the seeds of a self-fertilising plant, there would be any regression at all, and (ii) whether there is any difference in the average seed weights of daughter plants grown from light and heavy seeds of the mother plant? Had Galton had these points in mind, he might have thrown light on controversies of a much later date. Again, does the size of the mother seed influence the daughter seed only by way of heredity? Galton's small seeds led to sickly and often sterile plants, and it is quite probable that this might affect the weight of their seeds (see our Vol. II, p. 181). Be this as it may, Galton found from his data* that there was a *linear* regression of daughter seed on maternal seed. He does not yet use the term "regression," but speaks of a "reverting" towards "what may be roughly and perhaps fairly described as the average ancestral type." But it is difficult to believe that this reversion was solely due to heredity; if the original seed had fully represented the maternal plant and that plant had been indefinitely self-fertilised, the Law of Ancestral Heredity would suggest no regression at

---

* He issued packets of seven sizes of seeds, each containing ten seeds, and nine friends grew the plants. Two crops failing, he had all the seed offspring of $7 \times 7 \times 10 = 490$ carefully weighed seeds.

all. It is not possible to say whether the observed "reversion" was due to the weight of a single seed not representing the true maternal character, to the hypothesis of self-fertilisation not being correct or to other causes. Theoretically the important point is that Galton reached linear regression as a first feature of his correlation table. The next point Galton reached was the homoscedasticity or equal variability of the arrays of daughter seeds corresponding to a given mother seed*. "I was certainly astonished to find the family variability of the produce of the little seeds to be equal to that of the big ones; but so it was, and I thankfully accept the fact; for if it had been otherwise, I cannot imagine, from theoretical considerations, how the typical problem could be solved" (p. 10).

The second logical stage in Galton's analysis is mathematical; he endeavours, assuming that the population is stable and is distributed normally, to find what relation must exist between the "reversion" coefficient and

* Thus far I have not been able to find Galton's data for the weights of sweet-peas in the *Galtoniana* here. It is not easy, however, to find a special topic in the mass of note-books and undated and unindexed papers. Quite possibly, however, he lent his measurements to somebody, as he lent many series of observations to myself. It would be interesting to see exactly the data from which he deduced the two fundamental principles of a normal bivariate distribution, i.e. the straight-line regression and the equivariability of the arrays. Galton gives the correlation table of filial and parental seeds in the Appendix, p. 226, of his *Natural Inheritance* for *lengths* not weights. This shows that the mean length and variability of the parent seeds were arbitrarily chosen, there being 70 of each. Further, in the table the offspring seeds are modified to show 100 in each array. We do not know therefore the true means or standard deviations of either parental or offspring populations. This does not, however, affect the determination of either means or standard deviations of arrays. I find in hundredths of an inch:

Diameter of Parent Seed	Mean Diameter of Array of Filial Seeds	Standard Deviation of the Array
21	17·26	1·988
20	17·07	1·938
19	16·37	1·896
18	16·40	2·037
17	16·13	1·654
16	16·17	1·594
15	15·98	1·763

My means do not agree with Galton's, possibly he found his before reducing his whole numbers to percentages. (It could not be by the distribution of the filial diameters "Under 15," as this would tend, I think, to reduce all his means below mine.) He does not give his array standard deviations nor the quartiles. However, on some such numbers as these Galton reached his results. The array means are not incompatible with a straight-line relation; the standard deviations suggest that the smaller parental seeds had offspring seeds of less variability than those of the larger seeds, rather than equivariability being the rule. This view might be modified if we knew the actual distribution of the filial seeds "Under 15." Many of these dwarf seeds I suspect were abortions, as their lumping up at the tail of the arrays really prevents the latter from being considered as "normal curves." Galton states (*loc. cit. supra*) that he had obtained confirmatory results for the foliage and length of pod; this indicates that his experiments must have been carried on for a second year, as he started only with the parental seed.

the variability constant of the equivariable arrays in order that the population may owing to the laws just stated repeat in the filial the parental distribution.

Now there are two points to be regarded here. Galton first states that he is going to suppose no sexual selection at work, and further he next supposes every female to be reduced to an equivalent adult male standard. It is true that he does this by the aid of percentiles, but what it really amounts to is this: If m_2 be the female mean character, σ_2 the standard deviation and Δ_2 the deviation of an individual female from type, m_1, σ_1 and Δ_1 corresponding quantities for the male, then Galton replaces the female $m_2 + \Delta_2$ by a male $m_1 + \Delta_1$, where Δ_1 has the same percentile value p for males as Δ_2 for females. This really amounts to taking $\Delta_1 = \dfrac{\sigma_1}{\sigma_2} \Delta_2$; it appears to me that this reduction of female to male value is more correct than that which he adopted later in his memoir of 1886 and in *Natural Inheritance* (see our p. 15). Having got his midparental value as the mean of the father's and mother's characters, the last reduced to male value, Galton correctly asserted that if there be no sexual selection and the original population followed a normal distribution, the midparental distribution also would be normal with a standard deviation $\dfrac{1}{\sqrt{2}} \sigma_1$. He next introduces an ingenious artifice; instead of supposing the offspring to "revert" he supposes the midparent to revert and then to have offspring whose type (i.e. mean value) is that of the original parentage. In other words, if X be the character in a midparentage, then $r'X$, where r' is the reversion coefficient, will be the same midparentage after reversion. This really signifies a uniform "squeeze" in the ratio of r' to 1 of the normal curve of midparentages, or the new curve of reverted midparentages will be a normal curve of standard deviation $\dfrac{1}{\sqrt{2}} \sigma_1 \times r'$. We have lastly to distribute the offspring of these midparentages about their mean values with a constant variability, which we will represent by Σ; thus the standard deviation σ' of the distribution of offspring will be given by

$$\sigma'^2 = \tfrac{1}{2} \sigma_1^2 r'^2 + \Sigma^2.$$

But, if this standard deviation of the final normal curve is to repeat the original population, σ' must equal σ_1, or we have

$$\Sigma^2 = \sigma_1^2 \left(1 - \tfrac{1}{2} r'^2\right).$$

Here r' is the "reversion" of the midparent and is equal to $\sqrt{2}\,r$, if r be the reversion on a single parent*. In other words, if r be the reversion of offspring on parent then the constant standard deviation of the array of offspring for a given parent must be $\sigma_1 \sqrt{1 - r^2}$, if the population starts with a normal distribution and when reproduced is to have the same normal

---

* If the standard deviation of the "reverted" single parent be $r\sigma_1$, then $\sqrt{2}\,r\sigma_1$ will be the standard deviation of the reverted midparent, but if this be taken as $r'\sigma_1$ clearly $r' = \sqrt{2}\,r$.

distribution. This is the earliest appearance of the symbol r as a coefficient of "reversion"; the reasoning by which the result is obtained is only true, if parental and offspring generations have the same variability; in that case r is what we now term the coefficient of correlation, and Galton here deduces the relationship between the constant array variability and this coefficient.

In the course of his work he introduces the ideas of natural selection and of differential fertility. This section of the discussion is somewhat difficult to follow. Galton further supposes selection to take place symmetrically round the population mean or type. Finally to obtain the above result Galton supposes the selection and the fertility to be non-differential, or gives them mere percentage values for all parents alike*.

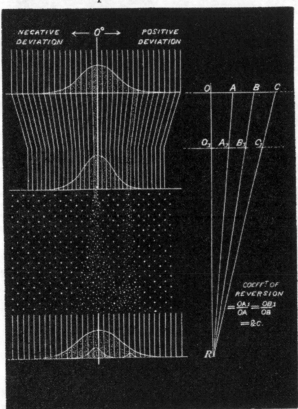

Fig. 2. Galton's Quincunx illustrating the nature of Regression.

The third point in this paper of Galton's is the ingenious "Quincunx" by which he illustrates the phenomenon of reversion and the continual maintenance by aid of inheritance of a stable population. Galton at first indicates how closely certain measured characters are given by a normal distribution and how such a normal distribution may be produced by a stream of pellets

* A paper in which this matter is more fully dealt with by the present writer will be found in *Biometrika*, Vol. VII, pp. 258–275, "On the Effect of a Differential Fertility on Degeneracy: A New Year's Greeting to Francis Galton, 1910."

falling vertically through a forest of horizontal pins. He next, starting with a normal distribution of variability σ_1, reduces the variability to $r\sigma_1$ by sloping his discharge tubes towards the type (see Fig. 2). This restriction of the tubes has the same effect as giving a uniform horizontal "squeeze" to his original distribution; he thus reaches his population of "reverted parents." If he now opens any single one of his tubes he will get a normal distribution, about the reverted parent character as type, which will have

Fig. 3. Galton's Quincunx illustrating the effect of Natural Selection.

the proper variability $\sigma_1\sqrt{1-r^2}$ if a suitable choice be made of the extent of "pin-forest" through which the pellets fall. Since this variability is the same for all parentages, the extent is constant, and if all the tubes be opened, all the "reverted" parentages contribute their share to building up again the population from which we started.

Those who hold the hypothesis of the pure line to be true, apparently overlook the fact that while the gametic distribution might be stable, they must appeal to a stringent natural selection, or a differential fertility, to maintain stability for two successive generations in somatic characters. This stability Galton achieved by aid of reversion.

In dealing with the problem of Natural Selection, Galton takes only the case of selection round type and assumes that those selected to live, not those selected to die, will follow a normal distribution. This limits to some extent its general applicability, but he illustrates his idea by a second ingenious Quincunx (see Fig. 3), in which the middle stage is formed by a vertical normal-curve diaphragm which cuts off from the descending pellets, uniformly distributed over the horizontal bases of their compartments in the top stage, the "selected pellets," which again are on the removal of the sliding floor allowed to run down into the third stage compartments where they form a normal distribution of much reduced variability.

Speaking of the principles of "reversion" and reduced variability in the offspring of a given parentage, Galton says:

> "The typical laws are those which most nearly express what takes place in nature generally; they may never be exactly correct in any one case, but at the same time they will always be approximately true and always serviceable for explanation. We estimate through their means the effects of the laws of sexual selection, of productiveness and of survival, in aiding that of reversion in bridling the dispersive effect of family variability. They show us that natural selection does not act by carving out each new generation according to a definite pattern on a Procrustean bed, irrespective of waste. They also explain how small a contribution is made to future generations by those who deviate widely from the mean, either in excess or deficiency, and they enable us to discover the precise sources whence the deficiencies in the produce of exceptional types are supplied, and their relative contributions. We see by them that the ordinary genealogical course of a race consists in a constant outgrowth from its centre, a constant dying away at its margins, and a tendency of the scanty remnants of all exceptional stock to revert to that mediocrity, whence the majority of their ancestors originally sprang." (*loc. cit.* p. 17.)

Thus Galton stated his law of reversion originally; we see that it really covers the most marked features of bivariate normal correlation, we have even the now-familiar symbol r. Whether, however, he was at that time justified in asserting reversion as a typical law of heredity on the basis of his sweet-pea results may be open to question. Is the weight or diameter of a single seed a fair representation of a parental somatic character? Was Galton justified in considering the variability of his offspring constant? These are points which have much bearing on later work and on what correlation the r really signified in the case of Galton's actual experimental data.

C. *Heredity in Stature of Man. Development of the Conception of Regression.* That Galton had some doubts himself is, I think, clear from the fact that for eight years he published nothing further on the subject of regression, but started by aid of his family records to collect data bearing on inheritance in man: see Vol. II, pp. 363 *et seq.* As soon as he had obtained enough data to deal with the inheritance of stature in man he returned to the subject, and in 1885 and 1886 published a number of papers dealing with the topic. The first of these is his Presidential Address to the Section of Anthropology of the British Association, Aberdeen Meeting, 1885*. He next published a more detailed paper in the *Miscellanea* of the *Journal of the*

* *B. A. Transactions*, 1885, pp. 1206–1214; *Nature*, Vol. XXXII, pp. 507–510.

Anthropological Institute*. He further took the subject as the topic of his Presidential Address at the Anniversary Meeting of that Institute in January, 1886†, having meanwhile again discussed it in a lecture at the Birmingham and Midland Institute entitled: "Chance and its Bearing on Heredity"‡. Finally we have the mathematical basis of Galton's work more fully provided in a paper on "Family Likeness in Stature" with an Appendix by J. D. Hamilton Dickson, presented to the Royal Society on January 1, 1886§. None of these papers is exclusive, each has something not in the others, but probably those in the *Miscellanea* of the *Journal of the Anthropological Institute* and in the *R. S. Proceedings* are the more important for those who have not time to read them all. We have throughout to remember that Galton was a pioneer, and could not see matters in the clearer light of to-day when we start from a knowledge of bivariate distribution with its two means, two variabilities and its coefficient of correlation; he did not yet clearly recognise the distinction between a coefficient of regression and a coefficient of correlation. It is difficult for the reader now-a-days to appreciate the paradox which Galton reached from his data and finds it needful to discuss at some length, namely: that the coefficient of regression for the offspring on a midparent is double what it is for the midparent on the offspring‖. A further difficulty is that Galton invariably thought in terms of grades, quartiles and the "ogive curve," and this I venture to think is by no means helpful for elucidating correlation, as the reader of the first ten pages of the Royal Society paper will find. It has always been a puzzle to me why Galton called in Mr Dickson and placed before him a somewhat artificial problem in probability the answer to which comes directly¶ from Galton's own two statements.

* Vol. xv, pp. 246–263.   † Vol. xv, pp. 489–499.

‡ Reported in the *Birmingham Daily Post*, December 7, 1886.

§ *Roy. Soc. Proc.* Vol. xl, pp. 42–73, 1886.

‖ Since the midparental standard deviation is, when the female is reduced to male equivalent, $\sigma_1/\sqrt{2}$ in our previous notation, the two regression coefficients are respectively: $\dfrac{\sigma_1/\sqrt{2}}{\sigma_1}\,r$ and $\dfrac{\sigma_1}{\sigma_1/\sqrt{2}}\,r$, that is, $r/\sqrt{2}$ and $\sqrt{2}\,r$, or one twice the other. I think Galton was slightly puzzled here, because he had not yet fully realised that the two variabilities not being the same, he must measure each variate in its own unit of variability in order to make both regressions the same.

¶ Galton had discovered that the offspring of parents of character deviation x vary about $(r\sigma_2/\sigma_1)x$ with a standard deviation $\sigma_2\sqrt{(1-r^2)}$. Hence if y be the deviation of the n offspring of the n' parents of deviation x, and we assume, as Galton, that parental and offspring generations both follow the normal law, the number of offspring of deviation y will be

$$\frac{nn'}{\sqrt{2\pi}\,\sigma_2\sqrt{1-r^2}}\,e^{-\frac{1}{2\sigma_2{}^2(1-r^2)}\left(y-\frac{r\sigma_2}{\sigma_1}x\right)^2}.$$

But $n'=\dfrac{N}{\sqrt{2\pi}}\dfrac{1}{\sigma_1}e^{-\frac{1}{2}\frac{x^2}{\sigma_1{}^2}}$, where N is the total population of parents, thus substituting for n' we have

$$z=\frac{nN}{2\pi\sigma_1\sigma_2\sqrt{1-r^2}}\,e^{-\frac{1}{2(1-r^2)}\left(\frac{x^2}{\sigma_1{}^2}-\frac{2rxy}{\sigma_1\sigma_2}+\frac{y^2}{\sigma_2{}^2}\right)}$$

as the frequency distribution of offspring and parents, the well-known result, which was not even written down by Mr Dickson!

The most noteworthy point, however, is this, that Galton having the correlation table before him of the statures of 928 offspring and of their midparents proceeded after smoothing the frequencies to determine the contour lines and found them to be:

(i) a system of concentric and similar ellipses about the common mean of the filial and midparental statures.

Further:

(ii) the regression straight lines were conjugate diameters to the two axes of stature.

He also determined from his contours the ratio of the axes of this ellipse system, and the inclination of the major axis to the horizontal. The ellipse, which served as type, is given in the accompanying diagram (see Fig. 4, p. 14), and the observed values on this ellipse and the values computed from Mr Dickson's Formulae are*:

	Galton from Contours	From Dickson's Formulae
Regression Slope	1 in 3	6 in 17·5
Major to Minor Axis	10 to 5·1	$\sqrt{7}$ to $\sqrt{2}$ or 10 to 5·35
Inclination of Major Axis	25°	26° 36′

It is needless to say that Galton was delighted with this accordance. He wrote† as follows with regard to it:

"I may be permitted to say that I never felt such a glow of loyalty and respect towards the sovereignty and magnificent sway of mathematical analysis as when his [Mr Dickson's] answer reached me confirming, by purely mathematical reasoning, my various and laborious statistical conclusions with far more minuteness than I had dared to hope, for the original data ran somewhat roughly, and I had to smooth them with tender caution‡."

We ought on no account to overlook the fact that the theory of linear regression and the associated homoscedasticity were evolved by Galton from his sweet-pea experiments, confirmed by his stature measurements, and resulted practically in the form of the normal surface for two variates with its elliptic contours, before the mathematical theory of correlated errors was known to him. It is one of the most striking lessons in what may be achieved by a patient analysis of even crude observations. Yet without being discouraged in our own attempts at similar discoveries, we do well to remember that only an exceptional mind has the insight to discriminate between the essential and the non-essential in a mass of statistical data, and to select those two principles which illuminate the manner in which a population reproduces itself stably by aid of heredity—and what is more in so doing to pave the way to the solution of many other problems of a wholly different character. Fig. 5, p. 16, shows the regression line of offspring on midparent for the case of stature; it is, I think, the second regression line ever drawn, and Galton indicates by the line at 45° exactly how much the offspring fall behind the stature of their individual midparent. He added to this regression diagram, a picture of his "Forecaster of Stature"—which might equally well be used to

* *Journal of the Anthropological Institute,* Vol. xv, p. 263.
† *Ibid.* p. 255.   ‡ *Ibid.* p. 255.

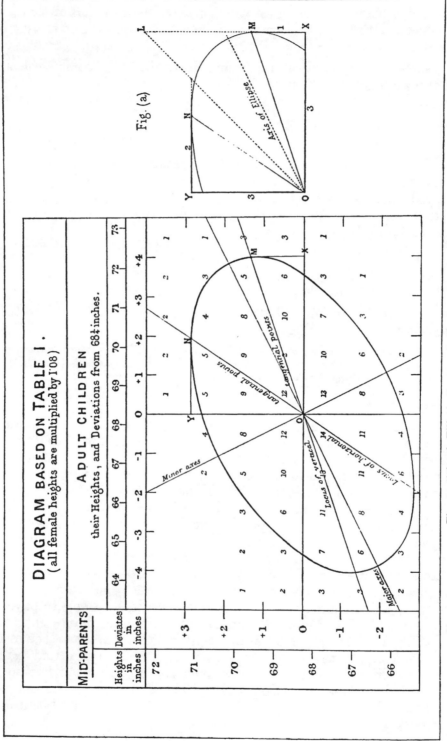

Fig. 4. Galton's Elliptic Contour drawn from his observations.

predict the probable value of any third variate from a knowledge of two others*. The working of the Forecaster is almost obvious on examination of the diagram, but for the benefit of those who come for the first time to the subject of regression I give Galton's own words:

"The weights M and F have to be set opposite to the heights of the mother and father on their respective scales; then the weight sd will show the most probable heights of a son and daughter on the corresponding scales. In every one of these cases it is the fiducial mark in the middle of each weight by which the reading is to be made. But, in addition to this, the length of the weight sd is so arranged that it is an equal chance (an even bet) that the height of each son or each daughter will lie within the range defined by the upper and lower edges of the weight on their respective scales. The length of sd is 3 inches = $2f$†; that is, 2 × 1·50 inch.

"A, B and C are three thin wheels with grooves round their edges. They are screwed together so as to form a single piece that turns easily on its axis. The weights M and F are attached to either end of a thread that passes over the movable pulley D. The pulley itself hangs from a thread which is wrapped two or three times round the grove of B and is then secured to the wheel. The weight sd hangs from a thread that is wrapped in the same direction two or three times round the groove of A, and is then secured to the wheel. The diameter of A is to that of B as 2 to 3. Lastly, a thread wrapped in the opposite direction round the wheel C, which may have any convenient diameter, is attached to a counterpoise.

"It is obvious that raising M will cause F to fall, and *vice versâ*, without affecting the wheels A, B, and therefore without affecting sd; that is to say, the parental differences may be varied indefinitely without affecting the stature of the children, so long as the mid-parental height is unchanged. But if the mid-parental height is changed, then that of sd will be changed to $\frac{2}{3}$ of the amount.

"The scale of female heights differs from that of the males, each female height being laid down in the position which would be occupied by its male equivalent. Thus 56 is written in the position of 60·48 inches, which is equal to 56 × 1·08. Similarly, 60 is written in the position of 64·80, which is equal to 60 × 1·08‡."

The last words indicate what is, I think, an important point: Galton obtains the female from the male stature by multiplying by the constant factor 1·08. This he obtained as the ratio of the male to the female mean value, and he practically assumes this ratio to be the same for all other statures.

In a certain sense I think this is, at least theoretically, a retrograde step from his suggestion of 1877. He then took the transmuted female mean to be the male mean plus the female deviation increased in the ratio of male to female variability. This appears to be theoretically a better process of transmutation. Practically the two methods will only agree, if the ratio of the two variabilities is equal to the ratio of the two means, i.e. if the so-called coefficients of variability of the two sexes are equal. This is approximately but not absolutely true for a number of human characters.

There are of course several other conditions which must be fulfilled to make Galton's definition of midparent valid, and some of these he discusses. In the first place the parents must mate at random with regard to the character dealt with, i.e. there must be no sexual selection in the form of assortative mating with regard to stature, tall must not tend to marry tall,

---

* It would only be needful to adopt scales in accordance with the constants of the bivariate regression formula.

† In this paper Galton uses the symbol f for the quartile deviate.

‡ *Journ. Anthrop. Institute*, Vol. xv, p. 262.

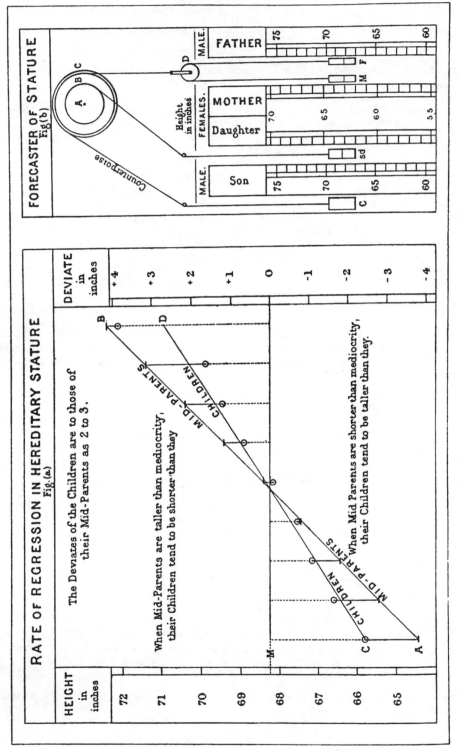

Fig. 5. Galton's Second Regression Line and his "Forecaster of Stature."

nor short, short. Galton discusses* the absence of assortative mating for stature and forms the following table, where the medium group embraces individuals of 67″ and up to 70″ stature for males or transmuted females:

Husband

		Short	Medium	Tall	Totals
Wife	Short ...	9	28	14	51
	Medium	25	51	28	104
	Tall ...	12	20	18	50
		46	99	60	205

He notes that there are 27 like marriages short with short and tall with tall, and 26 contrasted marriages† short with tall, and argues that there is no assortative mating in stature. In a fuller treatment of the same data by the present writer the coefficient of resemblance between husband and wife was found to be ·093 ± ·047‡, which might just be significant. Later work has shown that there is sensible assortative mating not only in stature (·280), but in span (199) and cubit (·198)§; in other words big men do tend to marry big women and small men small women. Galton's data show, however, so little assortative mating that his results were not sensibly influenced by disregarding it.

Galton now turns to another point, namely: Does the difference in stature of parents influence the stature of the offspring? He was clearly conscious that this was an important point, for on it depends whether his value for the midparental stature is or is not to be considered correct. As we should now express it, he was really asking whether the stature in the offspring was equally correlated with the statures of the two parents, or rather, that is the question he would have been asking had he transmuted his female deviations to male deviations by aid of the ratio of the two variabilities and not of the two means‖. If the two correlations be not equal, then Galton's "Forecaster," based on his conception of midparent, would give incorrect results. Galton indicates in a table (*Journ. Anthrop. Instit.* Vol. xv, p. 250) that the differential influence of the parents should not be very great, but he does not really

* *Journ. Anthrop. Instit.* Vol. xv, p. 251.
† Printed in *loc. cit.* 32 instead of 26.
‡ *Phil. Trans.* Vol. 187 A, p. 270, 1896.
§ *Biometrika*, Vol. ii, p. 373.
‖ If r_{13} be the paternal, r_{23} the maternal coefficient of correlation and r_{12} that of assortative mating, the bivariate formula shows us that to give equal weight to father and mother we must have equality of the two expressions

$$\frac{r_{13} - r_{12}r_{23}}{1 - r_{12}^2} \quad \text{and} \quad \frac{r_{23} - r_{12}r_{13}}{1 - r_{12}^2}$$

(*Roy. Soc. Proc.* Vol. viii, p. 240, 1895), and this involves $r_{13} = r_{23}$, i.e. the equality of the parental influences.

determine it quantitatively. Actually for his data we have the following correlations[*]:

	Father	Mother
Son	·396 ± ·024	·302 ± ·027
Daughter	·360 ± ·026	·284 ± ·028

There was thus really quite a well-marked prepotency of the father in the case of stature. Later results on ampler and better material have failed to confirm this prepotency[†]; I think it may well have been due to amateur measuring of stature in women, when high heels and superincumbent chignons were in vogue; it will be noted that the intensity of heredity decreases as more female measurements are introduced. Daughters would be more ready to take off their boots and lower their hair knots, than grave Victorian matrons. As we have not since succeeded in demonstrating any sex prepotency in parentage, Galton's assumption that such did not exist justifies his theory. But this assumption was not justified by his actual data and affects seriously the values of the constants he reached, which are all too low in the light of more recent research. I think we should be inclined to say now that the regression of the offspring deviate[‡] is on the average nearer to $\frac{4}{5}$ than to Galton's $\frac{2}{3}$ of the midparental deviate. Galton, however, recognised very fully that his numerical values were only first approximations. He writes:

"With respect to my numerical estimates, I wish emphatically to say that I offer them only as being serviceably approximate, though they are mutually consistent, and with the desire that they may be reinvestigated by the help of more abundant and much more accurate measurements than those I have at command. There are many simple and interesting relations to which I am still unable to assign numerical values for lack of adequate material, such as that to which I referred some time back, of the relative influence of the father and the mother on the stature of their sons and daughters.

"I do not now pursue the numerous branches that spring from the data I have given, as from a root. I do not speak of the continued domination of one type over others, nor of the persistency of unimportant characteristics, nor of the inheritance of disease, which is complicated in many cases by the requisite concurrence of two separate heritages, the one of a susceptible constitution, the other of the genus of the disease. Still less do I enter upon the subject of fraternal deviation and collateral descent[§]."

Galton's reasons for making a special study of stature are dealt with at considerable length and summarised as follows:

"The advantages of stature as a subject in which the simple laws of heredity may be studied will now be understood. It is a nearly constant value that is frequently measured and recorded, and its discussion is little entangled with consideration of nurture, of the survival of the fittest, or of marriage selection. We have only to consider the midparentage and not to

---

[*] *Phil. Trans.* Vol. 187 A, p. 270, 1896.

[†] See *Biometrika*, Vol. II, p. 378, 1902.

[‡] Galton in this paper introduces the term "deviate": "I shall call any particular deviation a 'deviate,'" *Journ. Anthrop. Instit.* Vol. xv, p. 252. The term was perhaps unnecessary considering the existence of "deviation," but it has come into general use, and is perhaps more justifiable in Galton's sense than "variate," which is now so often used, not for a particular variation, but for the "variable" itself.

[§] *Journ. Anthrop. Instit.* Vol. xv, p. 258.

trouble ourselves about the parents separately. The statistical variations of stature are extremely regular, so much so that their general conformity with the results of calculations based on the abstract law of frequency of error is an accepted fact by anthropologists. I have made much use of the properties of that law in cross-testing my various conclusions and always with success*."

Galton considers the fact that stature is not a simple element, but a compound of the accumulated lengths or thicknesses of more than a hundred parts, to be a distinct advantage and a source of the beautiful regularity of its frequency distributions†. He does not see that this may tend to screen some fundamental law which may be obeyed by the simple components. Thus we note that as a rule the parental correlations decrease as we take characters based on fewer elements, e.g. the parental correlations for span are less than those for stature, and those for forearm are less than those for span. There might be—I on my part do not assert that there is—an alternate inheritance in the simple components, which is screened in the complex compound‡. To this Galton might well have replied: Why should a single bone be looked upon as an ultimate element, if it develops from a number of centres of ossification, and pushing the matter further may we not be driven to find the simple component ultimately in a cell? The "simple components," which obey some equally simple law of inheritance, are still to find in the bony skeleton of man.

Two further terms defined by Galton may here be considered.

He recognises that the individuals in what we now term an array (a column, or row) of the correlation table are not in themselves blood kindred, they are not, for example, all sons of the same parents, or all brothers of the same individual. Their link is that they are all sons of a set of parents having the same small range of any character, or again all brothers who have a brother within the same small range of character. Thus these individuals probably differ in both ancestry and nurture. Galton proposes to call them "co-kinsmen" or more definitely according to the array type "co-filials" or "co-fraternals." By such terms he only means that their correlated variable (e.g. stature in parent, brother or collateral) has the same value, or limited range of values. Galton was thus fully aware that the variability within a family group of brethren, a fraternity, was not the same as the variability within such an array or co-fraternity, or co-kinship. Galton's terms have not come into general use, it is, perhaps, awkward to call individuals co-kinsmen who are not kinsmen at all. But the failure to distinguish between a fraternity in the true sense, and a co-fraternity in Galton's sense, has not been unfruitful of error§. It is, perhaps, best to stick to the words "filial array" or "fraternal

* *Journ. Anthrop. Instit.* Vol. xv, p. 251.   † *Ibid.* p. 249.

‡ Those who assert that stature or cephalic index "mendelises," have not explained how the bones on the dimensions of which they are formed themselves react to inheritance. If these simpler elements "mendelise," how comes it that the compounds do, and what becomes of the correlations between these components?

§ If r be the correlation coefficient of offspring on midparent and R be the multiple correlation coefficient of offspring on the whole of its ancestry, then, σ being the standard deviation of offspring, $\sigma \sqrt{1 - r^2}$ is the variability of a co-fraternity and $\sigma \sqrt{1 - R^2}$ the variability of a fraternity, or group of blood brothers.

array," the word array suggesting that we are dealing with a wider group than a single family.

The next idea raised by Galton is very important for later researches. He goes to the root of his law of regression when he states that the somatic character of the parents does not fully define the somatic character of the offspring. The somatic character of the parents is not the full representative measure of the germ plasm of the stirp. This is represented by a long series of ancestors, who become so numerous as we go backward, that their mean value for a generation cannot differ from mediocrity. Regression in Galton's view is the result of the influence of parental heredity pulling the offspring so to speak towards the parental value and the mediocrity of the more distant ancestry pulling towards its own value of the character.

Now we may or we may not know something of the ancestry behind the first midparent. If we know nothing absolutely then the fact that the first midparent has a certain character value enables us to predict a certain probable value for the next midparent and so on. If we did know completely the ancestry, we might replace the whole ancestry by a single midancestor. To this midancestor, we may give the name "generant." Again we had better cite Galton's own words, because although the idea is suggestive he does not define it in a manner which enables us to determine mathematically its nature. From what we said above it is clear that we may have a true generant and a probable generant based on only a partial knowledge of the ancestry*.

Galton's Conception of the Generant.

"The explanation of it [Regression] is as follows : The child inherits partly from his parents, partly from his ancestry. Speaking generally the further his genealogy goes back, the more numerous and varied will his ancestry become, until they cease to differ from any equally numerous sample taken at haphazard from the race at large. Their mean stature will then be the same as that of the race ; in other words, it will be mediocre. Or, to put the same fact into another form, the most probable value of the midancestral deviates in any remote generation is zero.

"For the moment let us confine our attention to the remote ancestry and the midparentages, and ignore the intermediate generations. The combination of the zero of the ancestry with the deviate of the midparentage is the combination of nothing with something, and the result resembles that of pouring a uniform proportion of pure water into a vessel of wine. It dilutes the wine to a constant fraction of its original alcoholic strength, whatever the strength may have been.

"The intermediate generations will each in their degree do the same. The middeviate in any one of them will have a value intermediate between that of the midparentage and the zero value of the ancestry †. Its combination with the midparental deviate will be as if, not pure water, but a mixture of wine and water in some definite proportion, had been poured into the wine. The process throughout is one of proportionate dilutions, and therefore the joint effect of all of them is to weaken the original wine in a constant ratio.

"We have no word to express the form of that ideal and composite progenitor, whom the offspring of similar midparentages most nearly resemble, and from whose stature their own respective heights diverge evenly above and below. If he, she or it, is styled the "generant" of the group, then the law of regression makes it clear that parents are not identical with the generants of their own offspring."

* *Journ. Anthrop. Instit.* Vol. xv, pp. 252–3.

† This sentence is not, I think, correct as it stands. A man might easily have four grandparents all taller than his parents. Galton probably meant to insert the words "on the average."

If U for any character be the deviate of the generant from the mean of the race, then the individual endowed with such U's for all characters would represent the stirp of any family. Unfortunately Galton does not give us any method for determining the U of the generant. I think, however, if we take the character U of the generant to be that linear function of the characters of all the ancestry which gives the highest correlation R with the character in the offspring, it throws light on Galton's idea. In this case U is simply proportional to the multiple regression expression. If we make the following hypotheses, which have considerable experimental evidence in their favour, namely :

(*a*) that the individual correlations of offspring with male and with female ancestors are equal,

(*b*) that such correlations with individual ancestors die out in a geometrical ratio, i.e. the correlations of the offspring with individual parents (father or mother), with individual grandparent (male or female), with individual great-grandparent, etc. form a series r_1, r_1a, r_1a^2, etc., where a is less than unity, then it can be demonstrated that the deviate U will be given by the formula*

$$U = \gamma\,(h_1 + \beta h_2 + \beta^2 h_3 + \ldots),$$

where h_1, h_2, h_3, etc. are the deviates of the midparental characters in the successive grades of ancestry and γ, β are constants, which can be found in terms of r_1 and a. Further, the fraternity of which U defines the stirp will vary round U with variability $\sigma\sqrt{1-R^2}$, where R (the "coefficient of multiple correlation") is known in terms of r_1 and a, or of γ and β.

The expression for U, or the deviate of the generant which defines the stirp, has been termed the *Law of Ancestral Inheritance*†. It is not a biological hypothesis, but the mathematical expression of statistical variates, which obey, as many measurable characters in man, certain forms of frequency distribution, these being maintained in successive generations. It can be applied with special values of γ and β to many biological hypotheses. We are, however, not concerned to discuss these matters here, but merely to point out that in the papers we are now dealing with Galton was feeling his way upwards towards this Law of Ancestral Inheritance, though I venture to think by a faulty stairway. The somewhat complicated mathematics of multiple correlation with its repeated appeals to the geometrical notions of hyperspace remained a closed chamber to him, necessary as multiple correlation now is for many practical problems of modern statistics. As I have said there is a true generant, i.e. one in which we insert the true values of the different ancestral midparental deviates, namely h_1, h_2, h_3, ... as above, and a probable generant for which we only know h_1 and put in probable values

---

* *Biometrika*, Vol. VIII, pp. 239–243.

† *Roy. Soc. Proc.* Vol. LXII, p. 386. For the fuller mathematical treatment see *Biometrika*, Vol. VIII, pp. 239–240 and Vol. XVII, pp. 129 *et seq*.

based on h_1 for h_2, h_3, ..., etc. Galton deals only with the latter. He writes as follows[*]:

"When we say that the midparent contributes two-thirds of his peculiarity of height to the offspring, it is supposed that nothing is known about the previous ancestor. But though nothing is known, something is implied, and this must be eliminated before we can learn what the parental bequest, pure and simple, may amount to. Let the deviate of the midparent be x (including the sign), then the implied deviate of the midgrandparent will be $\frac{1}{3}x$, of the mid-ancestor in the next generation $\frac{1}{9}x$ and so on. Hence the sum of the deviates of all the midgenerations that constitute the heritage of the offspring is $x(1 + \frac{1}{3} + \frac{1}{9} + \text{etc.}) = x\frac{3}{2}$."

Now I think this result erroneous because it assumes that the quantities γ, γa, γa^2, ... of the generant above can be found from the simple regression formula of parent on offspring. This we know to be very far from the fact, the multiple regression coefficients have no such simple relations to parental regression. The fallacy lies, I think, in this: we could imply that value of the grandparental from the parental deviate by means of the simple regression formula, but to do this is to assert that all the remaining h's, h_3, h_4, etc., are put zero, i.e. are given every conceivable value, with the mean value zero. But we are going to imply other than zero values for these h's, hence our system of implied ancestral values is not consistent and this, I think, is indicated by the total heritage coming out $x\frac{3}{2}$. To get over this difficulty Galton proceeds "to tax" each contribution to the heritage. He takes as two extreme cases (a) a uniform taxation of all ancestral contributions of $\dfrac{1}{n}$,

and (b) a taxation geometrical in amount, supposing $\dfrac{1}{m}$ of the total only to be transmitted from one generation to a second. He thus reaches the following expressions:

$$x\left(\frac{1}{n} + \frac{1}{3n} + \frac{1}{9n} + \text{etc.}\dots\right) = \frac{1}{n}\,x\,\frac{3}{2},$$

$$x\left(\frac{1}{m} + \frac{1}{3m^2} + \frac{1}{9m^3} + \text{etc.}\dots\right) = x\,\frac{3}{3m-1}.$$

But x being the midparental character the heritage of the offspring is, Galton says, $\frac{2}{3}x$, thus $\dfrac{1}{n} = \dfrac{4}{9}$, and $\dfrac{1}{m} = \dfrac{6}{11}$. From this he draws the conclusion that both may be taken to be $\frac{1}{2}$ approximately. But here the reasoning seems at fault, for the offspring heritage of $\frac{2}{3}x$ is based on all the other midparental deviates h_2, h_3, ... taking their average or zero values. The regression coefficient would not be two-thirds, if they took the values $\frac{1}{3}h_1$, $\frac{1}{9}h_1$, etc.

Finally from what Galton has just said it would appear that we might have two series for determining ancestral contributions, the one in n, i.e. $\frac{1}{2}$, $\frac{1}{2}$, $\frac{1}{2}$, ..., or the one in m, i.e. $\frac{1}{2}$, $\frac{1}{4}$, $\frac{1}{8}$, ... But this is clearly not what he

---

[*] *Roy. Soc. Proc.* Vol. LXII, p. 61, and compare *Journ. Anthrop. Instit.* Vol. XV, pp. 260 *et seq.*

understands, for having determined the midparental contribution to be $\frac{1}{2}$ from either series, he now writes* of the values of $\dfrac{1}{n}$ and $\dfrac{1}{m}$:

"These values differ but slightly from $\frac{1}{2}$, and their mean is closely $\frac{1}{2}$, so that we may fairly accept that result. Hence the influence, pure and simple, of the midparent may be taken as $\frac{1}{2}$, of the midgrandparent $\frac{1}{4}$, of the midgreatgrandparent $\frac{1}{8}$ and so on. That of the individual parent would be $\frac{1}{4}$, of the individual grandparent $\frac{1}{16}$, of an individual in the next generation $\frac{1}{64}$ and so on."

Thus Galton reaches his *Separate Contribution of each Ancestor to the Heritage of the Child*, a principle which is often spoken of as his Law of Ancestral Heredity. In reaching it he apparently drops his $\dfrac{1}{n}$ series altogether and follows his $\dfrac{1}{m}$ series with its geometrical system of taxation. This is distinctly more in keeping with the expression for the generant deviate U above, which runs in a geometrical series. If we assume all the ancestors to have the same deviation h, we have $U = \dfrac{\gamma}{1-\beta} h$, and, if the offspring value might in such a uniform breed be also taken as h, it follows that $\gamma = 1 - \beta$. Hence if we take the first midparents' contribution to be $\frac{1}{2}$, i.e. $\gamma = \frac{1}{2}$, with Galton, it follows that $\beta = \frac{1}{2}$, and our series is Galton's geometrical series with his radix value, a half. But I venture to think it was inspiration rather than correct reasoning which led him to a geometrical series for U.

On the other hand his multiple regression coefficients $\frac{1}{2}, \frac{1}{4}, \frac{1}{8}, \dots$ suffice to determine what the correlations between an individual ancestor in any generation and the offspring *ought* to be. They take the values for parents $\cdot3$, for grandparents $\frac{1}{2} \times \cdot3$, for great-grandparents $\dfrac{1}{2^2} \times \cdot3$ and so on. Galton found his midparental regression $\frac{2}{3}$ and took his parental to be $\frac{1}{3}$†. This is not so far from $\cdot3$, that we could say it confutes Galton's Ancestral Law. But we find Galton taking the grandparental regression and therefore the correlation $\frac{1}{9}$, the great-grandparental $\frac{1}{27}$ and so on. These values form a series $a, a^2,$ $a^3, \dots$ for the individual ancestral correlations and lead to $\gamma = 1, \beta = 0$, or to the generant U being solely determined by the parents, the higher ancestry contributing nothing to the generant‡. Hence it follows that Galton's Ancestral Law is not in keeping with the values he has taken for his individual ancestral correlations. The reasoning by which he has reached one or the other is defective. As I have said Galton's guess at a geometrical relation for the coefficients of U was an inspiration, but his idea that a grandson is the son of a son and so his regression (and with a stable population his correlation) must be $\frac{1}{3} \times \frac{1}{3} = \frac{1}{9}$ is fallacious. Regression coefficients cannot be obtained from each other in this manner.

* *Roy. Soc. Proc.* Vol. LXII, p. 62.

† This will be equal to the correlation, for the variabilities of both variates are taken to be the same.

‡ See *Phil. Trans.* Vol. 187, A, p. 306, 1896.

Galton, by means of seeking the slope of the regression line, found the regression of brother on brother to be $\frac{2}{3}$ and this accordingly would be the fraternal correlation; he then said: a nephew is the son of a brother, *therefore* his regression on his uncle $= \frac{1}{3} \times \frac{2}{3} = \frac{2}{9}$. Again I do not believe that regressions can be built up in this manner. It appears to be multiplying together probabilities that are not independent, but correlated; for all a regression provides is a probable deviation, and we cannot apply independent probabilities to a correlated triplet. Why may not a brother be considered as the son of a midparent and so have regression $\frac{2}{3} \times \frac{2}{3} = \frac{4}{9}$ instead of Galton's observed value $\frac{2}{3}$? Why might we not equally well argue that a nephew is the grandson of a midparentage, which gave rise to his uncle and thus the nephew-uncle regression be $\frac{1}{3} \times \frac{2}{3} \times \frac{2}{3} = \frac{4}{27}$ instead of $\frac{2}{9}$? Why should cousins* be considered the offspring of two brothers $\frac{1}{3} \times \frac{2}{3} \times \frac{1}{3}$ rather than as the grandsons of one midparentage $\frac{1}{3} \times \frac{2}{3} \times \frac{2}{3} \times \frac{1}{3}$? Even if we are always to take the "shortest way round," no argument is given in favour of it, and it could only be satisfactorily demonstrated by actual data.

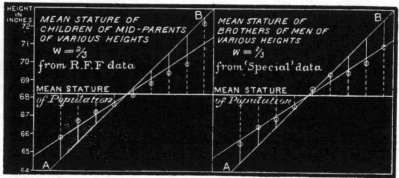

Fig. 6. Galton's Filial and Fraternal Regression Lines.

I do not think Galton's method of deducing the degrees of resemblance between kinsmen of various degrees of blood relationship from the single datum of the regression of a filial array on its midparent will pass muster; it is extraordinarily suggestive—no one had thought before of giving a quantitative measurement to the various types of kinship. Galton indicated how it could be done by aid of correlation tables and gave at this time two such tables†, those for midparent with offspring and for brother with brother. These are both from his *R. F. F.* (*Records of Family Faculties*), but he also provided another correlation table giving the distribution for a special series of pairs of brothers. In Fig. 6 will be found his regression lines for offspring on midparents, and for brother on brother. His method of reduction was, however, very different from any we should adopt to-day. When he wanted a mean he determined a median, and he did this by roughly proportioning (graphically) the total in the cell in which it lies, he worked not with the

---

* The value $\frac{1}{3} \times \frac{2}{3} \times \frac{1}{3} = \frac{2}{27}$ is given by Galton: *Natural Inheritance*, p. 133.

† If we include the earlier one for the seed-weights in mother and daughter plants for the case of sweet-peas (see our p. 4) we have here the four earliest correlation tables and regression lines ever published.

standard deviations, but the probable errors, and he determined these from the quartiles by rough proportioning as before. When he wanted a regression coefficient he plotted the medians of the arrays and fitted these with a straight line, presumably by testing with a straight edge. The slope of this straight line is Galton's regression coefficient. If we assume the standard deviation of the two marginal columns to be the same, then this regression coefficient is the coefficient of correlation, but that term was not used by Galton in the group of memoirs at present under discussion.

It will be remembered that Galton transmuted all his females to their male equivalent, and then found his regression for offspring on midparent to be $\frac{2}{3}$ and therefore on a single parent to be $\frac{1}{3} = \cdot 3333$. Reworking the whole material ten years later I found the mean of the four possible parental correlations to be $\cdot 3355*$, in singular accordance with his rougher methods, which, however, had largely screened the significant inequalities of the parental correlations in his case.

Turning to Galton's data for brothers I note that he nowhere tells us how he gets his regression coefficient of $\frac{2}{3}$. In the *R. S. Proc.* paper (*loc. cit.* p. 55) there is a small graph for the "Special" data for brothers, none for the *R. F. F.* data for brothers. The slope of the regression line Galton has run through the array medians is, as near as I can judge it, 34°, or the regression would be 6745, which Galton would call $\frac{2}{3}$. In the *Natural Inheritance*, p. 109, there are small charts for the regression lines of both the *R. F. F.* and the "Special" data, the former (which does not go truly through the mean) has an angle of 24° giving a correlation of $\cdot 4452$ and the latter an angle of about 33°, or a slope of $\cdot 6494$. Actually forming tables myself on Galton's data I found for the *R. F. F.* Regression of Brother on Brother $\cdot 4547$, and for the "Special" data $\cdot 5990$, not so violently diverse from Galton's results, when we consider the difference of methods, and personal equation in selecting pairs of brothers for tabular entry. But there is a point in which I find it needful to differ from Galton in the value of his material. I believe that the "Special" data were really heterogeneous; they contained pairs of brothers measured in an Essex volunteer regiment, who taken alone gave a regression of no less than 7175, while the remainder had only a value of $\cdot 5574$. I am inclined to think therefore that we need to throw out the volunteers, and if we do so the mean of Galton's two results, $\frac{1}{2}(\cdot 4452 + \cdot 5574) = \cdot 5013$, is very close to the mean value $\cdot 50$ which has since been found on more satisfactory and ampler data for a variety of characters in man. I doubt whether it is possible to accept Galton's original estimate of $\frac{2}{3}$ for fraternal regression and correlation, and believe that he may have been led to select the higher value of the two he had obtained by an idea that fraternal should equal midparental regression.

Anyhow in these numbers we find the first attempt to obtain a numerical measure of the degree of resemblance in brothers, just as in another part of the paper he has provided us with the first measure of filial resemblance. Galton knew quite well that his values were not final, but here, as so often, he blazed the track for others to build a highway.

* See our p. 18.

There is another suggestion in the Royal Society paper which has ultimately been followed up to great profit, namely that the variability within the family could be ascertained by considering the difference in the same character of pairs of brothers. Let R be the multiple correlation coefficient of an individual on all his ancestors or his correlation with his "generant," then since two brothers have the same ancestry the variability in a family of brothers is $\sigma \sqrt{1 - R^2}$, where σ is the standard deviation of brothers. Now if x_1 and x_2 be the characters in a pair of brothers, for example their statures, we have $\frac{1}{2}(x_1 + x_2)$ for their mean and $\frac{1}{4}(x_1 - x_2)^2$ for their standard deviation squared, or so-called variance. If this be taken for a large number of pairs, then it may be shown that

Mean variance for pairs of brothers $= \frac{1}{2}\sigma^2(1 - r) = \frac{1}{2}\sigma^2(1 - R^2)$,

where r is the simple correlation of brothers*.

These results have really been given as early as 1886 by Galton. He does not use R, and instead of standard deviations, speaks of quartile values, i.e. probable errors. He writes b for our $\cdot 67449\, \sigma \sqrt{1 - R^2}$, p for our $\cdot 67449\, \sigma$, and our r is his regression of brother on brother or his w. Thus in his symbols:

Mean (probable error)^2 of pairs of brothers $= \frac{1}{2}p^2(1 - w) = \frac{1}{2}b^2$.

These results are given on pp. 58–59 of the *R. S. Proceedings'* memoir, and demonstrated by methods which appeal only to the most elementary conceptions. When we come to actual numerical values, Galton finds a series of values for b (the probable deviation in a group of brothers) which ranges from $0''\cdot98$ to $1''\cdot38$—a result which might be anticipated from the rather heterogeneous nature of his material. If for the reasons already stated we do not trust to the "Special" data only, but use also the *R. F. F* results, the mean value (Table, p. 59) found by the various processes for b is $1''\cdot179$. For p I find from Galton's table on his p. 69, $1''\cdot684$, and thus deduce for R the value $\cdot7140$, comparing not badly with the value $\cdot7284$ obtained recently for brothers from probably better data†. Clearly with these values for p and b that for w, the regression of brother on brother or the correlation of brothers, is $\cdot5096$ and not $\frac{2}{3} = \cdot6667$ as Galton assumed it, trusting to his "Special" data; this is a result agreeing far better with later determinations of fraternal heredity‡.

The whole paper is a most remarkable one, not only for the wealth of new ideas it contains, but for the insight it shows Galton had into many problems which have only been recently, or are only at present, under

---

* *Biometrika*, Vol. XVII, pp. 130–1.   † *Ibid.* p. 138.

‡ A further point worth recording occurs on p. 58 of the *R. S. Proc.* memoir. Suppose samples of size n are taken from a normal distribution. Then the mean square standard deviation of these samples, $\bar{\mu}_2$, is given in terms of the standard deviation squared, Σ^2, of the sampled population, by $\bar{\mu}_2 = \dfrac{n-1}{n}\Sigma^2$. Galton puts this in probable deviation form as $d^2 = \dfrac{n-1}{n}b^2$ and putting $n = 4, 5, 6, 7$ applies it to find b^2 from mean square probable deviation (in his terminology the quartile) of brothers in families of different sizes. Thus anticipating more recent work on small samples.

discussion. He perceived for the first time that the problem of multiple correlation when solved would give the closest prediction possible to the probable value of the character in an individual from known characters in the kinsfolk, but he also recognised that long selection could not indefinitely reduce variability, that 30% reduction in variability was about as much as could be hoped for (i.e. p to b in his notation).

"The possible problems are obviously very various and complicated, I do not propose to speak further about them now. It is some consolation to know that in the commoner questions of hereditary interest, the genealogy is fully known for two generations, and that the average influence of the preceding ones is small.

"In conclusion it must be borne in mind that I have spoken throughout of heredity in respect to a quality that blends freely in inheritance. I reserve for a future inquiry (as yet incomplete) the inheritance of a quality that refuses to blend freely, namely the colour of the eyes. These may be looked upon as extreme cases, between which all ordinary phenomena of heredity lie*."

These words show that Galton fully recognised that his theory applied only to continuously varying and blending characters.

The paper in the Anthropological Journal *Miscellanea*, while less replete with ideas requiring mathematical interpretation than that in the *R. S. Proceedings*, contains two matters which deserve notice. Over and over again we meet with the statement that more able men are born from undistinguished parents than from parents of marked ability. In the year 1927 it formed the subject of a series of controversial letters in *The Times* newspaper, in which neither side seemed to have any statistical ammunition, nor appeared to be aware that they were dealing with a forty year old paradox, which Galton had refuted in 1885 :

"Let it not be supposed for a moment that any of these statements invalidate the general doctrine that the children of a gifted pair are much more likely to be gifted than the children of a mediocre pair. What they assert is that the ablest child of one gifted pair is not as likely to be as able as the ablest of all the children of very many mediocre pairs†."

In 1900‡ the biographer gave exact numbers for the production of ability on the assumption that one man in twenty may be treated as "able." It turned out that in 10,000 matings the 52 pairs of exceptional parents produced 26 exceptional sons, while the 9948 non-exceptional pairs produced 474 exceptional sons, thus the rate of production of exceptional sons by exceptional parents was 10 times greater than the rate by non-exceptional parents, but the latter produced more than 18 times as many exceptional sons as the former. The result flows merely from the fact that a rate of 10 times the production in the case of exceptional parents is counteracted in *total* output, by the fact that there are some 200 times more non-exceptional than exceptional pairs of parents. It is distressing to note how such distinguished scientists as Dr Leonard Hill are unable to grasp the interpretation of this simple statistical paradox first provided by Galton in 1885!

The second point is the publication of a diagram illustrating the variability of a stable population in the parental generation, for the midparentages, for the generants, and for the filial generation. The diagram (see our p. 28) is

* *R. S. Proc.* Vol. LXII, pp. 62–63.   † *Journ. Anthrop. Instit.* Vol. xv, p. 254.
‡ *Phil. Trans.* Vol. 195, A, p. 47.

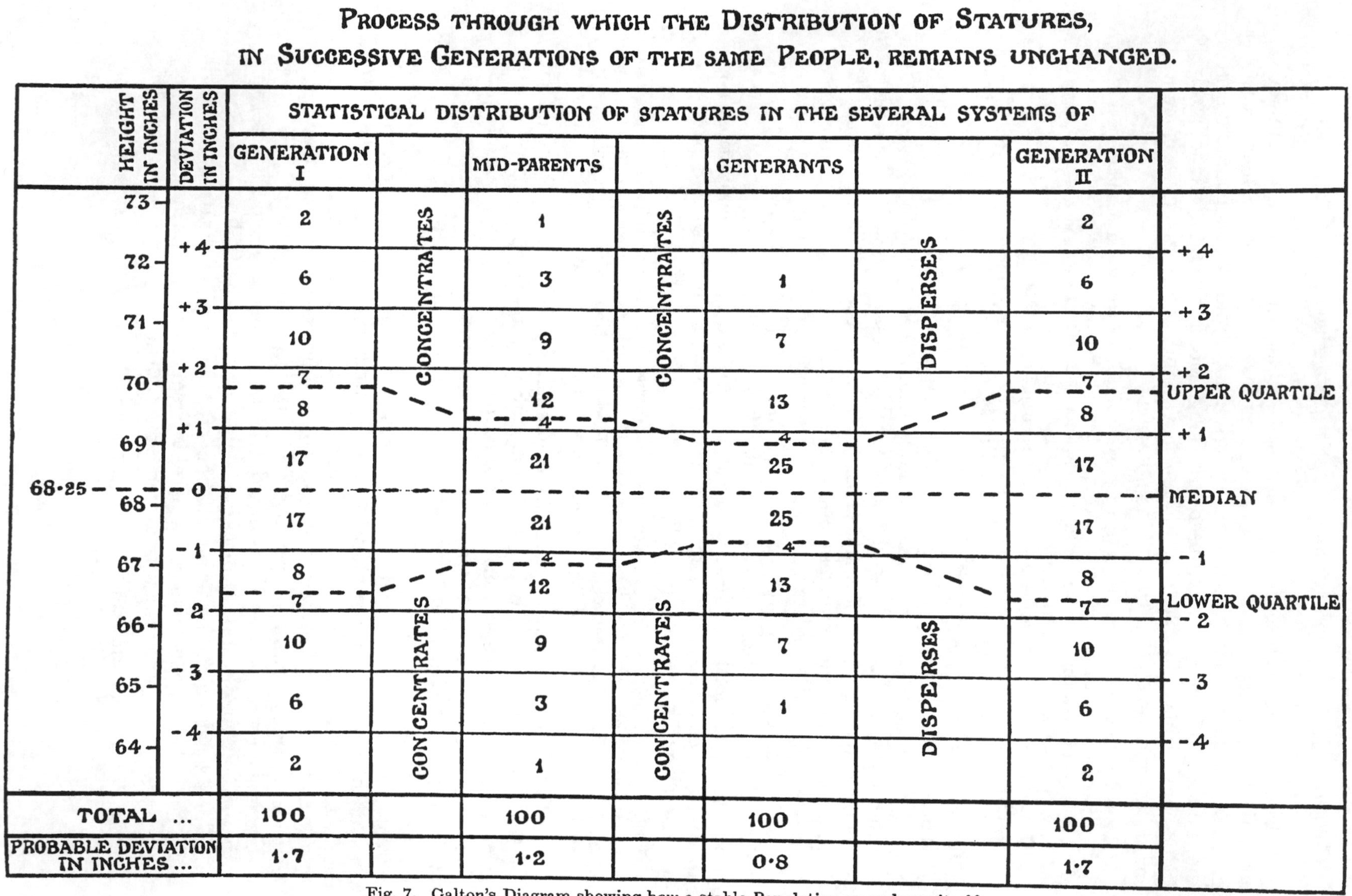

STATISTICAL DISTRIBUTION OF STATURES IN THE SEVERAL SYSTEMS OF

HEIGHT IN INCHES	DEVIATION IN INCHES	GENERATION I	MID-PARENTS	GENERANTS	GENERATION II	
73	+4	2			2	+4
72	+3	6	1		6	+3
71	+2	10	3	1	10	+2
70		7	12 (4)	7 (4)	7	UPPER QUARTILE
69	+1	8	21	13	8	+1
68·25 — 68	0	17	21	25	17	MEDIAN
		17	12 (4)	25	17	
67	−1	8	9	13 (4)	8	−1
66	−2	7	3	7	7	LOWER QUARTILE
65	−3	10	1	1	10	−2
		6			6	−3
64	−4	2			2	−4
TOTAL …		100	100	100	100	
PROBABLE DEVIATION IN INCHES …		1·7	1·2	0·8	1·7	

(Between columns: CONCENTRATES · DISPERSES · CONCENTRATES)

Fig. 7. Galton's Diagram showing how a stable Population reproduces itself.

not described at length*, and I have ventured to modify it in one or two directions, which I believe will make it somewhat clearer. The main difficulty I have is to interpret what Galton meant by the column headed "Generants." If he meant by " Generant " the hypothetical individual that I have represented by U above, a sort of "midancestor," replacing the whole stirp, then I think the variability of this midancestor should be given by $\sigma_1 \sqrt{1 - R^2}$, where σ_1 is the standard deviation of the population for the given character. For stature, using not σ_1 but the quartile, this would be Galton's b or $1''\!\cdot\!179$, or the value which Galton selects for b, i.e. $1''\!\cdot\!06$. In his diagram, we have under the "Generants" column " Probable deviation " $0''\!\cdot\!8$; this number does not occur, as far as I can see, anywhere else in the paper. One solution I can suggest is that Galton was thinking of the variability of pairs of his new population; in this the variability of these paired generants would be $b/\sqrt{2}$, and $\cdot8 \times \sqrt{2} = 1\!\cdot\!131$, almost the mean between the above values of b. Another explanation may be that Galton had not reached the comprehensive idea of the single midancestor, which I have defined by the " generant " above, but that his generant depended solely on the midparent and was to be taken as an individual with $\frac{2}{3}$ of the character of the midparentage. In this case the variability of the generant group would be $\frac{2}{3}$ of that of the midparental group, i.e. $\frac{2}{3}(1''\!\cdot\!2) = 0''\!\cdot\!8$. If this be true the generant would be only a hypothetical individual who produced offspring varying about his own, and not about a regressed type. I trust this latter solution may be erroneous, as I should like Galton to have conceived the idea of a single individual— not one depending only on the parents—who would represent the whole stirp or ancestry. At any rate let us preserve in future the good word " generant " for the hypothetical individual who possesses, in the manner indicated by the function U, all the midancestral characters which are capable of showing a blending inheritance. Such a generant is a sort of mean man for the stirp, who for statistical purposes represents the whole ancestry. If Galton had not this idea, he provided at least the origin from which it sprung! If his generants are the receded midparents, let us ourselves use generants for the midancestry, who will not of necessity involve regression at all.

Of the Birmingham Lecture on "Chance and its Bearing on Heredity" little need be said, it only adds to what we have already discussed, emphasis on the point that in a stable population the whole inheritance of any blending character depends on the knowledge of three constants: (i) the mean character in any generation, (ii) the corresponding variability, and (iii) a single hereditary correlation.

Galton gave a further account of his researches on regression in stature

* Galton is explaining how the new generation is a reproduction of the old and writes: " the process comprises two opposite sets of actions, one concentrative and the other dispersive, and of such a character that they neutralise one another and fall into a state of stable equilibrium (see Diagram [on our p. 28]). By the first set, a system of scattered elements is replaced by another system which is less scattered; by the second set, each of these new elements becomes a centre whence a third system of elements is dispersed" (*loc. cit.* p. 256). This is the only reference to the diagram or its interpretation I have noticed.

in his Presidential Address to the Anthropological Institute on January 26, 1886*. One or two points from this address may be noted. On pp. 491–3 he describes the working model which he exhibited to indicate how the *probable* stature of any man could be ascertained from that of a kinsman in any degree. Since the regression is constant all we have to do is to make use of the property of similar triangles. *AB* is a scale of stature, where *M* is the mean stature of the population.

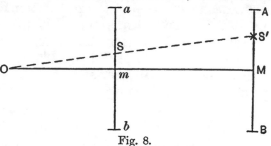

Fig. 8.

S' is any particular stature, *O* a point on the horizontal through *M*, so that *OM* = 10 units, then if *Om* = 10*r*, where *r* is the correlation of the particular grade of kinship, a string from *O* to *S'* will cut a vertical line through *m* in a point *S*, such that the point *S* gives the probable stature of the kinsman of the grade *r* of correlation. Galton put on a number of lines to determine probable stature in sons, nephews, grandsons, etc. He also constructed scales based on the standard deviation ($\sigma \sqrt{1 - r^2}$) showing the percentile distribution for each grade of kinship. These scales could be shifted up and down on their respective lines *ab*, so that the probability could be measured of any deviation from the probable stature *S*. As Galton's numerical values for the regressions were somewhat doubtful, I constructed at his suggestion some ten years later a life-size "Geniometer" on this plan with the revised values we had then determined for the hereditary correlations. It is reproduced on Plate I. The original figures which are in brilliant colours† gave Galton and I hope my audience some amusement.

In a presidential address of this kind, it is legitimate to let one's thoughts run freely, there is no need sternly to demonstrate each step as may be thought fitting in a Royal Society paper. Accordingly Galton "let himself go." Some quotations will illustrate for the reader what opinions were forming in his mind, they are not demonstrated judgments—it is doubtful if some are demonstrable at all.

(i) *On the Normal Distribution or Law of Error* (pp. 494–5).

"I know of scarcely anything so apt to impress the imagination as the wonderful form of cosmic order expressed by the 'law of error.' A savage, if he could understand it, would worship it as a god. It reigns with severity in complete self-effacement amidst the wildest confusion. The huger the mob and the greater the anarchy the more perfect is its sway. Let a large sample of chaotic elements be taken and marshalled in order of their magnitudes, and then, however wildly irregular they appeared, an unexpected and most beautiful form of regularity proves to have been present all along. Arrange statures side by side in order of their magnitudes, and the tops of the marshalled row will form a beautifully flowing curve of invariable proportions; each man will find, as it were, a pre-ordained niche, just of the right height to fit him, and if the class-places and statures of any two men in the row are known, the stature that will be found at every other class-place, except toward the extreme ends, can be predicted with much precision."

* *Journ. Anthrop. Instit.* Vol. xv, pp. 487–499, 1886.

† The actual artist, who was then a member of my staff, is now a distinguished man of science, a grave and learned professor, and might not be too pleased if I gave his name away!

Take the red thread through any value on the scale of stature, say 74″, then the average stature of persons having all the kinsfolk described below of that stature would be obtained by drawing a vertical line through the mark indicated by the kinsfolk till it meets the red thread, and carrying through this meeting point a horizontal line back to the scale of stature, which provides the average desired. For example the average nephew of *two* uncles, one paternal and one maternal, each 6 ft. 2 in. would be 6 ft. 0·6 in., but the average nephew of *four* uncles, two on each side, each 6 ft. 2 in. will be 6 ft. 1·8 in.* Again if both parents and paternal and maternal grandparents were of this stature, the grandsons would have progressed and be on the average 6 ft. 3·1 in. The statures are recorded for males, the corresponding female statures may be obtained by subtracting $\frac{8}{100}$ths from the male statures. In starting with females the male stature equivalent to that of the female must first be obtained by adding $\frac{2}{23}$rds of its value to the female stature. Thus a woman of 5 ft. 9 in. counts as a man of 6 ft. 3 in.

The reduction from my life-size diagram to the present small dimensions costs much in accuracy of reading, but serves to bring out the point that the regression ultimately changes to a progression.

* The regression coefficient used on the genometer was ·9614.

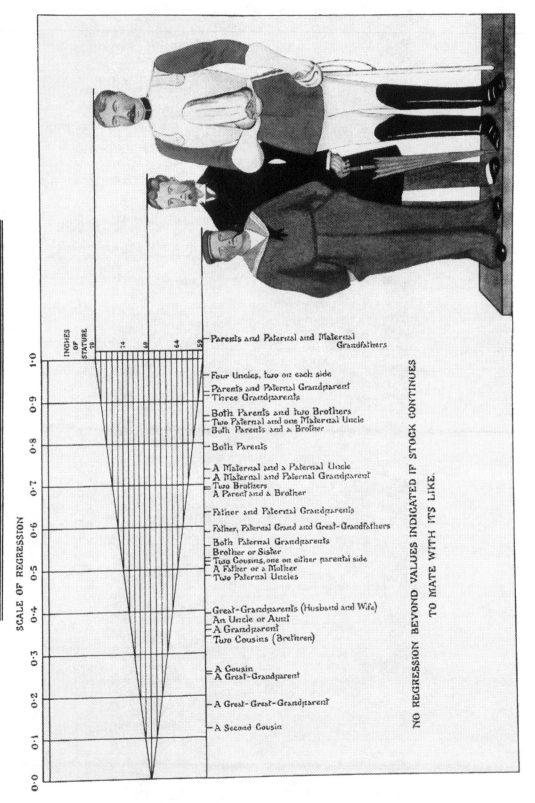

GENOMETER AFTER AN IDEA OF SIR FRANCIS GALTON

SCALE OF REGRESSION

0·0   0·1   0·2   0·3   0·4   0·5   0·6   0·7   0·8   0·9   1·0

INCHES
OF
STATURE

74   69   64   59

Parents and Paternal and Maternal
                              Grandfathers

Four Uncles, two on each side
Parents and Paternal Grandparent
Three Grandparents
Both Parents and two Brothers
Two Paternal and one Maternal Uncle
Both Parents and a Brother
Both Parents
A Maternal and a Paternal Uncle
A Maternal and Paternal Grandparent
Two Brothers
A Parent and a Brother
Father and Paternal Grandparents
Father, Paternal Grand and Great-Grandfathers
Both Paternal Grandparents
Brother or Sister
Two Cousins, one on either parental side
A Father or a Mother
Two Paternal Uncles
Great-Grandparents (Husband and Wife)
An Uncle or Aunt
A Grandparent
Two Cousins (Brethren)
A Cousin
A Great-Grandparent
A Great-Great-Grandparent
A Second Cousin

NO REGRESSION BEYOND VALUES INDICATED IF STOCK CONTINUES
TO MATE WITH ITS LIKE.

PLATE II

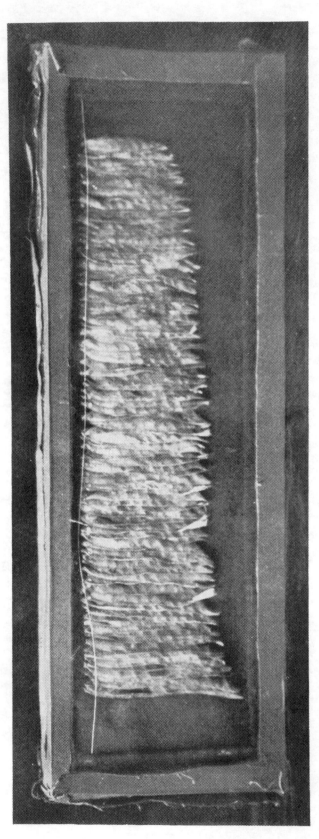

Galton's "Ogive Curve" as exhibited by a marshalled series of Bean Pods. Unfortunately in the many years since Galton built up this illustration several tips have been broken off and in other cases some of the pods have burst open and the shell has curled round.

Galton was wont to illustrate the beauty of the "pre-ordained niche" on a marshalled series of bean pods which he had many years before prepared. This series is reproduced on Plate II. Unfortunately the tips of some of the pods have bent back, but the general scheme survives.

(ii) *The Phenomenon of Regression, a great Hindrance to the Establishment of Breeds* (pp. 495–6).

"It will be seen from the large values of the ratios of regression how speedily all peculiarities that are possessed by any single individual to an exceptional extent, and which blend freely together with those of his or her spouse, tend to disappear. A breed of exceptional animals, rigorously selected, and carefully isolated from admixture with others of the same race, would become shattered by even a brief period of opportunity to marry freely. It is only those breeds that blend imperfectly with others and especially such of these as are at the same time prepotent, in the sense of being more frequently transmitted than their competitors, that seem to have a chance of maintaining themselves when marriages are not rigorously controlled—as indeed they never are, except by professional breeders. It is on these grounds that I hail the appearance of any new and valuable type as a fortunate and most necessary occurrence in the forward progress of evolution."

Galton admits that the precise manner in which a new type comes into existence is unknown, but suggests that a multitude of petty causes may contribute to reshape the grouping of the germinal elements and so lead to a new and fairly stable position of equilibrium, which admits of hereditary transmission. In favour of this view he cites the frequent experience of "sports," useful, harmful and indifferent and therefore without teleological intent. These, he considers, have various degrees of heritable stability, and form fresh centres towards which some at least of the offspring have a tendency to revert. He considers that such sports, by refusing to blend freely, may be transmitted almost in their entirety.

"On the other hand, if the peculiarity blends easily, and if it was exceptional in magnitude, the chance of inheriting it to its full extent would be extremely small...*. I feel the greatest difficulty in accounting for the establishment of a new breed in a state of freedom by slight and uncertain selective influences, unless there has been one or more abrupt changes of type, many of them perhaps very small, but leading firmly step by step, though it may be along a devious track, to the new form."

* Galton gives in a footnote the percentage of sons who are as tall or taller than their fathers. I have recalculated this table on somewhat better data than Galton had available (*Biometrika*, Vol. II, p. 381). It now runs as follows:

Father's Stature	Probable Stature of Son	Percentage of Sons taller than Father	Father's Stature	Probable Stature of Son	Percentage of Sons taller than Father	Father's Stature	Probable Stature of Son	Percentage of Sons taller than Father
67″·5	68″·56	67·4 °/₀	72″·0	70″·88	31·6 °/₀	77″·0	73″·46	6·4 °/₀
68″·0	68″·82	63·7 °/₀	73″·0	71″·40	24·5 °/₀	78″·0	73″·98	4·2 °/₀
69″·0	69″·33	55·6 °/₀	74″·0	71″·91	18·4 °/₀	79″·0	74″·49	2·6 °/₀
70″·0	69″·85	44·0 °/₀	75″·0	72″·43	13·4 °/₀	80″·0	75″·01	1·6 °/₀
71″·0	70″·37	39·4 °/₀	76″·0	72″·95	9·5 °/₀	81″·0	75″·52	0·9 °/₀

The considerable changes from Galton's percentages arise from the facts: (i) that the sons in our data had a mean stature 1″ greater than their father's, (ii) that our regression was ·516 against Galton's ·333.

Whatever we may think of Galton's arguments, it is clear that in 1886 he did not believe in the influence of natural selection as producing new forms by acting on continuously varying small deviations. This may have been due to the influence which the idea of *perpetual* regression * had upon his mind. Whatever its source, Galton was in 1886 and later a firm believer, as the above passage indicates, in evolution by mutation. He was a mutationist before De Vries published his first paper on mutations (1900).

(iii) *On the Inheritance of Ability and its Application to the Upper House of Legislature* (pp. 497–9).

Galton inquires how far the results for heredity in stature may be applied to heredity in ability. He holds that considerable differences have to be taken into account, and he classifies them under three heads:

"*Firstly*, after making large allowances for the occasional glaring cases of inferiority on the part of the wife to her eminent husband, I adhere to the view I expressed long since as the result of much inquiry, historical and otherwise†, that able men select those women for their wives who are not mediocre women, and still less inferior women, but those who are decidedly above mediocrity. Therefore, so far as this point is concerned, the average regression in the son of an able man would be less than one-third."

On better data‡ than Galton had at his command the regression of son's stature on father's stature is about ·52 instead of ·33, and, allowing for assertative mating, about ·82 on the midparent instead of Galton's ·67. When we introduce the grandparents the regression is not large. I think these points will explain Galton's difficulty as to ability without resort to the theory that extreme ability does not blend, which he suggests in his second statement:

"*Secondly*, very gifted men are usually of marked individuality, and consequently of a special type. Whenever this type is a stable one, it does not blend easily, but is transmitted almost unchanged, so that specimens of very distinct intellectual heredity frequently occur."

Unfortunately Galton gives no illustrations, and without statistical evidence it is difficult to interpret his meaning.

"*Thirdly*, there is the fact that men who leave their mark on the world are very often those who, being gifted and full of nervous power, are at the same time haunted and driven by a dominant idea, and are therefore within a measurable distance of insanity. This weakness will probably betray itself occasionally in disadvantageous forms among their descendants. Some of these will be eccentric, others feeble-minded, others nervous, and some may be downright lunatics."

The same point has been made frequently since Galton's day, but although isolated cases can of course be cited, the statement demands statistical demonstration. We require to know first whether the men "who leave their

---

* The theory of multiple regression shows us that if an individual mates with his like, he may regress on exceptional parents, but his offspring will not regress on him, nor further descendants either. A breed may be established if we select only parents and grandparents; the regression is thus of minor importance compared with the homogamy.

† See Vol. II, p. 105. ‡ *Biometrika*, Vol. II, p. 381.

mark on the world" are really always the able men, and if so, how many of them are "driven by a dominant idea." Again having defined this class, do statistics indicate that their offspring more often suffer from some form of nervous breakdown than the sons of men of lesser ability? Ryk ud med dine tal, bygmester! Talene på bordet!

I think Galton did not really believe that ability was inherited in a manner widely different from stature, for he now proceeds to suggest how a fitting House of Peers might be based on the knowledge gained by his inquiry. He supposes that in some new country it is desired to institute an Upper House of life-peers which shall be largely governed by the hereditary principle.

"The principle of insuring this being that (say) two-thirds of the members shall be elected out of a class who possess specified hereditary qualifications, the question is: What reasonable plan can be suggested of determining what those qualifications should be?

"In framing an answer we have to keep the following principles steadily in view: (1) The hereditary qualifications derived from a single ancestor should not be transmitted to an indefinite succession of generations, but should lapse after, say, the grandchildren. (2) All sons and daughters should be considered as standing on an equal footing as regards the transmission of hereditary qualifications. (3) It is not only the sons and grandsons of ennobled persons who should be deemed to have hereditary qualifications, but also their brothers and sisters, and the children of these. (4) Men who earn distinction of a high but subordinate rank to that of the nobility, and whose wives had hereditary qualifications, should transmit these qualifications to their children. I calculate roughly and very doubtfully, because many things have to be considered, that there would be about twelve times as many persons hereditarily qualified to be candidates for election as there would be seats to fill. A considerable proportion of these would be nephews, whom I should be very sorry to omit, as they are twice as near in kinship as grandsons*. One in twelve seems a reasonably severe election, quite enough to draft off the eccentric and incompetent, and not too severe to discourage the ambition of the rest. I have not the slightest doubt that such a selection out of a class of men who would be so rich in hereditary gifts of ability, would produce a senate at least as highly gifted by nature as could be derived by ordinary parliamentary election from the whole of the rest of the nation. They would be reared in family traditions of high public services. Their ambitions, shaped by the conditions under which hereditary qualifications could be secured, would be such as to encourage alliances with the gifted classes. They would be widely and closely connected with the people, and they would to all appearance—but who can speak with certainty of the effects of any paper constitution?—form a vigorous and effective aristocracy." (pp. 498–9.)

Galton does not state how he would start his Upper House *ab initio*, nor take into account the possible need of recruiting its stock from outside ability. His scheme would certainly introduce improved and better planned marriages among the peers, as they would be anxious to preserve the peerages within their own families. Here as elsewhere† he points out to our hereditary peers how little justification there is for their position, while at the same time he indicates that there is a basis in heredity for a really effective aristocracy. Such doctrines would scarcely appeal even now to either Tory or Democrat. Among the many proposals put forward for reforming the British House of Lords, none has endeavoured like Galton's to place it on a

---

* I think this is incorrect for reasons stated above (see pp. 22 and 24). The observed correlations between a man and his grandson and a man and his nephew are about equal.

† See our Vol. II, p. 93.

scientific basis by suggesting that the hereditary honour should follow ability in the stock and not be granted to a preordained individual.

D. *Attempt to demonstrate the Law of Ancestral Heredity on Eye-Colour.* In 1886 Galton published in the *Proceedings** of the Royal Society a paper on "Family Likeness in Eye-Colour." The only earlier paper I know which deals with this topic is that by Alphonse de Candolle†. That paper has no adequate statistical treatment, and suffers from two fundamental errors. The material was collected not only from Switzerland with its mixed races, but from Sweden, Germany and France, so that beyond the immediate parents, there must have been great differences in the eye-colours of the unrecorded earlier ancestry, and secondly the contributors were especially requested to leave out offspring of "doubtful" eye-colour, and also those of definite eye-colour whose parents had doubtful eye-colour. I do not think that in de Candolle's paper any results of real scientific value are reached. Galton's method of approaching the problem is entirely different. He starts from his Law of Ancestral Heredity, and endeavours to apply it to eye-colour, which he says does not usually blend. Accordingly he proportions the ancestral contributions not in the character of the individual but among the whole group of offspring. As Galton believed he had deduced from his mid-parental regression of $\frac{2}{3}$ the system $\frac{1}{2} + \frac{1}{4} + \frac{1}{8} + \ldots$ for contributions to the individual character in the case of stature, so he now supposes that an individual parent's eye-colour will determine on the average that of $\frac{1}{4}$ of the offspring, that of a grandparent $\frac{1}{16}$ of the offspring, and so on.

"Stature and eye-colour are not only different as qualities, but they are more contrasted in hereditary behaviour than perhaps any other simple qualities. Speaking broadly parents of different statures transmit a blended heritage to their children, but parents of different eye-colours transmit an alternative heritage. If one parent is as much taller than the average of his or her sex as the other parent is shorter, the statures of their children will be distributed in much the same way as those of parents who were both of medium height. But if one parent has a light eye-colour and the other a dark eye-colour, the children will be partly light and partly dark, and not medium eye-coloured like the children of medium eye-coloured parents. The blending of stature is due to its being the aggregate of the quasi-independent inheritances of many separate parts, while eye-colour appears to be much less various in its origin. If then it can be shown, as I shall be able to do, that notwithstanding this two-fold difference between the qualities of stature and eye-colour, the shares of hereditary contribution from the various ancestors are in each case alike, we may with some confidence expect that the law by which these hereditary contributions are governed will be widely, and perhaps universally applicable‡."

Galton starts his paper by considering whether there has been a secular change in eye-colour in the four generations to which his *Records of Family Faculties* extended. He started with those who ranked as "children" in the pedigree as Generation I; their parents, uncles and aunts were Generation II; the grandparents and their collaterals were Generation III, while the great grandparents and their collaterals were Generation IV. He gives the

* Vol. xl, pp. 402–416. Read May 27, 1886.

† "Hérédité de la couleur des yeux dans l'espèce humaine." *Archives des Sciences physiques et naturelles*, 3ième Période, T. xii, pp. 97–120, Geneva, 1884.

‡ *Roy. Soc. Proc.* pp. 402–3.

accompanying chart for the percentages of these eye-colours in the various generations, and concludes that there has been in these four generations

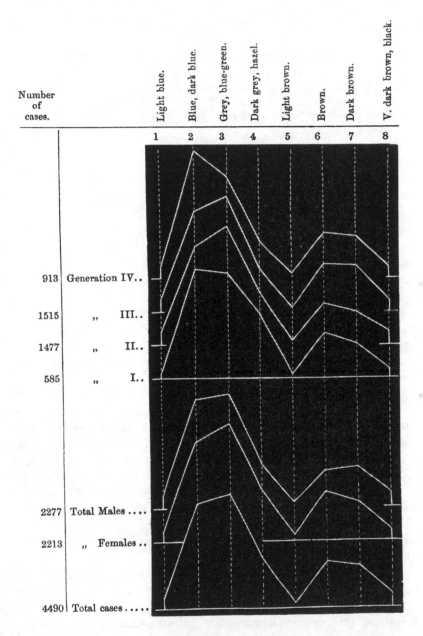

Fig. 9. Percentages of Eye-Colour in Successive Generations.

little secular change in eye-colour. It should, I think, be noted that the Generations III and IV are likely to be much older than Generations I and II when their eye-colours were recorded. Galton's data give the following percentage values on considerable numbers in the groups combined of Light and Dark Blue, Grey, Blue Green:

Generation	Male				Female			
	I	II	III	IV	I	II	III	IV
Percentages ..	58·2	58·4	62·9	70·4	58·0	58·1	58·6	56·2
Probable Errors	± 1·92	± 1·22	± 1·33	± 1·43	± 1·97	± 1·23	± 1·22	± 1·58

It will be seen that, while there is no significant change in the percentage of light eyes in the women, there is really such a change in the light eyes of the men; the grandparental and great grandparental generations have more bluish eyes. Were it not for the fact that there is no change in the women, we might attribute this not to a racial change going on, but to men's eyes growing lighter with extreme age. I have no statistical data to produce, but my impression of the marked frequency of very light colour in old men's eyes is strong. At the same time I know no physiological reason why men's and not women's eyes should grow lighter with greater age.

On the basis of his diagrams Galton considers that he may disregard "a current popular belief in the existence of a gradual darkening of the population, and can treat the eye-colours of those classes of the English race who have contributed to the records, as statistically persistent during the period under discussion" (p. 406).

Galton next states that he considers that there are only two fundamental types of eye-colour, the light and the dark, but under this supposition the medium tints are troublesome. Such tints he has classified under "Dark Grey and Hazel." In these cases the outer portion of the iris is usually of a dark grey colour, and the inner of a hazel. The proportions of grey and hazel vary, and the eye is called "dark grey" or "hazel" according to the colour which happens most to arrest the attention of the observer. Galton's attempt to deal with these medium eyes, of which there are in the population about 12·7 °/₀, is to me unconvincing; yet the fact that he recognises their existence is more satisfactory than the Mendelian treatment which disregards them entirely!

Galton for conciseness terms all these eyes "hazel." He defines a hazel-eyed family to be one in which there is at least one hazel-eyed child, and he proceeds to inquire into the constitution and ancestry of such "hazel-eyed" families or sibships. He obtains the results tabulated on p. 37.

Now it is clear from the table that when there is a hazel-eyed child in a sibship, the percentage of dark eyes in the sibship is only very slightly reduced, but the number of light-eyed brothers and sisters is 16 °/₀ below that of the general population. Again in the parental generation, there are 12 °/₀ fewer light-eyed parents of hazel-eyed parents, and this 12 °/₀ is transferred to the hazel-eyed group, the dark-eyed parents remaining at

Constitution and Ancestry of Hazel-Eyed Sibships.

Generations	Total cases observed in 168 families	Percentages		
		Light Eyed	Hazel Eyed	Dark Eyed
I. Siblings ...	948	45	32	23
II. Parents ...	336	49	25	26
III. Grandparents	449	60	13	27
General Population	4490	61·2	12·7	26·1

the general population percentage. The distribution of the grandparents of a hazel-eyed person is practically the same as that of the general population. From these data Galton concludes as follows:

"The total result in passing from Generation III to I, is that the percentage of the light eyes is diminished from 60 or 61 to 45, therefore by one quarter of its original amount, and that the percentage of the dark eyes is diminished from 26 or 27 to 23, that is to about [? by about] one-eighth of its original amount, the hazel element in either case absorbing the difference. It follows that the chance of a light-eyed parent having hazel offspring is about twice as great as that of a dark-eyed parent. Consequently since hazel is twice as likely to be met with in any given light-eyed family as in a given dark-eyed one, we may look upon two-thirds of the hazel eyes as being fundamentally light and one-third of them as fundamentally dark. I shall allot them rateably in that proportion between light and dark and so get rid of them. M. Alphonse de Candolle has also shown from his data that *yeux gris* (which I take to be equivalent to my *hazel*) are referable to a light ancestry rather than to a dark one, but his data are numerically insufficient to warrant a precise estimate of the relative frequency of their derivation from each of these two sources." (pp. 407–8.)

I find it very difficult to follow this reasoning, or to see from the table above its validity. It would seem to be essential to follow up the particular ancestry of each hazel-eyed individual, before we can draw the conclusions that Galton does from the *massed* numbers of children, parents and grandparents. Galton and de Candolle at least admit the difficulty of the hazel eyes; many Mendelian writers speak only of "brown" and "blue" eyes; others speak of hazel-eyed persons as heterozygotes*.

Galton having thus disposed of his *yeux gris*, now turns to the same multiple regression formula as he has used for stature, namely he makes the regression coefficient $\frac{1}{4}$ for a parent, $\frac{1}{16}$ for a grandparent and so on to higher ancestry. He also makes use of what is, I believe, an erroneous hypothesis, at any rate one inconsistent with his multiple regression coefficients,

* Sometimes a definition is given of pure blue eyes as being those without anterior pigment. According to one ardent Mendelian this can always and only be tested with a lens; another accepted relatives' statements, and came to the same conclusion without a lens. From twelve cases in which both eyes were carefully examined with a lens and thus found to be without anterior pigment, the excised eye when sectioned and examined microscopically showed quite clearly anterior pigment. Hitherto I have failed to come across any eye, however blue, which is without some anterior pigment when sectioned. At what degree of pigmentation does the recessive character cease?

namely, that if an individual has h of a certain character, the most probable value of the character in his parent will be $\frac{1}{3}h$, and in his grandparent $\frac{1}{3^2}h$ and his great grandparent $\frac{1}{3^3}h$ and so on.

Consequently, if we know nothing beyond the one parent of character h, the expected heritage is

$$h\left\{\frac{1}{4}+2\left(\frac{1}{3}\times\frac{1}{2^4}\right)+4\left(\frac{1}{9}\times\frac{1}{2^6}\right)+\ldots\right\}=h\times0{\cdot}30.$$

When one grandparent only is known to have h then the corresponding parent has $\frac{1}{3}h$, and the two great grandparents $\frac{1}{3}h$, the four great great grandparents $\frac{1}{3^2}h$ and so on. Thus the formula is

$$h\left\{\left(\frac{1}{3}\times\frac{1}{2^2}\right)+1\times\left(\frac{1}{2^4}\right)+2\left(\frac{1}{3}\times\frac{1}{2^6}\right)+4\left(\frac{1}{3^2}\times\frac{1}{2^8}\right)+\ldots\right\}=h\times\left(\frac{1}{12}+\frac{3}{40}\right)=h\times0{\cdot}16,$$

i.e. actually $0{\cdot}1583\,h$.

If a parent and the corresponding two grandparents be known Galton says the parent will contribute $\frac{1}{4}$ of his character and the two grandparents and their ancestry $\frac{3}{40}$ as above. But I do not think this is correct, even on Galton's assumptions. In the previous case we predicted the great grandparents and higher ascendants from a knowledge of the grandparents *only*. But in this case we have not only these two grandparents, but also the knowledge of their offspring, the parent, to predict from, and accordingly Galton's $\frac{1}{80}$ for the rest of the ancestry is not satisfactory. As he is working in round numbers, Galton puts $\frac{3}{40}\left(=\cdot075\right)$ as equal to $\cdot08$.

Three cases are now dealt with: I, both parents only known; II, four grandparents only known; and III, both parents and four grandparents known. I gives $2\times\cdot30=\cdot60$ of heritage with a residue of $\cdot40$ undetermined. Galton distributes this residue in the general population proportions of light to dark eyes after distributing the hazel eyes $\frac{2}{3}$ to light and $\frac{1}{3}$ to dark eyes, which give $70\,^{\circ}/_{\circ}$ and $30\,^{\circ}/_{\circ}$ of those eyes. Thus the residue $\cdot40$ is to be given $\cdot28$ to light and $\cdot12$ to dark eyes. The corresponding residues for cases II and III are $\cdot36$ and $\cdot18$, which Galton distributes as $\cdot25$ and $\cdot11$, $\cdot12$ and $\cdot06$* respectively.

Galton now combines all these results in a table from which with knowledge of the ancestry as far as parents and grandparents are concerned he considers prediction of eye-colour in offspring can be ascertained (p. 39).

Let me illustrate the use of this table. A family of 12 given by Galton had both parents light-eyed, 3 grandparents light-eyed and 1 hazel-eyed. If we predict from parents only we should have

$$12\times(2\times\cdot30+\cdot28)=12\times\cdot88=10{\cdot}56\text{ light-eyed.}$$

If we predict from grandparents only we should have

$$12\times(3\times\cdot16+1\times\cdot10+\cdot25)=9{\cdot}96\text{ light-eyed.}$$

* More accurately the latter pair should be $\cdot13$ and $\cdot05$.

And if from all our information

$$12 \times (2 \times \cdot25 + 3 \times \cdot08 + 1 \times \cdot05 + \cdot12) = 10\cdot92 \text{ light-eyed.}$$

Thus the best prediction gives 11 out of 12 children light-eyed. Actually all 12 were light-eyed. Take again another family 2 parents hazel, 2 grandparents light, 1 hazel and 1 dark. Total family, 7 children. The prediction is $7 (2 \times \cdot16 + 2 \times \cdot08 + 1 \times \cdot05 + \cdot12) = 4\cdot55$ light-eyed, the observed number was 4. Of course Galton only claims to give the average family, and some of the results he gives from his Table of 78 individual families are not good. But his Table III in which he deals with 16 groups of different ancestries is, considering what appears to me the doubtful character of his assumptions, really surprising. Out of 827, 629 were observed to be lighteyed. Predicted from parents only 623 were light-eyed, and from parents and grandparents 614. As a rule, however, III gives a better result than I; for example, out of 183 children, all of whose parents and grandparents were light-eyed (none hazel), 174 were observed to be light-eyed; here III predicts 172, and I only 161.

Prediction Table for Eye Colour in Offspring.

	Both Parents I		Four Grandparents II		Both Parents and Four Grandparents III	
	Light	Dark	Light	Dark	Light	Dark
Light-eyed Parent	0·30	—	—	—	0·25	—
Hazel-eyed Parent	0·20	0·10	—	—	0·16	0·09
Dark-eyed Parent	—	0·30	—	—	—	0·25
Light-eyed Grandparent ...	—	—	0·16	—	0·08	—
Hazel-eyed Grandparent ...	—	—	0·10	0·06	0·05	0·03
Dark-eyed Parent	—	—	—	0·16	—	0·08
Residue to be rateably assigned	0·28	0·12	0·25	0·11	0·12	0·06

It is certainly remarkable that the predictions should be even as accurate as they are—and they are indeed not perfect—considering the contradictory assumptions on which they are based[*]. Perhaps in the first glow of finding such an amount of accordance Galton was justified in writing:

"A mere glance at Tables III and IV will show how surprisingly accurate the predictions are, and therefore how true the basis of the calculations must be....My returns are insufficiently numerous and too subject to uncertainty of observation to make it worth while to submit them

[*] In particular Galton's assumption that the correlations of the offspring with the individual parent, grandparent, great grandparent, etc., form the series r, r^2, r^3, etc., is incompatible with his multiple regression coefficients $\frac{1}{4}$, $\frac{1}{16}$, $\frac{1}{64}$, etc. Any such series causes all those coefficients except the first or parental coefficient to vanish, and reduces the ancestral multiple regression to a simple biparental inheritance. Thus the parental characters determine completely those of the offspring, as in the well-known case of the Mendelian theory of gametic characters.

to a more rigorous analysis, but the broad conclusion to which the present results irresistibly lead, is that the same peculiar hereditary relation that was shown to subsist between a man and each of his ancestors in respect of the quality of stature, also subsists in respect to that of eye-colour." (pp. 415–6.)

The essential fact to be remembered here is that Galton supposes the ancestral contributions which blend in the case of the stature of the individual, will be found as alternative eye-colours in the same proportions as for stature in the total group of descendants. For example, if an ancestor contributes $1/p$th of his stature deviation to his descendant in the final generation, he will contribute his eye-colour to $1/p$th of his descendants in the same generation.

It would be of great interest to rework Galton's proportions with the actual correlations found from his data, and with the corresponding and consistent multiple regression coefficients, and ascertain whether accordance was not sensibly improved. His parental correlation $\frac{1}{3}$ is too small for his data, and his regression coefficients want considerable modification.

E. *Law of Ancestral Heredity applied to Basset Hounds.* Galton having applied his Law of Ancestral Heredity to Eye-Colour in Man sought for additional material to illustrate it. He found this eleven years later in Sir Everett Millais' large pedigree stock of Basset Hounds. This material reached him at the very time he was himself planning an extensive experiment with fast breeding small mammals*. One can but regret that that experiment was never undertaken. The Bassets are dwarf bloodhounds, and there are only two varieties of colour, they are either white with blotches from red to yellow technically termed "lemon and white," or they have in addition to this "lemon and white" black markings; in which case they are termed "tricolour." Galton had thus only two types to deal with, which he terms "tricolour" (T) and "non-tricolour" (N). A full report of his statistical reduction of Millais' data is given in a paper read before the Royal Society, June 3, 1897†.

Galton's material was contained in *The Basset Hound Club Rules and Studbook*, compiled by Everett Millais, 1874–1896, but with this valuable addition, that Sir Everett Millais had added the registered colours of nearly 1000 of the hounds (this copy is now in the Galton Laboratory). In this record are 817 hounds, the colour of whose parents are given, and 567 hounds in which the colours of the two parents and the four grandparents are known, and lastly in 188 cases in addition the colour of all the eight great grandparents.

Galton starts with the same idea as in the paper last dealt with, namely that each parent contributes $\frac{1}{4}$, each grandparent $\frac{1}{16}$ and so on, of the heritage taken as a whole to be unity. Here as in the case of eye-colour, the heritage is

---

* An extensive series on moth-breeding had been undertaken but had unfortunately failed to give any satisfactory results, partly owing to the diminishing fertility of successive broods, and partly to the disturbing effects of food differences and change of environment in different years.

† See *Roy. Soc. Proc.* Vol. LXI, pp. 401–413. An abstract appeared in *Nature*, July 8, 1897, Vol. LV, p. 235.

not taken to be that of an individual, but as represented by percentages of the total offspring, the coat colours being exclusive, i.e. there is no attempt to measure the degree of melanism. Galton gives some reasons for his law being a probable one:

"It should be noted that nothing in this statistical law contradicts the generally accepted view that the chief, if not the sole, line of descent runs from germ to germ and not from person to person. The person may be accepted on the whole as a fair representative of the germ, and, being so, the statistical laws which apply to the persons would apply to the germs also, although with less precision in individual cases. Now this law is strictly consonant with the observed binary subdivisions of the germ cells, and the concomitant extrusion and loss of one-half of the several contributions from each of the two parents to the germ cell of the offspring. The apparent artificiality of the law ceases on those grounds to afford cause for doubt; its close agreement with physiological phenomena ought to give a prejudice in *favour* of its truth rather than the contrary. Again, a wide though limited range of observation assures us that the occupier of each ancestral place *may* contribute something of his own personal peculiarity, apart from all others, to the heritage of the offspring. Therefore there is such a thing as an average contribution appropriate to each ancestral place, which admits of statistical valuation, however minute it may be. It is also well known that the more remote stages of ancestry contribute considerably less than the nearer ones. Further it is reasonable to believe that the contributions of parents to children are in the same proportion as those of the grandparents to the parents, of the great grandparents to the grandparents, and so on; in short, that their total amount is to be expressed by the sum of the terms in an infinite geometrical series diminishing to zero. Lastly, it is an essential condition that the total amount should be equal to 1, in order to account for the whole of the heritage. All these conditions are fulfilled by the series of $\frac{1}{2} + \frac{1}{2^2} + \frac{1}{2^3} +$ etc., and by no other*. These and the foregoing considerations were referred to when saying that the law might be inferred with considerable assurance à *priori*; consequently, being found true in the particular case about to be stated, there is good reason to accept the law in a general sense." (*loc. cit.* p. 403.)

Modern research shows that the "binary subdivisions of the germ cells, and the concomitant extrusion and loss of one-half of the several contributions from each of the two parents to the germ cell of the offspring" may have other interpretation than that put upon it by Galton. Objections may also be raised to Galton's proportioning of the "heritage" among the offspring, and to his allowance for ancestors whose characters are not known directly. But the criticisms of the "ancestral law," made chiefly by Mendelians, have failed to attack these weaknesses. They have been generally based on citing *individual* matings†, as if these had any application to a statistical law

---

* This seems incorrect: the conditions would appear to be equally well satisfied by

$$(1 - a)(1 + a + a^2 + a^3 + ...),$$

which series leaves a constant a to be determined by observation of one kind or another. By putting $a = \frac{1}{2}$, Galton excluded his ancestral law from describing Mendelian gametic inheritance, which corresponds to $a = 0$ or the parents' gametic constitutions *alone* determining the offspring.

† Occasionally hybridisations are cited. Galton in a letter to *Nature*, October 21, 1897, writes:

"Permit me to take this opportunity of removing a possible misapprehension concerning the scope of my theory. That theory is intended to apply only to the offspring of parents who, being of the *same variety*, differ in having a greater or less amount of such characteristics as any individual of that variety may normally possess. It does *not* relate to the offspring of parents of different varieties; in short it has nothing to do with hybridism, for in that case the offspring of two diverse parents do not necessarily assume an intermediate form."

Whether the limit to offspring assuming "an intermediate form" is needful is another question, and might raise a discussion as to whether the law could be applied to alternate describing what happens on the *average* in the case of a race or community mating at random. What Galton's critics have not seen is that the degree of accordance between his predictions and observed facts, if not perfect, is yet so considerable, in the cases of both eye-colour in Man and coat-colour in Basset Hounds, that it is not possible simply to put it for all characters on one side as of no importance. No entirely erroneous hypothesis could, I think, lead to such accordance as Galton shows in his Tables V and VI of this memoir!

I have already pointed out when dealing with Galton's views on eye-colour, that, because r is the regression coefficient of child on parent[*], it does not follow that r^2 will be that of child on grandparent or of grandparent on child. Galton drops this manner of allowing for the unstated characters of the higher ascendants when he comes to the coat-colour of Bassets. He argues as follows: Out of 1060 parents of 530 offspring with tricolour coats 836 were tricolour (T) and 224 were lemon and white (N), i.e. 79 °/₀ and 21 °/₀; he accordingly says that the chance that a tricolour offspring has a tricolour parent is ·79. He concludes that if a dog has a tricolour parent, but nothing is known of the grandparents, these will be ·79 °/₀ tricolour, and the parents of these grandparents will be $(·79)^2$ °/₀ tricolour and so on. I am inclined to doubt the accuracy of this method of correction for the past ancestry of the tricolour for two reasons: (i) if both parent and grandparent were tricolour, then it seems to me there would be a greater probability of the great grandparent being tricolour than ·79, for we know that not merely one, but two generations of the offspring of these ancestors have been tricolour[†]; (ii) further, in each ascending generation besides the ·79 °/₀ tricolour of a tricolour animal there will be ·21 °/₀ non-tricolour, but these non-tricolour dogs will have also a percentage of tricolour ancestry, namely 56 °/₀ according to Galton's Table III, and I cannot see that he has allowed for the non-tricolour ancestors' contribution of additional ancestral tricolours in his method of reckoning his tricolour "coefficients" of tricolour grandparents. Noting that Galton calls A_s the ancestry of the sth generation and a_0 the offspring, we may cite his words from p. 406:

"Suppose all the four grandparents, A_2, to be tricolour, then only 0·79 of A_3 will be tricolour also, $(0·79)^2$ of A_4, and so on. These several orders of ancestry will respectively contribute an average of tricolour to each a_0 of the amounts of $(0·5)^3 \times 0·79$, $(0·5)^4 \times (0·79)^2$, etc. Consequently the sum of their tricolour contributions is

$$(0·5)^3 \times (0·79) \{1 + (0·5) \times (0·79) + (0·5)^2 \times (0·79)^2 + \text{etc.}\}$$

which equals 0·1632. The average tricolour contributions from *each* of the four tricolour grandparents must be reckoned as the quarter of this, namely, 0·0408."

characters in either eye-colour or coat-colour; but Galton's disclaimer, made with regard to Professor Henslow's criticisms of the law (see *Gardeners' Chronicle*, September 25, 1897) based on plant hybridisations, has been overlooked by those who more recently have cited hybridisations as disproving the law.

[*] $r = 0·3$ according to Galton.

[†] Thus from Galton's Table I we find that if parents and grandparents were all tricolour the percentage was 89, and not 79, tricolour offspring. Galton treats really correlated relationships as independent probabilities.

Now I think this does not involve all the tricolour ancestry of the four tricolour grandparents, for 0·21 of the great grandparents are non-tricolour, and there will be $(0·21) \times (0·56) \times (0·5)^3 \times (0·79) \times (0·5)^2$ of the great great grandparents tricolour. At each stage a non-tricolour branch will split off, showing in the next ascending generation some tricolour. It appears to me that Galton has overlooked the sum of all these ancestral tricolour contributions in estimating the tricolour in a_0. They may be considerably less than those retained, but I do not think they can be disregarded without justification.

"By a similar process," Galton writes, "the average tricolour contribution from the ancestry of *each* non-tricolour grandparent is found to be 0·0243." (p. 406.)

It would seem that this is obtained from:

$$(0·5)^3 \times (0·56) \{1 + (0·5) \times (0·56) + (0·5)^2 \times (0·56)^2 + \text{etc.}\} = ·0972,$$

for one-fourth of this is 0·0243.

But the above expression is not, I think, correct, for after the great grandparental 0·56 of tricolour we must surely use not 0·56 but 0·79 to pass from tricolour to tricoloured ancestry. Thus the result should be

$$(0·5)^3 \times (0·56) \{1 + (0·5) \times (0·79) + (0·5)^2 \times (0·79)^2 + \text{etc.}\} = ·1157,$$

of which the fourth part is ·0289.

Here as before the non-tricoloured ancestors of earlier generations who would themselves have tricoloured parents, etc., are neglected.

Taking Galton's illustration (p. 406) of both parents tricolour, three grandparents tricolour, and one lemon and white, Galton's factor of ·8342 is only changed to ·8388 by the above correction, but this gives 100 tricolour hounds out of a total of 119 offspring in this category, while the observed tricolours were 101, a remarkably close accordance.

I illustrate the sort of accordance obtained in the following examples:

Both Parents Tricolour	Number of Tricolour Grandparents				
	4	3	2	1	Totals
Tricolour Offspring:					
Observed ...	106 (119)	101 (119)	24 (28)	8 (11)	239 (277)
Calculated ...	108	100	21	8	237

Both Parents and three Grandparents Tricolour	Number of Tricolour Great Grandparents				
	8	7	6	5	4	Totals
Tricolour Offspring:						
Observed ...	—	17 (18)	19 (21)	14 (16)	6 (6)	56 (61)
Calculated ...	—	16	18	13	5	52

The numbers in brackets denote total offspring.

Two cases give rather poor results, those for 1 parent and 3 grandparents tricolour, no great grandparents or higher ancestry known (92 calculated for 79 observed in 158) and 1 parent, 3 grandparents and 5 great grandparents tricolour with no higher ancestry known (18 calculated for 8 observed out of 31). In the latter case especially it is the observations which seem to me questionable, because for one parent tricolour and the other lemon and white, whatever be the more remote ancestry we get 139 tricolour to 122 non-tricolour, while with 3 grandparents and 5 great grandparents tricolour, the observations only give us 8 tricolour to 23 non-tricolour or a *drop* from 50 °/₀ to 26 °/₀ in tricolour, with an increase of tricolour ancestry. If we can trust the classification, then no simple Mendelian hypothesis will provide a formula to fit the data, because neither tricolour × tricolour nor non-tricolour × non-tricolour breeds true. I have said, if we can trust the classification, because as Galton points out there is a strange prepotency of sire over dam*, the ratio of sire colour to dam colour in offspring being of the order of 6 to 5. A more important fact bearing on the classificatory accuracy arises from an investigation by an entirely different method from Galton's[†], where it appeared that the resemblance of the offspring to the sire was far less than to the dam. This suggested that the parentage was more certain in the case of the dam than in that of the sire, a difficulty not unlikely to arise from the carelessness of kennel attendants.

In the opinion of the present biographer the Law of Ancestral Heredity has been shown by Galton to be at least approximate in two very different cases, and this justifies further attempts to deal with it, either in Galton's or a more generalised form, on more satisfactory material and with possibly more accurate methods of computing the corrections for the unknown characters of the higher ancestors.

F. *Representations of the Ancestral Law.* Several graphical representations of Galton's form of the Ancestral Law have been provided. Perhaps the best is that devised by A. J. Meston of Pittsburg[‡], which was modified by Galton himself in a communication to *Nature*, January 27, 1898.

The diagram (p. 45) is of the following nature.

It is based on a square of unit edge; 2 and 3 represent the parents; 4, 5, 6 and 7 the grandparents; 8, 9, 10, 11, 12, 13, 14, 15 the eight great grandparents, and so on. All even numbers represent males and uneven numbers females. $2n + 1$ is the female mate of the male $2n$. The father and mother of n are always $2n$ and $2n + 1$ respectively. Every ancestor in whatever line has now got a definite number, and every number denotes a definite ancestor. For example:

(i) What is the proper number to represent a child's mother's mother's

* In the *Roy. Soc. Proc.* paper, p. 404, Galton says the dam is prepotent. But on this page and in Table II, p. 410, sire and dam should be interchanged. This slip is acknowledged by Galton himself in a letter to *Nature*, October 21, 1897, on the *Hereditary Colour in Horses*, to which we shall refer later. It does not affect his work as he has made no use of this prepotency in his calculations.

† *Roy. Soc. Proc.* Vol. LXVI, p. 158. January, 1900.

‡ See *The Horseman*, December 28, 1897, Chicago.

father's father's mother's father's father's father's mother? The child's mother is 3, her mother $2 \times 3 + 1 = 7$, her father $2 \times 7 = 14$, his father 28, 28's mother $= 2 \times 28 + 1 = 57$, 57's father is 114, 114's father is 228, 228's father is 456 and lastly 456's mother is 913, which is the number signifying the required ancestor.

Fig. 10.

(ii) What ancestor does 253 represent? 253 is odd and therefore the mother of $\frac{1}{2}(253 - 1) = 126$, who being even is the father of 63, who being odd is the mother of 31 who is the mother of 15 who is the mother of 7, who is mother of 3 the child's mother. Accordingly 253 is the child's mother's mother's mother's mother's father's mother.

This numerical nomenclature is not due to Meston, but to Galton himself, appearing in his paper of 15 years' earlier date on "Arithmetic Notation of Kinship*."

We may, to use Galton's notations, say that:

$$m \, m \, f \, f \, m \, f \, f \, f \, m = 913, \quad \text{and} \quad 253 = m \, m \, m \, m \, m \, f \, m.$$

The ancestral lines of 913 and 253 separate off at the parents of the child's maternal grandmother.

G. *Experiments in Moth-Breeding.* Before we turn to a number of papers and projects directly arising from the "Law of Ancestral Heredity," it is desirable to say a few words on the abortive moth-breeding experiments (see our p. 49). At first sight the idea of breeding moths seems exceedingly hopeful. They breed rapidly and apparently could be fairly successfully reared and bred in captivity. Accordingly Galton in January, 1887, six months after the reading of his paper on Human Eye Colour, issued for

* *Nature*, September 6, 1883.

private circulation a circular entitled *Pedigree Moths*. He had already enlisted the assistance of an able entomologist, Mr F. Merrifield, of Brighton, who had suggested working with the Purple Thorn Moth (*Selenia illustraria*). The circular consisted of two parts, one by Galton stating the purpose of the experiments, and an Appendix by Merrifield asking his fellow entomologists for advice and help. After referring to the number of pupae he needed, Merrifield asks his colleagues for information as to the number of eggs, best means of mating and laying of fertile eggs, preservation of moths, cages, possible feeding, stupefying, etc., feeding of larvae, and preserving pupae. It will be seen that the experiments were not rashly entered upon by mere lay workers. Most careful inquiries were made and the plan of operations well thought out with no undue haste. I think it needful to emphasise this as I know of two later laborious experiments in moth-breeding which also failed to attain any satisfactory results, and of which we might possibly say that Galton's experience was not turned to profit by the undertakers. Too often such experience is overlooked, or the investigator trusts to a belief in his own greater skill, and the use of a different species*.

Galton in his section of the circular, after referring to his work on regression in stature and to the Law of Ancestral Heredity as exhibited in eye-colour, states that he thinks it desirable to obtain data providing more than the three to four generations he has been able to deal with in these cases. He considers that moths would form suitable material, and that the time needful would be shortened by taking a species which bred twice a year. He proposed to measure the size of wing for six generations, and in order to measure the effect of selection to breed from large male and female, from small male and female, and from mediocre male and female. Thus he would establish three lines, and from the largest he would again pick the very large male and female, and from the smallest the very small male and female, and from the mediocre line the mediocre were again to be chosen; these latter were to act as a control series whereby to standardise the large and the small lines. After six generations Galton proposed to reverse the process, and return by selection to his original wild moth. The whole experiment would have taken at least six years. In his circular Galton makes two statements. One is that his Law of Regression leads to *his* Ancestral Law; this I believe to be incorrect. There is a relation between the ancestral correlations and a Law of Ancestral Heredity, but the numerical values given by Galton for his regression and his ancestral contributions are incompatible with each other (see p. 39 above). In the second place Galton makes the following statement:

" It is, however, highly probable from other considerations that though this simple formula may be closely true for the parents and nearly true for the grandparents, it may become sensibly and increasingly different for remoter progenitors. It is this fact that I want to investigate, because all theory concerning the nature of stability of type, and of much else, must be based on the facts of Regression, which such experiments as those proposed can alone, so far as I see, be likely to declare in a trustworthy way."

* It is possible that moths held in captivity for generations, like the silk-worm moth, would show less erratic results than wild moths reared under what must after all be very artificial conditions.

Now what is Galton's difficulty when he thus wishes to modify the contributions of the earlier progenitors?

I think his difficulty can be elucidated from passages in his other writings. In the first place, on February 2, 1887, Galton and Merrifield read papers to the Entomological Society. These are respectively entitled:

Pedigree Moth-breeding, as a means of verifying certain important Constants in the General Theory of Heredity, and

Practical Suggestions and Enquiries as to the Method of Breeding Selenia illustraria for the purpose of obtaining Data for Mr Galton, and were published in the *Transactions**.

These papers consist of an enlargement of the proposals in the Circular of January and a fuller account of the methods to be adopted in obtaining, feeding, breeding, and measuring the moths. Between January and February apparently the measurement of length of wing had been definitely fixed upon. Galton in his paper says that the laws of simple heredity as he has propounded them involve only five constants.

"These admit of being separately determined, and they are at the same time connected by an equation that serves to verify their observed values. The equation depends on the fact alluded to, that successive generations of the same population yield identical biological statistics, although each family or brood is full of variations, and although the 'median' of each characteristic in each brood is on the average *always more mediocre* than the corresponding characteristic in the mean of the two parents. The first of these events, 'fraternal variability,' increases the variability of the population as a whole, and the latter event, which I call 'Regression,' decreases it; the two can be shown to counterbalance each other and give rise to a position of stable equilibrium. The five constants are (1), the Median of the race; (2), the Quartile of the race; (3), the Quartile of the broods of the same parents, i.e. brothers and sisters; (4), the Quartile of the broods of a large number of like parents, mixed together in a single group; (5), the coefficient of Regression." (p. 28.)

Before we go further let us endeavour to interpret this important passage in terms of more modern notation and more modern conceptions of multiple correlation. Corresponding to (1) we have the mean of the race M; to (2), the standard deviation of the race σ; to (3), the variability of the family, i.e. $\sigma_f\sqrt{1-R^2}$, where R is the multiple correlation coefficient of an individual on all his ancestry and σ_f is the standard deviation of the totality of offspring; to (4), $\sigma_f\sqrt{1-r^2}$, where r is the correlation coefficient between parent and offspring; and to (5), $\sigma_f r/\sigma_p$, where σ_p is the standard deviation of the parental generation in its totality. Now Galton, I think, throughout supposes (after reducing female to male values) that $\sigma_f = \sigma_p = \sigma$, or he supposes his parental variability to be the same as that of his general population and again equal to that of the total offspring population. Further he supposes M to be the same for every generation, and this is the most stringent limitation of all, for it hinders the possibility of a continuous (or discontinuous) change of type. We can illustrate this from a statement in a second paper of Galton's, that on the Coat Colour of Basset Hounds (see our p. 40 and p. 402 of the memoir itself). Therein he writes as follows:

"The law may be applied either to total values or to deviations, as will be gathered from the following equation. Let M be the mean value from which all deviations are reckoned, and

* *Trans. Entomological Soc.* London, 1887, Part i, pp. 19–34.

let D_1, D_2, etc. be the means of all the deviations, including their signs, of the ancestors in the 1*st*, 2*nd*, etc. degrees respectively; then

$$\tfrac{1}{2}(M + D_1) + \tfrac{1}{4}(M + D_2) + \text{etc.} = M + (\tfrac{1}{2}D_1 + \tfrac{1}{4}D_2 + \text{etc.}).''$$

This is sufficient evidence that Galton had not at the time under consideration reached the full meaning of multiple regression. The Ancestral Law is nothing but the principle of multiple regression applied to ancestral inheritance, but in this case the deviations must all be measured not from a general mean, but from the mean of the corresponding generation. The Law of Ancestral Heredity is therefore independent of the change of type, if such is taking place; it can tell us nothing of the laws ruling that change of type, which is something wholly independent of it. Galton's statements that the law may be applied either to total values or deviations is only true for a population stable through the whole ancestry, whereas the application to deviations (with the proper ancestral coefficients, i.e. the multiple regression coefficients) is generally true, and if Galton had recognised this, it would have saved him from doubts as to the compatibility of his law with evolutionary changes.

That Galton recognised the difference between the Quartile of the single brood and the Quartile of the clubbed broods of like parents shows that he fully appreciated the difference between R and r. I do not think, however, that he recognised that his Ancestral Law, i.e. the values he had chosen for his coefficients, actually enforced a definite relation between R and r. But he fully realised the relation between the regression coefficient and r, his "index of correlation*." We can now continue our citation from the Entomological Society paper, which brings out Galton's difficulty:

"The laws in which these constants play a part give calculated results that prove to be closely true to observation in the ordinary cases of simple heredity, where there has been no long-continued selection, but it does not at all follow that they will hold true for the descendants of a long succession of widely divergent parents. It is this that I want to test. The point towards which Regression tends cannot, as the history of Evolution shows, be really fixed. Then the vexed question arises whether it varies slowly or by abrupt changes, coincident with changes of organic equilibrium which may be transmitted hereditarily; in other words, with small or large changes of type. Moreover the values of the Quartile in (3) and (4) cannot be strictly constant and are probably connected in part with the value of the Median and require a modified treatment by using the geometrical law of error instead of the arithmetical one (*Proc. Royal Soc.* 1879). Again the diminution of fertility and of vitality that accompany wide divergence from racial mediocrity have yet to be measured, by comparing the A [selected large size] and Z [selected small size] broods with the M [mediocre size] broods. It was assumed not to vary in the approximate theory of which I spoke." (p. 28.)

These words bring out the difficulty which arose in Galton's mind from treating regression as taking place towards a *fixed* racial value, instead of supposing it to arise from measuring deviations from the means of their groups. In this way a rather mysterious entity "the racial centre of regression" was created, which was given biological significance, when it really was only a factor in the purely statistical description of mass phenomena. Once recognise that in each generation the deviation is measured from the

---

* He speaks of his five constants being connected by "an equation."

mean of its generation and we find no incompatibility of the Ancestral Law with any change of type. What we obviously must do is to study the change of type or of successive means; regression is a wholly independent matter, and "the racial centre of regression" something which has no essential existence, biologically or statistically.

The next point that Galton makes is that the variabilities $\sigma_f \sqrt{1-R^2}$ and $\sigma_f \sqrt{1-r^2}$ cannot be quite constant; it is not clear whether he only means by this that σ_f and therefore σ, the population variability, changes with the course of evolution. This is very possible, though there is small likelihood of its being discoverable in breeding only six generations of moths, *if kept under the same environment.* Selection might equally well change type and variability; but if the distributions of frequency were normal, the type and variability would be uncorrelated, and the selection of one would not necessarily affect the other. Hence I do not see why Galton says the change in the Quartiles is probably connected with the value of the Median; least of all do I grasp why he should refer at this point to Macalister's curve for the geometric mean. Whatever application that curve may have to variation in sensations, this is the only occasion on which I have seen it suggested that it has any claim to be used for bodily measurements. It might be as justifiably used for physical measurements on man as for those on moths, but I can hardly imagine profit coming from such an application.

The last point made by Galton, namely that the fertility and vitality of stocks widely divergent from the mediocre are likely to be affected, is a very important one and is probably the reason why it is not possible to carry size selection far, at any rate by rapid strides. This has been demonstrated not only on the moth material at present under discussion, but by more recent endeavours to modify small mammals by selecting for size.

The reader who is interested in this matter would do well to refer at least to Merrifield's first report * on the moth-breeding experiments. He will then quickly understand why they failed to satisfy Galton's thirst for data! The spring and autumnal broods were really dimorphous, the males appeared to be larger in one and the females in the other; the wing lengths were not the same in the two. Thus the fact of two broods a year would certainly not expedite matters. Further, the fertility of the largest and the smallest was reduced below that of the mediocre, and when Merrifield took steps to obtain by forcing under higher temperatures more frequent broods, not only did he increase the size of his moths' wings, but the "giant" line and the "dwarf" line became sterile and he had to start again from the mediocre. In fact artificial means had to be used to get the moths from the pupae near enough in time to breed with one another. Further, changes in environment or food had to be made to hasten the larvae to the pupal stage because food supplies were getting low. And all these changes appear to have been associated with variations in size so that finally the irregularities were too widespread for any statistical treatment of the data, or as Galton himself

---

* *Trans. Entomological Soc.* London (Dec. 7, 1887), 1888, pp. 123–136.

expressed it ten years later: "No statistical results of any consistence or value could be obtained from them*." Thus ended what had at first sight appeared to be a hopeful series of experiments, experiments upon which much thought and labour had been expended.

H. *Correlations and their Measurement.* As I have already pointed out the conception that the regression coefficient for inheritance could be applied to a measure of the relationship of associated variates, provided each was measured in terms of its own scale of variability, first occurred to Galton while he was taking a walk in the grounds of Naworth Castle in the year 1888 (see p. 393 of Vol. II). On December 5, 1888, Galton sent to the Royal Society a paper read fifteen days later and entitled: "Co-relations and their Measurement, chiefly from Anthropometric Data†." The twentieth of December is therefore the birthday of the conception of correlation in biometric data as apart from the idea of regression in heredity which Galton had reached some years earlier, without perceiving at once its capacity for wide generalisation in the treatment of associated variates in all living forms.

Like so much of Galton's work the present paper reaches results of singular importance by very simple methods; his methods are indeed so simple that we might almost believe they must lead to a fallacy had not Galton deduced thereby the correct answer. It is the old experience that a rude instrument in the hand of a master craftsman will achieve more than the finest tool wielded by the uninspired journeyman.

The first three paragraphs of this memoir define Galton's method of considering correlation, and indicate that in 1888 even the spelling of the word had not been fixed‡:

"'Co-relation or correlation of structure' is a phrase much used in biology, and not least in that branch of it which refers to heredity, and the idea is even more frequently present than the phrase; but I am not aware of any previous attempt to define it clearly, to trace its mode of action in detail, or to show how to measure its degree.

"Two variable organs are said to be co-related when the variation of the one is accompanied on the average by more or less variation of the other, and in the same direction. Thus the length of the arm is said to be co-related with that of the leg, because a person with a long arm has usually a long leg, and conversely. If the co-relation be close then a person with a very long arm would usually have a very long leg; if it be moderately close then the length of his leg would only be long, not very long; and if there were no co-relation at all then the length of his leg would on the average be mediocre. It is easy to see that co-relation must be the consequence of the variations of the two organs being partly due to common causes. If they were wholly due to common causes, the co-relation would be perfect, as is approximately the case with the symmetrically disposed parts of the body. If they were in no respect due to common causes, the co-relation would be *nil*. Between these two extremes are an endless number of intermediate cases, and it will be shown how the closeness of co-relation in any particular case admits of being expressed by a simple number.

"To avoid the possibility of misconception it is well to point out that the subject in hand has nothing whatever to do with the average proportions between the various limbs in different

---

* *Roy. Soc. Proc.* Vol. LXI, p. 402.

† *Ibid.* Vol. XLV, pp. 135–145.

‡ Five years later in 1893 when the volume containing the letter *C* of the *Oxford English Dictionary* was issued, the Galtonian or biometric sense of " correlation " was not given.

races*, which have been often discussed from early times up to the present day, both by artists and by anthropologists. The fact that the average ratio between the stature and the cubit is as 100 to 37† or thereabouts does not give the slightest information about the nearness with which they vary together. It would be an altogether erroneous inference to suppose their average proportion to be maintained so that where the cubit was, say, one-twentieth longer than the average cubit, the stature might be expected to be one-twentieth greater than the average stature, and conversely. Such a supposition is easily shown to be contradicted both by fact and theory." (*loc. cit.* pp. 135–6.)

Let us now describe Galton's procedure. In the first place Galton does not use means, he uses throughout medians, both for his marginal totals and his arrays. Further he does not use standard deviations, he makes use of the quartile measurements. Thus if Q_1, M and Q_3 be the measurements at first, second and third quartile divisions, he takes M as his median and $\frac{1}{2}(Q_3 - Q_1)$ as his measure of variation. Thus his results, unlike our modern treatment, depend essentially on assuming that all his data follow a normal (or "curve of errors") distribution‡. If M_c be the median of any character c and $_bM_c$ the median of an array of this character for a given value b of a second character c', then Galton plots:

$$\frac{_bM_c - M_c}{\frac{1}{2}(Q_3 - Q_1)_c} \quad \text{to} \quad \frac{b - M_{c'}}{\frac{1}{2}(Q_3 - Q_1)_{c'}}.$$

In other words he reduces the deviation of an array median from the population median to its unit of variation obtained from the quartiles, and plots this to the deviation of the second character from its median reduced likewise to its own unit of variation. Then he plots:

$$\frac{_aM_{c'} - M_{c'}}{\frac{1}{2}(Q_3 - Q_1)_{c'}} \quad \text{to} \quad \frac{a - M_c}{\frac{1}{2}(Q_3 - Q_1)_c},$$

where a is a value of the first character and $_aM_{c'}$ the median of the corresponding array of the second character, and thus gets a second series of points. He takes six or seven values of a and of b, plots two sets of six or seven points and notes that the first and second series of points are nearly on one and the same straight line§. He draws this straight line as closely as he can to the points and through the median, and reads off its slope. This slope is Galton's measure of co-relation. If we take the mean deviation of c' for a given value of c, Galton calls c the "Subject" and c' the "Relative," but perhaps it would be best to call the latter the "Co-relative." Galton's data consisted of about 350 males of 21 years and upwards, of whom the majority were young students, measured in his Laboratory in 1888. He deals with

---

* [The *variation* in the ratio of stature to cubit does, however, provide a means of determining the correlation. K.P.]

† [Rather 100 to 27 or thereabouts on Galton's numbers, i.e. 67·20″ for stature and 18·05″ for cubit. K.P.]

‡ In the table given on p. 52 for the correlation of Stature and Left Cubit it is very difficult to see any approximation to normality in the distribution of stature.

§ In order to get the *same* straight line, if c be the subject and c' the co-relative, and the "subject" axis horizontal, then it is needful when c' is subject and c co-relative to plot c' along the *same* axis as was used in the first case for c. In other words the character axes must be interchanged.

six characters: Head Length, Head Breadth, Stature, Length of Left Middle Finger, Left Cubit and Height of Right Knee. But he only provides as illustration one table such as we now term a correlation table, and one diagram illustrating how he found what we now term the correlation co-efficient. The table and diagram dealing with the co-relation of stature and cubit are given below. Readings were made to one-tenth of an inch.

Correlation Table for Stature and Cubit.

Length of Left Cubit in inches, 348 adult Males.

	Under 16·45	16·45— 16·95	16·95— 17·45	17·45— 17·95	17·95— 18·45	18·45— 18·95	18·95— 19·45	19·45 and above	Totals
Above 70·45	—	—	—	1	3	4	15	7	30
69·45—70·45	—	—	—	1	5	13	11	—	30
68·45—69·45	—	1	1	2	25	15	6	—	50
67·45—68·45	—	1	3	7	14	7	4	2	38*
66·45—67·45	—	1	7	15	28	8	2	—	61
65·45—66·45	—	1	7	18	15	6	—	—	47*
64·45—65·45	—	4	10	12	8	2	—	—	36
63·45—64·45	—	5	11	2	3	—	—	—	21
Below 63·45	9	12	10	3	1	—	—	—	35*
Totals	9	25	49	61	102	55	38	9	348

(Stature in inches — left axis label)

* Printed as 48, 48, and 34 respectively in the *Roy. Soc. Proceedings.*

Diagram illustrating the Graphical Process of finding the Slope of the Regression Line, i.e. the Correlation Coefficient of to-day's terminology.

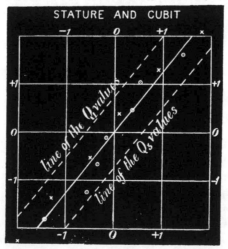

Fig. 11.

Galton says he constructed tables and diagrams like the above. "It will be understood that the Q value is a universal unit applicable to the most varied measurements, such as breathing capacity, strength, memory, keenness of eyesight, and enables them to be compared together on equal terms not-withstanding their intrinsic diversity. It does not only refer to measures of length, though partly for the sake of compactness, it is only those of length that will be here given as examples" (*loc. cit.* p. 137). Galton already saw clearly that his new method enabled comparison to be made on equal terms between variates with such intrinsic diversity as acuity of vision and head breadth*.

I have endeavoured to check Galton's work. I expect he found his medians and quartiles by plotting an "ogive curve" (see our p. 31 and Plate II) and smoothing it. The process of checking is rendered difficult by the following statements on p. 138:

"It is unnecessary to extend the limits of Table II [that of stature and cubit reproduced above] as it includes every line and column in my MS. table that contains not less than twenty entries. None of the entries lying within the flanking lines and columns of Table II were used."

The first statement seems to suggest that the whole table has not been printed, the second leaves one in doubt as to how to find the medians of the arrays, or indeed of the marginal totals, if none of the entries in the flanking lines and columns had been used. Unfortunately I have not succeeded in discovering the original work and manuscript tables for this memoir among Galton's papers†. Putting aside the possibility of re-examining Galton's own work by more modern methods, we can, I think, indicate how closely his semigraphic median, quartile and regression slope methods accord with those obtained from much longer series by more accurate processes. First let us consider the correlation coefficients:

| Character Pair | Correlation Coefficient | |
	As found by Galton from 350 Male Adults	As found by Macdonell from 3000 Criminals
Stature and Cubit	0·80 {0·8290}	0.7999
Stature and Head Length ...	0·35	0·3399
Stature and Middle Finger ...	0·70	0·6608
Cubit and Middle Finger ...	0·85	0·8464
Head Length and Head Breadth	0·45	0·4016
Stature and Height of Knee ...	0·90 {0·8665}	—
Cubit and Height of Knee ...	0·80 {0·8028}	—

The values in the first column of this table were the first organic correlations ever published, and on that account are of great historical interest.

* It is not without interest to note that more than a quarter of a century later, Major Leonard Darwin could assert that the influences of environment and heredity could not be compared, because there was no common unit of measurement applicable to them both! He appeared still ignorant of Galton's use of Q. See *Eugenics Review*, Vol. v, p. 152.

† My colleague, Miss E. M. Elderton, has taken out the first 348 entries for male adults 21 years and upwards from Galton's Laboratory records, and the resulting values from her tables, computed by modern methods, are given in brackets in the above and the following tables. Our table for stature and cubit differs somewhat from Galton's but with a probable error of ·0113 the correlation is hardly significantly different from Galton's value. Both Knee Height and Cubit are measured in the Anthropometric Laboratory at University College, but the former is measured to the lowest point of the patella with the subject standing at rest, while Galton measured to the top of the knee with the subject sitting. Galton deducted the measured heel, we measure with boots off. Our correlation for male students of Knee Height and Cubit is only 0·66.

Macdonell's values, obtained by a far more refined and accurate method, indicate —especially when we remember that they are for a very different population—how successfully Galton solved his problem. Doubtless he was somewhat aided by the fact that anthropometric physical measurements are far more nearly normal than many other variates. Had his distributions been more skew, his median estimates would not have given as accurately the correlation coefficients. We can now compare the mean or median values and the standard deviations as found from the quartiles with later results:

Character	Means			Standard Deviations		
	Galton (Adults 21 and upwards)	Macdonell (Criminals)	Schuster (Oxford Students)	Galton (Adults 21, and upwards)	Macdonell (Criminals)	Schuster (Oxford Students)
Stature (cm.) ...	170·69*	166·46	176·50 {170·25}	6·58	6·45	6·61 {6·80}
Cubit (cm.)	45·70	45·06	— {45·85}	2·11	1·96	— {2·04}
Height of Knee (cm.)	52·00	—	— {52·15}	3·01	—	— {2·62}
Middle Finger (mm.)	115·32	115·24	—	5·63	5·48	—
Head Length (mm.) ...	193·55	191·66	196·05	7·11	6·05	6·23
Head Breadth (mm.)	152·40	150·04	152·84	6·82	5·01	4·92
Cephalic Index ...	78·74	78·28	78·02	·—	—	2·92

Considering the difference of social class in the three series, Galton's results can hardly have exception taken to them, except in the case of the variabilities of Head Length and Head Breadth. These are excessive, but as we have not the original tables from which the quartiles were determined, it is not possible to investigate wherein they are anomalous†.

The degree of accordance reached by Galton's process may be illustrated by his tables for Stature and Knee Height:

Stature in inches	Mean of corresponding Knee Heights		Height of Knee in inches	Mean of corresponding Statures	
	Observed	Calculated		Observed	Calculated
70·0	21·7 (30)	21·7	22·2	70·5 (23)	70·6
69·0	21·1 (50)	21·3	21·7	69·8 (32)	69·6
68·0	20·7 (38)	20·9	21·2	68·7 (50)	68·6
67·0	20·5 (61)	20·5	20·7	67·3 (68)	67·7
66·0	20·2 (49)	20·1	20·2	66·2 (74)	66·7
65·0	19·7 (36)	19·7	19·7	65·5 (41)	65·7
—	—	—	19·2	64·3 (26)	64·7

The figures in brackets give the numbers of individuals upon whom the observed medians of the arrays were determined. It will be observed that the accordance between observation and theory is again very good.

* For a general hospital population: Stature = 170·59 (*Biometrika*, Vol. IV, p. 126).

† Galton says "The head length is the maximum length measured from the notch between and just below the eyebrows" (p. 137). Is this the glabella?

Table V on Galton's p. 143 is noteworthy. In Column 3 we have the co-efficients of correlation tabled under the now familiar symbol r. In Column 4 we have the values of $\sqrt{1-r^2}$, to enable the Quartile of the arrays to be found. In Column 5 we have, placed one under the other, the two regression coefficients, and in Column 6 in the same manner the Quartiles of the arrays (i.e. $\cdot 67449\,\sigma_x\sqrt{1-r^2}$ and $\cdot 67449\,\sigma_y\sqrt{1-r^2}$)*. Throughout, without referring directly to the matter, Galton assumes linear regression and homoscedasticity, i.e. he is thinking in terms of the bivariate *normal* surface. Next he draws attention to the relation of his present work to his former work on heredity.

On the fifth line of p. 144, he has the words: "from $\dfrac{1}{1\cdot 7}$ to $\dfrac{1}{3}\times\dfrac{1}{1\cdot 2}=1$

to $0\cdot 44$, which is practically the same." This should read "from $\dfrac{1}{1\cdot 7}$ to $\dfrac{1}{3}\times\dfrac{1}{1\cdot 2}=1$ to $0\cdot 47$, which is identically the same," as it should be since it expresses the coefficient of correlation found from the second regression line. Galton emphasises the importance of the reduction in the variability of the array, as measured by $\sqrt{1-r^2}$, and points out how this affects the efficiency of Bertillon's system of identification by anthropometric measurements. Bertillon had asserted that his measurements were independent variates. A reference to Plate LII of our second volume will show that Galton had chosen several of Bertillon's "independent" measurements and determined their actual correlation.

Galton next outlines a method by which the influence of n variates on another might be determined. He suggests that after transmuting the variates we should sum them, when the probable error of the sum would " be $\sqrt{n}$, if the variates were perfectly independent, and n if they were rigidly and perfectly related. The observed value would be almost always somewhere intermediate between these extremes, and would give the information that is wanted" (p. 145).

This would not, I believe, be a feasible method of approaching multiple correlation; it neglects the possibility of negative correlations, and does not provide for the influence on one variate of all the remainder. It is an attempt to obtain a sort of average value of the interlinkage of a system of n variates†. I do not think that at this time Galton had realised the existence and importance of negative correlation.

* A large proportion of values in the 5th and 6th columns have rather serious numerical errors, corrected by Galton on a copy of the paper in my possession. He also states thereon that he wishes to change the symbol r to ρ, presumably because he was thinking of it as the "correlation coefficient," not as the regression coefficient, when units are reduced to respective variabilities. The regression coefficient without reduction he had termed r in his memoir on stature.

† Let $x_1, x_2, \ldots x_s, \ldots x_n$ be the n variates, and $\sigma_1, \sigma_2, \ldots \sigma_s, \ldots \sigma_n$ their standard deviations, $\bar{x}_1, \bar{x}_2, \ldots \bar{x}_s, \ldots \bar{x}_n$ their means. Then if $\chi = \overset{n}{\underset{1}{S}}(x_s-\bar{x}_s)/\sigma_s$, we have:

$$\sigma_\chi{}^2 = n + 2\,S'(r_{ss'})$$
$$= n, \text{ if all the correlations } r_{ss'} \text{ are zero,}$$
$$= n + 2\tfrac{1}{2}\,n\,(n-1) = n^2, \text{ if all the correlations are } plus \text{ one.}$$

Hence $\sigma_\chi = \sqrt{n}$ and n in the two cases respectively, as Galton says. But the actual value of σ_χ

Galton sums up his results as follows*. Let x be the deviation of the subject, and y_1, y_2, y_3, etc. the corresponding deviations of the correlative, all deviations being reduced to their proper unit of variability, and also let the mean of the y deviations for the given x be $\bar{y}_x$, then we find:

(1) That $\bar{y}_x = rx$ for all values of x; (2) that r is the same, whichever of the two variables is taken for the subject; (3) that r is always less than 1; (4) that r measures the closeness of correlation.

It will be seen at once that we have here the first fundamental statement as to the correlation coefficient and its properties. Probably Galton did not recognise that $r = 0$ does not signify independence of the two variates, only the independence of *means* of arrays. In addition to this, complete independence involves the arrays being similar and similarly placed curves. It was not till normal distributions were seen to be non-universal that the distinction between the vanishing of r and the absolute independence of variates was fully recognised. For the same reason the idea of non-linear regression did not cross Galton's mind. He got as far as an acceptance of the normal frequency distribution permitted. Only when we look at what has happened since 1888, do we realise the importance of that short paper on "Co-relations"! Thousands of correlation coefficients are now calculated annually, the memoirs and text-books on psychology abound in them; they form, it may be in a generalised manner, the basis of investigations in medical statistics, in sociology and anthropology. Shortly, Galton's very modest paper of ten pages from which a revolution in our scientific ideas has spread is in its permanent influence, perhaps, the most important of his writings. Formerly the quantitative scientist could only think in terms of causation, now he can

would not be proportional to the sum of the $r_{ss'}$ even if they were all positive. Perhaps a better measure of the same type would be to use σ_χ^2, where

$$\chi = \overset{n}{\underset{1}{S}}(x_s - \bar{x}_s)^2/\sigma_s^2 \text{ and } \bar{\chi} = n;$$

hence:

$$\sigma_\chi^2 = \text{mean } (\chi - \bar{\chi})^2$$

$$= \text{mean } \left\{ \overset{n}{\underset{1}{S}}(x_s - \bar{x}_s)^4/\sigma_s^4 + 2S'(x_s - \bar{x}_s)^2(x_{s'} - \bar{x}_{s'})^2/\sigma_s^2\sigma_{s'}^2 \right.$$

$$\left. - 2nS(x_s - \bar{x}_s)^2/\sigma_s^2 + n^2 \right\}$$

$$= 3n + 2S'(1 + 2r_{ss'}^2) - 2n^2 + n^2$$

$$= 2n + 4S'(r_{ss'}^2),$$

if the variates follow normal distributions, and thus σ_χ^2 lies between $2n$ and $2n^2$. This at any rate would present no difficulty arising from the existence of negative correlations. We see, however, from this result that possibly the best measure, u, of the total correlativity in a system would be simply to take

$$u = \frac{2S(r_{ss'}^2)}{n(n-1)},$$

for in this case u will always lie between 0 and 1, the former value corresponding to no association in the variates of the system, and the latter to perfect correlation of all of them.

* Galton has interchanged his x and y variates. The paper shows here as elsewhere signs of haste in preparation.

think also in terms of correlation. This has not only enormously widened the field to which quantitative and therefore mathematical methods can be applied, but it has at the same time modified our philosophy of science and even of life itself. The words which I have cited at the beginning of this chapter show that Galton, if he expressed himself modestly, still realised the importance of his work. The root idea at the bottom of correlation must not be treated as merely rebuilding on a securer mathematical basis statistical science. It is a much greater innovation which touches in its philosophical aspects the epistemology of all the sciences.

I have already referred (Vol. II, pp. 380–386) to Galton's attempt to introduce the conception of correlation* to anthropologists in 1889. It was a hopeless task! Most physical anthropologists in this country lack a. thorough academic training, and statistical methods will only penetrate here after they have been adopted in Germany and France as they are being adopted in Russia, Scandinavia and America. English intelligence is distributed according to a very skew curve, with an extremely low modal value; we have produced great men, who have propounded novel ideas, but our mediocrity fails to grasp them or is too inert to turn them to profit. Years later these ideas come back to England, burnished and luring, through foreign channels, and mediocrity knows nothing of their ancestry!

In 1889 Galton read at the British Association (Newcastle-upon-Tyne) a note entitled: "Feasible Experiments on the Possibility of transmitting Acquired Habits by means of Inheritance"; it is published in the *B. A. Report*, p. 620, also in *Nature*, Vol. XL, p. 610, October, 1889. Galton considers that creatures reared from eggs would be most satisfactory and suggests fish, fowls and moths. He considers that fish may be taught to adopt habits not conformable to their nature (Möbius' experiment with pike and minnows). Fowls have an instinctive dread of certain insects, but might be taught to eat mimetic and harmless insects. Larvae are fastidious in their diet, but can be induced to take food which they naturally avoid, and which is found perfectly wholesome. Would acquired habits of this kind be in any case transmitted to their offspring?

I. *Natural Inheritance.* The ideas on heredity and correlation which had been working in Galton's mind during the decade of the 'eighties found final expression in his book entitled *Natural Inheritance*, published in 1889 when Galton was 67 years of age. It may be said that this publication created Galton's school; it induced Weldon, Edgeworth and the present biographer to study correlation and in doing so to see its immense importance for many fields of inquiry. It is idle to overlook the haste with which it was prepared and the many slips and positive errors to be found in its pages, but no one who studied it on its appearance and had a receptive and sufficiently trained mathematical mind could deny its great suggestiveness, or be other than grateful for all the new ideas and possible problems which it provided. The methods of

---

* Spelled thus in the Presidential Address of Jan. 2, 1889, and, I think, ever afterwards.

Natural Inheritance may be antiquated now, but in the history of science it will be ever memorable as marking a new epoch, and planting the seed from which sprang a new calculus, as powerful as any branch of the old analysis, and valuable in just as many fields of scientific research.

In its application to inheritance the work suffers from the same misinterpretation of "regression" that I have several times referred to, namely making the regression of offspring of given parentage a great biological law, when it really arises from the clubbing together of all offspring of given parentage *without regard to their earlier ancestry.* Given selected parentage and grandparentage alone, then with our present numerical values of the multiple correlation constants, it seems highly probable that the progeny of selected offspring would progress rather than regress on their parents and grandparents. In other words, given a line in which by chance or artificial selection there has been marked ancestry for two or three generations, and which is then isolated or inbred, there is reason to believe it would progress even beyond its ancestry rather than regress. Statistical investigations of heredity since 1889 seem to indicate a progressive evolution in selected lines, rather than a general regression to a population mean*. That would only arise from the far too frequent mating with mediocrity or worse than mediocrity. If Galton's misinterpretation of regression runs through *Natural Inheritance*, and makes him appeal to "sports" for evolutionary changes; if the reader is puzzled to know why Galton should study "variations proper," which according to him have no permanent value for evolution; still the book is a great book, for it applies a wholly new calculus—if one still in its infancy—to an important biological problem.

I think, however, that Galton fully grasped how much more important was his method than its special application. He writes that his conclusions

"depend on ideas that must first be well comprehended, which are now novel to the large majority of readers and unfamiliar to all. But those who care to brace themselves to a sustained effort, need not feel much regret that the road to be travelled over is indirect, and does not admit of being mapped beforehand in a way that they can clearly understand. It is full of interest of its own. It familiarises us with the measurement of variability, and with curious laws of chance that apply to a vast diversity of social subjects. This part of the inquiry may be said to run along a road on a high level, that affords wide views in unexpected directions, and from which easy descents may be made to totally different goals to those we have now to reach. I have a great subject to write upon, but feel keenly my literary incapacity to make it easily intelligible without sacrificing accuracy and thoroughness." (Chapter i, pp. 2–3.)

Galton in his Introductory Chapter states that there are three problems with which he will be principally concerned. The first problem is to determine how a population can, under the laws of heredity, keep stable from generation to generation. The second problem regards the *average* share contributed to the character in the offspring by each ancestor severally. The third problem is to measure numerically the nearness of kinship in

---

* There has always been this element of truth in Johansen's theory of "pure lines," that selected lines do not regress if they are isolated or inbred. The doubtful dogma of that theory is that exceptional members of a "pure line" are only "fluctuating variations," and so no further selection is of any value within a "pure line."

various degrees (pp. 1–2). Such are the three fundamentally novel problems which Galton set himself in *Natural Inheritance*; we shall endeavour to show the extent to which he has solved them, or at least has suggested methods of solving them, in the following discussion of that work.

Chapters II and III are general in character, expressing Galton's own views on heredity, and erring, if at all, in rather too much appeal to analogy. In the first of these chapters Galton states his opinion as to "natural" and "acquired" characters, indicating that he considers the inheritance of the latter extremely doubtful; he emphasises the importance of closely criticising the evidence offered in each case to prove the transmission of acquired faculties, citing especially the possibilities of intra-uterine influence*. He refers to the difficulty of combining male and female measures, and states that:

"Fortunately we are able to evade it altogether by using an artifice at the outset, else, looking back as I now can, from the stage which the reader will reach when he finishes this book, I hardly know how we should have succeeded in making a fair start. The artifice is never to deal with female measures as they are observed, but always to employ their male equivalents in place of them. I transmute all the observations of females before taking them in hand, and thenceforward am able to deal with them on equal terms with the observed male values." (p. 6.)

Galton for stature multiplied every female stature by 1·08 to reach its male equivalent, or added about one inch to every foot of female stature. He does not tell us how he demonstrated that equivalence, whether from the ratio of the mean values in men and women, or more adequately by finding it held (approximately) for all grades†. The true method is to reduce each deviation from the mean by dividing by its standard deviation, or other measure of variability, and it was an inspiration on Galton's part that led him to recognise that at any rate for the case of stature, the ratio of variabilities in male and female was close to the ratio of their mean values. See our p. 15 above.

On p. 7 Galton deals with what he terms *Particulate Inheritance*. He recognises that an individual may possess characters, which are known to have existed in an ancestor, but were not in the immediate parents. From this idea of latent characteristics Galton reaches the conception of inheritance in the individual as a "mosaic" of ancestral factors, and illustrates his views by two analogies, that of a builder's yard, with fragments of old buildings ready to be used again (p. 8), and the vegetations on two islands which spread to adjacent islets (pp. 10–12). I think he would have done better to have retained his earlier conception of the "stirp" (see our Vol. II,

---

* The complexity of this latter source must be borne in mind, if we can accept Galton's statements on pp. 15–16, that not a drop of blood passes from mother to child, and yet that a mother's system may be "drenched with alcohol and the unborn infant alcoholised" during all its intra-uterine existence.

† Probably in this latter way; see his p. 42, where he says we are to transmute female to male measures by comparing their respective "schemes," and devising a formula which will change one to the other. A "scheme," supposed normal, depends on two constants, the mean and the variability. Galton does not point this out, or state the inference which follows from his use of the factor 1·08.

p. 185), that is of the continuity of the germ-plasm. It is only in a figurative sense that we can look upon the inheritance of the individual as a mosaic, and speak of the contribution of an ancestor to the result. The individual is the product of the germ-plasms that go to his production, not of the individual ancestors. The study of the characters of the individual ancestors is only ancillary to a study of the possibilities of those germ-plasms. The correlation of a somatic character in a great grandparent, say, and great grandchild is not in any sense a real measure of what the former contributes to the latter, nor is the corresponding multiple regression coefficient such a measure. We are testing what *on the average* we can predict of the somatic characters of the offspring from a knowledge of what the germ-plasms of the "stirp" have produced in the past. In other words the term "contribution of an ancestor" should be interpreted as, or be replaced by, "contribution of the ancestor to the prediction formula." *It is in no sense a physical contribution to the germ-plasms on which the somatic characters of the offspring depend.* I do not think that anyone acquainted with the theory of multiple correlation would interpret the Law of Ancestral Heredity in any other sense; but Galton's use of the terms "particulate inheritance," "mosaic," "heritage from distant progenitors," must be admitted to be easily capable of mis-interpretation.

Galton then deals with the "heritages that blend and those that are mutually exclusive," citing as an illustration of the former, skin-colour in crosses between white and negro, and of the latter eye-colour. He does not here, any more than in his fundamental paper on eye-colour (see our p. 34), explain for what reason he assumes the distribution of eye-colour in the array of offspring due to a definite ancestry will be in the same proportions as in the case of a blended character in an individual offspring. Galton concludes that:

"There are probably no heritages that perfectly blend, or that absolutely exclude one another, but all heritages have a tendency in one or the other direction, and the tendency is often a very strong one....A peculiar interest attaches itself to mutually exclusive heritages, owing to the aid they must afford to the establishment of incipient races." (pp. 13–14.)

So far, however, as the struggle for existence and evolution are concerned, this last sentence must mean that a mosaic of the characters of two distinct races is for some environmental reasons more fitting than either pure race, and what is more, that the characters in the new mixed race will be stable and not segregate out again.

In the concluding paragraph we read:

"The incalculable number of petty accidents that concur to produce variability among brothers, makes it impossible to predict the exact qualities of any individual from hereditary data. But we may predict average results with great certainty, as will be seen further on, and we can also obtain precise information concerning the penumbra of uncertainty that attaches itself to single predictions. It would be premature to speak further of this at present; what has been said is enough to give a clue to the chief motive of this chapter. Its intention has been to show the large part that is always played by chance in the course of hereditary transmission, and to establish the importance of an intelligent use of the laws of chance and of the statistical methods that are based upon them, in expressing the conditions under which heredity acts." (pp. 16–17.)

Galton in his Chapter III deals with the theory of *Organic Stability*, illustrating it by the model of a polygonal slab, which has positions of stable equilibrium with various degrees of stability, i.e. which may require large or only small displacements to pass from one position of equilibrium to a second. He considers that his model (see Fig. 12) shows how the following conditions

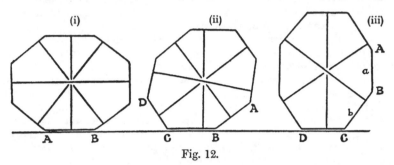

Fig. 12.

may co-exist: (1) Variability within narrow limits without prejudice to the purity of the breed (i); (2) Partly stable sub-types (ii); (3) Tendency, when much disturbed, to revert from a sub-type to an earlier form; (4) Occasional sports which may give rise to new types (iii) (pp. 27–30). Again the whole argument is one of analogy, and the reader may be pardoned a little vexation when he finds such important topics as the *Stability of Sports* and *Infertility of Mixed Types* only discussed (pp. 30–32) by reference to the analogy of hansom cabs and the impossibility of their useful blend with four-wheelers*!

The fact, I think, is that Galton's own ideas at this time were obscured by his belief that the ancestors actually did contribute to the heritage; he regarded the incipient structure of the new being to be the result of a clash of elements contributed from many ancestral sources, and the resulting building up out of more or less opposing elements of a particulate individual inheritance as the result of chance†. A further source of difficulty to Galton in his interpretation of hereditary phenomena lay in his mistake as to the nature of regression. This forced on him the conception of positions of stable equilibrium, each with its own centre of regression, and led him to the view that evolution must generally proceed by sports, and not by minute steps. It is true that on p. 32 he draws a distinction between the two views that the steps *may* be small and that they *must* be small, but as he has elsewhere applied his view of regression to indicate that small steps *cannot* be the source of evolution, the distinction is not really much of a concession (see our pp. 31–2). The following words of Galton deserve, however, to be quoted not only

* I find my copy of the *Natural Inheritance*, read and annotated forty years ago, defaced by many marginal notes expressing anger at Galton's analogies in this Chapter. But these notes were written before I had read and grasped the value of much of the later work in the book.

† Of course the Mendelian appeals to the same doctrine of chance to explain the variation in the members of an individual brood or litter, but he does so on the basis of homogeneous germ cells having a heterogeneous factor formula. I am inclined to believe that the germ cells of the same individual are not always and absolutely homogeneous, at any rate in the higher organisms, and that the clash of elements to be determined by chance need not lie in the factors of the formulae of the gametes, but in the fertilising germ cells themselves.

because they express his own strong convictions, but also because they may serve as a warning that we must appeal with caution to the continuity of the palaeontological record :

"An apparent ground for the common belief [that evolution proceeds by minute steps only*] is founded on the fact that whenever search is made for intermediate forms between widely divergent varieties, whether they be of plants or of animals, of weapons or utensils, of customs, religion or language, or of any other product of evolution, a long and orderly series can usually be made out, each member of which differs in an almost imperceptible degree from adjacent specimens. But it does not at all follow because these intermediate forms have been found to exist, that they are the very stages that were passed through in the course of evolution. Counter evidence exists in abundance, not only of the appearance of considerable sports, but of their remarkable stability in hereditary transmission. Many of the specimens of intermediate forms may have been unstable varieties, whose descendants had reverted; they might be looked upon as tentative and faltering steps taken along parallel courses of evolution and afterwards retraced. Affiliation from each generation to the next requires to be proved before any apparent line of descent can be accepted as the true one. The history of inventions fully illustrates this view. It is a most common experience that what an inventor knew to be original, and believed to be new, had been invented independently by others many times before, but had never become established. Even when it has new features, the inventor usually finds on consulting lists of patents, that other inventions closely border on his own. Yet we know that inventors often proceed by strides, their ideas originating in some sudden happy thought suggested by a chance occurrence, though their crude ideas may have to be laboriously worked out afterwards. If, however, all the varieties of any machine that had ever been invented, were collected and arranged in a museum in the apparent order of their evolution, each would differ so little from its neighbour as to suggest the fallacious inference that the successive inventors of that machine had progressed by means of a very large number of hardly discernible steps." (pp. 32–3.)

In concluding this chapter Galton apologises for largely using metaphor and analogy, on the ground that he wished to avoid any "entanglements with theory," as no complete theory of inheritance had yet been propounded that met with general acceptance (p. 34). This seems to me to show that Galton looked upon his statistical analysis of inheritance not as a theory of heredity, but as a description of hereditary facts, which it undoubtedly is.

Chapter IV deals with Galton's "ogive curve" (see our pp. 30–31) by which he represents a frequency distribution by aid of grades or percentiles. Galton had discussed this manner of representation in numerous earlier papers, and we may refer to Plate II for a graphic representation of his curve. The only novel point in Chapter IV is the suggestion, not very fully worked out, that the scheme of grades or percentiles might be applied to "inexact measures," i.e. to our present so-called "broad categories," and that these may be measures of a great variety of characters including relative professional success. He cites on this latter point Sir James Paget's analysis of the successes of 1000 of his pupils at St Bartholomew's Hospital. Sir James made five classes: (a) Distinguished, (b) Considerable, (c) Moderate, (d) Very limited success, and in the fifth class (e) he put Failures. Galton made the numbers in each 28, 80, 616, 151, and 125 respectively. Among the foremost were the three professors of anatomy in Cambridge, Edinburgh and

* It is a strange but widely spread notion that those who believe in continuous variation of a non-fluctuating character, must *ipso facto* suppose evolution to proceed by "minute steps." Given a race with mean cephalic index of 75 and a range in index from 65 to 85, there is nothing to prevent by isolation the establishment of a brachycephalic race of cephalic index 82— a spring as great as from Englishman to Jew—without transition through all the small intermediate steps from 75 to 82.

Oxford, and the three last were two men who committed suicide under circumstances of great disgrace and Palmer, the Rugeley murderer, who was hanged. There is possibly little knowledge to be obtained from the result for a single medical school, but comparative statistics for several would be of considerable value.

Chapter V deals with *Normal Variability*, and Galton shows how the distribution depends only on the two constants, the median and the quartile, and further that if two individuals whose grades are known be actually measured, then the median and quartile, and so the whole distribution of variation, can be discovered (p. 62, footnote, and cf. our Vol. II, p. 385). The origin of the normal distribution is illustrated mechanically by aid of the " quincunx " (see our pp. 9 and 10). Nor is Galton able to avoid becoming poetically enthusiastic in a paragraph headed *The Charms of Statistics*, for he writes :

"It is difficult to understand why statisticians commonly limit their inquiries to averages and do not revel in more comprehensive views. Their souls seem as dull to the charm of variety as that of the native of one of our flat English counties, whose retrospect of Switzerland was that, if its mountains could be thrown into its lakes, two nuisances would be got rid of at once. An average is but a solitary fact, whereas if a single other fact be added to it, an entire Normal Scheme, which nearly corresponds to the observed one, starts potentially into existence.

"Some people hate the very name of statistics, but I find them full of beauty and interest. Whenever they are not brutalised, but delicately handled by the higher methods, and are warily interpreted, their power of dealing with complicated phenomena is extraordinary. They are the only tools by which an opening can be cut through the formidable thicket of difficulties that bars the path of those who pursue the Science of Man." (pp. 62–63.)

Galton at the end of his Chapter V gives the two fundamental propositions on which his normal surface for the distribution of characters in two relatives depends. He envisages it in the following manner.

"(1) Bullets are fired by a man who aims at the centre of a target, which we will call its *M*, and we will suppose the marks that the bullets make to be painted red, for the sake of distinction. The system of lateral deviations of these red marks from the centre *M* will be approximately Normal, whose *Q* [Probable Error] we will call *c*. [This is the distribution of the first relative.] Then another man takes aim, not at the centre of the target, but at one or other of the red marks, selecting these at random. We will suppose his shots to be painted green. The lateral distance of any green shot from the red mark at which it was aimed will have a Probable Error, that we will call *b*. Now if the lateral distance of a particular green mark from *M* is given [*a*], what is the *most probable* distance from *M* of the red mark at which it was aimed? It is $\frac{c^2}{c^2 + b^2} a$*.

"(2) What is the Probable Error of this determination? In other words, if estimates have been made for a great many distances founded upon the formula in (1), they would be correct on the average, though erroneous in particular cases. The errors thus made would form a normal system whose *Q* [Probable Error] it is desired to determine. Its value is $\frac{bc}{\sqrt{b^2 + c^2}}$ †."

(pp. 69–70.)

* Unfortunately Galton has the value $\sqrt{\dfrac{c^2}{c^2 + b^2}}$, which is very liable to confuse the reader.

† In more modern notation, this may be looked upon as the variability of the array of the second relative $= c^2 (1 - r^2)$; therefore $r = \sqrt{c^2/(c^2 + b^2)}$. Hence the regression of first relative on second relative $= rc/\sqrt{c^2 + b^2} \times a = \dfrac{c^2}{c^2 + b^2} \times a$. Again the variance of the difference in character between the two relatives $= c^2 + (c^2 + b^2) - 2c\sqrt{c^2 + b^2}\, r = b^2$, or *b* has for physical meaning the probable error of the distribution of the difference in character between the two relatives.

It was by the help of these propositions that Galton discussed the action of inheritance in stable populations. Assuming normal distribution of characters, as he did, then the above relations really involve the fundamental properties of bivariate regression, stated with a truly amazing minimum of algebra.

In Chapter VI Galton describes his data. After referring to the moth-breeding experiments then in progress, and to his much earlier experiments on the characters of sweet-peas, he passes to his *Records of Family Faculties* obtained by the offer of £500 in prizes. He obtained the records of 150 families, 70 by male and 80 by female recorders. The records contained data as to Stature, Eye-Colour, Temper, the Artistic Faculty, and some forms of Disease. As a measure of the amount of material thus obtained, we find 205 couples of parents and 930 adult children of both sexes. A further set of *Special Data* was obtained by circulars requesting measurements of the stature of pairs of brothers. The constants for this material differ considerably from those for the Family Records. I think Galton thought the former material more reliable, but in working through his data in 1895* I came to the conclusion that the Special Data, owing to the heterogeneity of their origin, were scarcely to be fully trusted.

The chapter on *Data* concludes with some account of Galton's work on the weight of sweet-pea seeds. He states that:

"The results were most satisfactory. They gave me two data, which were all that I wanted in order to understand, in its simplest form, the way in which one generation of a people is descended from a previous one; and thus I got at the heart of the problem at once." (p. 82.)

Galton had thus first learnt of the nature of regression in 1875 from his sweet-pea experiments. He gives in Appendix C, pp. 225–6, of the *Natural Inheritance*, the first correlation table for inheritance, that of the diameters of parental and filial plants. The regression is about $\frac{1}{3}$. I have drawn the regression line (see our p. 4). Galton also states that he had made confirmatory measurements on foliage and length of pod, but he does not enter into details.

Chapter VII contains the *Discussion of the Data of Stature*. This chapter covers the same ground as the papers dealt with in our pp. 11–20, but there is some amplification and some attempt to simplify the mathematical reasoning†. The table on p. 133 is, as I have indicated on our pp. 23–4, very doubtful as far as the numerical values are concerned. In particular Galton terms the mean *regression w*, and then says that the probable deviation of the regressed array is $p\sqrt{1-w^2}$, where p is the probable deviation of

* See *Phil. Trans.* Vol. 187, A, pp. 283–4.

† Certain corrections should be made. On p. 127, formula (2), there should be no radical before $c^2/(b^2+c^2)$. This is a relic of an error on p. 70, where $\dfrac{c^2}{c^2+b^2}a$ should be read for $\sqrt{\dfrac{c^2}{c^2+b^2}}$, see p. 224. The numerical value for b deduced from (2) is correct. On p. 128, the numerical value for b should be ·96 not ·98, and this value, ·96, should be inserted in the table on p. 129 instead of the 1·10 given under the (3) heading. The mean is then 1·03 instead of 1·06.

the population. This is not generally correct; Galton is confusing the regression coefficient with the correlation coefficient. As long as both relatives have equal variability, which we may suppose to be the case with father and son or uncle and nephew, the two coefficients are numerically equal ; but when the two variates have not equal variability, this formula is of course incorrect. In the first entry in the table we have the regression of sons on midparent given as $\frac{2}{3}$, and Galton calculates from $p\sqrt{1-w^2}$ the probable deviation of the array of sons to be 1·27. The variability of midparents is, however, not equal to that of sons, but is in the ratio of 1 to $\sqrt{2}$; accordingly $r = w/\sqrt{2}$ must be used here instead of w, and the probable deviation of the array of sons is 1·50 and not 1·27.

Further the equality of the regressions of sons on midparents and of brothers on brothers is made by Galton to be $\frac{2}{3}$ in both cases. I think this value is too low in the case of midparents and too high in the case of brothers, the regressions being much more nearly in the ratio of 1·0 to 0·5 than in a ratio of equality. Other regressions entered in this table are very doubtful. We have to look upon the numerical values given as suggestions of the relative degrees of resemblance of various kinsmen, rather than conclusive values founded on observation of adequate numbers (see our pp. 23–4). The main result of Galton's work was to indicate the mechanism by which a population could remain stable notwithstanding variation and inheritance. It was a great direct achievement, and in the indirect light it cast on the general idea of correlation of still greater importance.

Chapter VIII contains the *Discussion of the Data of Eye-Colour*. This corresponds to the Royal Society paper, which I have already analysed on pp. 34–40 above. The same criticisms must be considered as still valid, and need not be repeated here.

Chapter IX deals with *The Artistic Faculty*. I do not think the contents of this chapter had been previously discussed by Galton. The data were deduced from the answers in *Records of Family Faculties* to the questions : " Favourite Pursuits and Interests ? " and " Artistic Aptitudes ? "

The object of this chapter is not to give a reply to the simple question, whether or no the Artistic Faculty tends to be inherited. A man must be very crotchety or very ignorant, who nowadays seriously doubts the inheritance either of this or of any other faculty*. The question is whether or no its inheritance follows a similar law to that which has been shown to govern Stature and Eye-Colour, and which has been worked out with some completeness in the foregoing chapters (p. 155). The conclusions

---

* It may be interesting with regard to these words to cite a few sentences from an obituary notice of Francis Galton which appeared in *Nature*, February 2, 1911 (Vol. LXXXV, p. 441). The writer says:

"Only once do I remember on a public occasion a slight severity in his usually gentle tone. A medical man of distinction [Dr Charles Mercier], speaking obviously without any knowledge of the literature of the subject, had asserted that the supposition that the children of parents with certain mental and moral peculiarities would reproduce these features, arose from a totally false conception of what the laws of heredity are. The mental and moral aptitudes were for the speaker outside the purview of hereditary investigation. Galton's reply was very simple: Much of what his critic had said 'might have been appropriately urged forty years ago, before accurate measurement of the statistical effects of heredity had been commenced, but it was quite obsolete now.'"

reached by Galton in this chapter are, I think, on the whole correct, but his handling of "broad categories" by means of percentages, in particular when no probable errors of the percentages are provided, is not to the modern statistician very conclusive. I think it would be labour well spent, should the opportunity arise, to work through his data afresh. Meanwhile we may arrange rather differently his tabulations and consider what flows from them. He tells us that he found it difficult to separate music from drawing, and finally classed both into a single group, the "artistic." Thinking also that parents were likely to overestimate the artistic capacity of young children, he excluded all but adults. Thus in the parental table the data chiefly refer to members of the second and third generations.

The first table I have deduced from Galton's data is that for Husband and Wife. It contains 894 couples and gives a percentage of 28 for males and of 32 for females with artistic temperament. The probable error of the difference 4 of these percentages is 1·46, or the difference is about 2·7 times its probable error, it may therefore be just significant. Galton concludes that:

"Part of this female superiority is doubtless to be ascribed to the large share that music and drawing occupy in the education of women, and to the greater leisure that most girls have, or take, for amusing themselves. If the artistic gifts of men and women are naturally the same, as the experience of schools where music and drawing are taught apparently shows it to be, the small difference observed in favour of women in adult life would be a measure of the smallness of the effect of education compared with that of natural talent." (p. 156.)

I should not have thought the experience of art schools was in favour of the equality of artistic gifts in the two sexes. Galton's data really tell us nothing as to the grade of artistic faculty in the two sexes, as for this we require grouping in at least three categories. But my impression is that a larger proportion of the prizes and studentships for creative work still goes to the men, even in those schools where the women are in a majority.

Assortative Mating in Artistic Faculty.

Husband

		Artistic	Non-Artistic	Totals
Wife	Artistic ...	107 {80}	179 {206}	286
	Non-Artistic	143 {170}	465 {438}	608
	Totals	250	644	894

Assuming the artistic faculty to be a continuous normal variate, we find from the above table the coefficient of correlation between Husband and Wife to be no less than ·2418 ± ·0376. This value for the mating of like with like for a mental temperament is singularly in accord with the intensity of assortative mating for physical characters*. It denotes a resemblance between

* Stature, ·2804 ± ·0189; Span, ·1989 ± ·0204; Forearm, ·1977 ± ·0205. Health as measured by Duration of Life: Wensleydale and Wharfedale, ·2200 ± ·0244; Oxfordshire, ·2500 ± ·0211; Society of Friends, ·1999 ± ·0212. See *Biometrika*, Vol. II, pp. 373 and 487.

Husband and Wife as great as that between cousins. I have placed in brackets after the observed numbers those that would arise in each category if the mating were purely random. It will be seen at once that the tendency for like to marry like is increased at the expense of the unlike marriages. I fail to understand how Galton interpreted his percentages; naturally if like marries like above the random allotment, there must be a reduction in the marriages of unlike individuals, the random 42 °/$_o$ of the latter being in fact reduced to 36 °/$_o$. Thus he writes:

"There is I think trustworthy evidence of the existence of some slight disinclination to marry within the same caste, for signs of it appear in each of the three sets of families with which the Table deals. The total result is that there are only 36 per cent. of such marriages observed, whereas if there had been no disinclination but perfect indifference, the number would have been raised to 42. The difference is small and the figures are few, but for the above reasons it is not likely to be fallacious. I believe the facts to be, that highly artistic people keep pretty much to themselves, but that the very much larger body of moderately artistic people do not. A man of highly artistic temperament must look upon those who are deficient in it, as barbarians; he would continually crave for a sympathy and response that such persons are incapable of giving. On the other hand, every quiet unmusical man must shrink a little from the idea of wedding himself to a grand piano in constant action, with its vocal and peculiar social accompaniments; but he might anticipate great pleasure in having a wife of a moderately artistic temperament who would give colour and variety to his prosaic life. On the other hand a sensitive and imaginative wife would be conscious of needing the aid of a husband who had enough plain common sense to restrain her too enthusiastic and frequently foolish projects*." (pp. 157–8.)

I have cited this passage, because, although it endeavours to explain a "slight disinclination to marry within the same caste," which Galton's data rightly interpreted show no evidence for, it yet throws light on some of his personal views of life. I can well picture what torture to him it would have been to be wedded to "a grand piano in constant action." While always exhibiting the best of old-fashioned courtesy to women, he had, when I first knew him, little belief in their intellectual strength; just as he held, that while women gifted with great physical strength existed, it was well for the repose of the other sex that they were rare (see our Vol. II, pp. 374–376). I think that later in life, when he came more in touch with academically trained women, and saw what work they could do on his own lines, his views suffered considerable modification. Again I am not content to pass without protest the rather sweeping statement that sensitive and imaginative persons, whether men or women, are apt to require restraining from "too enthusiastic and frequently foolish projects"; it denies that such persons often combine their sensitiveness and imagination with a rational power of control. It does not seem to me that the three factors, reason, sensitiveness and imagination, are incompatible, but that the success of truly great minds lies in the just combination of the three.

---

* Galton has written in pencil against this passage in his personal copy of *National Inheritance*, that it must be corrected, and I have also found some printed lists of *Errata*, in which the passage is stated to be incorrect. But none of the half-dozen copies I have examined of the work contains this *Errata* slip, and thus it is desirable to draw the attention of possible readers to a misinterpretation, which would certainly have been corrected in a second edition.

But let us return to less exciting questions. Galton does not, it is sad to record, classify his data in four fundamental parental tables, and till the material is reworked we must be content with the following arrangement.

Midparent and Child, Artistic Faculty.

		Both Parents Artistic	Only one Artistic	Neither Artistic	Totals
Child	Artistic ...	95	201	173	469
	Non-Artistic	53	319	666	1038
	Totals	148	520	839	1507

Unfortunately there is no distinction of sex in the offspring. Working out the correlation of this table in three different ways[*] I find the mean correlation coefficient to be ·4405 with a probable error of the order of ·024. There appears little doubt accordingly of the resemblance of offspring in artistic faculty to their parents, but the problem which Galton was investigating was not the existence of this resemblance, but whether its intensity might be taken as practically identical with those he had found for eye-colour and stature. The reader for whom the following remarks may be too technical is recommended to pass to the conclusions at the end of this paragraph. Galton assumes (i) equal inheritance from both parents, we will represent this by the correlation coefficient r; (ii) he does not correct by reducing female to male measure, we will suppose this done; (iii) he neglects the assortative mating, we will represent this by the correlation coefficient ϵ, in the present case this being equal to ·2418. The following results can be easily demonstrated:

$$(a) \quad \frac{\text{Variability of Midparent}}{\text{Variability of Offspring}} = \sqrt{\frac{1+\epsilon}{2}},$$

$$(b) \quad \text{Correlation of Offspring and Midparent} = \frac{r\sqrt{2}}{\sqrt{1+\epsilon}} = ·4405.$$

$$(c) \quad \text{Regression of Offspring on Midparent} = \frac{2r}{1+\epsilon} = \frac{\sqrt{2} \times ·4405}{\sqrt{1+\epsilon}},$$

or substituting the value of ϵ:

$$\text{Regression on Midparent} = ·559 = \tfrac{2}{3} \times 0·84.$$

$$\text{Parental Correlation, } r = ·3471 = \tfrac{1}{3} \times 1·04.$$

Now Galton deduced for regression of offspring on midparent for both stature and eye-colour the value $\tfrac{2}{3}$, and for parental correlation $\tfrac{1}{3}$. For the

---

[*] Treating the degrees of artistic faculty in the midparents as 1, 0·5, and 0, a biserial correlation coefficient after correction for class index gives ·4523 ± ·0138. The two possible divisions giving fourfold tables provide ·4655 ± ·0240 and ·4039 ± ·0298. The three results are thus in reasonable accord.

artistic faculty the former value is therefore 16°/₀ in defect and for the latter value 4°/₀ in excess. Galton, using what he admits to be a very crude method of percentages, shows that a regression of ⅔ would give him:

Percentage of Artistic Offspring.

	Both Parents Artistic	One Parent Artistic	Neither Artistic
Theory ...	40 °/₀	38·5 °/₀	17 °/₀
Observation	36 °/₀	39 °/₀	21 °/₀

Observation differs by 10°/₀ in the first case and 23·5°/₀ in the last case from theory. Galton says that the first values are "in very happy agreement," that the second "agree excellently well" and that the third give "a very fair accordance," and concludes:

"that the same law of Regression, and all that depends upon it, which governs the inheritance both of Stature and Eye-colour, applies equally to the Artistic Faculty." (pp. 161–2.)

But if the best value we can find from Galton's data for the Regression differs 16°/₀ from the value he assumes*, it is clear that we cannot assert that the accordance of percentages between theory and observation given in the above table justifies us in assuming on the present material that the Regression is the same for Artistic Faculty and Stature. Nevertheless while it may be impossible to accept on the basis of Francis Galton's data in *Natural Inheritance* that agreement between the constants for the inheritance of Stature and Artistic Faculty—that is between physical and psychical characters—which he thought he had found, we have yet to bear in mind two points: First that since 1889 more refined tools and better and more ample data have distinctly tended to confirm the equality of inheritance of mind and body; and secondly that Galton was foremost in the endeavour to obtain statistically a quantitative measure for the strength of resemblance in psychical characters.

Before we pass to the subject of Disease, it seems fitting here to note that Galton dealt with a second psychical characteristic, that of Temper. He refers to this on p. 85, where he deals with Marriage Selection, and also in Appendix D, pp. 226–238, which is a reprint with slight revision of a paper which first appeared in the *Fortnightly Review*, July, 1887†, under the title of "Good and Bad Temper in English Families." Galton found Temper in his Family Records described under 15 "Good" epithets and 46 "Bad" epithets, and he divided these into five classes, the first two corresponding to his "Good," and the remainder to his "Bad." These were

---

* If the midparental regression had been deduced from a correlation coefficient found in the ordinary way, its probable error would have been ·0096; the difference of ·667 and ·559 is ·108, more than 10 times that probable error. I think we must conclude that ·559 cannot be treated as ⅔.

† Vol. XLII, N.S. pp. 21–30.

(i) Mild, (ii) Docile, (iii) Fretful, (iv) Violent, and (v) Masterful. I have already discussed this classification in Vol. II, p. 271. I would only add that if we follow Galton's classification of Good and Bad Temper, we find a slight *negative* correlation between Husband and Wife. If it be considered significant, the mild temper of one mate may be due to the experience that control is needful or at least advisable in the environment of a violent consort. On the other hand the fourfold table for siblings, i.e. offspring of the same parents, is:

Temper in Siblings.

1st Sibling

		Good Temper	Bad Temper	Totals
2nd Sibling	Good Temper Bad Temper	330 {264} 255 {321}	255 {321} 454 {388}	585 709
	Totals	585	709	1294

The numbers in curled brackets give the frequencies which would occur in each category on the basis of independent chance. It will be seen that observation shows a heaping up in the like categories at the expense of the unlike categories. The correlation coefficient is ·3167 for this fourfold table; there is thus a considerable degree of resemblance between the temper of siblings, but I believe this measure would be considerably increased if sullen and fiery tempers were not included in one group.

Chapter X deals with the subject of *Disease*. This is a most interesting and suggestive chapter, but the data were too sparse to provide definite conclusions of any kind.

Galton states (p. 165) first (by again appealing to an analogy!) his *Preliminary Problem*. We know, he tells us, the ages at death and the causes of death of the population as a whole. We know the proportions at each age of those who die of diseases *A*, *B*, *C*, etc. He would assume—which I think is somewhat doubtful—that the proportions of persons dying of these diseases at various ages in two successive generations are the same. If now we limit ourselves to persons dying at a certain age of disease *A*, how, if at all, does this affect the distribution of deaths from the diseases *A*, *B*, *C*, etc., in the offspring generation?

The problem is an exceedingly difficult, if an extraordinarily important one, for it requires an immense mass of data. In the first place the proportional death distribution is a function of social class, and of occupation; it is as we have seen a function of age; it influences fertility; and in more than one way is affected by sex*. Anyone who has seriously faced the problem, and seen the number of groups into which the various affecting factors compel him to sort the material, will recognise how hopelessly

---

* The male in many cases, as by foreign travel or by military or naval service, runs greater risks than the female.

inadequate must be the schedule-series which can be collected by any single investigator however energetic. Galton's deduced schedule extracted from his data was a good one, but I should like to see added two columns, one for occupations and one for domiciles, under which latter heading I understand such descriptions as "rural," "urban," "India," "Nigeria," etc., stating years of life in each domicile. I reproduce one of Galton's working schedules, in

Sample of one of Galton's Schedules for Heredity of Disease

Father's Name ....James Gladding
Mother's Maiden Name......Mary Claremont

Initials	Kin	Principal Illnesses and Ailments	Cause of Death	Age at Death
J.G.	Father	Bad rheum. fever; agonising headaches; diarrhœa; bronchitis; pleurisy	*Heart Disease*	54
R.G.	Brother	Rheum.; gout	*Apoplexy*	56
W.G.	Brother	Good health except gout; paralysis later	*Apoplexy*	83
F.L.	Sister	Rheum. fever; rheum. gout	*Apoplexy*	73
C.G.	Sister	Delicate (inoculated and died)	*Smallpox*	?
M.G.	Mother	Tendency to lung disease; biliousness; frequent heart attacks	*Heart Disease* and *Dropsy*	63
A.C.	Brother	Good health	*Accident*	46
W.C.	Brother	Led a wild life	*Premature Old Age*	62
E.C.	Brother	Always delicate	*Consumption*	19
F.R.	Sister	Smallpox three times	*General Failure*	85
R.N.	Sister	Bilious; weak health	*Cancer*	50
L.C.	Sister	?	*Fever*	21
	Offspring			
M.G.	Brother	Inflam. lungs; rheum. fever	*Heart Disease*	17
K.G.	Brother	Debility; heart disease; colds	*Consumption*	40
G.L.	Sister	Bad headaches; coughs; weak spine; hysteria; apoplexy	*Paralysis*	50
F.S.	Sister	Bad colds; inflam. lungs; hysteria	....	Living
R.F.	Sister	Infantile paralysis; colds; nervous depression	....	Living
L.G.	Sister	Inflam. brain, also lungs; neuralgia; nervous fever	....	Living
		Space left for remarks		

Suggested additions, columns for occupation and environment. Also transfer the word "Living" from "Age at Death" to "Cause of Death" column, and if living state age at time of record.

which he has changed the real names. It is of interest as showing a case in which inoculation was followed by smallpox and death, and a second case in which one person had smallpox three times, both being phenomena of which the possibility raised heated controversy in the 18th century.

Now if we remember that we can hardly form less than 15 principal disease groups nor fewer than 10 age groups, and that we have two parents, it will be seen that we require to divide our material to start with into 300 categories. It would be of little service, if we are to reach really definite conclusions, unless we had 50 to 100 parents in each of these groups, that is to say, records of 15,000 to 30,000 of the first generation, and it might be hoped five times as many in the second generation, a total say of 100,000 to 200,000 recorded deaths, and we must assume these cases to be in a fairly uniform social class and with a fairly like environment. Probably it would be best to work with one social class, and weed out cases having very differentiated environments. Galton had 160 usable family records, with an average of 75 individuals, so that he might hope to reach 12,000 records of disease and perhaps 6000 of deaths. Actually he had only about 2000 causes of death recorded, which might correspond to some 300 groups of two parents and five children. On the average this would give for each special age group and each special disease group about *two* parents of the same sex, mustering *ten* children of two sexes and all ages, from which to determine whether and to what extent a parent dying at a given age of a specific disease influences the offspring dying of that or other specific diseases at given ages*. The problem is one of probabilities and we shall not have data enough to answer it. Suppose a man to die of cancer between 65 and 75; then we may further be supposed to know the chance that a man of 35 to 45 will die of cancer, but how are we to determine, supposing the latter man is son of the former, whether the relationship in disease is one of chance or heredity? We can only do it, if we have enough pairs of fathers and sons like the above to calculate from the observed frequency the probability of sons dying round the given age of cancer, and to determine if it differs significantly from the probability of deaths in general from cancer of men between the ages of 35 to 45, whether they have or have not a cancerous parent. I have enlarged on these points, because the measurement of heredity in disease is a fundamental problem of eugenics, but its complexity in the general form is rarely recognised; the labour and great cost of such investigations are in most cases prohibitive. Galton spent £500 in getting his Family Records, but although inheritance of disease was to be an essential part of it, he obtained practically nothing of value. Thus he writes in this connection:

" I had hoped even to the last moment, that my collection of Family Records would have contributed in some small degree towards answering this question, but after many attempts I find them too fragmentary for the purpose. It was a necessary condition of success to have the complete life-histories of many Fraternities who were born some seventy or more years ago, that is during the earlier part of this century, as well as those of their parents and all their uncles and aunts. My

---

* It is most important to bear this age factor in mind, as the relative proportions of the diseases of which an individual may die vary in life from age to age.

records contain excellent material of a later date, that will be valuable in future years, but they must bide their time; they are insufficient for the period in question. By attempting to work with incomplete life-histories the risk of serious error is incurred." (pp. 166–7.)

And farther on Galton sees what is the kernel of the matter:

"*Data for Hereditary Diseases.* The knowledge of the officers of Insurance Companies as to the average value of unsound lives is by the confession of many of them far from being as exact as is desirable*....

"Considering the enormous money value concerned, it would seem well worth the while of the higher class of those offices to combine in order to obtain a collection of completed cases for at least two generations, or better still for three....They would have no perceptible effect on the future insurances of descendants of the families, even if these were identified, and they would lay the basis of a very much better knowledge of hereditary disease than we now possess, serving as a step for fresh departures. A main point is that the cases should not be picked and chosen to support any theory, but taken as they come to hand. There must be a vast amount of good material in existence at the command of the medical officers of Insurance Companies*. If it were combined and made freely accessible, it would give material for many years' work to competent statisticians, and would be certain, judging from all experience of a like kind, to lead to unexpected results." (pp. 185–6.)

Still from his "fragments" Galton drew certain suggestive conclusions. He tested the trustworthy character of his data by determining whether deaths due to cancer, consumption, drink and suicide appeared less frequently on his records than in ordinary tables of mortality and found that they did *not*. He concluded that his correspondents had entered with interest into what was asked for, and had freely trusted him with their family histories. Galton throws out a curious suggestion: Namely suppose that one parent has a disease A and the other a disease B; if the child inherits a tendency to both diseases, how far are they mutually exclusive, how far do they blend or how far does the blend change them into a third form of disease? I think, for example, there is evidence to show that such hereditary diseases as phthisis and rheumatism are largely antagonistic. What effect on offspring results from the marriage of mates from rheumatic and tuberculous stocks?

Galton considers that there was evidence in his records of two obvious hereditary tendencies in stocks, the one to disease and the other to the absence of

* It is worth noting that Mr W. P. Elderton (now Actuary and Manager of the Equitable Life Assurance Society) fifteen years later, speaking at a meeting of the Sociological Society where Galton had read a paper on Eugenics, made the following statement with regard to heredity of disease: "An important item in the study of heredity is the heredity of disease, and I think life assurance offices might be able to give useful statistics. When a person whose life is assured dies, a certificate of death is given to the office and is put away with the papers that were filled up when the assurance was taken out. These original papers state the causes of death of parents, brothers and sisters and their ages at death, or their ages if they were alive when the assurance was effected. These particulars give information for the study of heredity in relation to disease, and from the same source light might be thrown on a question of great importance —the correlation between specific disease and fertility. One point in conclusion. Dr Hutchinson spoke of the greater importance of environment, but in that he would hardly get actuaries to agree with him. Their observation, judged by life offices, experience and practice, would seem to show that environment operates merely as a modifying factor after heredity has done its work." *Sociological Papers*, 1904, Macmillan & Co., 1905, p. 62.

Fifteen years passed after Galton threw out the suggestion that material might be available in assurance offices, before an actuary told us it did actually exist. Twenty-five further years have rolled by and still nothing has been done!

disease. This seems to be confirmed by a strong inheritance of general physical health independent of any special disease, which has been established since Galton's inquiry. He purposely adopts in order to cover many popular expressions the term "consumption." But beside actual consumption he graded in three additional classes (for which he gives the rather vague descriptions used), the context of the record also being considered*. These are (i) Highly suspicious, (ii) Suspicious and (iii) Somewhat suspicious. He reckoned at the rate of 4 of (i) to three actually consumptive, 4 of (ii) to two actually consumptive and 4 of (iii) to one such. Dividing a total of consumptives thus formed by the total offspring he formed a ratio, which multiplied by 100 he termed the "consumptivity" of the fraternity. For example, in a fraternity of which one member was actually consumptive, two suspicious and four somewhat suspicious, Galton would reckon three consumptive members, and the taint, or consumptivity, would be 43 %. To his surprise he found on making frequency distributions of consumptivity in fraternities, whether for one brother or one parent consumptive, that low and high degrees of consumptivity were both maxima, and moderate degrees gave a minimum or "anti-mode." Thus Galton, as far as I am aware, reached the first U-shaped distribution of frequency. He himself, notwithstanding his great belief in the normal curve, says it is not possible to torture the figures so as to make them yield the single-humped normal curve :

"They make a distinctly double-humped curve whose outline is no more like the normal curve than the back of a Bactrian camel is to that of an Arabian camel. Consumptive taints reckoned in this way are certainly not 'normally' distributed. They depend mainly on one or other of two groups of causes, one of which tends to cause complete immunity and the other to cause severe disease, and these two groups do not blend freely together. Consumption tends to be transmitted strongly or not at all, and in this respect it resembles the baleful influence ascribed to cousin marriages, which appears to be very small when statistically discussed, but of whose occasional severity most persons have observed examples." (pp. 175–6.)

Galton shows on pp. 177 and 179 by aid of very slender data, namely 14 fraternities with a "high" degree of consumption, which signified about 50 % deaths from lung trouble, and nine fraternities severely affected as to the heart, that the parentages in the two cases were of a very different character. In the latter case there was practically no distinction between the diseases from which the father and mother died; in the former no more deaths than those of two fathers could be associated with lung trouble, while some nine mothers out of fourteen were consumptive. This led Galton to take the view that consumption, while partly due to the inheritance of a tuberculous diathesis, which may be transmitted equally by either parent, is also transmitted by infection, and that in this respect the mother is by her closer contact far more a source of infection than the father. Is this differential influence of parents for tuberculosis confirmed by later investigations? I have taken the unpublished results for some 400 phthisical patients in King Edward VII's Sanatorium, and classified their parents into definitely phthisical and "suspicious," where owing to mention of their ailments there was suspicion

* See his pp. 172–3. Something of the same kind is still undertaken by tuberculosis officers in grading the families of the admittedly tuberculous.

of lung trouble. In 413 cases where information as to the father was given, he was definitely phthisical in 7·02 °/$_{\circ}$ and there were suspicions of phthisis in 17·19 °/$_{\circ}$. In 420 cases of mother the corresponding numbers were 6·90 °/$_{\circ}$ and 13·81°/$_{\circ}$. Thus Galton's view of the greater influence of the mother, whether by infection or by heredity, is not confirmed on large numbers*. Notwithstanding Galton's suggestion as to the fundamental part played by infecting mothers he proceeds on pp. 181–185 to discuss consumption on the basis of heredity. Although we may not feel this justifiable, his method is so suggestive and generally applicable that it must be discussed here. He starts by assuming that the distribution of resistance or immunity in the population may be supposed to have a normal distribution of mean M and—to use modern notation—a standard deviation σ. Now according to Galton's data 16 °/$_{\circ}$ of the deaths of his general population were from consumption, hence $M - ·9945\sigma$ is the level of immunity at which consumption begins its ravages, and the mean immunity of those who die from consumption is $M - 1·5207\sigma$. But, if we accept Galton's figures for stature, the parental regression (and correlation) is $\frac{1}{3}$, or the marriage in which only one parent is consumptive gives rise to a "co-fraternity" (modern "array") of mean $M - \frac{1}{3} \times 1·5207\sigma = M - ·5069\sigma$, with a variability or standard deviation of $\Sigma = \sigma\sqrt{1 - \frac{1}{9}} = \sigma \times ·9428$. Accordingly the centre of this array is at a distance $·4876\sigma$ from the limit to immunity and the ratio of this to the standard deviation of the array $= ·5172$. The table of the Probability Integral shows that this is only very slightly over 30 °/$_{\circ}$. Galton, disregarding the fact that by choosing his regression, he has *ipso facto* chosen the variability of his array, tries values for it which he thinks reasonable and which give him 31 °/$_{\circ}$, 29 °/$_{\circ}$ and 27 °/$_{c}$ of consumptives in the offspring of a consumptive parent. These are not far from the value 30 °/$_{\circ}$ we have obtained. Galton by his different methods obtained 26 °/$_{\circ}$ and 28 °/$_{\circ}$ of consumptive offspring of a consumptive parent, but this is only a minimum limit, as it does not appear that he confined himself to families all the members of which were already dead, or had passed practically through the age zone of really lethal tuberculosis. Of course the method supposes that within reasonable limits the degree of immunity of each individual remains constant, and that, within reasonable limits again, this degree of immunity is not affected by the size of the dose.

The importance of the method is greater than that of its application, which is rendered doubtful by the use of the special values, not confirmed, for stature, and by the fact that Galton had already attributed much of the result to infection. What, however, the method indicates is, that if we know the frequency of a particular type of disease in the community and its

---

* One curious result does seem to flow from my data. If we divide our patients into male and female, then of the 423 parents of the female subjects 8·75°/$_{\circ}$ were definitely phthisical and 18·20°/$_{\circ}$ were suspected; but of the 410 parents of male subjects only 5·12°/$_{\circ}$ were definitely phthisical and 12·68 °/$_{\circ}$ suspected of phthisis. This suggests either that the parentage was more influential in the case of the female, or that women knew more or were less reserved than the men about the diseases of their parents.

frequency in the case of the offspring of a parent suffering from this disease, then by a series of approximations we can readily obtain the value of the correlation between offspring and parent, or the intensity of heredity in the case of that disease. Galton himself states:

"Too much stress must not be laid on this coincidence*, because many important points had to be slurred over, as already explained. Still, the *primâ facie* result is successful, and enables us to say that so far as this evidence goes, the statistical method we have employed in treating consumptivity seems correct, and that the law of heredity found to govern all the different faculties as yet examined, appears to govern that of consumptivity also, although the constants of the formula differ slightly." (p. 185.)

The penultimate chapter of *Natural Inheritance* is termed *Latent Elements*. The main point to which Galton appeals here is that the parents contributing on the average a definite amount to the heritage of a child, according to Galton each $\frac{1}{4}$, the residue of the stock of either parent can on the average only contribute a definite amount, i.e. $\frac{1}{4}$ on this view, to the child, or only $\frac{1}{4}$ of the characters of the ancestry can lie latent in the parent, and be contributed to the child. Galton argues that "either parent must contribute on the average only one quarter of the Latent Elements, the remainder of them dropping out and their breed becoming absolutely extinguished" (p. 188). He illustrates this by the selection of 13 out of a pack of 52 cards; any card may be chosen but actually 39 are rejected, yet if a great many sets of 13 are chosen, i.e. a great many individuals be taken, every card in the original pack will ultimately appear. "No given pair can possibly transmit the whole of their ancestral qualities; on the other hand there is probably no description of ancestor whose qualities have not been in some cases transmitted to a descendant" (p. 189). The throwing out of half the latent (as well as half the patent) elements at each crossing is really part of Galton's idea of all inheritance being particulate, a mosaic of ancestral characters. Even his idea of a blend is not a summation of continuous contributions, but a summation so to speak of quanta from individual ancestors.

In his next paragraph Galton deals with a *Pure Breed*, and again his error as to regression appears to come to the surface. He discovered regression simply as a statistical result, i.e. because he took parents of given characters, whose earlier ancestry might be anything whatever, he naturally found the offspring nearer than the parents to mediocrity. But unfortunately this idea of regression fixing itself in his mind became for him a biological fact, and he considered that he had discovered in stability of types, i.e. in groups each with their own focus or centre of regression, the source of evolution, or change from one type to a second. He now tells us that in the case of pure breed in which there has been long selection, the influence of a large quantity of mediocre ancestry would disappear, and so would the tendency to regression, except in so far as it is "connected with the stability of different types" (p. 189). In other words we have now *two* sources of regression, while Galton's original deduction of regression was purely statistical and depended

* That of the above percentages.

on the presence of the ancestral mediocrity. He does not state what experimental evidence there is for this other type of regression, and my impression is that it arose in his mind from a belief that regression was always in action and so evolution impossible by mere selection of continuous variations.

Under the same heading of *Pure Breed* Galton also considers the variation within a sibship or group of brethren. If, as he defines a pure breed, it be merely a line in which the ancestors have been given the same selection value for a large number of generations, then on Galton's theory of normal distribution of variates, the theory of multiple regression shows us that the variability will be the same within the sibship, whether the ancestry have been selected or not. Galton, to whom that theory was not familiar, deduces by a rough approximative method that the ratio of variability in the pure breed is to that in the mixed breed as ·98 to 1·00; but actually on his hypothesis they are equal; the variability of the sibship is independent of whether the characteristics of the progenitors are alike or unlike. Of course the reader must understand that by pure breed and mixed breed Galton is only referring to sibships which have their progenitors alike and their progenitors unlike in character respectively. All these progenitors are supposed of the same race, and he was not dealing with cross-breds, or mixture of races, in his "mixed" breeds.

In the final paragraph of this chapter Galton gives the results of his experience. He considers that for practical prediction you need to know not only the obvious somatic characters of the two parents, but the latent characters of their germ-plasms. These latter he considers can be respectively determined with a fair degree of approximation from the paternal and maternal uncles and aunts, if they exist in considerable numbers. Also what may be ascertainable of grandparents and their sibships will be of value. But he considers that if he were to start collecting family records again, he would limit himself to families having at least six adult children, and with as many members in both paternal and maternal sibships. There is much that is true in this view, yet at the same time, where a stirp occasionally throws a noteworthy individual, it may be doubted whether a sample of 12 in the first generation and six in the second is large enough to bring out all the latent possibilities which may be of importance. The desire of the Eugenist must always be for as complete a family pedigree as possible. It would not be feasible on a sample of 18 to say whether a single occurrence showed insanity to be a latent character of the stock or not.

Galton's final chapter contains a brief summary of the work, of which our present section is a more complete one. Only two points may be referred to. On p. 196 he writes:

" There are no means of deducing the measure of fraternal variability [i.e. variability in the sibship from the same pair of parents] solely from that of the co-fraternal [i.e. the array of individuals who all have one parent of the same character value]. They differ by an element of which the value is thus far unknown."

We need no longer admit this ignorance. If R be the multiple correlation of an individual on all his ancestry, or on his " generant," then $\sigma \sqrt{1 - R^2}$ is the variability of the sibship or fraternity proceeding from that generant, where σ is the standard deviation or variability of the general population. If r be the correlation of brothers in the ordinary sense, then $\sigma\sqrt{1-r^2}$ is the variability of an array or co-fraternity of brothers. The connecting link missed by Galton is : $R^2 = r$[*].

The second topic is :

> "the fundamental distinction that may exist between two couples whose personal faculties are naturally alike. If one of the couples consist of two gifted members of a poor stock, and the other of two ordinary members of a gifted stock, the difference between them will betray itself in their offspring. The children of the former will tend to regress ; those of the latter will not. The value of a good stock to the well-being of future generations is therefore obvious, and it is well to recall attention to an early sign by which we may be assured that a new and gifted variety possesses the necessary stability to easily originate a new stock. It is the refusal to blend freely with other forms. Some among the members of the same fraternity might possess the characteristics in question with much completeness, and the remainder hardly or not at all." (pp. 197–8.)

It will be perceived from this paragraph that Galton does *not* hold the absence of regression in the "gifted" stock to be due to less mediocrity in the ancestry, but to the creation of a "new" stock by some trick of falling into a fresh position of stability, which enables the stock, at any rate in some of its members, to breed true. That is, he appeals to mutations for the source of "gifted" stocks.

Whether this be true or not, Galton I think reached his views owing to a misinterpretation of the statistical phenomena of regression. It was a misfortune that he really did not get beyond the idea of regression in two variates, because to be clear as to the true relation between his "midparental heredity" and his "Law of Ancestral Heredity" a knowledge of multiple regression is essential. But it was the greatest good fortune that he got as far as he did; he blazed the track, which many have followed since, and if he left unsolved or half-solved problems, his disciples ought to be grateful that the master has provided the problem as well as the tool, rather than be stern critics of his pioneer work[†]. *Natural Inheritance* is a great book even if it has its obvious blemishes.

The work concludes with the reproduction of tables from the memoirs on percentiles, on stature and on eye-colour, etc. Also with a series of Appendices. A gives particulars of Galton's own works and memoirs. B reprints Hamilton Dickson's paper (see our p. 12). C describes the experiments on sweet-peas, never fully dealt with. D reprints the *Fortnightly Review* paper on Temper (see our pp. 69–70 and Vol. II, p. 271). E reproduces Galton's paper on the Geometric Mean (see Vol. II, pp. 227–8). F reprints Galton and Watson on the Probable Extinction of Families (see Vol. II, pp. 341–343). G deals with the orderly arrangement of hereditary data, in particular with

---

[*] See *Biometrika*, Vol. XVII, p. 131.

[†] We have in a case in the *Galtoniana* of the Galton Laboratory the first map of Damaraland. Is it of less value because it is not an Ordnance map of what was once German South-West Africa?

the case of recording disease. Much of this the reader who wishes to go farther than our pages will find more easily here than in the original papers. Certain numerical misprints in the tables require that they should be carefully examined before use.

J. *Discontinuity in Evolution.* In 1894 appeared William Bateson's *Materials for the Study of Variation, treated with especial regard to Discontinuity in the Origin of Species.* One of the strange misconceptions with regard to Galton's views and to his work lies in the fact that he has been over and over again considered as the propounder of the view that evolution has taken place by the selection of slight or continuous variations. As a matter of fact Galton had for some years before the appearance of Bateson's book been preaching emphatically the doctrine of *discontinuity* in evolution. Indeed his opinions on the manner of evolution date back to 1872 : see our Vol. II, pp. 84, 170–174 and 190. They are more clearly expressed in the preface to the 1892 reprint of *Hereditary Genius.* There we read :

"Another topic would have been treated more at length if this book were rewritten—namely the distinction between variations and sports. It would even require a remodelling of much of the existing matter. The views I have been brought to entertain since it was written, are amplifications of those which are already put forward in pp. 354–5*, but insufficiently pushed there to their logical conclusion. They are that the word variation is used indiscriminately to express two fundamentally distinct conceptions: sports and variations properly so called. It has been shown in *Natural Inheritance* that the distribution of faculties in a population cannot possibly remain constant if on the average the children resemble their parents. If they do so the giants (in any mental or physical particular) would become more gigantic and the dwarfs more dwarfish, in each successive generation. The counteracting tendency is what I called 'regression.' The *filial* centre is not the same as the *parental* centre but it is nearer to mediocrity; it regresses towards the *racial* centre. In other words the filial centre (or the fraternal if we change the point of view) is always nearer, on the average, to the racial centre than the parental centre was. There must be an average 'regression' in passing from the parental to the filial centre." (pp. xvii–xviii.)

The flaw in Galton's argument is again one that we have had several times to notice, namely that he is overlooking the fact that he has clubbed together parents of all possible types of ancestry, and the " regression " of his sons is solely due to the large number of such parents who have sprung from an ancestry mediocre or below mediocrity. The amount of filial regression depends entirely on the amount of this mediocrity, and there will be no regression if two or three generations above the parents are of like deviation from mediocrity. Thus although it may still be a matter for experiment and discussion, whether evolution proceeds by variations proper or by sports, whether it be continuous or advance by jerks, the reason which made Galton the pioneer in advocating discontinuous evolution was a misinterpretation of his own discovery of " regression."

* These pages deal with Galton's idea of the stability of types: see our p. 61 and Vol. II, p. 113. It is quite reasonable to suppose that by successive selection of extreme variations proper we might reach a position of unstable equilibrium of the parts of an organism. But there does not exist experimental evidence at present to indicate that such instability would lead to a sport breeding truly rather than to non-viable forms of the organism. See our pp. 93–4.

This is expressed so definitely in the following paragraph that I must cite it:

"It is impossible briefly to give a full idea, in this place, either of the necessity or the proof of regression; they have been thoroughly discussed in the work in question*. Suffice it to say, that the result gives precision to the idea of a typical centre from which individual variations occur in accordance with the law of frequency, often to a small amount, more rarely to a larger one, very rarely indeed to one that is much larger, and practically never to one that is larger still†. The filial centre falls back further towards mediocrity in a constant proportion to the distance to which the parental centre has deviated from it whether the direction of the deviation be in excess or in deficiency. All true variations are (as I maintain) of this kind, and it is in consequence impossible that the natural qualities of a race may be permanently changed through the action of selection upon mere variations. The selection of the most serviceable *variations* cannot even produce any great degree of artificial and temporary improvement, because an equilibrium between deviation and regression will soon be reached, whereby the best of the offspring will cease to be better than their own sires and dams." (p. xviii.)

The flaw in the argument here is that Galton uses "filial centre" in *two* senses. In the first sense it refers to all the offspring of pairs of parents of the same character values, *whatever their parental ancestries may be.* Hence there must always be regression. In the second sense it is used of the offspring of an individual pair of parents, and interpreted to mean that the offspring of a given individual stock always regresses to the population mean, more and more in each generation. This is not true, the stock may with assortative mating even progress. The misfortune arose from Galton not having reached the formulae for multiple regression. Had he done so, he would have seen the contradiction between his "Law of Ancestral Heredity" and his interpretation of "Regression." Whether continuous or discontinuous evolution, or partly one and partly the other, expresses the truth, it is quite certain that Galton in 1892 supported evolution by mutations owing to an error of interpretation. His views on the subject undoubtedly contributed to directing attention to discontinuous evolution. He writes:

"The case is quite different in respect to what are technically known as 'sports.' In this a new character suddenly makes its appearance in a particular individual, causing him to differ distinctly from his parents and from others of his race. Such new characters are also found to be transmitted to descendants. Here there has been a change of typical centre, a new point of departure has somehow come into existence, towards which regression has henceforth to be measured and consequently a real step forward has been made in the course of evolution. When natural selection favours a particular sport, it works effectively towards the formation of a new species, but the favour that it simultaneously shows to mere variations seems to be thrown away, so far as that end is concerned. There may be entanglement between a sport and a variation which leads to a hybrid and unstable result, well exemplified in the imperfect character of the fusion of different human races. Here numerous pure specimens of their several ancestral types are apt to crop out, notwithstanding the intermixture by marriage that had been going on for many previous generations." (pp. xviii–xix.)

Unfortunately the only method of settling points of such fundamental importance—that of critical experiment—was not adopted‡. Some biologists

---

* [*Natural Inheritance*, 1889. See our pp. 57 and 65.]

† [This sentence is lacking in Galton's usual precision of statement.]

‡ Galton here first indicates what for years he believed to be the right experimental method for solving problems in heredity; his scheme, however, failed because he endeavoured to work

poured scorn on statistical methods even while they rejoiced in being ignorant of the mathematical processes, which would alone have enabled them to understand and criticise them effectively. Other biologists contented themselves with asserting that material collected by "non-biologists" could not possibly be of biological value. Many rash statements were made which would hardly now be maintained by the most ardent mutationist or Mendelian*. The controversy over Galton's method of dealing with heredity became a logomachy, or as some would say a tauromachy, and contributed little of permanent value to science. It was idle because the fundamental questions as to whether "variations proper" could serve as a basis for selection, and whether and to what extent sports bred true, were not investigated by agreed critical experiments. No one who has tried or even thought over such experimental work—bound to be of a secular nature—will be in the least likely to minimise the difficulty of devising and carrying through a crucial experiment. Nevertheless that was and remains the sole satisfactory method of settling a scientific dispute as to natural phenomena. The opinion that no real conclusion could be reached, except by direct experiment, was the actual reason why Galton's lieutenants ultimately retired from the controversy concerning the application of his methods to the measurement of heredity. Galton himself for another decade endeavoured to provide means for secular experimentation. What was the outcome of his attempts we shall see later on.

Again when Galton came to study finger prints, he was struck by the scarcity of transitional types; further his evidence indicated that there was little if any correlation between type and any bodily or mental characteristics, or that the types were peculiar to any human races.

"It would be absurd therefore to assert that in the struggle for existence, a person with, say, a loop on his right middle finger has a better chance of survival, or a better chance of early marriage, than one with an arch. Consequently genera and species are here seen to be formed without the slightest aid from either Natural or Sexual Selection, and these finger patterns are apparently the only peculiarity in which Panmixia, or the effect of promiscuous marriages, admits of being studied on a large scale. The result of Panmixia in finger markings corroborates the arguments I have used in *Natural Inheritance* and elsewhere, to show that 'organic stability' is the primary factor by which the distinctions between genera are maintained; consequently the progress of evolution is not a smooth and uniform progression, but one that proceeds by jerks, through successive 'sports' (as they are called), some of them implying considerable organic changes; and each in its turn being favoured by Natural Selection.

"The same word 'variation' has been indiscriminately applied to two very different conceptions, which ought to be clearly distinguished; the one is that of 'sports' just alluded to,

by a committee of incompatibles. I shall return to his attempts later, but their first foreshadowing appears in the 1892 preface to *Hereditary Genius*:

"It has occurred to others as well as myself, as to Mr Wallace and to Professor Romanes, that the time may have arrived when an institute for experiments on heredity might be established with advantage. A farm and garden of a very few acres, with varied exposure, and well supplied with water, placed under the charge of intelligent caretakers, supervised by a biologist, would afford the necessary basis for a great variety of research upon inexpensive animals and plants. The difficulty lies in the smallness of the number of competent persons who are actually engaged in hereditary inquiry, who could be depended upon to use it properly." (p. xix.)

* For example, that two-factor dominant and recessive Mendelian hypotheses would account for the heredity of coat-colour or eye-colour. Or that albinotic eyes were those without any granular pigment, and individuals possessing them would breed true.

which are changes in the position of organic stability, and may through the aid of Natural Selection, become fresh steps in the onward course of evolution; the other is that of the variations proper, which are merely strained conditions of a stable form of organisation, and not in any way an overthrow of them. Sports do not blend freely together; variations proper do so. Natural Selection acts upon variations proper, just as it does upon sports, by preserving the best to become parents, and eliminating the worst, but its action upon mere variations can, as I conceive, be of no permanent value to evolution, because there is a constant tendency to 'regress' towards the parental type. The amount and results of this tendency have been fully established in *Natural Inheritance*. It is there shown, that after a certain departure from the central typical form has been reached in any race, a further departure becomes impossible without the aid of these sports. In the successive generations of such a population, the average tendency of filial regression towards the racial centre must at length counterbalance the effects of filial dispersion; consequently the best of the produce cannot advance beyond the level already attained by the parents, the rest falling short of it in various degrees*."

The views of Galton here summarised show that the view he took in the *Natural Inheritance* of 1889†, that evolution was largely carried out by "sports" or in jerks, i.e. was chiefly discontinuous, was not the outcome of reading Bateson's work, although in that work he found support for his ideas. It will be seen at once also that he had divided, years before later controversies, "variations" into "sports"—now termed "mutations"—and "variations proper," which Galton held (and had indeed demonstrated) were inherited, but believed could not be of permanent value, because of what he termed the "constant tendency to regress." The fact that they are inherited distinguishes Galton's "variations proper," and very definitely distinguishes them, from the "fluctuating variations" of Mendelian writers, which are asserted by them to be non-inheritable. How far the theory of discontinuous variation—with all its contradiction in many cases of the palaeontological record‡—was really forced on the attention of biologists by Galton's writings it is not possible to say. We do know that both De Vries and Bateson were at one time enthusiastic students of Galton's works. However this may be, what is now clear is that there is no "unexpected law of universal regression" as Galton supposed, it is merely a misinterpretation of his own data and the constants based upon them.

It is important to examine this point, not only with regard to Galton's views on Discontinuity in Evolution, but also owing to the many biological misinterpretations of the statistical conception of regression. Galton found that the *average* value of the stature of sons of fathers having an excess h in stature above the population mean had only an excess of $\frac{1}{3}h$ above that same mean. Practically all his conclusions are based on this single fact and the statement that the array of such sons varies about this regressed mean with a variation about 6 °/₀ less than the variation of the general population. The reduction of variation is so small, that it is possible practically to select sons of the same character deviation as their parents possessed. In order to

---

* *Finger Prints*, 1892, pp. 19–21.

† See Chapter III, "Organic Stability," and compare our pp. 60–62 above.

‡ "The distinctive feature of palaeontological evidence is that it covers the entire pedigree of variations, the rise of useful structures not only from their minute, apparently useless, condition, but from the period before they occur." HENRY F. OSBORN, 1889.

illustrate what Galton overlooked let us take his Ancestral Law coefficients as if they represented the absolute truth and investigate what would be the mean stature of sons if their parents and grandparents were by natural or artificial selection raised to a deviation h above the population value.

The mean of the sons would now be

$$\tfrac{1}{4}(h+h)+\tfrac{1}{16}(h+h+h+h)=\tfrac{3}{4}h,$$

the offspring have accordingly regressed $\tfrac{1}{4}h$ from the parental deviation. Now suppose selection to cease, and owing to isolation or other cause the offspring to interbreed; then their offspring will have the average value

$$\tfrac{1}{4}(\tfrac{3}{4}h+\tfrac{3}{4}h)+\tfrac{1}{16}(h+h+h+h)+\tfrac{1}{64}(h+h+h+h+h+h+h+h)$$
$$=\tfrac{3}{8}h+\tfrac{1}{4}h+\tfrac{1}{8}h=(\tfrac{3}{8}+\tfrac{1}{8})h+\tfrac{1}{4}h=\tfrac{3}{4}h.$$

In other words there is *no further regression*, or what these offspring lose in the regression of their parents *is compensated by the exceptionality of their grandparents*. Applying the formula once more we have for the offspring average

$$\tfrac{1}{4}(\tfrac{3}{4}h+\tfrac{3}{4}h)+\tfrac{1}{16}(\tfrac{3}{4}h+\tfrac{3}{4}h+\tfrac{3}{4}h+\tfrac{3}{4}h)+\tfrac{1}{64}(h+h+h+h+h+h+h+h)$$
$$+\tfrac{1}{256}(h+h+h+h+\text{to sixteen times})$$
$$=\tfrac{3}{8}h+\tfrac{3}{16}h+\tfrac{1}{8}h+\tfrac{1}{16}h=(\tfrac{3}{8}+\tfrac{1}{8})h+(\tfrac{3}{16}+\tfrac{1}{16})h=\tfrac{3}{4}h,$$

or the exceptional great grandparents make up for the loss of the regressed parents and the exceptional great great grandparents for the loss of the regressed grandparents; and so on. In other words there is no "unexpected law of universal regression." Regression in Galton's sense arises solely from the fact that by clubbing into a single array the offspring of all fathers of a given character deviation he has given them not only mothers whose average stature will be mediocre, but also a mediocre ancestry. But if there be isolation and inbreeding what Galton treated as a regression is a permanently progressed value for the offspring. Indeed if we continue to select, not with increased deviation, but with the same deviation (h), there is, so far from a regression, a continuous progression towards the selection value. For example if we select for

	1 generation	2 generations	3 generations	4 generations	5 generations
we progress:	$\tfrac{16}{32}h,$	$\tfrac{24}{32}h,$	$\tfrac{28}{32}h,$	$\tfrac{30}{32}h,$	$\tfrac{31}{32}h,$

and then inbreeding due to isolation or other cause after any one of these generations maintains the group at the progressed value.

Shortly there is no law of "universal regression," and we can deduce from Galton's own theory that his "variations proper," if selected and inbred, would establish a breed with a "new centre of regression." It is of course more than probable that our new centre of regression, i.e. the type of our new breed, may be unsuited to survive, that is to say in Galton's sense may be unstable. One cannot alter one character in an organism without modifying all the correlated characters, and some of those altered are likely to have survival value. But Galton's own data and Galton's own theory rightly interpreted lead to no "universal regression," still less to an argument that "variations proper" cannot be the subject of selection and the formation of new breeds.

This does not prove that "variations proper" have been the basis of evolution, but it removes Galton's chief reason for belief in evolution by discontinuity, that is by sports or mutations. The law of "universal regression"—over which Galton undoubtedly stumbled—is only true when we neglect ancestry beyond the parents and suppose mating at random, but these are not the conditions which exist when intense selection is taking place and the selected interbreed.

Having prefaced Galton's views on Discontinuity with some criticism, which I think is needful, of his theory of regression, we may turn to his paper on "Discontinuity in Evolution," which was published in *Mind*, Vol. III, N.S. pp. 362–372, July, 1894. Galton begins by saying that students of the laws of variation need not be disheartened by the impossibility of learning what is the cause of variation. Galton, who, as we have seen, believed in individuality in the numerous germ cells of an organism, and that germ cells were subject to selection, found no difficulty in attributing variation to the effect of interacting germinal elements*. He considered that the actual cause of any particular variation might be put on one side by those who study the degree and character of variation generally.

We are next provided with a definition of race based upon the idea of a typical centre of regression. As I understand him A and B are two different races, if the offspring of the members of A and the offspring of the members of B regress to two different centres of regression. But how can we practically demonstrate this? If we take the offspring of a pair of individuals of race A, the degree in which they differ from their parental mean will depend upon the long line of ancestry of those parents (to adopt Galton's own views); if the parents were relatively small in stature, say, for their ancestry, the offspring average may exceed the parental stature; if they were relatively tall, the offspring average may fall short of the parental. If we choose such a large number of parents of given statures, that we may assume the ancestors of the parents have for average value that of the general population, then the offspring average will regress to the population mean, and should we know the regression coefficient accurately, this will provide the population mean or "typical centre of regression." Similarly we might determine the typical centre of the race B, and ascertain whether the two centres were or were not significantly different. But I cannot see that this is any more than inquiring whether the populations A and B have different

* I must confess to feeling it extremely difficult to accept the view that the population of germ cells belonging to an individual organism are like atoms, identical in character, and have a germinal capacity defined by absolutely the same formula. Such a population of germ cells is, if parasitical, still an organic population, and one continually in a state of reproduction and change. No other organic population that we know of is without variation among its members, and I find it extraordinarily hard to believe that it is a matter of complete indifference which individual spermatogonium of an organism is the ultimate source for fertilisation of an individual ovum of a second organism.

means, and we might proceed at once to do this without introducing at all the ideas of inheritance and regression. Galton's definition might be of service if we could determine from the regression of the offspring of a single pair of parents, or a few pairs, the typical centre, but this is no more feasible than to determine from a few individuals the population mean; the very backbone of Galton's conception of parental regression is that the ancestors of the parents cover all the possible pairs in the community, or are on the average mediocre.

Having defined his races A and B to be those having different centres of regression, which if the races are stable simply connotes different population means, Galton concludes that if A and B are stable then intermediate types are less stable. I think this is only a theory, not necessarily a demonstrable fact. It may be that races A and B have not diverged from a common ancestral race C by continuous variation, but there is nothing in Galton's theory of regression to prevent A and B arising from C or even A from B by continuous variation. The idea of "stability" as a source of organic evolution is one that Galton was very fond of; when a race has been largely selected, it topples over, so to speak, into a new form of organic equilibrium with a new centre of regression. In this way Galton would account for "sports" and the prepotency and permanency of certain sports, and he considers that *most* breeds have arisen from sports. He then refers to various kinds of sports as in peacocks, peaches, and the appearance of remarkable intellectual gifts in man. Under the latter category he cites Sebastian Bach. "Can anybody believe that the modern appearance in a family of a great musician is other than a sport?" (p. 368). He also refers to Inaudi the mental arithmetician, who started as an illiterate Piedmontese boy. In the latter case, however, the Inaudi stock may well have possessed similar, if less intense powers which were never called into activity, while in the former case we now possess the pedigree of the Bach family, and their remarkable musical power is certified for five or six generations. All variation is discontinuous when examined in small groups such as families, and the extreme deviations in such small groups may be easily interpreted as sports. Newton again may well have been a sport, but till we know more than we do at present of his mother's ancestry, it is hardly wise to hold that he was such. Nor again if some of these men are to be considered "sports," can we dogmatically assert that they might, like the "japanned" or black-shouldered peacocks, have produced offspring regressing to a new typical centre.

"The phrase organic stability must not as yet be taken to connote more than it actually denotes. Thus far it has been merely used to express the well-substantiated fact that a race does sometimes abruptly produce individuals who have a distinctly different typical centre, in the sense in which those words were defined. The inference or connotation is that no variation can establish itself unless it be of the character of a sport, that is, by a leap from one position of organic stability to another, or as we may phrase it, through '*transilient*' variation. If there be no such leap the variation is, so to speak, a mere bend or divergence from the parent form, towards which the offspring in the next generation will tend to regress; it may therefore be called a '*divergent*' variation. Thus the unqualified word variation comprises and confuses what I maintain to be two fundamentally different processes, that of transilience and that of divergence, and its use destroys the possibility of reasoning correctly in not a few important matters. The interval leapt over in a transilience may be at least as large as it has been in any hitherto observed instance, and it may be smaller in any less degree. Still whether it has been large or small, a leap has taken place into a new position of stability. I am unable to conceive the possibility of evolutionary progress except by transilience, for if they were merely divergences, each subsequent generation would tend to regress backwards towards the typical centre, and the advance that had been made would be temporary and could not be maintained." (p. 368.)

We see that Galton only differed from the mutationists by supposing that their not-inherited "fluctuating variations" were really inherited, although they were of no permanent account, being rendered nugatory by his principle of regression. In view of the inheritance he had found for grades of stature, he could hardly hold otherwise than to suppose them inherited, but he coupled this inheritance with a misinterpretation as I have shown of his own statistical theory of regression, which left him practically in the ranks of the mutationists—a strangely inconsistent position for one who has been looked upon as the founder of the Biometric School! A little farther on Galton writes:

"These briefly are the views that I have put forward in various publications during recent years, but all along I seemed to have spoken to empty air. I never heard nor have I read any criticism of them, and I believed they had passed unheeded and that my opinion was in a minority of one. It was, therefore, with the utmost pleasure that I read Mr Bateson's work bearing the happy phrase in its title of discontinuous variation,' and rich with many original remarks and not a few trenchant expressions. ...It does not seem to me by any means so certain as is commonly supposed by the scientific men of the present time, that our evolution from a brute ancestry was through a series of severally imperceptible advances. Neither does it seem by any means certain that humanity must linger for an extremely long time at or about its present unsatisfactory level. As a matter of fact, the Greek race of the classical times has surpassed in natural faculty all other races before or since*, and some future race may be at least the equal of the Greek, while it is reasonable to hope that when the power of heredity and the importance of preserving valuable 'transiliences' shall have been generally recognised, effective efforts will be made to preserve them." (pp. 369 and 372.)

What direction those "effective efforts" should take Galton does not indicate. He tells us that human sports of considerable magnitude in both the moral and intellectual fields assuredly occur. But when we face the question of increasing the number of their offspring we soon recognise that endowment of parenthood will achieve little in the case of a rare mutation; we find ourselves led into the thorny field of speculating on the eugenic as distinguished from the social value of monogamy and on the possible utility of endogamy in the perpetuation of human sports. The elimination of animal passions still strong in man would have to be carried much farther than it has yet been, before the tribal customs as to marriage and family life could be safely called into question even in the case of individuals of surpassing intellectual or moral eminence, were it feasible, indeed, to determine such individuals with anything like unanimity. The difficulty of the problem should not discourage all consideration of it, for it is clearly fundamental if we are consciously to use heredity to elevate mankind; on the other hand the very difficulty of the problem forbids hasty solutions being adopted and proclaimed as essentials of the eugenic programme†.

* [See, however, my remarks, Vol. II, pp. 107–109.]

† See on this point, *The Times* (December 31st) report of a discussion under the auspices of the Eugenics Society at the Educational Congress on December 30, 1927.

Galton's interest in Discontinuous Evolution was further manifested in the same year by a circular which will be found in the *Transactions of the Entomological Society of London*, 1895 (April 3rd). It consists of three questions addressed to breeders and others, not only entomologists but to those who pursue any branch of natural history. The questions are for information on the following topics:

"(i) Instances of such strongly-marked peculiarities, whether in form, in colour, or in habit, as have occasionally appeared in a single or in a few individuals among a brood; but no record is wanted of monstrosities, or of such other characteristics as are clearly inconsistent with health and vigour.

"(ii) Instances in which any one of the above peculiarities has appeared in the broods of different parents. [In replying to this question, it will be hardly worth while to record the sudden appearance of either albinism or melanism, as both are well known to be of frequent occurrence.]

"(iii) Instances in which any of these peculiarly characterised individuals have transmitted their peculiarities, hereditarily, to one or more generations. Especial mention should be made whether the peculiarity was in any case transmitted in all its original intensity, and numerical data would be particularly acceptable that showed the frequency of transmission: (*a*) in an undiluted form, (*b*) in one that was more or less diluted, and (*c*) of its non-transmission in any perceptible degree."

The context attached to the questions shows that Galton was still troubled by the question of regression: "Regressiveness and stability are contrasted conditions and neither of them can be fully understood apart from the other." As I have endeavoured to indicate regression is merely a statistical result, which holds for a population, not for an individual, when we table the former with a knowledge of only a limited number of the kinsfolk of individuals and *assume the mean of each generation to remain the same**. The biological problem is to determine how this mean changes and is quite independent of the statistical idea of regression. As I have indicated above (p. 83) the offspring of selected ancestry on Galton's own theory do *not* regress to the population mean, and in this respect the only contrast that could be drawn between the offspring of a "sport" and of such selected ancestry is the question of the extent to which a sport breeds true without having even a limited amount of selected ancestry. This is really the point which Galton's third question would tend to answer†.

K. *Eugenics as a Religious Faith.* I have already pointed out that a very fundamental characteristic of Galton's mind was his desire that our progressive knowledge of natural law should at once be turned to practical service in attempts to elevate the race of man. He could not think of the doctrine of

---

* This assumption is made by Galton, but it is not in the least needful to the statistical theory of regression, which measures each generation from its own mean.

† I am not certain whether it was in reply to this circular that Galton received information about a singular family of lunatic cats. He described the family in a letter to *The Spectator* (April 11, 1896), entitled: *Three Generations of Lunatic Cats*. The sires of the kittens were unknown, but may be assumed to have been normal. Nevertheless the lunacy, which may be considered as a sport, was transmitted by the mother to all her offspring and grandchildren with undiluted strength. The only doubt that can be raised is whether the sire of "Phyllis," who was brought from Ewart Park, Northumberland, might possibly have been a wild cat. It is a pity the family could not have been preserved for the study of hereditary lunacy.

evolution merely as a contribution to academic biology; for him the type of "sport" of greatest interest and value was that embracing the human moral or intellectual "sports," and he desired at once to know how we might perpetuate for the service of mankind such supermen as might appear. Evolution according to him was providing for the survival of the physically and mentally more vigorous members of the race, and he desired to see this achieved with greater rapidity and less pain to the individual. In 1894 a book entitled *Social Evolution*, written by Benjamin Kidd, was published, and created for a time some stir as dealing with the relation of supernatural, or at least ultra-rational religion to the social evolution of man. It was not written from the scientific standpoint and contained little of permanent value. It led Galton, however, to publish a rather remarkable article on "The Part of Religion in Human Evolution*." Kidd's thesis may be briefly summarised as follows: Intra-group struggle for existence is the *sine quâ non* of social progress; this beneficent working of the struggle for existence is so painful to existing men that they would not, if they were rationalists, pay the price for it; to check the anti-social and anti-evolutionary force of reason religion has been evolved to provide an ultra-rational sanction for moral conduct.

Galton starts his paper by suggesting that superstitions in barbaric times, such illusions as totems, tutelar deities, and we may add tribal and national gods, gave cohesive force and compactness to a group and tended to render it successful against other groups, which on rational grounds had begun to question such illusions. Galton recognises the important part religion may play in determining national stability. Even after men of education have realised the irrationality of a national creed, it may be unwise precipitately to destroy it.

"The social system of every nation, including its religion, whatever that may be, has adjusted itself into a position of stability which is dangerous to disturb. Deep sentiments and prejudices, habits and customs, all more or less entwined with the established religion of each nation, are elements of primary importance to its social fabric. It is true that vast changes become obvious in the social system of every progressive people, whenever its habits and customs at one period are compared with those of another long after, but, as a rule, the changes are piecemeal. Each change is primarily confined to a single part, the remainder adapting itself to the new condition with a comparatively small shift of the position of the centre. Commonsense teaches how much can be thus done with safety at any given time. Great and sudden changes in religion are hardly to be attempted except when the stability of the existing system is tottering and on the point of falling." (p. 758.)

Whatever views we may take about religion, whether we regard it as a supernatural revelation or not, we can agree that one of its chief functions is to curb selfishness in the individual, to inculcate altruism, and by restraining human passions to help the stabilisation of society. With this end in view religion from the earliest times has been the guardian of tribal custom in regard to marriage, birth and death. It has therefore concerned itself with matters which from our present knowledge of the laws of natural selection and heredity we recognise as bearing on human evolution. It is impossible—

* *The National Review*, August 1894, pp. 755–765.

and this the Church is now beginning to recognise—to place the scientific doctrine of evolution and the moral conduct of man as inspired by religious belief in separate idea-tight compartments. Slowly, but nevertheless surely, this aspect of religion is taking possession of the minds of the more thoughtful clergy. It has long been seen by many men of science that it formed the most hopeful field for co-operation between the old supernaturalism and the new scientific knowledge. It is from this conception that Galton, as an agnostic, starts to bring religion into touch as a living force with our belief in human evolution.

Galton cites three definitions of religion. (*A*) that of John Stuart Mill: The essence of Religion is the direction of the emotions and desires towards an ideal object, recognised as rightly paramount over all selfish objects of desire. (*B*) that of Kant: Religion consists in our recognising all our duties as Divine commands. And (*C*) that of Gruppe: A belief in a State or Being which, properly speaking, lies outside the sphere of human striving or attainment, but which can be brought into this sphere in a particular way, namely by sacrifices, ceremonies, prayers, penances and self-denial.

Gruppe's definition is historical, indicating religion only by its past outward and nigh outworn forms. Kant's definition of religion as a recognition of supernatural sanction for *all* duties is too narrow in its sanction and too wide in its duties*; it demands also a continuous revelation as duties continuously change with human progress. It was not unnatural therefore that Galton selected Mill's definition of religion. He points out that any guiding idea that takes passionate possession of the mind of a person or a people is an adequate adversary to purely selfish considerations, and that such would be religious in Mill's sense but not in that of Kant or Gruppe.

"Many of the ordinary emotions which influence conduct admit of being excited to so high a pitch that the merely self-regarding feelings do not attempt to withstand them, but yield themselves unresistingly to be sacrificed to the furtherance of a cause. That the emotions can be so excited, whether in a party or in a nation, easily and often irrationally, is one of the common teachings of history." (p. 757.)

No supernatural command or sentiment is needful. Religious enthusiasms in the sense of Kant or Gruppe may give great help, but they are not indispensable.

"The ambitions, loves, jealousies, and hates of nations, families, and persons, seem fully strong enough to force men who are under their influence, to disregard what is commonly understood by the phrase selfish desires." (p. 758.)

Galton, under a conviction of its truth, then makes the following affirmation:

"The direction of the emotions and desires towards the furtherance of human evolution, recognized as rightly paramount over all objects of selfish desire, justly merits the name of a religion." (p. 758.)

---

* To render unto Caesar what is Caesar's may be the dictate of a great religious teacher, but is scarcely a *religious* duty—even if Caesar be not a foreign war-lord!

Thus, I think, he sympathises with the Victorian scientific criticism of religion as defined by Kant and Gruppe, but he desires to see a religion in Mill's sense built up to replace the formal religions. He holds that:

"the destructive task is a necessary though painful preliminary, because until obstructions are thoroughly cleared away, and the view is quite open, the character and exigencies of the vacant space cannot be rightly understood, nor can a judgment be formed as to how far and in what way rebuilding is needful. It is also pardonable enough that the work of destruction should be over zealously indulged in by some who have long chafed under what they consider to be the irrationality of one or other of the many conflicting creeds.

"All earnest inquirers recognize the awful mysteries that surround human life, but they are angered by theosophies that attempt to solve part of the problems by means of hypotheses that are improbable in themselves, while they introduce gratuitous complications. For instance if we strip from Milton's fable and from the *dramatis personae* of *Paradise Lost* all the glamour thrown over them by his superb diction, a grotesquely absurd framework remains behind. His high undertaking to justify the ways of God to man becomes ludicrously inadequate. The same spirit under another guise that moved our ancestors in the days of the Reformation to shatter the authority of Rome, is abroad again but is now directed against the dogmas of the time. The spirit is that of a determination to face and view the grand and terrible problem of life in the clear light of day, and not through artificial mediums that partly hide, partly colour and partly refract it." (pp. 758–9.)

Galton, while desiring a reformation of religion in the sense of Erasmus, was perfectly conscious that the bulk of our people, who may be weary of the old superstitions, are in such a backward state that they will be more ready to accept new superstitions* than to seek a rational basis for a national religion. Granted the discredit of the long accepted ultra-rational faith, granted that a nation "be suffering in a still more acute form than our own from poverty, toil, and an unduly large contingent of the weakly, the inefficient, and the born-criminal classes, and that the existing social arrangements are acknowledged to be failures," what will follow? Galton held that socialistic experiments on various scales and in various ways will be largely tried and will be admitted to be ineffective *owing to the moral and intellectual incompetence of the average citizen*†.

"There would then be a widely-felt sense of despair; there would be ominous signs of approaching anarchy and of ruin impending over the nation, while a bitter cry would arise for light and leading. A state of things like this is by no means impossible in the near future, even here in England, and therefore, it deserves some consideration as being something more than a merely academic question. In the imagined event, preachers of all sorts of nostrums would abound, mostly fanatics who could see only one side of a question, and on that account they would be all the more earnest in their opinions and persuasive to the multitude." (p. 759.)

Thus Galton uses the probable ineffectiveness of socialistic experiments as an argument in favour of the acceptance of eugenics as a social and at

* Salvationism, Spiritualism, Theosophy, Christian Science, to say nothing of the resurrection of the urge, dating from the Neolithic period, to the sacrifice of the people's deity or its totem, and a communal feast on the remains.

† I do not remember any other reference of Galton's to Socialism. The present passage indicates that it was not in its ideas antipathetic to him, but that he conceived it would fail owing to the moral and mental feebleness of the average citizen. The present experiments in Russia and China will serve to test his opinions in the eyes of sympathetic onlookers. Their failure would convince men according to Galton that racial progress in the eugenic direction must precede or accompany social reconstruction.

the same time a religious programme for the nation. He puts into the mouth of a supposed agnostic and "somewhat fanatic" preacher opinions which were undoubtedly his own—even though he states that, with much sympathy for them, he would not commit himself to them without serious reservations the statement of which "would merely distract the argument*." It would indeed be a loss, if Galton's views thus boldly expressed should perish in an ephemeral review article. He himself has added in a list of his papers in my possession a note to the effect that this article suggests Eugenics—although that name is not mentioned—as a religion in accordance with our modern views on human evolution. We note here the beginning of Galton's last period in which he devoted his activities to eugenic propaganda. I cite the following characteristic passages:

"The mystery is unfathomed as to whence the life of each man came, whether it pre-existed in any form or not. The mystery is equally great as to what will become of his life after the death of the body; whether it will be perpetuated in a detached form as some creeds say, whether it will be absorbed into an unlimited sea of existence, as other creeds assert, or whether it will cease entirely. As regards this life, there are also mysteries. Every act may or may not have been determined by previous conditions, but man has the sense of being free and responsible: he is accustomed to do and to be done by as if he were so, therefore we may provisionally believe that he is free and should act on that supposition. There is a further mystery as regards the cosmic conditions under which we live, for no assurance can as yet be obtained of any supernatural guidance, the facts alleged in evidence of its existence being more than counterbalanced by those that point the other way. We cannot, in consequence, tell with certainty whether human life is subject to an autocracy, or whether, at least for practical purposes, it exists as an isolated republic; but the latter appears at present to be the more probable, and should, therefore, guide our conduct. Each man's destiny during his life may then be viewed with propriety as depending entirely on his own physiological peculiarities and on his surroundings. He has, consequently, to conduct himself as a member of a free executive committee during his brief life, guiding his actions by whatever he can learn of the tendencies of the cosmos, in order to co-operate intelligently with what he cannot in the long-run resist. The sense of responsibility that is imposed by this view would sober, brace, and strengthen the character, just as that of dependence on an autocratic power effeminates and enfeebles it....

"On the foregoing basis our agnostic might say: 'Let us consider what is peculiarly profitable and proper for man to attempt. One of the most prominent conditions to which life has been hitherto subject, is the newly discovered law of the survival of the fittest, whose blind action results in the progressive production of more and more vigorous animals. Any action that causes the breed or nature of man to become more vigorous than it was in former generations is therefore accordant with the *process* of the cosmos, or, if we cling to teleological ideas, we should say with its *purpose*.

"It has now become a serious necessity to better the breed of the human race. The average citizen is too base for the every day work of modern civilization. Civilized man has become possessed of vaster powers than in old times for good or ill, but has made no corresponding advance in wits and goodness to enable him to direct his conduct rightly. It would not require much to raise the natural qualities of the nation high enough to render some few Utopian schemes feasible that are necessarily failures now. Conceive, for the sake of argument, the nation to be divided in the imagination into three equal groups L, M, N, in order of their natural civic capacities. At present the production of the forthcoming generation is chiefly effected by L and M, the lowest and the middle; if it were hereafter effected by M and N, the middle and the

---

* What these reservations may be we do not know, they probably related to evolutionary, as opposed to revolutionary change. The opinions of the "supposed agnostic" are so akin to those which Galton has himself expressed in other passages, but never more briefly or forcibly, that we may well be certain they are really his own.

highest, a distinct gain would be achieved in the lifetime of many of those who initiated the reform, for it is probable that the inefficient multitude of weaklings in brain, character, and physique would be sensibly diminished in thirty years.

"Our agnostic preacher might go on to say that this terrible question of over-population and of the birth of children who will necessarily (in a statistical sense) grow into feeble and worse than useless citizens must be summarily stopped, cost what it may. The nation is starved and crowded out of the conditions needed for healthy life by the pressure of a huge contingent of born weaklings and criminals. We of the living generation are dispensers of the natural gifts of our successors, and we should rise to the level of our high opportunities. The course of nature is exceedingly wasteful in every way. It is careless of germs, tens of thousands of pollen grains perishing of which none have had the chance of effecting fertilization, by being transported to the proper spot at the proper moment, by the blind agency of an insect ferreting among the flowers for food. It is equally careless of the microbes whose part in the animal world is ana-logous to the pollen of the flowers; they are produced in myriads, though only one is needed for fertilization. It is no exaggeration to say that the number of them which is produced each year by an average male of any of the larger animals, would suffice to fertilize a million of females, if every one of them were utilized. The course of nature is also indifferent and ruthless towards our own lives, but reason can teach us to effect with pity, intelligence, and speed many objects that nature would otherwise effect remorselessly, unintelligently and tediously. By its action, suffering may be minimized and waste diminished. Wherever intelligence chooses to inter-vene, the struggle for existence ceases, that struggle being by no means so absolute a necessity in evolution as Mr Kidd assumes it to be. ...Horses are bred in the number and of the stamp required, within the limits of excellence that experience has taught to be possible. A general high level of the qualities that make a good horse has been attained without any aid from natural selection, artificial selection having superseded it.

"Before, however, as even a fanatic must allow, any form of artificial selection could be ap-plied to the human race, other than such moderate, yet not ineffective, reforms as might produce the results mentioned a little way back, much is needed. Accurate knowledge has to be obtained on numerous details connected with productiveness, of which we are now curiously ignorant and careless to study, while national customs would have to be profoundly modified. The fanatic might, however, fairly urge that in considering what is feasible, and what not, the three following canons ought to be freely accepted:

"1. The customs of every nation are liable to change to an extent that is barely credible to those who do not bear history in mind; therefore the existing customs of any nation may be lightly regarded while discussing future possibilities.

"2. No custom can be considered seriously repugnant to human feelings that has ever pre-vailed extensively in a contented nation, whether barbarous or civilized.

"3. Any custom established by a powerful authority soon becomes looked upon as a duty, and, before long, as an axiom of conduct which is rarely questioned.

"Fortified by these three canons, an anthropologist who is necessarily familiar with the cus-toms of many nations will find abundant elbow-room for his wildest speculations. There is hardly any proposition, however monstrous it may seem to us now, that is thereby precluded from consideration....

"It is quite credible that a nation whose old religious notions and social practices, whatever they were, have avowedly failed, who have been aroused to the knowledge that man possesses vast and hitherto unused powers over the very nature of unborn generations, who have learnt to realize the dilatoriness, ruthlessness, and pain that accompany the evolution of man, when it is left as now to cosmic influences, and who have satisfied themselves that the present low state of their race might be materially improved by concerted national action, should seize with irresistible ardour upon the idea of utilizing their power.

"That is to say, the nation might devote its best energies to the self-imposed duty of carrying out, in its manifold details, the following general programme: (1) Of steadily raising the natural level of successive generations, morally, physically, and intellectually, by every reasonable means that could be suggested; (2) of keeping its numbers within appropriate limits; (3) of developing the health and vigour of the people. In short, to make every individual efficient, both through nature and by nurture.

"A passionate aspiration to improve the heritable powers of man to their utmost, seems to have all the requirements needed for the furtherance of human evolution, and to suffice as the basis of a national religion, in the sense of that word as defined by J. S. Mill, for, though it be without any ultra-rational sanction, it would serve to 'direct the emotions and desires of a nation towards an ideal object, recognized as rightly paramount over all selfish objects of desire.'" (pp. 761–3.)

I trust this long citation will not have wearied the reader; for his biographer it contains some of the most important lines Galton ever wrote. There is no reason to be afraid of plain words. Man has learnt how to breed plants and most inferior forms of life that are of service to him. He has yet to learn how to breed himself. When he has studied heredity and environment in their influences on man, the application of the laws thus found to the progressive evolution of the race will become the religion of each nation. Such is the goal of Galtonian teaching, the conversion of the Darwinian doctrine of evolution into a religious precept, a practical philosophy of life. Is this more than saying that it must be the goal of every true patriot* ?

L. *Miscellaneous Papers on Evolution, Heredity, etc.* We may now turn to a series of short papers by Galton, chiefly published in *Nature*, and dealing with hereditary and evolutionary topics from 1897 onwards.

The first paper we note is entitled: "Rate of Racial Change that accompanies Different Degrees of Severity in Selection," and will be found in *Nature*, April 29, 1897 (Vol. LV, p. 605). This is an important paper, because it deals with the effect of continued selection in modifying a variate continuously distributed in a population. Galton starts with his two-thirds regression of the offspring on the midparent for stature and the reduction of the variability of the offspring of such midparents in the ratio of 1·5 to 1·7 inches. He then continues to select both parents at the 99th, 95th, 90th, 80th and 70th grades for 1, 2, 3, 4, 5 and an infinite number of generations in order to determine the progression there would be in stature by such continuous selection. Galton, unfortunately ignorant of the formulae of multiple regression, makes three erroneous assumptions, namely: (i) that the regression between each generation is the same, namely $\frac{2}{3}$, notwithstanding the earlier ancestry being as we advance more and more selected; (ii) that the variability of the array of selected ancestry remains for the later generations the same as for the first selected generation; this is of course incorrect, the variability steadily diminishing towards a finite limit; (iii) that if the selected race be now left to itself, it will regress indefinitely to the old general population mean:

"It must be borne in mind, that there is no stability in a breed improved under the supposed conditions; but that as soon as selection ceases it will regress to the level of the rest of the population through stages in which the deviation, at starting, sinks successively to $w, w^2 ... w^n$ of its value†. It may, however, happen that a stable form will arise during the process of high

---

* Some may question whether we have more here than in Comte's *Religion of Humanity*. I think so, because it is freed of the ceremonialism which Comte and Gruppe demanded as a factor of religion, and it is essentially based on the acquirement of knowledge in a field of science, which had little if any existence in Comte's day.

† w is Galton's regression coefficient in the case of selected midparent, *with no selected previous ancestry*.

breeding, that shall afford a secondary focus of regression, and become the dominant one, if the ancestral qualities that interfere with it be eliminated by sustained isolation and selection. Then a new variety would, as 1 conceive, arise; but into this disputable topic there is no need to enter now." [See, however, my footnote, p. 79.]

We now know that on the theory of multiple regression, this indefinite regression has no existence; there is a slight regression in the first generation of breeding from the selected stock, but it ceases with this generation. We have again in the cited passage evidence that Galton was obliged to appeal to "sports" to account for evolutionary progress, because he had misinterpreted the theory of regression. If w be the regression of the offspring of the first generation of selected midparentage, the regression of the offspring of parents of the first generation, *who have also selected midgrandparents*, is not to be taken w again. Thus the formulae, the numerical table and the conclusions drawn from it in this paper are I think in error. But the idea at the back of it that the more intense the selection, the more rapid is the relative progress, is true; as also the idea that there may in each case be a limiting value. Probably no such continued selection is really feasible; too many characters in the organism are highly correlated, so that if it were possible to carry under conditions of viability an individual character to a height much above the population mean, some one or other of the correlated characters would be almost certain to be incompatible with the continued efficiency of the organism in relation to its environment or its functions*.

I have not recalculated Galton's table, because with the data at present available, I am inclined to believe that selection for two or three generations and then inbreeding would be followed, at any rate in some characters, by a progression rather than a regression. In other words the strength of inheritance is such that with a very brief period of selection followed by isolation a continuous differentiation will proceed—so far as it is not checked by a counter natural selection. This suggests that we may have to seek in heredity itself for the basis of progressive evolution; a variation maintained for a couple of generations, followed by an isolation of the offspring, will continue to progress. If this be true we surmount the difficulty of why variations to which the environment is not hostile, or indeed may be favourable when they are sufficiently developed, can reach the stage of development at which they become important as a new factor of efficiency in the individual. We see that it is not natural selection, but the mere force of heredity, which leads in isolated groups to the genesis of variations of sufficient importance to have survival value to the individual. We may term this theory of the genesis of remunerative variations the "Heredity Theory of Progressive Evolution." It seems at first sight in flat contradiction to Galton's views on continuous regression when selection ceases, unless the selection has led to the creation of a sport. Yet it really flows

---

* Nor is this confined only to the functions of the individual, but may concern the functions of other members of its race. Thus breeders of bull-dogs have gone on continuously selecting the size of the head until the mortality of puppies and bitches at littering has become so serious as to threaten even the survival of the breed.

from a more complete view of multiple regression, and the more accurate values we now have for the heredity constants.

The second paper to be referred to was published in *The Gardeners' Chronicle* for May 15, 1897, and is entitled "Retrograde Selection." In this Galton asks from horticulturists advice as to the cultivation of a plant or plants existing in an original stock R and a stable variety V. For example V might be a dwarf variety. Galton proposes to endeavour by selection to pass back from V to R. If the plants in progress of selection be X, then Galton proposes to pass towards R by selecting the plants above the quartile of X on the R side of the character. He describes very fully how the experiments could be made in an orderly fashion and the needs as to soil and methods of growth; he refers to his paper of April (see our p. 93) for a measure of the rate at which changes might be supposed to take place. No doubt much might be learnt from such experiments—if only, that "retrograde selection" is impossible. I am not certain that Galton had not this in mind, for if in his view stable varieties could only originate in sports, we could not select back to the original stock, unless selection itself conduces to sporting. I do not think the paper led to any actual experiments on Galton's part, although the Editor of the *Chronicle* wrote strongly in favour of such experiments in a leader in the following week, and there were some suggestions on May 29, 1897*.

In *Nature*, November 4, 1897 (Vol. LVII, p. 16), Galton gave a brief account of E. T. Brewster's paper on "A Measure of Variability and the Relation of Individual Variation to Specific Differences†." Brewster is really comparing what we now term Interracial with Intraracial Variability, measuring his variability by what is practically the coefficient of variation V. His thesis is that if for any two interracial characters A and B, V_A is $>V_B$, then for the corresponding intraracial characters in the "allied races" v_a will be $> v_b$. There does not seem any obvious theoretical reason for this, but Galton holds that Brewster "has provisionally established his thesis that whenever any special character varies much in individuals of the same race, it is probable that it will be found to vary much in 'allied races' and conversely."

The next paper to be considered is entitled: "Hereditary Colour in Horses," and appeared in *Nature*, October 21, 1897 (Vol. LVI, pp. 598–9). Galton tabulated his data from material collected and published by "Tron Kirk" in the Chicago journal, the *Horseman* (Christmas Number, 1896). In the fundamental table which Galton gives there are 3025 matings of bay sires, but as "Tron Kirk" informs us that 3100 foals were born to no more than 46 different bay sires‡, or an average of 67 foals to the sire, it is clear that

---

* The interpretation put by the practical gardeners was that Galton wanted to go back from an *improved* form to a *poor* original. But I do not think this by any means the chief purport of his paper; he wanted if possible to go back from a specialised variety to the form, not necessarily inferior, from which it had been obtained.

† *Proceedings, American Academy of Arts and Sciences*, May, 1897.

‡ It is not said that the matings of bay sires cover these 3100 foals, but presumably they do. The difference in numbers may be due to the omission of grey foals or to twinning.

there will be a great bias in the returns owing to the limited number of gametic constitutions in the sires. The following table gives the results as

Table of Colour Inheritance in Horses.

	No. of Observations	Colour of Dam	Colour of Sire	Percentages of Colour in Offspring			
				Chestnut	Bay	Brown	Black
A (i)	68	Chestnut	Chestnut	100	—	—	—
(ii)	1900	Bay	Bay	10	81	6	3
(iii)	19	Brown	Brown	—	42	52	5
(iv)	25	Black	Black	—	4	28	68
B	407	Chestnut	Bay	33	61	4	2
	366	Bay	Chestnut	30	63	3	4
C	52	Chestnut	Brown	—	86	11	2
	69	Brown	Chestnut	16	65	10	9
D	72	Chestnut	Black	6	76	15	3
	57	Black	Chestnut	30	40	—	30
E	221	Bay	Brown	1	79	14	6
	450	Brown	Bay	6	66	18	10
F	156	Bay	Black	3	60	30	7
	268	Black	Bay	7	53	16	24
G	55	Brown	Black	—	22	38	40
	6	Black	Brown	—	16	50	33

Percentages taken only to whole numbers.

published by Galton. In the first line of the series of rows, A (i), we see that for 68 cases of chestnut mated with chestnut all the offspring were chestnut. Galton does not comment on this, but it was the source of considerable later controversy. A certain number of matings of chestnut with chestnut taken from Wetherby's *Thoroughbred Studbook* gave the same result as the first row of Galton's matings; but a longer series, wherein it was pointed out that the *Studbook* did contain some instances of chestnut mated with chestnut not producing chestnut, was rejected on the ground that these instances must be due to error of record, a most circular process of reasoning.

It is clear from B where we are dealing with a fairly adequate number of crossings both ways that (i) there is not a predominance of sire or dam for chestnut with bay matings, and (ii) bay may contain a factor of chestnut. If we work out from B the number of bays with a factor for chestnut we find them to be 31·5 °/$_{o}$, while 68·5 °/$_{o}$ lack that factor. Applying these percentages to the 1900 bay and bay matings in A (ii) we should

PLATE III

Drawn with the colouring matter as in black human hair.

H. C. Sorby 1878.

Reproduction in close colour facsimile of Dr Sorby's painting of a tree with the pigment from black human hair. [Colouring matter largely from melanin pigment granules?]

PLATE IV

Drawn with the colouring matters as in dark red human hair.

H. C. Sorby 1878

Reproduction in close colour facsimile of Dr Sorby's painting of a tree with the pigment from dark red human hair. [Colouring matter largely from the diffused pigment of the fibrillae?]

anticipate 9·9 °/₀ chestnut and 90·1 °/₀ non-chestnut foals, a result almost exactly that observed.

If we judged by A (iii), A (iv), and G, we should conclude that black and brown had no factor for chestnut. In that case chestnut and bay crossed with black and brown would produce no chestnut foals. This is flatly contradicted by C, D, E and F, which indicate that browns and blacks can contain a factor for chestnut. The only explanation is, perhaps, a rather forced one, namely that the matings in A (iii), A (iv) and G were few in number and possibly made from a very few sires of brown and black colour without factors for chestnut, while C, D, E and F, providing far more numerous matings, contained blacks and browns with such factors. C, D, E and F, indeed, tend to confirm this, for when the sire is black or brown there are far fewer chestnuts produced than when the dam is black or brown; in the latter case the larger number of dams used would give a greater chance of their carrying factors for chestnut.

Galton himself by *averaging* up the likenesses in coat-colour of foal to dam and to sire concluded that as some 32·83 °/₀ of foals followed the colour of their dam and 31·75 °/₀ that of their sire, there was no prepotency, but such an averaging method misses the possibility of discovering a prepotency due to the presence or absence of "factors." The equality of male and female hereditary influence is borne out by the long series B, but hardly by the other and shorter series such as C and D*. Galton was very fully aware of what he terms the "rudeness" of the data. He had been troubled with much the same problems in considering hair colour; but he had then obtained an analysis of the pigments in human hair from Professor Sorby and the latter investigator had shown the existence of two distinct pigments, one red and one black. Sorby painted two trees in these two pigments extracted from human hair, pictures which used to hang in Galton's dressing room and are now in the Galton Laboratory. More recent microscopic investigation seems to show that the same two pigments occur in the hair of horses and dogs, but that the red pigment is diffused in the hair, i.e. "the whole ground substance of the fibrillae is impregnated with it." On the other hand the black or melanine pigment occurs in the form of pigment granules†. On examination of a number of specimens of horse hair in samples from ribs and mane of chestnut horses, ranging from the golden chestnut of the Trakehnen stud to the black chestnut, the diffused red pigment was found in all, but the pigment granules varied from scarcely any in the golden and light chestnuts to close packing in the dark and black chestnuts. This corresponds very closely to the range of granular pigmentation found in passing from light red to dark auburn hair in man. Such results suggest that it is very desirable to study microscopically the distribution of the two pigments in the hair of horses'

---

* For example in 124 matings of chestnut dam by brown or black sire there were only 6 chestnut foals, but in 126 matings of chestnut sire with black or brown dam there were 46 chestnut foals.

† Pearson, Nettleship and Usher: *Monograph on Albinism*, Part II, pp. 319–345, Cambridge University Press.

coats, and to remember that granular pigmentation varies enormously within the range of coat-colours described as chestnut by hackney breeders. Galton assumes that full red pigmentation counts for 1·0 and takes chestnut to be 0·8; bay, 0·7; brown, 0·4; and black, 0·1. Then by using the results of the several lines in A, he concludes that each chestnut parent contributes 40 units to the offspring, each bay 33·7 units, each brown 25·3, and each black 10·4. He is now able to deal with the crosses in B, C, D, E, F and G. He finds that there should be in the offspring of the matings:

	B	C	D	E	F	G	
Theoretically	74	65	50	59	41	36	units of red
While observations give	70	64	60	61	54	35	units of red

But is this really conclusive? It is possible that there are almost the same amounts of diffused pigment in the different coats and that the visible colours arise from the relative amounts of granular pigmentation. Had Galton put the amount of red pigment $= 0·8$ for all coat-colours, he would have got his theoretical and observational numbers in *perfect* agreement. But I do not think this would justify the assumption that the amount of red pigment is the same for all coats. I think then that we cannot assume his far rougher agreement is any proof of the numbers he has selected, or indeed of his theory of average parental contributions[*]. After the experience we now have had in the coat-colour of mammals, I feel fairly convinced that it is necessary to supplement the macroscopic classification by microscopic examination, for the categories formed in the former manner contain a great range of both diffuse and granular pigmentation.

Prepotency in Trotting Horses. Still another paper of the same year occurs in *Nature*, July 14, 1898 (Vol. LVIII, pp. 246–7). It is entitled "The Distribution of Prepotency," and deals with data for American Trotting Horses. Galton was much occupied at this time with the idea of the prepotency of individuals. He believed that some favoured individuals had a power of impressing their exceptional characters on their offspring, and that this prepotency was of the nature of a "sport." The American Trotting Horse data provided, he considered, a method of testing this belief. Wallace's *Year Books* give lists (i) of the sires of offspring any one of which has succeeded in trotting one mile in 2 minutes and 30 seconds or less, or who has "paced" (ambled) the same distance in 2 minutes and 25 seconds or less; (ii) of the dams of two such offspring, or else of one such offspring and one such grandchild. Galton selected from these lists of sires and dams those foaled before 1870 and therefore who would be at least 25 years of age in the *Year Book* for 1896, which he was using. He considered that this would give at least 20 years of breeding age to the parents and 5 years of attempted

[*] I took the relative proportions of red to be r_1, r_2, r_3, r_4, and determined their values to fit B, C, D, E, F, G by least squares instead of guessing their values; the ratio of the r's was $1·00 : 1·04 : 1·14 : 1·06$, almost a ratio of equality!

record making at least to the foals. In this way Galton obtained 716 sires and 494 dams who had produced offspring satisfying the above conditions, and he classified them by the number of times they had produced such marked foals. Reduced to percentages of sires and dams the following table resulted:

Percentage numbers of Standard Performers produced
by a single Sire or Dam.

	1	2	3	4	5	6—10	11—20	21—30	31—40	41—50	51 and over
Sire	46	17	10	7	3	9	4	1	1	1	1
Dam	50	35	10	3	1	1	—	—	—	—	—

Galton explains the difference between sire and dam by remarking that while the sire produces some 30 foals annually, the dam produces only one, and therefore the chance of a large number of standard performers is much less for her. He even allows that some of the exceptionally noteworthy performances of the sires (Blue Bull, 60; Strathmore, 71; George Wilkes, 83; Happy Medium, 92; and Electioneer, 154 standard performers) may be due in part to the best mares being sent to famous sires. But he concludes that the extraordinary "tail" of high-class offspring of the sires must be due to some prepotency in some of the sires which enables them to impress their character on their offspring, and he remarks:

"My conclusion is that high prepotency does not arise through normal variation, but must rank as a highly heritable sport, or aberrant variation; in other words its causes must partly be of a different order, or else of a highly different intensity, to those concerned in producing the normal variations of the race. In a sport the position of maximum stability seems to be slightly changed. I have frequently insisted that these sports or "aberrances" (if I may coin the word*) are probably notable factors in the evolution of races. Certainly the successive improvements of breeds of domestic animals generally, as in those of horses in particular, usually make fresh starts from decided sports or aberrances, and are by no means always developed slowly through the accumulation of minute and favourable variations during a long succession of generations."

Here, I think, Galton has forgotten two things:

(i) The average difference between the first and second individuals in a group of 100 tabled to any character is no less than ·36 of the variability of the group, and in a random sample may be still higher, but this is no adequate reason for treating the first individual (or the last) as a sport because he is not, like mediocre individuals, practically continuous with his neighbours.

(ii) That the number of distinguished offspring any individual gives rise to must be considered in relation to his total output. Galton merely says that a sire produces "some 30 foals annually." I do not think this is adequate.

Many years ago I saw a good deal of the working of a large thoroughbred stud; the stud contained a number of stallions, some famous for their racing

---

* The word is quite good English, if Joseph Glanvill and Sir Thomas Browne are authorities.

achievements or for those of their progeny, others less famous. The service lists of the former were always full up with external and home mares; this could not be said of the latter. I think it would be safe to say that the former stallions served annually at least double the number of mares the latter did. I hold therefore that to really demonstrate even a relative superiority in producing standard performers Galton ought to have taken the number of standard performers per total foals produced and this for both sire and dam. Owing to one cause or another one mare may fail more frequently than another to produce her annual foal. Out of four viable foals she might produce three standard performers. Galton's method would make her a less exceptional mare than one that produced four standard performers out of ten foals. Thus his conclusions may be correct, but they cannot be said to be proven until we know the relation of exceptional to total offspring. The distribution of standard performers to sires looks like a "*J*"-shaped frequency curve, and I do not understand why it is not as justifiable as a *J*-shaped curve for cricket scores, nor do I believe that anything is deducible from the deviation of its tail from normality*.

Foundation of "Biometrika." In October 1900 the present biographer sent in a paper to the Royal Society; that paper was printed in the *Philosophical Transactions* and was published in November of the *following* year. Meanwhile William Bateson, who had read the paper as one of the referees, wrote a sharp criticism of it, which the then Secretary of the Royal Society printed and issued in slip to the Fellows, before the latter had any opportunity of studying the criticised paper itself†. Michael Foster, notwithstanding the remonstrances of the biometricians, failed to see any objection to a referee criticising a paper before its publication, and as a result of his attitude, it was determined early in 1901 to found a Journal for the publication of biometric papers. Weldon and the biographer were to be Acting Editors with Galton as Consulting Editor. It is all past history now, and with twenty volumes issued of *Biometrika*, one can afford to smile, when one thinks of Bateson and Michael Foster as unwitting parents of what they would have considered an unviable hybrid! *Biometrika* appeared in October 1901, and Galton contributed an introductory notice entitled "Biometry" (Vol. I, pp. 7–10). A good deal of that paper would now be unintelligible without the light

* There is a long review in the same number of *Nature* (Vol. LVIII, pp. 241–2) by Galton of Alexander Sutherland's *The Origin and Growth of the Moral Instinct*. Galton praises the book highly, as extremely original and extending and confirming the masterly sketch by Darwin in Chapters IV and V of his *Descent of Man* of the evolution of the moral instinct. Galton does not, however, contribute any special views of his own, except the remark that "it would be very interesting to trace and describe the origin and purport of superstitious fears in human nature and their bearing on moral instinct." Galton, it must be remembered, was always appreciative and generous in reviewing; there is, even allowing for this, much information collected in Sutherland's book, which should give it a permanent place in the evolutionist's library.

† Shortly afterwards a resolution of the Council was conveyed to me, requesting that in future papers mathematics should be kept apart from biological applications. βίος was an admissible topic, μέτρον also, but their combination was anathema, and that at a time when statistical theory had to be worked out step by step as the biological applications demanded.

that the above historical facts throw upon it. Galton's tale of Sir Joseph Banks and the young geologists was the parable which he provided in order that he who runs might read.

"Now that nearly a century has slipped past since the event, there can be no harm in digging up and bringing to light a buried but amusing historical fact."

Then follows the inner story of the foundation of the Geological Society. "But," continues Galton,

"it is not in the least my intention to insinuate that Biometry might be served by any modern authority in so rough a fashion, but I offer the anecdote as forcible evidence that a new science cannot depend on a welcome from the followers of old ones, and to confirm the former conclusion that it is advisable to establish a special Journal for Biometry."

Speaking of those early difficulties of Biometry, Galton writes:

"The new methods occupy an altogether higher plane than that in which ordinary statistics and simple averages move and have their being. Unfortunately the ideas of which they treat, and still more the technical phrases employed in them, are as yet unfamiliar. The arithmetic they require is laborious, and the mathematical investigations on which the arithmetic rests are difficult reading even for experts; moreover they are voluminous in amount and still growing in bulk. Consequently this new departure in science makes its appearance under conditions that are unfavourable to its speedy recognition, and those who labour in it must abide for some time in patience before they can receive much sympathy from the outside world. It is astonishing to witness how long a time may elapse before new ideas are correctly established in the popular mind, however simple they may be in themselves. The slowness with which Darwin's fundamental idea of natural selection became assimilated by scientists generally is a striking example of the density of human wits. Now that it has grown to be a familiar phrase, it seems impossible that difficulty should ever have been felt in taking in its meaning. But it was far otherwise, for misunderstandings and misrepresentations among writers of all classes abounded during many years and even at the present day occasional survivals of the early stage of non-comprehension make an unexpected appearance. It is therefore important that the workers in this new field who are scattered widely though many countries, should close their ranks for the sake of mutual encouragement and support. They want an up-to-date knowledge of what has been done and is doing in it. ...

"This Journal, it is hoped, will justify its existence by supplying these requirements either directly or indirectly. I hope moreover that some means may be found, through its efforts, of forming a manuscript library of original data. Experience has shown the advantage of occasionally rediscussing statistical conclusions, by starting from the same documents as their author. I have begun to think that no one ought to publish biometric results without lodging a well arranged and well bound manuscript copy of all his data, in some place, where it should be accessible under reasonable restrictions, to those who desire to verify his work. But this by the way*.

"There remains another urgent reason of a very practical kind for the establishment of this Journal, namely that no periodical exists in which space could be allowed for the many biometric memoirs that call for publication. Biometry has indeed many points in common with Mathematics, Anthropology, Botany and Economic Statistics, but it falls only partially into each of these. An editor of any special journal may well shrink from the idea of displacing matter which he knows would interest his readers, in order to make room for communications that could only interest or be even understood by a very few of them." (pp. 7–8.)

Thus Galton in his eightieth year heartened his young lieutenants for their task, and his words have been through some 28 years a guide to the

* It is noteworthy that Galton's suggestion of a store of data (which has been provided in the archives of the Galton Laboratory for all papers worked out there) has recently been revived by Professor Julian Huxley, and suggestions made for storing measurements in the British Museum (Natural History).

surviving editor of *Biometrika*, never in the first place to expect recognition too quickly, and always if possible to give opportunities for publication to the younger men, whose work and enthusiasm might elsewhere meet with a cool reception.

Gifted Sons of Gifted Fathers. On November 28, 1901 (*Nature*, Vol. LXV, p. 79) Galton published a paper entitled: "On the Probability that the Son of a very highly-gifted Father will be no less gifted."

"Here we meet again with the specious objection which is likely to be adduced, as it has already been urged with wearisome iteration, namely, that the sons of those intellectual giants whom history records, have rarely equalled or surpassed their fathers*. In reply I will confine myself to a single consideration and, ignoring what Lombroso and his school might urge in explanation, will now show what would be expected if these great men were as fertile and as healthy as the rest of mankind.

"The objectors fail to appreciate the magnitude of the drop in the scale of intelligence, from the position occupied by the highly exceptional father down to the level of his *genetic* focus (as I have called it), that is the point from which his offspring deviate, some upwards, some downwards. They do not seem to understand that only those sons whose upward deviation exceeds the downward drop can attain to or surpass the paternal level of intelligence, and how rare these wide deviations must be."

Galton points out that besides the exceptional quality of the father there are three other factors influencing the position of the offspring's genetic focus: (i) the quality of the mother, (ii) the quality of the father's ancestry and (iii) that of the maternal ancestry. The problem is—if we do not discuss it from an individual case—what weight to give to these three additional factors. Now it is a well-recognised fact that while exceptional parents produce exceptional sons at a much higher rate than non-exceptional parents do, the pairs of the latter are so much more numerous than those of the former that it is far more probable that an exceptional man is the son of non-exceptional than of exceptional parents. Hence when we are dealing with average results (ii) and (iii) will not be highly contributory. On the other hand many exceptional men have wives much above the average, and we ought to reckon something for the influence of the mother. Let us take her influence to be measured by an exceptionality one-fifth that of her husband and suppose him to be one man in a thousand†. If we have somewhat over-estimated the average exceptionality of the wife, as one woman in two hundred, we have done so purposely partly to account for possibly neglected paternal and maternal ancestry, and partly to give the son a better chance of reaching to his father's exceptionality. On these assumptions we may treat the problem on rather more modern lines than Galton has done. The "genetic centre" or mean of the array of offspring of our exceptional man and his wife will be at a distance $2.086 \times \sigma$, where σ is the standard deviation or variability of the population for the given character. This supposes the parental correlations to be equal and of intensity $.46$, and the coefficient of assortative mating to be $.25$, both reasonable average values. The *average* son of our

---

* See what has already been said regarding this point on our p. 27 above.

† Assuming that the coefficient of assortative mating to be $.25$ then the average wife of an exceptional husband (1 in 1000) would only be 1 in 40.

exceptional parents is only one man in fifty-four. We have now to determine what is the variability of this array of sons about the mean and how many sons in that array will equal or exceed the father in exceptionality. The variability of such sons is $\cdot 8133\sigma$, and the deviation from the filial mean of the father is $(3\cdot 100 - 2\cdot 086)\,\sigma = 1\cdot 014\sigma$, or the deviation is $1\cdot 014/\cdot 8133 = 1\cdot 25$ nearly in terms of the sons' variability. From which we ascertain that only 11 times in 100 occasions would the son equal or exceed his father's exceptionality, i.e. one in nine sons. Granting an average of three sons to each father we have to examine the cases of three exceptional fathers before we come across a son equal to his father in ability.

But Galton was considering a much higher degree of exceptional ability ; he suggests seven or eight times the quartile for his excess above mediocrity. Let us take one man in 100,000, and suppose a nearer approach in the mother to Galton's view, say she was one woman in fifty, then the deviations of the parents would be $4\cdot 27\,\sigma$ and $2\cdot 05\sigma$ respectively. The genetic centre would be $2\cdot 32\sigma$, and the deviation of the filial mean from the father in terms of filial variability would be $(4\cdot 27 - 2\cdot 32)/\cdot 8133 = 2\cdot 40$ nearly, or 8 sons out of 1000 would reach or exceed their fathers' level or 1 son in 125, or allowing three sons to a father only 1 son would arise in the case of 40 fathers of distinction who would be at least his father's equal.

In a population of 10,000,000 adult men there would be 100 of this exceptional ability each producing able sons at the rate of $\cdot 025$ apiece. The remaining 9,999,900 produce $97\cdot 5$ or nearly at the rate of 1 per 1000, or $\cdot 001$ apiece. That is to say 39 exceptional men are produced by non-exceptional fathers to one produced by exceptional fathers, but the latter produce exceptional sons at 25 times the rate of the former. This is the paradox which Galton tried in vain to make people understand. It has quite recently been again confusing the minds of Professors Raymond Pearl and Leonard Hill, who cannot grasp how great ability is inherited, because the majority of distinguished men have not distinguished fathers.

Pedigrees. Few of those who have had the task of making pedigree charts have not been worried by the unmanageable size to which they are apt to grow, but still more by the difficulty of indexing in a connecting form the material on which pedigrees are ultimately to be based. Galton in a paper entitled "Pedigrees," published in *Nature*, April 23, 1903 (Vol. LXVII, pp. 586–7), suggests a method of what may be termed an "index pedigree"—or as he himself termed it a "pedigree based on fraternal units." This consists in giving a numbered page to each family group. The family group consists of: (i) Father and (ii) Mother with reference to their family group numbers; (iii) their offspring, with any facts the purpose of the pedigree is to illustrate stated about them; thus the main information is to be found on the page, where an individual is one of the offspring, i.e. under his family group number ; (iv) the wife or husband of each child with their family group numbers; and (v) the family group number which gives the offspring of each marriage in the first family group. The birthdays of the *parents* in every family group are

given, in each case for the purpose of distinguishing couples with the same
names. The following is a slightly modified reproduction of Galton's illustration. The whole "index pedigree" will of course have an index, the main
family group of any individual and the family group founded by him being
recorded. For example we look up Frank Gore in this index and find against
his name 205, 340. The latter entry will give his birthday and confirm that

Family Group*.

John Gore	29 October 1822		31 *d*	101
Amy Myers	4 May 1826		43 *c*	

		Characterisation			
Fred. Gore	101 *a*		Mary Drew	144 *a*	205
George Gore	101 *b*		Jane Boyle	136 *e*	211
Ellen Gore	101 *c*		John Piers	105 *b*	207
Susan Gore	101 *d*		*Unmarried*	—	—
Steph. Gore	101 *e*		*Unmarried*	—	—
Fanny Gore	101 *f*		Harry Pitt	163 *f*	223

George Drew	27 March 1827		51 *d*	144
Eliz. Patten	9 May 1830		62 *a*	

Harry Drew	144 *a*		Rose Spry	123 *e*	315
Mary Drew	144 *b*		1. Fred. Gore	101 *a*	205
,, ,,	144 *b*		2. George Lewis	165 *c*	340

Fred Gore	26 November 1858		101 *a*	205
Mary Drew	4 October 1862		144 *b*	

Frank Gore	205 *a*		Anne Fox	218 *a*	340
Amy Gore	205 *b*		James More	265 *e*	344
Anne Gore	205 *c*		*Unmarried*	—	—
Alex. Gore	205 *d*		Eva Sully	241 *d*	370
Rose Gore	205 *e*		Stephen Bell	270 *b*	315

* Slightly modified from Galton's form. Letters to individual numbers need only be attached when there are no names.

he is the man we are seeking. It will give us information as to all his children
and by reference to the number in the last column on the right we can find
out the characteristics of his grandchildren and so on to lower descendants.
By reference to the other number 205, we find the full particulars of all his
brothers and sisters, and can trace by the numbers 344, 370, 315 all his
nephews and nieces, and then upwards to their ancestors or downwards to
their descendants. The family numbers 101 and 144 lead us to the particulars
of his father and mother, and to those of all his paternal and maternal uncles
and aunts respectively. 101 and 144 also give us the family group numbers
of his paternal grandparents. The numbers 211, 207 and 223 will enable us
to find all his paternal, while 315 will give us his maternal cousins. 340 will
lead us to his half-brothers and sisters.

It is fairly clear that if the General Registry were indexed in this way, or even special registries like those of the Society of Friends, pedigree making would be easy work. The Family Group system becomes somewhat more cumbersome in the case of rapidly breeding mammals, for example, dogs. In this case it is needful to replace the family group by the dam, sire and single litter, even if the mating be repeated, as the material becomes too unwieldy. For very small mammals—guinea-pigs, rats or mice—where names are not given, it is the index number of the individual which needs careful thought, especially if it is desired to provide in that index number some indication of the generation to which an individual belongs. A small letter may be given to each individual in the litter attached to the family group index number, and F_s may be added to denote the sth generation from foundation stock, but it is difficult if, say, an F_2 sire has been mated with an F_3 dam to indicate this relation briefly. The difficulty is greater when such a mating is some distance back in the ancestry. If such matings have occurred in considerable numbers the use of generation marks in the index number of animals becomes a doubtful blessing, and we may well fall back on Galton's Family Group numbers plus a small letter.

Nomenclature of Kinship. Galton, still thinking over various methods of expressing kinship, turned from the numerical expression of it to seek a brief nomenclature, and published in *Nature*, January 28, 1904 (Vol. LXIX, pp. 294–5) a paper entitled: "Nomenclature and Tables of Kinship." In this he endeavours to give a self-explanatory, brief and euphonious name to each grade of kinship, which he had in earlier papers provided with an appropriate literal or numerical symbol (see our Vol. II, pp. 354–5, and the present volume, pp. 44–5). He does it in the form of a schedule, here reproduced (see p. 106), for recording in all his known relatives some character X known to exist in A.B. This schedule is practically what he used in the same year to obtain the distribution of successes in the kinsfolk of Fellows of the Royal Society, a topic to which we shall return shortly. The schedule is republished here because it may form a starting point for those desirous of making similar inquiries. One of the most important points in it is the insistence (by the presence of a separate column) on the importance of enumerating the *total* number of relations of each class. Even up to the present year I have seen disease schedules drafted in which the question is asked: How many brothers (sisters, cousins, uncles, aunts, etc.) have been subject to the disease? without the slightest consciousness that such information is idle unless accompanied by the statement of the total number of relatives, affected and not affected, in each class.

There is only one way and that a rather incomplete one in which such imperfect data can be somewhat inadequately utilised. That is by ascertaining the *average* number of relatives of each class in the population at large. Galton often pressed the present writer for data on this point, but there arose considerable difficulties in the way of obtaining them, perhaps the chief of which were the secular changes in the size of families and the infant death-rate*.

* There are also difficulties with regard to the "weighting" of the large families both in the collecting of the data, and in the actual use of them when obtained.

Distribution of the Peculiarity X in the Family of A.B.

fa = Father, or father's, according to its place; similarly, *me* = Mother; *bro* = Brother; *si* = Sister; *so* (or *son* where more euphonious) = Son. The links in the chain of kinship are to be read as leading outwards from A.B. Thus, *me da* signifies "A.B.'s mother's daughter is." *fa bro son* means "A.B.'s father's brother's son is."

Ordinary names for generalised kinships	Titles showing the precise chain of kinships	Adults alone		Titles showing the precise chain of kinships	Adults alone		Names in full of those whose initials appear in the preceding column
		Total No. of sons and daughters	Initials of those whose X deserves record		Total No. of sons and daughters	Initials of those whose X deserves record	
Grandfather	*fa fa*	1		*me fa*	1		
Grandmother	*fa me*	1		*me me*	1		
Uncles ...	*fa bro*			*me bro*			
Aunts ...	*fa si*			*me si*			
Father ...	*father*	1		— —	—	—	
Mother ...	*mother*	1		— —	—	—	
Brothers ...	*brother*			— —	—	—	
Sisters ...	*sister*			— —	—	—	
Half-brothers	*fa son*			*me son*			
Half-sisters	*fa da*			*me da*			
Nephews ...	*bro son*			*si son*			
Nieces ...	*bro da*			*si da*			
First cousins Male ...	*fa bro son* *fa si son*			*me bro son* *me si son*			
First cousins Female ...	*fa bro da* *fa si da*			*me bro da* *me si da*			

Maiden name of the wife	Year of marriage	Number who survived infancy		Initials of those whose X deserves record
		Sons	Daughters	

Number of Kinsfolk. This question of the "Average Number of Kinsfolk in each Degree" was raised by a paper with this title published in *Nature*, September 29, 1904 (Vol. LXX, p. 529). Galton tells us that the simplest conditions for a general theory are those which suppose (i) the population to be stable, i.e. its numbers statistically constant in successive generations; (ii) that the generations do not overlap; (iii) that they are completed by passing wholly into history; and lastly (iv) that any individual is taken into account at whatever age he or she may have died. Galton further supposes that the numbers of the two sexes may be taken as roughly equal. Thus he considers it only needful to work out the results for a single sex. Suppose the average number of females born to a woman who is a mother to be d, then he says that on the average only one of her female children will be fertile of female children, or the chance that any one of these females will be fertile of females is $1/d$. Any mother has d female and d male children and therefore any one of these children will have $d - \frac{1}{2}$ brothers and $d - \frac{1}{2}$ sisters *on the average.* Galton uses a dash to denote a female relative who is fertile of females. Thus the number of sisters (si) is $d - \frac{1}{2}$, but the number of fertile sisters (si') is only $(d - \frac{1}{2})/d$, and each of these produces d daughters (da). Accordingly the number of sisters' daughters (sororal nieces) of a woman will be $(d - \frac{1}{2})/d \times d = d - \frac{1}{2}$. In this way the following table is reached:

	Specific Kinships	Average Number in each	
Ancestry :	me' (mother)	1	1
	$me'\ me'$ (mother's mother)	1×1	1
	$me'\ me'\ me'$ (mother's mother's mother) ...	$1 \times 1 \times 1$	1
Collaterals :	si (sisters)	$d - \frac{1}{2}$	$d - \frac{1}{2}$
	$me'\ si$ (maternal aunts)	$1 \times (d - \frac{1}{2})$	$d - \frac{1}{2}$
	$me'\ me'\ si$ (maternal grandmother's sisters)	$1 \times 1 \times (d - \frac{1}{2})$	$d - \frac{1}{2}$
	$si'\ da$ (sisters' daughters)	$(d - \frac{1}{2})/d \times d$	$d - \frac{1}{2}$
	$me'\ si'\ da$ (mother's sisters' daughters) ...	$1 \times (d - \frac{1}{2})/d \times d$	$d - \frac{1}{2}$
	$si'\ da'\ da$ (sisters' daughters' daughters) ...	$(d - \frac{1}{2})/d \times 1 \times d$	$d - \frac{1}{2}$
Descendants :	da (daughters)	d	d
	$da'\ da$ (daughters' daughters)	$\left(d \times \frac{1}{d}\right) \times d$	d
	$da'\ da'\ da$ (daughters' daughters' daughters)	$\left(d \times \frac{1}{d}\right) \times \left(d \times \frac{1}{d}\right) \times d$	d

Further explanatory letters were published by Galton, October 27, 1904, November 10, 1904, and January 12, 1905. These note one or two misprints and also reply to an objection raised by Professor G. H. Bryan. The reader will find an interesting paradox to solve, if he asks why his wife's sisters' daughters are on the average slightly less numerous than those of his own wife!

Kinsfolk of Fellows of the Royal Society. The main purpose of several of the notes by Galton discussed above becomes clear when we read the paper he communicated to *Nature* on August 11, 1904 (Vol. LXX, pp. 354–6) entitled: "Distribution of Successes and of Natural Ability among the Kinsfolk of Fellows of the Royal Society." Galton received more than 200 replies to a circular with a blank schedule (see our pp. 105–6) which he had sent to the Fellows. In this paper he deals with the 110 which arrived up to a certain date, and contained one or more noteworthy kinsfolk of the Fellow. Galton introduces a slightly arbitrary system of marking, namely 3, 2, 1 or 0 marks to measure more or less noteworthiness, but gives lists of what sort of positions and honours he paid attention to. All Fellows of the Royal Society were given the highest or starred class with 3 marks. In many cases the judgment as to noteworthiness depended on the opinion of the F.R.S. who filled in the schedule, more especially when it concerned the women of his family. Those who will take the trouble to examine the book later published by Galton and Schuster (see our pp. 113–121) will see how differently various Fellows rated "noteworthiness" in their own families; some consider success as merchant or solicitor, or even the becoming an advocate, as a noteworthy achievement, while others would probably never for a moment suppose such occurrences in their family as more than the ordinary routine of middle-class professional life. Galton for obvious reasons does not provide the marks he allotted to such noteworthiness, and he probably marked it low, but the fact that he gave the highest number of marks to *every* Fellow of the Royal Society makes his present biographer somewhat sceptical as to the value of his system in grading ability; at the one end you may have a born scientific genius who revolutionises men's ways of thinking of nature, at the other the professional scientist, not known outside his own country, scarcely beyond his own university, and in no way more able than the normal man in any profession who makes a living by his calling. Admittedly Galton's task was a very difficult one and probably his method may have been, if rough, sufficiently accurate to demonstrate the results he considers to flow from it. Let us consider some of these results:

In the first place he gives a Table, we may notice, for the successes of *male* kin of Fellows of the Royal Society through *A* (Male) and *B* (Female) lines. In this Table the columns headed "Index of Success" are the *total*

Successes of Kinsmen of Fellows of the Royal Society.

A. Through Male Lines		B. Through Female Lines	
Kinship	Index of Success	Kinship	Index of Success
fa fa bro	26	*me me bro*	5
fa bro son	45	*me si son*	31
fa fa	67	*me fa*	58
fa bro	66	*me bro*	64
Total	204	Total	158

marks for "noteworthiness" obtained by the *total* corresponding grade of kinsmen of the 110 Royal Society Fellows. It is important that the reader should bear this in mind as it is not an index in the usual sense of a ratio or percentage.

On this Galton comments:

"A popular notion that ability is mainly transmitted through female lines is more than contradicted by these figures."

A first impression might be that this result is due to overlooking ability in the women, but Galton had on his schedule a list referring only to women*. Even if the Fellows did overlook the capacity of their mothers (which is not usual with sons of ability) this does not account for their overlooking the achievements of their mothers' relatives. I think the explanation is to be sought in other directions. We find that our 110 Fellows had 57 fathers of distinction, but only 16 mothers. The fathers are credited with 136 marks or 2·4 marks apiece and the mothers with 24 or 1·5 marks apiece. It is clear that more than half the fathers of the Fellows were "noteworthy" in a fairly high degree and not more than 16 of these noteworthy fathers, possibly none of them, married a woman of special ability. That is to say they handicapped their sons by not marrying women of marked *ancestral* distinction. Had they chosen wives with equal *ancestral* distinction to that of their own line these 57 fathers would probably have had a still larger number of noteworthy sons. I particularly emphasise the word ancestral because an examination of the above table shows that our 57 fathers did not simply marry mediocrity. The *me bro* group is sensibly equal to the *fa bro* group in noteworthiness, and the *me fa* group (i.e. that corresponding to the fathers-in-law of the fathers) is not so far behind the *fa fa* group. In other words it would appear that our fathers of distinction were thrown by circumstance or inclination into the society of women (from whom they chose their wives) who were the sisters or daughters of men of distinction, but that in making their choice they paid no attention to their wives' earlier ancestry. The point is a somewhat subtle one and wants testing on more ample data, but it seems to me the real explanation of the results in Galton's table. We cannot conclude from it that ability in men is mainly transmitted through the male line. If the above interpretation be correct then the eugenist must ever bear in mind that it is not enough from the standpoint of offspring to marry a woman of ability, he marries so to speak also her stock†.

The next point Galton makes is that the families of the Fellows must be fertile, because the number of brothers, whether of selves or of fathers, comes out 2·43. This would correspond, since $d - \frac{1}{2} = 2\cdot43$, to $2d = 5\cdot86$ or practically nearly 6 members in the family. But although this is a measure of the observed fertility in the 110 Fellows, I think a word of warning is needful.

* Suggested categories: Social leader, Great force of character, Reputed very clever, Artistic (in any way) to an exceptional degree, Successful worker in educational, civic and philanthropic matters, Brilliant prize winner at school or college, etc.

† This is of course only repeating the biological fact, that the genetic potentialities of an individual are only very partially determined by his or her somatic characters, and the only way to obtain a wider appreciation of them (in the case of man where experimental breeding is impossible) is to examine the whole stirp as fully as possible.

Fifty-seven fathers of these Fellows were themselves distinguished men. Now let us start from distinguished fathers; we have seen (see our pp. 102–3) that they are likely to have a higher percentage than mediocre fathers of distinguished sons, but the probability of a distinguished son occurring to a distinguished father depends on the number of his male offspring. Hence if we start by selecting distinguished men, we are likely to find that their fathers had families above the average, especially if those fathers were themselves distinguished. I do not think therefore that we can reach a measure of the fertility of distinguished men from the number of their brothers, nor indeed from the number of their fathers' brothers, if a large number (upwards of 50 °/₀) of those fathers were themselves distinguished. We require the number of children of the Fellows themselves, and this has not been provided.

The next point raised by Galton is of very considerable interest, namely the relative intensity of heredity in the direct line and in the collaterals of this line. I am a little puzzled to follow Galton here. In the direct line of male ancestors there is only one representative in each generation, and there is no necessity to divide by 110 the total marks obtained by each grade of ancestry if we are dealing only with relative measures of noteworthiness. In the one case of *fa fa* and *me fa*, the grandparents, Galton does divide by two. Yet when he comes to the collateral kinsmen, he puts down the *total* marks gained by brothers, and these number not 110 but 110 × 2·43 brothers, and therefore it is not legitimate to compare the *total* marks obtained by brothers with those obtained by "selves" or fathers. In the same way Galton does divide by two the sum of the total marks obtained by paternal and maternal uncles, but forgets that uncles are more numerous than fathers or selves! I have therefore ventured to recompute Galton's Table III, adding to it one or two additional items, but giving in each case the average number of marks obtained by a kinsman of the given grade instead of Galton's total marks of the class. The following will illustrate the method by which my average has been obtained. There are four kinds of male cousins: (i) *fa bro son*; their average number is $1 \times \{(d - \frac{1}{2})/d\} \times d$, for there is only one father; his average number of brothers $= d - \frac{1}{2}$, and $1/d$ (see our p. 107) is the probability that a brother will have sons and d is the number of his sons. Accordingly (on Galton's theory) $d - \frac{1}{2}$ is the number of male cousins that a man will have of the class *fa bro son*; (ii) *fa si son* will also provide $d - \frac{1}{2}$ male cousins; (iii) *me bro son* and (iv) *me si son* will give the same number, or the average number of total male cousins is $4 (d - \frac{1}{2})$. Galton gives 2·43 as the average number of brothers in the self generation and the father generation. Hence $d - \frac{1}{2} = 2\cdot43$ and $d = 2\cdot93$, and therefore on Galton's theory 9·72 is the average number of male cousins or 19·44 the average number of cousins of both sexes combined*. Now the following are the total marks obtained by the cousins of 110 men, i.e. 2138·4 cousins: *fa bro son*, 45; *fa si son*, 25; *me bro son*, 46; and *me si son*, 31; total marks, 147. Average marks of a cousin: ·07.

* This seems to me rather a low average number of cousins for the individual, but I think it is the number which results from supposing the population stable; probably no population ought to be considered as such.

Proceeding in this manner we obtain the table below. I am inclined to think the average marks for Sons and Nephews too low, as it is possible that many of them would not have had time to reach full noteworthiness. As I have noted there is something defective in the earlier generations through the female line, and I have contented myself with using *fa fa bros* and *fa fa fa* as representatives of their grade. Galton does not even give *fa me fa* so that we cannot tell whether they got zero marks or he omitted to classify them. It is quite probable that many men know more of their father's paternal than of his maternal grandfather, a result of the old habit of tracing descent only through the male line.

Average Noteworthiness of Kinsmen in Direct and Collateral Lines of 110 Fellows of the Royal Society.

Generation	Kinship	Numbers	Total Marks	Average Marks	Kinship	Numbers	Total Marks	Average Marks
I	Self	110	330	3·00	Brothers	267·3	170	0·64
II	*fa*	110	136	1·24	*fa bros* ⎱ *me bros* ⎰	534·6	130	0·24
III	*fa fa* ⎱ *me fa* ⎰	220	125	0·57	*fa fa bros*	267·3	26	0·10
IV	*fa fa fa*	110	11	0·10	—*	—	—	—
Additional	Sons	322·3	49	0·15	—	—	—	—
	—	—	—	—	Nephews ⎱ = *bro sons + si sons* ⎰	534·6	48	0·09
	—	—	—	—	Cousins	2138·4	147	0·07

* No entries have been made by Galton for the father's great uncles.

From this revised table Galton's main conclusion flows as definitely as, perhaps more definitely than, from his own Table. The ancestor in the direct line is far more noteworthy than the average collateral in the same grade. To be in the direct line from distinguished ancestry amounts to much, but to be merely the collateral of a great man means very little. Examining the numbers in the first three lines of the table we see that the collaterals of a man of distinction have on the average only $\frac{1}{5}$ of the noteworthiness of his direct ancestor in each generation. As Galton puts it elsewhere, to be the cousin of a man of ability means little if the kinsman gets only the cousin's average share; it might mean a good deal if the character did not blend and the kinsman ran the chance, if a small one, of getting the whole of his cousin's exceptionality.

Galton next discusses the "Relation of Success to Natural Ability." He proceeds by stating that success is due to the combined effect of Natural Gifts and Circumstances. His method is to record success in terms of 1, 0, and − 1 marks to each division of a third of the frequency distribution for ability† and he marks each grade of circumstance ("healthy rearing, family

† The mean values of the thirds, 1·09, 0, and − 1·09, would be more legitimate.

and social influences, education, money, leisure, and surroundings that en-courage work or idleness") in the same way. Galton then assumes that if S be the measure of success, A of ability and C of circumstance,*

$$S = \tfrac{1}{2}(A + C),$$

and he then points out that the regression of success on ability will be just one half, if ability and circumstance be uncorrelated. But I see absolutely no reason for assuming the above form of relation between Success, Ability and Circumstance†.

Galton considers the intensity of the relationship of Ability to Environment at some length. He suggests that "a bright attractive boy receives more favour, and thereby has more opportunities of getting on in life, than a dull and unpleasing one, but these advantages are not without drawbacks; attractiveness leads to social distractions, such as have ruined many promising careers." Then he cites Henry Taylor's couplet:

> "Me, God's mercy spared from social snares with ease,
> Saved by the gracious gift, ineptitude to please."

But I fear that no generalities, only numerical observations, can lead us to a true appreciation of the value of r_{AC}. Researches since Galton's day show how small is the correlation of Ability and Environment. Galton suspected this and wrote that he believed home influences were much less potent than might be supposed. Galton states that the result of his inquiry was "to prove the existence of a small number of more or less isolated hereditary centres, round which a large part of the total ability of the nation is clustered, with a closeness that rapidly diminishes as the distance of kinship from its centre increases."

He further held that these exceptionally gifted families were an asset to the nation. "It must suffice for the present to mention the existence of at least nine gifted families connected with Fellows of the Royal Society, two or three of whom are exceptionally gifted." He concludes (as he has done elsewhere: see Vol. II, pp. 120–2) that it would be both feasible and advan-tageous to make a register of gifted families. Such a register Galton started for other fields of noteworthiness than the scientific, and fragments of this boldly outlined scheme still lie in the archives of the Galton Laboratory.

I have given considerable space to this paper of Galton, partly because it forms the basis of the later book on Fellows of the Royal Society, but

---

* The regression of success on ability would be $\tfrac{1}{2}(\sigma_A + \sigma_C r_{AC})/\sigma_A$, where σ_A and σ_C are the variabilities and r_{AC} the correlation of ability and circumstance. Clearly the regression of success on both ability and circumstance $= \tfrac{1}{2}$, if $r_{AC} = 0$.

† Preserving the type of symbols used in the last footnote the better form of relationship would be

$$\frac{S - \bar{S}}{\sigma_S} = \frac{r_{SA} - r_{SC} r_{AC}}{1 - r^2_{AC}} \frac{A - \bar{A}}{\sigma_A} + \frac{r_{SC} - r_{SA} r_{AC}}{1 - r^2_{AC}} \frac{C - \bar{C}}{\sigma_C},$$

where a bar denotes a mean value. Short of determining from actual observation the three correlations r_{SA}, r_{SC}, and r_{AC}, I do not see that we can profitably guess at values (such as $\tfrac{1}{2}$) for the multiple regression coefficients.

chiefly because, although I doubt the accuracy of some of the processes adopted, it is highly suggestive for kindred researches, and appears to have attracted little of the attention it deserved at the time of its publication in *Nature*.

Closely associated with the material on which the above memoir was based is a letter Galton published in *The Times*, November 17, 1904, with regard to the character and ancestry of Lord Northbrook, who had died on the 15th of the same month. Galton was in a position to comment on the character of Lord Northbrook, for he had served on a council* with him for two years and noted his "rare combination of thoroughness and quickness," which were reported family characteristics of the Barings. Galton was also well acquainted with the family history of the Barings for Lord Northbrook as a Fellow of the Royal Society had replied to Galton's schedule very amply and sympathetically. A full pedigree of the Barings as a noteworthy family would be well worth working up. Like many families of distinction in Great Britain, the Barings in the direct male line show foreign blood.

Noteworthy Families.

We now turn to the work which embraces the data on which the preceding two communications were based. The material was collected by schedules issued by Galton which were filled in by about half the Fellows and returned to him. From these Mr Edgar Schuster† selected the families in which there were at least three noteworthy kinsmen, and formed lists of their achievements on Galton's model. He thus compiled the brief biographical notices of sixty-six noteworthy families which fill about two-thirds of the volume. The book is entitled:

Noteworthy Families (Modern Science). *An Index to Kinships in Near Degrees between Persons whose Achievements are honourable and have been publicly recorded.* By Francis Galton, D.C.L., F.R.S., Hon. D.Sc. (Camb.) and Edgar Schuster, Galton Research Fellow in National Eugenics. Vol. I of the Publications of the Eugenics Record Office of the University of London. John Murray: London.

The intention was to collect similar material in other fields and publish corresponding volumes for Literature, Art, Politics, etc. Some of this material was actually collected‡.

If we consider briefly the material compiled by Schuster one is bound to confess that it is disappointing. As only about half the Fellows replied, and the families of only 63 are discussed, it is clear that we cannot look upon the results as representative of the Royal Society, much less of British

* Probably that of the Royal Geographical Society of which Lord Northbrook was at one time President.

† Mr Schuster had, in October 1904, been elected to the first Research Fellowship in National Eugenics founded by Francis Galton in connection with the University of London: see Chapter XVI below.

‡ "This volume is the first instalment of a work that admits of wide extension." Galton's *Preface*, p. ix.

Science. We cannot assume that the bulk of those who did not reply omitted to do so because their families presented no noteworthy members. We thus obtain no wholly trustworthy general picture of the frequency with which noteworthy men of science arise from noteworthy or commonplace families. Further in the 63 families dealt with as noteworthy we feel the definition is too arbitrary, several scarcely reach real distinction, and for those that do and are well worthy of record a trained genealogist could have given a truer picture and more interesting account of the family (with a pedigree chart!) from fairly accessible sources. We have indeed no certainty that our sample is a "random" one. Galton in his *Preface* of xliii pages, which forms the more valuable part of the book, admits that the facts given are "avowedly bald and imperfect," but considers that they lead to certain important conclusions, for example he considers they show "that a considerable proportion of the noteworthy members in a population spring from comparatively few families" (p. ix). This is very likely true, but it is difficult to accept it on evidence which does not indicate how many noteworthy persons there are in the population or how many we are to expect in a family, and deals only with what is probably not a truly random sample of even the men of science in the population, i.e. 63 out of a total which in 1914 was fixed at 1729 for the British Empire*.

Galton notes several important points, which may be of value as cautions to future circularisers. I cite some of them:

"The questions were not unreasonably numerous, nor were they inquisitorial; nevertheless, it proved that not one-half of those addressed cared to answer them. It was, of course, desirable to know a great deal more than could have been asked for or published with propriety, such as the proneness of particular families to grave constitutional disease. Indeed the secret history of a family is quite as important in its eugenic aspect as its public history; but one cannot expect persons to freely unlock their dark closets and drag forth family skeletons into the light of day." (pp. ix–x.)

Galton accordingly only asked for information on points which "could be stated openly without the smallest offence to any of the persons concerned."

One matter astonished Galton; he found it extraordinarily difficult to obtain even for near kin the number of kinsfolk of each person in each specific degree of kinship. Sometimes the omission was no doubt due to oversight or inertia, but Galton was surprised to find in how many cases the number of near kin was avowedly unknown.

"Emigration, foreign service, feuds between near connections, differences of social position, faintness of family interest, each produced their several effects, with the result as I have reason to believe, that hardly one-half of the persons addressed were able, without first making inquiries of others, to reckon the number of their uncles, adult nephews, and first cousins. The isolation of some few from even their nearest relatives was occasionally so complete that the number of their brothers was unknown." (pp. x–xi.)

Galton (p. xiii) states that he uses the epithet "noteworthy" to correspond in all branches of effort to that which would rank with an F.R.S. among scientific men. He considers that the term covers all those who appear in the *Dictionary of National Biography*, and about half those who appear in

* *Who's Who in Science*, 1914.

Who's Who. No attempt, he tells us, is made in *Noteworthy Families* to deal with the transmission of ability of the highest order. Galton here repeats what he has suggested elsewhere, namely that genius is akin to insanity: "the highest order of mind results from a fortunate mixture of incongruous constituents, and not such as naturally harmonise. Those constituents are negatively correlated, and therefore the compound is unstable in heredity" (p. xv). I do feel it impossible to accept this view; it is quite easy to cite the names of men, to whom the world accords the title of genius, who have had a strain of madness. But one is apt to exaggerate their number and possibly their greatness. Galton states that "the highest imaginative power is dangerously near lunacy." He tells us that he once heard Bonamy Price narrating how as a young man he had asked Wordsworth what was the exact meaning of the lines in the famous *Ode to Immortality**:

> "Not for these I raise
> The song of thanks and praise;
> But for those obstinate questionings
> Of sense and outward things,
> Fallings from us, vanishings;
> Blank misgivings of a Creature
> Moving about in worlds not realised," etc.

Wordsworth had replied that he had had not unfrequently to exert strength, as by shaking a gatepost, to gain assurance that the world around him was a reality. Galton concludes that at such times the mind of Wordsworth could not have been wholly sane; indeed he goes further and considers that such conduct suggests temporary insanity. Yet it seems to me that to every contemplative man, or at least to every contemplative child, such slipping away from their momentary environment, even in a crowded gathering, will not be unfamiliar, and that they can remember instances when they have experienced a distinct effort to recall themselves to their space and time relations; even if they do not need to shake a gatepost, they may require to shake themselves. It is very curious that Galton, who was so essentially a psychologist and attributes much to the subconscious mind†, should have been unfamiliar with the states in which the mind seems switched off from external reality, although conscious that it is still continuing. It is, I think, unreasonable on this ground to associate Wordsworth with insanity. I cannot

---

* This is not Wordsworth's title, which is: *Ode, Intimations of Immortality from Recollections of Early Childhood*—a title which suggests when the "obstinate questionings" arose, and with them the remedy.

† Galton indeed held genius to be something akin to inspiration, and supposed that the powers of unconscious work possessed by the brain are abnormally developed in those who exhibit it. "The heredity of these powers has not, I believe, been as yet especially studied. It is strange that more attention has not been given until recently to unconscious brainwork, because it is by far the most potent factor in mental operations. Few people, when in rapid conversation, have the slightest idea of the particular form which a sentence will assume into which they have hurriedly plunged, yet through the guidance of unconscious cerebration it develops itself grammatically and harmoniously. I write on good authority in asserting that the best speaking and writing is that which seems to flow automatically shaped out of a full mind." (See pp. xvii–xviii of the work under discussion.) Lagrange when listening to music or at social gatherings would sink into deep reveries and lose all consciousness of his environment.

help regretting that the greater authority of Galton was thrown into the scale which was already weighted with Lombroso and Ellis.

In the following chapter Galton discusses the proportion of noteworthies to the generality, but his final conclusion that "the proportion of one noteworthy person to one hundred of the generality who were equally well circumstanced as himself does not seem to be an over-estimate" requires perhaps more evidence than is provided.

Chapter V deals with "noteworthiness as a measure of ability," and discusses on the lines of his paper in *Nature* (see our pp. 108 *et seq.*) the interrelation of Success, Ability and Environment. I have already commented on Galton's treatment of this topic. It seems to me that his discussion is based solely on classification and nothing can be predicted of the correlation between these three factors until their relative frequencies in the several classes have been determined by observation.

Chapter VI deals with Galton's convenient nomenclature for kinship (see our p. 106). In Chapter VII we have the vital question investigated of the number of kinsfolk to be expected in each degree. I say this is vital, for without it we cannot possibly obtain any measure of the strength of heredity. Galton does not here adopt the method of his paper described on our p. 107, where he worked with a stable population, but he makes his returns for each class of kinship on the basis of the F.R.S.'s returns, the schedule containing an inquiry as to the number of kinsfolk in each degree, who *survived childhood*. Hence Galton's previous results do not strictly apply as they were based on all children born, as well as on a theoretically stable population. His Table V (p. xxx) gives only the data for 100 Fellows. I looked at the schedules and found a rather larger number available as schedules appear to have come in later after his Table V was completed. But as the averages were not essentially altered, I will reproduce Galton's numbers, citing them in a different form. The problem wants answering on far more extensive material, but I do not know where else to find even a rough approximation to the average number of relatives a man may expect.

Average Number of Kinsfolk in each Degree.

Class	Kin ♂	Average Number	Kin ♀	Average Number	Size of Family
Brothers and Sisters	*bro*	2·06	*si*	2·07	5·13
Uncles and Aunts	*fa bro*	2·28	*fa si*	2·07	5·35
	me bro	2·19	*me si*	2·38	5·57
Totals	Uncles	4·47	Aunts	4·45	—
First Cousins	*fa bro son*	2·65	*fa bro da*	3·02	
	fa si son	1·84	*fa si da*	2·08	Total
	me bro son	2·36	*me bro da*	2·66	First
	me si son	2·37	*me si da*	2·46	Cousins
Totals	♂ Cousins	9·22	♀ Cousins	10·22	19·44

Clearly the number of nephews and nieces is also contained in the table. A man may expect on the average 4·49 nephews and 5·10 nieces, while a woman would have on the average 4·73 nephews and 5·12 nieces. The later generation seems to give a slightly smaller family than the earlier generation. Since the families include only those who have reached adult age, and the infant death-rate was certainly greater in the older generation, the decrease in size of family is probably larger than appears. The calculations show that an individual has on the average about one fertile relative in each specific type of kinship. Galton now says that he proposes to make "the reasonable and approximate assumption" that "the number of fertile individuals is not grossly different to that of those who live long enough to have an opportunity of distinguishing themselves".... "Thus if a group of 100 men had between them 20 noteworthy paternal uncles it will be assumed that the total number of their paternal uncles who reached mature age was about 100, making the intensity of success as 20 to 100 or as 1 to 5. This method of roughly evading the serious difficulty arising from ignorance of the true values in the individual cases is quite legitimate, and close enough for present purposes" (p. xxxiii).

The argument is not easy to follow. Galton, for example, has (p. xxx) shown that the number of paternal uncles who *survived childhood* in the case of 100 F.R.S.'s is 228, and he now says we must consider this as only 100, and so we see the above number reduced to less than one half. But I think he is contrasting those who survived childhood with those who lived long enough to have an opportunity of distinguishing themselves. He considers that only one individual in each grade of kinship can on the average be fertile in a stable community, and such an individual would probably live to an age at which he would have had an opportunity of distinguishing himself. But it is difficult to see why those who have an opportunity of distinguishing themselves are limited to the fertile. The unmarried uncle may equally with the married have a chance of distinguishing himself. Assuming that "survived childhood" meant to the Royal Society Fellows the surviving 15 years of age —and Galton refers to competitive success at school—and that by 40 years any man has had an opportunity of distinguishing himself, then only some $\frac{1}{5}$ of those alive at 15 are dead before 40. Thus our 228 paternal uncles would scarcely be reduced to 182 if they had died at the rate of the total *average* community. Probably their lives were considerably better than the average, and it would be safe to suppose nearly 200 lived to forty years. This is 100 °/₀ more than Galton proposes to take. I should therefore be prepared to double Galton's number of candidates for distinction in each collateral grade of kinship* (but this will not affect his conclusions, if we are discussing only relative, not absolute degrees of noteworthiness) and to suppose the same number of relatives in each grade, which is approximately true (see our p. 107).

In Chapter VIII Galton limits his inquiry to males. He says that:

"Women have sometimes been accredited in these returns by a member of their own family circle, as being gifted with powers at least equal to those of their distinguished brothers, but definite facts in corroboration of such estimates were rarely supplied." (p. xxxiv.)

* This does not apply to the direct line, in which the number who lived to bear offspring is known exactly. Of course any direct ancestor may have died without reaching the age when he could obtain noteworthiness, but Galton does not consider the effect of this.

It may be difficult to get adequate appreciation of women's noteworthiness, but it is still more difficult to measure heredity in ability, unless we have some direct measure of whether ability can be transmitted through the mother with strength equal to that of transmission through the father. We know whether the father was or was not noteworthy, but if we have no measure of the ability of the mother, we cannot determine whether an able maternal stock transmits its ability equally through an able and through a mediocre woman member. Further Galton does not discuss the sons of Fellows as many might not have reached maturity; 467 persons were addressed, 207 of these sent serviceable replies, of which only 65 are treated in Schuster's list of noteworthy families of F.R.S.'s (pp. 1–79). Galton's data are numerically based on the 207 cases. He states that the real crux of the problem lies in what the remaining 260 were like. Abstention might be due to dislike of publicity, to inertia, or to pure ignorance; such causes would hardly affect the randomness of the sample, but if the 260 did not reply because they had no noteworthy kinsfolk this would influence the sample, and badly influence it. The two extremes are that (*a*) we suppose the 260 to share the richness of the 207 in noteworthy kinsfolk, (*b*) we consider that the 260 had no noteworthy kinsfolk. Galton says he cannot guess which of these hypotheses is the more remote from the truth, but considers that actuality cannot be very far removed from their mean value. I cannot find, however, that this is what Galton has really used. For example the F.R.S.'s had 81 noteworthy fathers. The percentage of noteworthy fathers on the first hypothesis is $81 \times \frac{100}{207} = 39 \cdot 13$ and on the second hypothesis is $81 \times \frac{100}{467} = 17 \cdot 34$; thus the mean of the two is $28 \cdot 24$. Galton, however, does not take the mean of the two hypotheses, but of the numbers 207 and 467, and gets 337; then he finds $\frac{81 \times 100}{337} = 24 \cdot 04$, and this is the percentage he actually uses. Taking 1 man in 100 as noteworthy—a somewhat arbitrary assumption—he states (p. xl) that F.R.S.'s have 24 times as many noteworthy fathers as the generality of men. Before we pass to Galton's final table we may cite one or two points he makes which are of distinct interest and importance for similar investigations. In Chapter IX he gives the result of marking individual degrees of noteworthiness; he made three categories and gave to them in degree of noteworthiness marks 3, 2, 1. He then reduced the total of marks for each degree of kinship (657) to the total number of cases of noteworthiness (329). As a first appreciation the two results differed very little; thus (p. xxxvii):

Comparison of Results with and without Marks in 65 Families.

	First Degree	Second Degree	Third Degree	First Cousins	Total
Number of marks assigned ...	225	208	102	122	657
Marks reduced by factor $\frac{329}{657}$ ...	113	104	51	61	329
Number of Cases of Noteworthiness	110	112	46	61	329

The reason for this approximate concordance lies in the distribution of triple, double and single marks being much the same in the different kinship groups. Galton concluded that marking for different degrees of noteworthiness would be a waste of energy in such a rough inquiry as that he was undertaking. But I think it would have been of great interest had Galton divided his material in another way, i.e. classified his F.R.S.'s into the three categories of noteworthiness, and tested whether their kinsmen had the same or different totals of marks. In other words he would have answered the question of whether ability leading to noteworthiness is inherited in quality as well as quantity.

The next point is very important. Most men know beside their own name that of their mother, i.e. her maiden name. Hence both the numbers and achievements of the uncles and aunts in both paternal and maternal lives are known and there is no difference of a sensible kind in Galton's totals. This holds also for the achievements of the grandparental generation. But when we come to the great grandparents and great uncles, there have been further changes of name in *fa me fa, me me fa, fa me bro* and *me me bro*, and Galton attributes the ridiculously low number of cases of noteworthiness compared with those for *fa fa fa* and *fa fa bro* with a loss of record owing to change of name. This probably has a good deal to do with it, but it does not account for the successes of *me fa fa* and *me fa bro*, who of course bear the mother's maiden name, being only half those of *fa fa fa* and *fa fa bro*, who bear the father's name. I am inclined to think that the factor of assortative mating to which I have referred on p. 109 is at least a contributory cause.

I now reproduce Galton's final table of results, to which I have added percentages[*]:

Numbers and Percentages of Noteworthy Kinsmen recorded in 207 Returns of F.R.S.'s.

Kinship	Numbers Recorded	Percentages	Kinship	Numbers Recorded	Percentages
fa	81	28·24	—	—	—
bro	104	32·26	—	—	—
fa fa	40	13·94	*fa fa fa*	11	3·83
me fa	42	14·64	*fa me fa*	2	0·70
fa bro	45	15·69	*me fa fa*	5	1·74
me bro	52	18·13	*me me fa*	1	0·35
fa bro son	30	10·46	*fa fa bro*	12	4·18
me bro son	19	6·62	*fa me bro*	2	0·70
fa si son	28	9·76	*me fa bro*	6	2·09
me si son	22	7·67	*me me bro*	2	0·70
Total Cousins	99	34·51	—	—	—
Male Cousins, each type	24·75	8·63	—	—	—

[*] Obtained from Galton's assumption that we may take the mean of the extreme cases, i.e., we multiply by $\frac{1}{2}\left(\frac{100}{207}+\frac{100}{467}\right) = \cdot 3486$. I prefer this to his actual method.

From this table we see how degree of noteworthiness diminishes as we pass from the near relatives of the noteworthy to more distant kinsmen. If we accept Galton's two hypotheses: (i) that only one relative in each class can on the average be considered as having lived and been mature enough to have had the opportunity of reaching noteworthiness (see our p. 116) and (ii) that one person in a hundred of the generality is noteworthy, then the above percentages express the numbers of times the F.R.S.'s have more noteworthy kinsmen than the generality of men*. It will be seen that the kinsmen with surnames different from those of the F.R.S.'s fathers and mothers have even a lesser percentage of distinction than the generality of men! Allowing that this may be to some extent due to ignorance of the names, and so of the achievements of these relatives, are we justified in holding that the percentage of noteworthiness in the generality is as high as 1 %? Galton himself says:

"The reader may work out results for himself on other hypotheses as to the percentage of noteworthiness among the generality. A considerably larger proportion would be noteworthy in the higher classes of society, but a far smaller one in the lower; it is to the bulk, say three-quarters of them, that the 1 per cent. estimate applies, the extreme variations from it tending to balance one another.

"The figures on which the above calculations depend may each or all of them be changed to any reasonable amount, without shaking the truth of the great fact upon which Eugenics is based, that able fathers produce able children in a much larger proportion than the generality."

(p. xli.)

Finally Galton refers to the fact that while there was a general high level of ability in the families of F.R.S.'s, some parents were in no way remarkable, so that the "Fellow" was simply a "sport," in respect of his taste and ability. "It is," he remarks, "to be remembered that 'sports' are transmissible by heredity, and have been, through careful selection, the origin of most of the valuable varieties of domesticated plants and animals. Sports have been conspicuous in the human race, especially in some individuals of the highest eminence in music, painting and in art generally, but this is not the place to enter further into so large a subject." Galton cited Bateson, De Vries and his own earlier writings (see our pp. 79 *et seq.*) for the treatment of this topic.

I find it very difficult to accept the view that a Fellow of the Royal Society, whose parents or even the whole of whose known kindred fail to be remarkable, or rather to have been recorded as remarkable, is a sport. In the first place when a pedigree like that of the musician Bach is fully worked out, he is seen to be very far from a sport; he is only the ablest member of a very able musical stirp. And in the next place, if we take a family every member of which for indefinite generations has been mediocre for any given character, we find the variability of an array of offspring is some 70 % of the variability of the population at large, which contains among its members the specially able. Hence although the specially able will not

---

* We might divide these numbers by two, if we assume that in collateral kinship, there will be two on the average who will reach an age when to be noteworthy is possible.

arise as frequently from the mediocre stirp as from the able stirp, they will occur albeit in smaller numbers. I see no reason for terming such occurrences "sports" (see our pp. 78–9, 102–3 above).

Galton's *Preface* was written when he was 84 years of age; it was written at a time when he was feeling keenly that he could no longer undertake the lengthy accumulation of data and their reduction. Nevertheless it is remarkable in its discovery of new problems to be solved and in the suggestions of how they may be solved. The rest of the book is somewhat ephemeral in character, and its judgments of noteworthiness open to criticism, but I think Galton's contribution deserves to be preserved, and I have therefore abstracted it at length here.

Miscellanea. Closely allied to the endeavour Galton made to obtain a register of noteworthy scientific families was a schedule he prepared entitled: "Register of Able Families," with a view to collecting material on a broader basis than that of the Royal Society. The object of the inquiry was "to collect information concerning a large number of exceptionally able families in *all ranks* of society." Ability and exceptionality are therein defined as follows:

"Ability refers to the powers of mind or body, to character, and to every quality which makes a person valuable to his country or to the society in which he lives. It is shown by an artisan who becomes a foreman or an employer, by a clerk who rises to a position of trust, by a private soldier who gains a commission, by a student who wins scholarships and university honours, by those who educate themselves in the absence of other opportunities of instruction, and by all who have fairly achieved honourable distinctions."

Exceptionality, we are told, refers to the middle classes:

"The same amount of ability that is exceptional among them would be very much more exceptional among the lower classes, but not very uncommon in the most distinguished circles of society. The interpretation of the word in each particular case is left to the judgment of the correspondent."

Then comes a characteristically Galtonian paragraph:

"The merit of a family as a whole falls under three distinct heads: (1) Its number, large families being more valuable than small ones when the individuals are of equal merit. (2) The average merit of the individuals. (3) The absence of serious drawbacks in respect to character or physique. *Civilised man being at present the worst bred of all animals, it is extremely rare to find families who are unstained by any moral or physical blemish**. Correspondents should, therefore, not err on the side of diffidence in proposing names; it will be the business of the office to examine the returns that are received and to select the best."

This circular was issued, but probably not in large quantities. What returns Galton obtained I do not know. At any rate no filled-in copies were among the papers that reached the Laboratory named after him. He may have destroyed what he received as worthless, or recognised before its issue that the circular must fail of its object. Exceptional ability is the last to recognise itself under that name, and if you ask mediocrity to register ability you will find that even if it can recognise its existence, it cannot appreciate its degrees, and will almost certainly underestimate its national importance.

* Italics the biographer's.

Galton, as I have often informed the reader, was ever young, ever believed that his fellow mortals had the same enthusiasm for the acquisition of knowledge that he himself had, and was always trustful that they would act as dispassionately in assessing their fellow mortals as he himself acted. Thus he launched his schedules and seemed never discouraged even when they brought little or no harvest!

In the January number of the *Monthly Review* for 1903 Sir Edward Fry published a paper entitled: "The Age of the Inhabited World." In this paper he endeavoured to show that Natural Selection is incapable of doing much that has been accredited to its agency especially citing the case of mimetic insects. He wrote:

"...useful deception will not take place until the protected form is nearly approached. Thus during the whole interval occupied in passing from the normal form of group *A* to near the normal form of group *B*, natural selection will have been entirely inoperative....Either birds are deceived by a small amount of imitation or they are not. If they are, natural selection cannot have produced perfect imitation; if they are not so deceived, then group *A* has passed over from its original form to something close upon the form *B* without any guidance from this principle."

Galton criticised this statement in *Nature*, February 12, 1903 (Vol. LXVII, p. 343) in a letter entitled: "Sir Edward Fry and Natural Selection." He writes:

"I deny this sharp dilemma and assert the existence of many intermediate stages. Two objects that are somewhat alike will be occasionally mistaken for one another when the conditions under which they are viewed are unfavourable to distinction. The light may be faint, only a glimpse of them may have been obtained, the surroundings may confuse their outlines*. While these conditions remain unchanged, the frequency of mistake serves as a delicate measure of even the faintest similarity....If one edible group *A* has individual peculiarities within the limits of variation, that give it a resemblance, however slight, to one of the noxious group *B*, it will occasionally be mistaken by a bird for a *B* and allowed to live unharmed. The similarity may be due to a characteristic attitude, to a blotch of colour, to a preference for resting on a part of the foliage to which its own form bears some likeness, or to other causes. In any case, it may well prove to be the salvation of 1, 2 or more per cent. of those who would otherwise have been seen and eaten. If so the thin edge of natural selection will have found an entrance, and its well-known effects must follow."

It will be noted that Galton says "within the limits of variation." That point is so often overlooked that I must again emphasise it. Few biologists have ever measured the blotch or spot on a butterfly's wing in the case of 400 or 500 members of the same species. They think in terms of a type specimen and suppose the type of one species has to be *gradually* shifted by small stages to the type of another. But the absolute *range* of variation may possibly be 25 °/$_c$ of the type value†. Stringent selection for one or two generations may easily raise the type 10 °/$_o$ or 15 °/$_o$. Such selection is not the same thing as proceeding by minute stages.

* I think Galton is here thinking of his own experimental work on degrees of resemblance and the use of blurrers: see our Vol. II, pp. 329–333.

† The mean length of thigh bone in the type Englishman is say 447 mm., but the *range* of English thigh bones runs from 381 to 513, a range practically covering the type of all existing races. If existence for man depended on the length of his thigh bone there is nothing to prevent severe selection—say the destruction of all individuals with thigh bones over 400 mm.—lowering the English thigh bone to the value, 411 mm., of the Fuegian even in a couple of generations.

Sir Edward Fry replied in *Nature*, March 5, 1903 (Vol. LXVII, p. 414), and falls at once into the fallacy of supposing that because variation in group *A* is continuous, it can only approach group *B* by converting *minute* points of likeness in the midst of unlikeness into such a preponderance of likeness as to produce deception. He holds, as so many others have held, that the theory of the accumulation of *minute* variations fails to account for the facts of mimetism. The error lies in supposing that because the organ varies "continuously," therefore evolution by natural selection involves a gradual accumulation of minute variations in a given direction. Let us suppose the edible group *A* to enter a new environment, where the protected group *B* exists, and that a small percentage of *A* differing *widely* from type has a sufficient resemblance to *B* to escape destruction at any rate to some smaller degree than its brethren. The bulk of *A* will be rapidly destroyed, but the widely divergent section will be, as it were, isolated by the destruction of their fellows, they will inbreed, and the tendency will be, according to the heredity theory of progressive evolution (see our p. 58), for the protecting character to continually increase in intensity, until in a larger and larger percentage it succeeds in deceiving its foes. Sir Edward Fry's appeal to the interspace that separates "the first minute change that deceives no one to the point of first deception," in which interspace he holds natural selection cannot operate, is clear evidence to my mind that he did not know how wide is the range of variation in nearly all organs of all organisms. Natural selection is not forced to choose an individual differing by a minute amount from the type. To hold this view is to think only in terms of the type, and not in terms of the whole population.

Some further communications very typical of Galton may be noted here.

He was far too human not to appreciate what the mass of men found of interest, and among other gatherings, he enjoyed great race meetings. Speaking of the Derby he writes in his *Memories* (p. 179):

"For my own part, I especially enjoy the start of the horses, for their coats shine so brightly in the sunshine, the jockeys are so sharp and ready, and the delays due to false starts give opportunities of seeing them well. I don't care much for its conclusion; but I used often after seeing the start to run to the top of the rising ground between the starting point and the stand, and sometimes got a good opera-glass view of much of the finish."

That Galton frequently went to the Derby is clear, and two instances deserve notice as characteristic of the man. On one of these occasions he persuaded Herbert Spencer and an Oxford clerical don to accompany him. We can imagine how Galton would enjoy this incongruous party who, however, he tells us, enjoyed each other's society. "All went off quite well, except that Spencer would not be roused to enthusiasm by the races. He said that the crowd of men on the grass looked disagreeable, like flies on a plate; also that the whole event was just what he had imagined the Derby to be."

Nevertheless Spencer was sufficiently fascinated to join Galton's Derby party again. We have unfortunately not the don's impressions of the philosopher, the statistician or the races! On another occasion Galton found it too hot to run to the hill, and facing the distant stand he watched the massed faces on the grand stand before the race and just as the horses approached the winning post. The result of his observations was communicated to *Nature**, and runs thus:

The Average Flush of Excitement.

"I witnessed a curious instance of this on a large scale, which others may look out for on similar occasions. It was at Epsom, on the Derby Day last week. I had taken my position not far from the starting-point, on the further side of the course, and facing the stands, which were about half a mile off, and showed a broad area of white faces. In the idle moments preceding the start I happened to scrutinise the general effect of this sheet of faces, both with the naked eye and through the opera-glass, thinking what a capital idea it afforded of the average tint of the complexion of the British upper classes. Then the start took place; the magnificent group of horses thundered past in their fresh vigour and were soon out of sight, and there was nothing particular for me to see or do until they reappeared in the distance in front of the stands. So I again looked at the distant sheet of faces, and to my surprise found it was changed in appearance, being uniformly suffused with a strong pink tint, just as though a sun-set glow had fallen upon it. The faces being closely packed together and distant, each of them formed a mere point in the general effect. Consequently that effect was an averaged one, and owing to the consistency of all average results, it was distributed with remarkable uniformity. It faded away steadily but slowly after the race was finished. F. G."

There is a notion still very current that gouty constitutions should avoid stoneless fruits, in particular strawberries. Galton's creed was that: "General Impressions are never to be trusted. Unfortunately when they are of long standing they become fixed rules of life, and assume a prescriptive right not to be questioned." What about gout and that noble fruit the strawberry? Galton (as well as his biographer) had come across instances, wherein belief dominating desire, enforced asceticism, and so deprived the believer of much harmless pleasure, by dogmatically asserting harmful consequences. Judge of Galton's joy while reading the biography of Linnæus, at discovering that the great naturalist, when the doctors failed to cure his gout, had got quit of his disease by large doses of strawberries! Galton wrote in 1899 a letter to *Nature†* on Linnæus' strawberry cure for gout. One can see the twinkle in his eye as he looked from his writing table towards Harley Street.

"The season of strawberries is at hand, but doctors are full of fads, and for the most part forbid them to the gouty. Let me put heart into those unfortunate persons to withstand a cruel medical tyranny by quoting the experience of the great Linnæus....Why should gouty persons drink nasty waters at stuffy foreign spas, when strawberry gardens abound in England?"

A further characteristic letter appeared in *Nature*, December 20, 1906 (Vol. LXXV, p. 173) regarding the "Cutting a Round Cake on Scientific Principles." The problem to be solved was clearly a personal one for Sir Francis and his niece, who averaged a small cake every three days. "Given a round tea-cake of some 5 inches across and two persons of moderate appetite to eat it, in what way should it be cut so as to leave a minimum of exposed surface to become dry?" The accompanying diagram shows

* June 5, 1879 (Vol. XX, p. 121).

† June 8 (Vol. LX, p. 125). See D. H. Stoever, *Life of Sir Charles Linnæus*, 1794 (Eng. Trans.), p. 416.

Sir Francis' solution. Broken lines show intended cuts; ordinary straight lines the cuts that have been made. The segments are kept together by an elastic band.

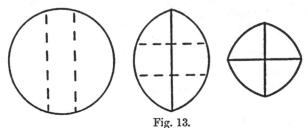

Fig. 13.

Always assuming, which I feel some doubt about, that both consumers of this cake ate their daily allotment of the circular rind, the method leaves an unconscionable amount of dry rind (some $\frac{4}{7}$th) for the third day's consumption! I rather suspect that the cook would have been instructed by the lady of the house to bake in rectangular tins in future.

Another amusing contribution: "Number of Strokes of the Brush in a Picture," was made to *Nature*, June 29, 1905 (Vol. LXXII, p. 198). Galton as I have already noted* sat in 1882 for his portrait (not a very successful one) to Graef. The source of the failure is, perhaps, revealed, for Galton finding it tedious to sit doing nothing counted the painter's slow methodical strokes per minute and then averaged them up. As he knew only too well the number of hours spent in the sittings, he obtained the total he desired to ascertain, some 20,000 strokes to the portrait. About 22 years later he was painted by Charles Furse†, whose method was totally different from that of Graef. He looked hard at Galton while mixing his colours, then he made dabs so fast that Galton found difficulty in keeping up his count. The difference of the two artists' work will be recognised, if the reader compares the Graef picture (Vol. II, Plate XI) with the Furse picture (Frontispiece to Vol. I). It may, however, destroy his pleasure in both, if he thinks of the two artists both having caught the aspect of Galton when silently counting! However to Galton's great surprise Furse's dabs came out about 20,000 to Graef's 20,000 strokes! Only we must remember that Furse did not fully complete his portrait. For comparative purposes Galton computed the number of stitches in an ordinary knitted pair of socks and found 102 stitches in the widest part to each row and 100 rows to 7 inches, whence he computed that the leg parts of a pair of socks would contain over 20,000 separate movements, or rather more than required for a portrait. Galton concludes:

"Graef had a humorous phrase for the very last stage of his portrait, which was 'painting the buttons.' Thus, he said, 'in five days' time I shall come to the buttons.' Four days passed, and the hours and minutes of the last day, when he suddenly and joyfully exclaimed, 'I am come to the buttons.' I watched at first with amused surprise, followed by an admiration not far from awe. He poised his brush for a moment, made three rapid twists with it, and three

* Vol. II, p. 99 and Plate XI.

† Furse died October 16, 1904, of phthisis. His unfinished portrait of Galton must have been one of his last works.

well painted buttons were thereby created. The rule of three seemed to show that if so much could be done with three strokes what an enormous amount of skilled work must go to the painting of a portrait which required 20,000 of them. At the same time, it made me wonder whether painters had mastered the art of getting the maximum result from their labour. I make this remark as a confessed Philistine. Anyhow I hope that future sitters will beguile their tedium in the same way that I did, and tell the results*."

Committee for the Measurement of Plants and Animals. It is impossible to pass over in Galton's *Life* the last decade of the nineteenth century without some reference to this Committee; it took up too much of Galton's energies and consumed too much of his valuable time to remain without some notice in his biography. But the time has hardly yet arrived, when it is possible to write fully about it, and cite at length the voluminous letters and other documents which indicate the parts played by various individuals in first hindering and then entirely perverting the original purposes of the Committee.

The Committee was appointed at Galton's suggestion by the Royal Society Council on January 18, 1894, and consisted of Francis Galton (Chairman), Francis Darwin and Professors Macalister, Meldola, Poulton and Weldon (Secretary), with the very definite purpose of "conducting Statistical Inquiries into the Measurable Characteristics of Plants and Animals." The first report was made in 1896, and consisted of a detailed account of Weldon's measurements on *Carcinus mœnas*, and also his "Remarks on Variation in Animals and Plants†." In the latter paper Weldon emphasised his own view that while "sports" in certain exceptional cases may contribute to evolution, ordinary "continuous" variations were a more probable source of change and further stated, what is almost self-evident, that "the questions raised by the Darwinian hypothesis are purely statistical, and the statistical method is the only one at present obvious by which that hypothesis can be experimentally checked." In asserting this he was only saying that heredity and selection in Nature are mass phenomena and must be treated as such. To those who have read the earlier pages of this chapter, it will be clear that Weldon's view as to the relatively small importance for evolution of "sports" was opposed to Galton's, but this divergence of opinion by no means caused friction between the Chairman and the Secretary of the Committee. It did, however, call forth reams of criticism and numerous letters of protest from William Bateson to the Chairman. The only addition to the Committee in 1896 was, however, that of the present biographer. That Weldon's paper admitted of criticism not only from the biological, but from the statistical side must be allowed, but the fatal mistake was the old one, the evil of attempting to work through a Committee. Had Weldon's paper been published

---

* Would the result be that many subjects would have the strained look of those practising mental arithmetic? The late Mr Hope Pinker told me that he was once modelling a bust of Jowett. The Master remained stolidly silent; Pinker found his task hopeless, and told Jowett that he must throw up his commission, unless the Master consented to talk. "I will try to be good, I will try," replied Jowett, and the portrait was completed. It is not always the artist's fault, if sittings end in a failure.

† See *Roy. Soc. Proc.* Vol. LVII, pp. 360–382.

as his independent contribution, it could have been criticised in the usual way ; he could have defended it, and its merits as well as the difficulties of its subject would have been amply recognised. As it was the reason given for the criticisms (which came from more than one quarter) was that of saving the Committee from making serious blunders*. The Chairman became the centre to which attack and rejoinder were directed, and in despair he wrote to Weldon on November 17, 1896 :

"Herewith is another paper from Bateson, and I enclose with this his accompanying letter to myself. We must talk over what is the fairest course to adopt when we meet (as we probably shall) before the meeting of the R. Soc. on Thursday.

"You see that he offers to print his four letters for circulation among members of the Committee. My greatest difficulty in thinking what should be done arises from the lengthiness.of these papers. I wish the issue could be stated in much more condensed language.

"It would in many ways be helpful, if Bateson were made a member of our Committee, but I know you feel that in other ways it might not be advisable†. The other members besides yourself hardly do enough."

In 1897 the Committee was enlarged by the addition of zoologists and breeders, some of whom had small desire to assist quantitative methods of research—Sir E. Clarke, F. D. Godman, W. Heape, E. Ray Lankester, E. J. Lowe, M. T. Masters, O. Salvin, W. T. Thiselton-Dyer and W. Bateson. It was further rechristened "Evolution (Plants and Animals) Committee of the Royal Society." For several years there was no dominant personality, who could effectively guide this very mixed assembly. Personally I ceased to attend its meetings, resigning in 1900, and was followed in that year by Weldon and later by Galton. Mr Godman then became Chairman and the Reports of the Committee were devoted entirely to the publications of Bateson and his school. The capture of the Committee was skilful and entirely successful‡. I think the feeling of the young biometricians towards Galton's enlarged Committee was more or less expressed by the letter to Galton I now quote, the date is February 12, 1897 :

"I wanted to write a few words to you about yesterday's meeting, but have hardly had, nor indeed hardly now have time to do so. I felt sadly out of place in such a gathering of biologists, and little capable of expressing opinions, which would only have hurt their feelings

* A paraphrase of some of these criticisms will indicate the spirit in which they were written. Vast labour, it was said, had been put into the work and its author no doubt thought himself justified in the conclusions put forward. Perhaps the Committee had thought too little of the responsibility it undertook in publishing such work. The author must know that many would accept his conclusions though few would be able to follow the paper or judge the matter for themselves. Nevertheless the critic found the evidence so inadequate and superficial that he could not understand how responsible people could entertain the question of accepting it. He very truly regretted the countenance given to such a production, etc. etc. Poor Galton! There are some people, whose unfortunate temperaments compel them to believe that as a matter of conscience they are born to be their brothers' keepers.

† Bateson had absolutely no sympathy with the statistical treatment of biological problems, the very work for which the Committee had been appointed.

‡ Perhaps the small understanding shown by the ruling spirits of the Royal Society of what had taken place, was evidenced in 1906, when inquiries were made as to whether the Society would accept the Weldon Memorial Medal and Premium, and the President wrote suggesting that the Evolution Committee would be an appropriate selecting body!

and not have been productive of real good. I always succeed in creating hostility without getting others to see my views; infelicity of expression is I expect to blame. To you I mean to speak them out, even at the risk of vexing you.

"All the problems laid down by you in your printed paper seem to me capable of solution, and nearly all of them *in one way only*, by statistical methods and calculations of a more or less delicate mathematical kind. The older school of biologists cannot be expected to appreciate these methods, *e.g.* Ray Lankester, Thiselton-Dyer, etc. A younger generation is only just beginning its training in them.

"I believe that your problems could be answered by direct and well devised experiments at a 'farm' or institute under the supervision of some two or three men who appreciate the new methods. I think you were entirely right in the idea of a committee to carry out such experiments. But I venture to think that the Committee you have got together is entirely unsuited to direct such experiments. It is far too large, contains far too many of the old biological type, and is far too unconscious of the fact that the solutions to these problems are in the first place statistical, and in the second place statistical, and only in the third place biological. It was the character of the Committee as now constituted which led me to support Michael Foster's motion that the Committee should not experiment, but assist experiment, and further to object to his words 'under the Committee.' Fancy the attempt to make real experiments on variation, correlation, or coefficients of heredity 'under a Committee' of which, I shrewdly suspect, only the Chairman and Secretary know the significance of these terms!

"Hence to sum up, your method seems to me a right one—a Committee to undertake experiments of a definite statistical character*. But your actual Committee is quite a wrong one. It might be a good Committee to press the public with subscription lists; but it is, I believe, a hopeless one to devise experiments which will solve in the only effective way these problems."

Meanwhile besides the criticisms already referred to, there were factors, other than the hope of peace, inducing Galton to enlarge his Committee and widen its programme. As early as February 3, 1891, Alfred Russel Wallace had written to Galton urging that the time was ripe for an experimental farm or institute to undertake researches which might decide disputed points in organic evolution.

Copy of Letter from Alfred Russel Wallace to Francis Galton†.

PARKSTONE, DORSET. *February* 3, 1891.

MY DEAR MR GALTON, Don't you think the time has come for some combined and systematic effort to carry out experiments for the purpose of deciding the two great fundamental but disputed points in organic evolution,—

(1) Whether individually acquired external characters are inherited, and thus form an important factor in the evolution of species,—or whether as you & Weismann argue, and as many of us now believe, they are not so, & we are thus left to depend almost wholly on variation & natural selection.

(2) What is the amount and character of the *sterility* that arises when closely allied but permanently distinct species are crossed, and then "hybrid" offspring bred together. Whether the amount of infertility differs between the hybrids of species that have presumably arisen in the *same area*, & those which seem to have arisen in very *distinct* or *distant areas*—as oceanic or other islands.

* The Royal Society had on Dec. 11, 1896, decided to retain the old name of the Committee, which contained the word "measurement." It was not till the following year, that with enlarged numbers and a wider programme, the Council acceded to Galton's request that the Committee be called "The Evolution (Plants and Animals) Committee."

† I have to thank Alfred Russel Wallace's son, Mr W. G. Wallace, for kindly permitting me to publish the following letters of his father.

Both these questions can be settled by experiments systematically carried on for ten or twenty years. The question is how is it to be done. Talking over the matter with Mr Theo. D. A. Cockerell, a very acute & thoughtful young naturalist, we came to the conclusion that a Committee of the British Association would probably be the best mode of carrying out the experiments, by the aid of a B. Ass^n. grant & a Royal Society grant, aided perhaps by subscriptions from wealthy naturalists. It seems to me that *one* paid observer giving his whole time to the work could carry out a number of distinct series of experiments at the same time,—and if the Zool. Soc. would allow some of the experiments to be made with their animals in their gardens much expense would be saved. To be really good however the hybridity experiments (and the others too) would have to be carried out with large numbers of animals, and thus some sort of small experimental farm would be required. Surely some wealthy landlord may be found to give a small tenantless farm for such a purpose. Then, using small animals such as *Lepus* and *Mus* among mammalia, some gallinaceous birds and ducks, and also insects, a good deal could be done even on a large scale, at a small cost. On the same farm a corresponding set of plant-experiments could be carried out; and an intelligent well educated gardener or bailiff, with a couple of men, or even one, under him, could superintend the whole operations under the written directions and constant supervision of the Committee.

Would you move for such a Committee at the next B. Ass^n. Meeting? *You* are the man to do it both as the original starter of the theory of non-inheritance of acquired variations, the only experimenter on pangenesis, & the man who has done most in experiment and resulting theory on allied subjects.

We thought first of a separate Society, but I doubt if a new society could be established & supported, whereas a Committee either of the B. Ass^n. or of the Royal Society could do the work quite as effectively & would probably receive as much support from persons interested in these problems. It seems to me a sad thing that years should pass away & nothing of this kind be systematically done. I feel sure you would meet with general support if you would propose the enquiry. Believe me, Yours very faithfully, ALFRED R. WALLACE.

FRANCIS GALTON, F.R.S.

P.S. It would of course be better still if a fund could be raised sufficient to establish an *Institute* for *experimental Enquiry into the fundamental Data of Biology*. This is surely of far higher importance than the anatomical, embryological, & other work for which the Plymouth Biological Station was founded. A. R. W.

42, RUTLAND GATE, S.W. *Feb.* 5/91.

MY DEAR MR WALLACE, The views you express so clearly & forcibly, agree with those I have often considered—ranging between a modest cottage with hutches & a bit of ground, up to an Heredity Institute. There was also a half move in this direction made last spring by Ray Lankester, Romanes & others. The difficulties I fear and which I hope you can remove, are as follows. Let us suppose that funds have been collected, a small farm procured and a sensible manager installed in it and that operations are ready to begin. Also I would suppose that the cost of conducting experiments would be met by those who devised them, who themselves had obtained a grant for the purpose from the R. Soc., Brit. Assoc. or otherwise.

Now (1) I doubt if it would be easy to devise a sufficiency of experiments to occupy the establishment of a sort that wd. generally be recognised as crucial. In the two groups of desiderata you mention, no one that I know of, has yet suggested an experiment, much less several experiments, that those who believe in and those who don't believe in the hereditary transmission of acquired characters would accept as fair. If a few such could be devised all my fears as to the utility of the establishment would vanish. If it could settle this one question pains and cost would be amply repaid.

(2) Similarly as regards the sterility question though in a much less degree. The uncertain and often large effects of confinement on fecundity would be a serious disturbing cause.

It then seems to be the first desideratum before making any move that a fairly long list of definite problems, that such an establishment might be set to work upon, ought to be drawn up. Would you put your views as to these on paper?

The number of experimenters is sadly small.

(3) Another difficulty is that the experiments are not likely to be so carefully tended & guarded in an establishment as they would be by oneself or by personal friends. I have had some very marked evidence of this in my own experience, which I don't like to put on paper for fear of causing annoyance.

If the difficulties I have mentioned can be shown to be small, all the rest would be plain sailing. The farm would bear a similar relation to Heredity both plant and animal that the Kew Observatory does to experimenters in Physical Science.

It might grow into a repository of stud books and all about domestic animal breeding, and pay its way well in this department. Also it might become a repository of family genealogies & facts about human heredity, and also pay its way here; the people love to have their genealogies put on record, photos of family portraits preserved, &c. & would pay for the trouble it might cost to keep them.

But the first thing is the experimental farm—in connection with Kew or Chiswick—the Zool. Society & Marine biological laboratories. It could be started moderately under the same roof, so to speak, as one of these, so as to avoid many expenses of a separate establishment, while an independent home was being prepared for it to be entered into if it succeeded.

I have much that would be helpful to say, if you can remove these initial difficulties of prospect. Very sincerely yours, FRANCIS GALTON.

Pray give our united kind remembrances to Mrs Wallace, & accept them yourself.

Copy of Letter from Alfred R. Wallace to Francis Galton.

PARKSTONE, DORSET. *Feby. 7th*, 1891.

MY DEAR MR GALTON, On receipt of your interesting letter I sat down & jotted the enclosed notes of the *kind* of experiments that it seems to me *would* test the theory of heredity or non-heredity of individually-acquired characters. Also a few as to fertility or sterility of hybrids, & as to the real nature of *some* of the supposed *instincts* of the higher animals. I do not myself see *much* difficulty in carrying out any of these, but then I am not an experimenter as you are. Still, I shall be glad to know exactly where the difficulty or insufficiency lies. If these, or any modifications of them, would be valuable & to the point, it seems to me that the mere keeping the plants and animals in health & properly isolated would fully occupy the keeper or keepers of the farm,—while the actual experiments—the deciding on the *separation without selection* of the various lots to experiment with,—which should be crossed & when, and other such matters, would recur only at considerable intervals & could be supervised by the members of the Committee, or some of them, by means of, say, a weekly inspection.

I have limited my notes to three points in which I feel most interest, but of course experiments in *variation* such as Mr Merrifield is carrying on for you, could be added to any extent if there were any danger of the keepers having too little to do!

All the experiments I suggest would require considerable numbers of individuals to be kept healthy and to be largely increased by breeding,—and they would all have to be continued during several years depending on the duration of life of the various species experimented with.

My wife and I are in pretty good health & beg to be kindly remembered to Mrs Galton. As everybody seems to come to Bournemouth we shall hope some day to have a call from you.

Yours very faithfully, ALFRED R. WALLACE.

F. GALTON, Esq., F.R.S.

This letter was accompanied by a detailed list of possible experiments.

42, RUTLAND GATE, S.W. *Feb.* 12/91.

MY DEAR MR WALLACE, I have thought much & repeatedly over your letter & have talked with Herbert Spencer & with Thiselton-Dyer, but cannot yet see my way. I hate destructive criticism,—for it is so easy to raise objections,—& want to offer constructive criticism & to help progress but have every point in view & in all the details I see serious difficulty without any considerable gain.

As an example of many others of the suggested experiments, take the first, viz. that of plants in windy & in still localities. Suppose (1) there was a difference in the seedlings from them, then the advocates of non-inheritance of acquired faculties would protest against its applicability saying that there *had* been selection, the lofty plants & the wide spreading ones would have been preferentially blown down and the weakly ones would have been killed by the rigour of conditions, therefore there had been selection in favour of the small & hardy. Now suppose (2) that there was no difference,—then the same people would say "I told you so." The expt would be for them a case of "heads you lose, tails I win."

Next, to produce any notable effect the expt must, as agreed by all, be protracted for many generations.

Lastly, nature affords an abundance of excellent examples, far superior to artificial ones. Thus take an (elevated) region swept with winds but with hollows in it which are sheltered and all of which is forest clad. The trees in the sheltered hollows will have been from time immemorial finer than those of the same kinds of the exposed places; collect the seeds and plant them under like conditions elsewhere.

During a (Swiss) tour a man might collect an abundance of such seeds of contrasted origin of many species of trees. Even a morning's walk would afford more data than a century of artificial experiment.

So again the seeds of plants originally of English stock but reared for some generations in various parts of the world might be collected and planted side by side.

[The last is Thiselton-Dyer's proposal.]

The only certain employment in the plant department of your proposed farm is to make experiments such as these, or rather to verify in a regular methodical way much that is known already, including expts on the opposite side such as graft-hybridism.

Dyer says that no *experimental* work is likely to succeed at such places as Kew in the ordinary course of work, where careful oversight is required. The men have much other work to do. It would require a man to be specially devoted to its oversight.

The animal experiments seem to be enormously costly.

The case you mention of hybrids & sterility would require many hundreds of animals at the lowest of the computations you give data for. Where the effects of disuse are concerned the animals should be, as a rule, underfed as regards their appetites and only eat just enough to keep them in health; then as there is a deficiency of material for growth, economy of structure would be effective. This would be *very* difficult to ensure. Some of the most interesting experiments are those of the Brown-Séquard type, but these must be put out of court in the present mood of the public & of the law.

Is not the bird nesting experiment continually the unconscious subject of experiment in those fowls who have been hatched from eggs in incubators?

Did you happen to see some remarks I made at Newcastle British Assoc/n, which are printed in the last Journal but one?

I suggested expts on those creatures which are reared from eggs apart from parents. Chickens in incubators, fish, & insects. The incubator industry is large in France & so is the silk-worm. But the naturalists present seemed not inclined to dwell on those views*.

Could anything be made of the following:

A farm for the verification of easy experiments, within easy reach of London.

Cordial relations between it and

(1) The Zoo., the Horticult., Kew, & Royal Agricult. Society.

(2) Private persons of various ranks who would agree to help in expts.

Library of reference on heredity got mostly by begging.

Log-book of daily work preserved (? in duplicate).

Publication of results in some one of the existing Scientific periodicals.

Superintendent (qualifications & Salary to be considered).

All under a c/ttee (? of the Royal Society).

In all this I am keeping the Kew Observatory in view as a somewhat analogous institution.

But before anything could be done, even before asking for its serious consideration, a few *carefully* and *fully* worked out proposals of experiment ought I think to be drawn up. I mean just as much as would have been done if the proposer handed them in to the Gov/t Grant or other committee, for a grant of money.　　　Very sincerely yours, FRANCIS GALTON.

* See our p. 57 above.

Copy of Letter from Alfred R. Wallace to Francis Galton.

PARKSTONE, DORSET. *Feby.* 13*th*, 1891.

MY DEAR MR GALTON, It will be I am afraid impossible to discuss the difficulties of experiment you urge by correspondence, and I will therefore confine myself to a short reply to the objections you have actually made, which seems to me very easily done.

Plants in windy and still air.

You say, "it might be said" there had been selection. But this is very easily obviated, & is the very point on which experiment is superior to observation of nature. In an ordinary open garden or field plants properly cultivated are *not* killed or prevented from flowering & seeding by wind. They grow healthily under it, and I feel sure that not *one* in a *hundred* plants would so suffer. The contrast wd. be produced not by the *violence* of the wind in the one case but by its absence in the other set, they having grown in a glass-covered (or glass-sided) garden. If a common perennial plant was grown—a mallow or a wallflower—for example—a set of 50 or 100 plants might be grown on for 3 or 4 years so as fully to establish whatever change could be produced in the *individuals* by the diverse conditions. Then at the end of that time take the *whole* of the seed produced by each lot,—take two samples of say the 100 smallest or lightest or better perhaps 100 of the average of each, and cultivate them side by side under *identical conditions*. It would not matter to *me*, or I think to *you*, what anybody *said*, but if there were— (*a*) a decided & measurable difference in the two lots of plants from which the seeds were taken, and—(*b*) there was *no* measurable or decided difference between the plants grown from these seeds under identical conditions, this would be *one definite fact against inheritance**,—while if there was a difference of the same nature & fairly comparable in amount it would be a decided *fact in favour of inheritance*. No doubt it might be urged that the effect would be minute but cumulative, & that might be admitted, & the experiment continued under exactly the same conditions for say ten generations. If then no differential effect were produced in the offspring the evidence would be strong against inheritance. Of course the fairest way would be for the advocates of inheritance to formulate the experiments they would admit to have weight, and the opponents of inheritance to do the same.

Then you say "nature affords an abundance of excellent examples, far superior to artificial ones." This I altogether demur to. In nature we *always & inevitably* have selection of various kinds, due to soil, aspect, winds, enemies, overcrowding, &c. &c. &c. & we cannot *possibly separate* the effects of these from any possible *inherited effects* due to diversity of conditions. But this is what we *can & do* do in cultivation.—We save plants from overcrowding & therefore from the struggle with other plants, we can give all the same soil & aspect, protect all alike from enemies, give both the same selection or the same absence of selection of seeds. In nature you cannot possibly tell whether any peculiarity in individuals is due to *conditions* or to *genetic variation*, while if you take those cases where the difference is clearly in *adaptation to conditions*— as the dwarfer plants at higher altitudes—you have the probability, *almost certainty*, of a considerable amount of nat. selectn. By experiment you are able to avoid all these uncertainties & determine the effects of certain definite modifications of environment on *individuals*,—& then ascertain whether the modifications thus produced are inherited.

In nature too, you have the uncertainty introduced by double parentage; each parent in all cross-fertilised plants, may have had *different characters & have* grown under *different conditions*. In experiment you eliminate this cause of uncertainty.

Of course the experiments with animals would involve expense, but with the smaller animals not very much,—& I understood you to say that *this* would not be an obstacle.

If you or any one else will point out the difficulties or uncertainties in the other experiments I suggested I will be glad to answer them, as I think I have done in the *one case* you have referred to.

It is only in this way that we can arrive at a satisfactory mode of procedure, & I regret that I cannot have the advantage of discussing the question with yourself & others who are well acquainted with the subject and with the special difficulties of experimentation.

Believe me, yours very truly, ALFRED R. WALLACE.

* [i.e. of acquired characters.]

P.S. Pray do not trouble to reply to this unless you think anything further from me may be of any use.   A. R. W.

Of course I have referred to the one experiment of *wind & no wind* as an example, not by any means considering it one of the best experiments.   A. R. W.

It will be seen that Wallace had a due appreciation of the necessity for "large numbers"; he recognised that the true method of approaching these problems was *statistical*. If the time was ripe for such experimental work forty years ago, what must we consider it now?

Apparently it was not till 1895 that Galton having got his Committee on the Measurement of Plants and Animals recurred to Wallace's idea of an experimental farm, which Wallace in 1896 termed a "Biological Farm." But a new possibility had arisen, that of acquiring the Darwin house at Down as a station for experimental evolution. Everything was favourable to such a desirable project. The Darwin family were prepared to part with the house for a national purpose on terms which meant a very large contribution from themselves. Galton named a large sum which an anonymous donor was willing to contribute towards the work of experimentation. There can be little doubt that had the scheme been pushed with energy, Down might thirty years ago have been obtained for a purpose urgently necessary and thoroughly in keeping with the spirit of Charles Darwin's work. But a bold scheme only appeals to the bolder minds, and these seemed to be entirely wanting among the men to whom Galton wrote with the hope of engaging their support for the proposed Biological Farm*, as it was termed in the circular issued by Galton on November 30, 1896. I reprint that document here:

*To*_____

The Committee appointed by the Royal Society, for the Measurement of Plants and Animals, proposes to hold an informal meeting at the Royal Society, on Friday, December 4th, at 4 p.m., which they hope you will favour with your presence.

The purpose of the meeting is to discuss the propriety of asking aid from the Council of the Royal Society in establishing and maintaining a Biological Farm, to supply materials (mostly zoological) appropriate to the investigations on which the Committee is occupied, and for undertaking experiments in breeding during many successive generations for the use of those who study the causes and conditions of Evolution.

The general idea that such a Farm would fulfil, somewhat resembles that which was present to the founders of the physical Institute known as the "Kew Observatory," which has been for many years under a Committee of Management appointed by the Council of the Royal Society. It was to procure a place where investigators could have experiments carried on at their own cost, subject, of course, to the permission of the Committee of Management, the cost being, in most cases, defrayed out of grants in aid to the investigators, made by the Royal Society or by the British Association.

It is likely that a farm-house with 20 acres of suitably varied land, and some running water, would amply suffice, so long as the experiments were chiefly confined to small animals. The farm would be in the charge of a resident caretaker under the direct authority of a scientific superior, who might hold the office of Secretary to the Committee of Management. It would be his duty to see that their instructions were duly carried out.

Independently of the farm, and perhaps preliminary to the attempt to raise money for its maintenance, the suggested Committee could accomplish a very important service in a similar direction, for the performance of which it is believed that funds would be immediately

---

.* Meldola, who was throughout warmly in favour of such an institution, actually termed it a "Biometric Station" in December, 1896.

available. That is, they might communicate with persons, many of high social position, who are breeders on a large scale in their own grounds, thereby initiating a widely spread system of co-operation in carrying out experiments desired by the Committee. It is not to be expected that the several results would be equally trustworthy with those made under specially trained management as in the proposed farm. On the other hand, whenever it was found that similar experiments made simultaneously at many different places led to the same results, those results would eminently deserve confidence. The incidental advantage of interesting influential persons in the work of the Committee would be great.

The cost of the complete scheme does not seem likely to be very formidable. It would be chiefly made up of the rental of the farm, the erection of enclosures, hutches, etc., the small initial cost of the animals, their feed, and the wages of the caretaker and assistants. The salary of the Secretary need not at first be large, since the duties of the office would not then be so onerous as to prevent his holding other appointments.

The meeting will be asked to consider this scheme, amending and altering it as desirable, to discuss its cost, and the ways of meeting that cost. If, after this, the prevalent feeling should be in favour of further proceedings, the meeting might appoint an Executive Committee, not consisting exclusively of Fellows of the Royal Society, to examine the subject closely in its various details, to consider the precise experiments that might be first undertaken, and to report to an adjourned meeting.

FRANCIS GALTON
(*Chairman of the Committee of the Royal Society for the Measurement of Plants and Animals*).

42, RUTLAND GATE, S.W.
November 30th, 1896.

The response was most heartrending. Even such warm friends of Galton as Sir J. D. Hooker and Herbert Spencer were not helpful. The former thought that experiments on plants could be undertaken at Kew, and no new station was needful; the latter thought the course suggested impolitic, the proposed purchase of the Darwin house was no doubt appropriate as a matter of sentiment, but most inappropriate as a matter of business. He would be disinclined to cooperate if any such imprudent step were taken*. Great matters must spring from small germs, which would only justify themselves by their success. Real encouragement came only from Adam Sedgwick, from Meldola, and from Weldon ("Surely £4000 can be raised somehow!"). The Darwin brothers it is needless to say wrote most generously and helpfully, but the scheme fell dead even among the biologists who thought it worth while to come to the meeting with the view of discussing it. There was among them no broad conception of what a station for experimental evolution might achieve for their science, and there was not the slightest chance of enthusiasm and energy being put into the project so that it might be carried to a successful issue. The money for the acquisition of Down was still to be found, but there was the sum of £2000 assured by the anonymous donor†, and one distinguished biologist, thinking a bird in the hand was worth two in the bush, asked, if they had not come to allot that sum for their experimental work, what had they come for? I never left a meeting with a greater feeling of despair, and this was shared by Weldon, and to a lesser extent by Galton, who was consoled to some extent by Francis Darwin's writing that, however much he regretted the Down project could not be worked, he was not going to

* As a matter of fact Spencer had not been consulted, but had heard of the matter indirectly through Adam Sedgwick, and had then written to Galton to know what it was all about!

† "There is assurance that a sum of £2000 would be available to start the undertaking, if a thoroughly satisfactory programme could be agreed to."

consider that scheme as finally dead. Now after thirty years it looks as if Down would be retained as a national possession. One may hope that it will be put to as good and fitting a purpose as Galton proposed for it. He has left a lengthy paper dealing with the work he considered the Biological Farm should undertake; it is based on the suggestions he received from many quarters, modified by his own ideas. It is a scheme for "Further accurate observations on Variation, Heredity, Hybridism, and other phenomena that would elucidate the Evolution of Plants and Animals." The matter is arranged under 16 headings, and it is sad to consider that, although more than thirty years have passed since the scheme was drafted, but little has been done to solve the problems therein suggested. It is impossible to print the full manuscript here, but some idea of what it deals with may be judged from its table of *Contents*:

"A. *Preparatory.* (1) Procedure (especially emphasising the need for continuity in observation and for secular experiments). (2) Cooperation (Institutions and Individuals). (3) Breeds suitable for Experiments (necessity for stores of pure stocks of small animals). (4) Place for Station (Down, and existing establishments). B. *Heredity as affected by and related to:* (5) Close interbreeding, Panmixia, Prepotency. (6) Hybridism. (7) Telegony. (8) Acquired modifications in parent. (9) Mental influence on Mother ("Jacobise" in a variety of ways). (10) Instinct (nest building by birds, who have never seen the nest of their species; directive instinct in dogs, taken to unknown place and watched from a distance by a stranger). (11) Variations, "Sports" and their intensity of inheritance. (12) Natural and Physiological Selection. (13) Parthenogenesis. (14) Fertility (many problems stated). (15) Sex and its causes. (16) Gestation."

The bundle of papers in which this and other schemes and letters from innumerable correspondents are included is labelled by Galton: "Old Papers concerning the Evolution Committee of the R. Soc. of probably no present value. Might be useful if a Darwinian Institute were ever founded." "Of probably no *present* value"—what a criticism of the biologists of 1890–1900!

Here, as in Experimental Psychology, Galton was ahead of his age, and few have recognised how much even by raising these questions, he stimulated that movement for experimental biology, which the present generation of biologists believes was unthought of by their Victorian predecessors. Thus came to an end Galton's plan for an experimental station for evolution; it was another illustration of the futility of working through ill-assorted committees. I say came to an end, but hardly in Galton's mind. It must I think have been in 1903, when in the summer vacation the biometricians were employed on their summer tasks at Peppard and Galton was of the company, that the matter again arose. One evening he asked his two lieutenants to prepare a draft scheme for a biological farm, to state its size, staff, equipment, its probable cost and annual expenditure for maintenance and experimentation. Weldon and I talked the matter over, and felt that although Galton was well-to-do, he was not so wealthy, that to run a biological farm might not deprive him of some of the easements necessary to his age. We therefore determined to estimate the cost of the farm on the scale of maximum effectiveness. It was a pious fraud, but the suggestion of a biological farm was never again referred to, and Galton's thoughts of increasing human knowledge soon turned to less expensive projects.

Appendix to Chapter XIV*.

"The Weights of British Noblemen during the last Three Generations," *Nature*, January 17, 1884 (Vol. XXIX, pp. 266–268).

This rather amusing paper is not included in any list of Galton's memoirs known to me, nor were any offprints of it to be found in the *Galtoniana*. It seems to have been forgotten by Galton himself and would have certainly been overlooked by me had I not stumbled across it in reading Romanes' review of Galton's *Record of Family Faculties* and *Life History Album* in the same number of the Journal. Galton—whom the Goddess of Chance certainly favoured—became acquainted with the fact that an old established firm of wine and coffee merchants had been since about 1750 in the habit of weighing their customers, and that upwards of 20,000 persons, many of whom were famous in English history of the eighteenth century, had for their use or amusement sought the firm's huge

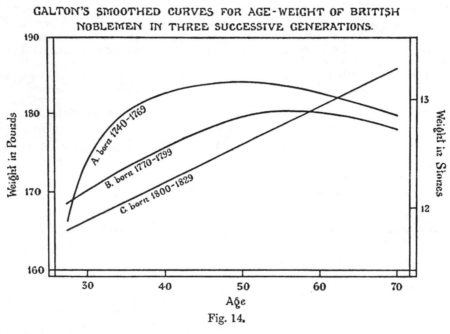

GALTON'S SMOOTHED CURVES FOR AGE-WEIGHT OF BRITISH NOBLEMEN IN THREE SUCCESSIVE GENERATIONS.

Fig. 14.

beam scales. Galton confined his attention almost entirely to noblemen as a well-rounded class, whose ages were easily ascertainable, and to their data in respect only of two characteristics, namely the degree of fluctuation in weight as exhibited by the age-weight curves of individual noblemen, and the difference in the average age-weight curves of noblemen born in the three periods 1740–1769, 1770–1799, 1800–1829. He found that the average annual fluctuation in the earlier group was about 7 lbs. and that in the latest group it was only 5 lbs. He concluded that this pointed to an

---

* Some notice of the following paper should have appeared in Section H of Chapter XIII (Vol. II), but its existence was then unknown to me.

irregularity in the mode of life that was greater two or three generations back than now. Further he found that the "prime" for weight was also earlier in age for the older generations, being hardly discoverable at all in those born in the first third of the nineteenth century or in the professional classes of the 'eighties. His three smoothed curves reproduced on p. 136, with the table of mean weights at each central age, indicate that noblemen of the generation which flourished about the beginning of last century attained their meridian and declined much earlier than those of the generation sixty years their juniors, or indeed than the mid-Victorian professional classes, where the culminating point was difficult to ascertain.

Galton's data were somewhat scanty as the following table will indicate, but his general conclusions appear to be justified:

Actual Mean Weights in pounds at Various Ages.

Class	Years of Age					
	27	30	40	50	60	70
Born 1740–1769	166 (13)	176 (18)	184 (24)	181 (21)	181 (18)	180 (12)
Born 1770–1799	168 (24)	171 (23)	172 (24)	184 (26)	178 (26)	178 (15)
Born 1800–1829	165 (35)	165 (44)	171 (43)	175 (37)	181 (22)	188 (7)
Mid-Victorian Professional Class	161	167	173	174	174	?

"There can be no doubt," he writes, "that the dissolute life led by the upper classes about the beginning of this century, which is so graphically described by Mr Trevelyan in his *Life of Fox*, has left its mark on their age-weight traces. It would be most interesting to collate these violent fluctuations with events in their medical histories; but, failing such information, we can only speculate on them, much as Elaine did on the dints in the shield of Launcelot, and on looking at some huge notch in the trace [for the individual], may hazard the guess, 'Ah, what a stroke of gout was there!'"

Although no great importance can be attached to Galton's results for this particular class of subject, yet the problems his paper suggests might be profitably studied on more ample material now extant. I am therefore glad to have brought to light once more this long forgotten paper.

CHAPTER XV

PERSONAL IDENTIFICATION AND DESCRIPTION

"It became gradually clear that three facts had to be established before it would be possible to advocate the use of finger-prints for criminal or other investigations. First it must be proved, not assumed, that the pattern of a finger-print is constant throughout life. Secondly that the variety of patterns is really very great. Thirdly that they admit of being so classified, or 'lexiconised,' that when a set of them is submitted to an expert, it would be possible for him to tell, by reference to a suitable dictionary, or its equivalent, whether a similar set had been already registered. These things I did, but they required much labour." Galton: *Memories of my Life*, p. 254.

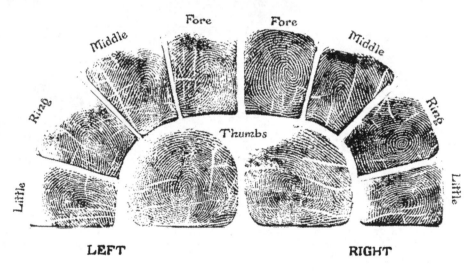

Fig. 15.

§ I. *History and Controversy.*

The writer must confess to having felt not a little puzzled when he had to determine in what order to present Galton's work on Personal Identification. It is not only that his work was scattered over very numerous publications, but that in order to make it effective Galton had to step into the public arena; and this had its usual consequences, namely controversy and misrepresentation, factors which had hitherto played but a small part in Galton's career. On the whole it is strange how little controversy intruded on Galton's long and quiet years of study; this was in part due to the peace-loving mind of the man, but there were also other causes at work. In the first place he was labouring most of his life in an entirely untilled field, and there could be no friction therefore with other pioneers. In the next place his fellow scientists were slow to realise that the new logical tools he was forging would ultimately be applied in their own fields of work, and when that application came, whether in anthropology, psychology, biology, sociology or medicine, there would be sure to be friction, and resulting controversy—heated and bitter. That experience was left to his lieutenants and their disciples.

In the matter of finger-print identification, however, Galton was not only sharpening a new tool, but urging on all and sundry its application to practical problems. The tool was soon seen to be so efficient that it had to be adopted, but its very efficiency raised jealousy and controversy, as to whom the merit of its introduction must be attributed. I shall endeavour in this chapter not only to put before my readers the history of the adoption of finger-print identification in this country, but the means Galton took to popularise the idea, and finally provide a *résumé* of his scientific contributions to the subject which form the foundation on which all later work in this field has been based.

Investigations with regard to finger-prints occupied much of Galton's time during the years from 1887 to 1895. I say advisedly from 1887, because soon after the opening of his second Anthropometric Laboratory, he began the collection of those thousands of finger-prints, on the study of which so many of his conclusions depended*. It may be safely said that no one had in the early 'nineties so vast a collection of finger-prints as Francis Galton, a collection covering not only our own English race, but also Welsh, Hindoos, Jews, Negroes, and special groups such as idiots, criminals, etc. That collection started by Galton has continued to grow to the present day, although now it is chiefly, but not entirely, confined to hereditary data. I have already indicated how Galton in the 'eighties was occupied with the problem of portraiture and personal identity (see Vol. II, Chapter XII), and it was from the standpoints (*a*) of ethnology, (*b*) of heredity, that he first approached the problem of the papillary ridges on the fingers. It is well known now that finger-prints may be classified into certain types, which Galton called "genera," and that variations appear to cluster round these typical forms. It would have been a great achievement to show that in certain human races only one "genus" occurs, or indeed that the "genera" occur in very different proportions. Galton failed in racial collections of fairly considerable numbers to detect any *marked* differentiation of this kind. I am not prepared to assert that with larger collections and more modern statistical methods *some* differentiation might not still be found; it is hard to believe that from the very origin of *Homo* these genera have

---

* This collection in large indexed cabinets exists in the *Galtoniana*, and I should be very grateful to any one whose finger-prints were taken thirty to forty years ago, if they would call in at the Galton Laboratory, University College, W.C.1, and allow their fingers to be reprinted. Galton in delivering his Presidential Address to the Anthropological Institute, January 22, 1889, refers to his lecture of 1888 as "of last spring" and mentions that he is taking the two thumb prints and describes the technique he has adopted. His work therefore began certainly in 1888 and I suspect experimentally in 1887 before his lecture of May, 1888. The earliest *dated* finger-prints that I can find in this laboratory are from March, 1888.

been scattered almost indiscriminately among the ten fingers; yet Galton failed to solve any anthropological problem by the aid of finger-prints*.

In the matter of heredity he was more successful, he produced evidence adequate enough to demonstrate that finger-prints were hereditary, but neither he nor any one since has produced a satisfactory account of the manner in which they are inherited.

In later years† Galton formed a considerable collection of family prints of the two forefingers only. These were tabled and reduced in 1920 by Miss Ethel M. Elderton, who demonstrated the general inheritance of the ridge patterns, but noted that two finger-prints were far from adequate to determine the intensity of heredity, as although a parental peculiarity of pattern might pass to the same finger in the child, or with less probability to the homologous finger, it might also pass to any one of the remaining eight fingers; this, if it happens to any individual finger with still less probability, may occur with equal or even greater probability when we take into account the total eight of them. While the existence of ten fingers in man is a distinct advantage in the matter of personal identification—or if we like a distinct misfortune to the criminal—it is also something of a misfortune to the geneticist. At any rate Galton's work left much to be done in determining the organic correlations between prints of the different fingers in the same individual‡ and the bearing of these organic correlations on the problem of heredity in the ridges. Thus it came about that while Galton did much pioneer work in the collection and co-ordination of material his chief contribution to the subject was in the matter of identification. He was the first to publish matter, largely due to Sir William Herschel, fully establishing the persistence of finger-print patterns; he was the first to show the nature of their variety and to classify them, and lastly he was the first to prove that it was possible to index them and rapidly to find, from a given set of prints, whether their owner was already in the index. All these problems were fundamental and must be definitely solved, if finger-prints were to be used for police purposes. None of this spade work had been achieved or at any rate published before Galton took up the subject. Before his day we have mere suggestions of the possible usefulness of these prints. Within ten years from his first study of the subject by the aid of his papers dealing with the prints from a scientific standpoint, by repeated letters to the press, by action through the British Association and by definite demonstrations in his Laboratory to the Commission appointed by Mr Asquith to consider the question of criminal identification in England, Galton had got not only bertillonage accepted in

---

* More recent researches, for example, those of Kubo (1918) and Collins (1915), seem to indicate that the Oriental races have a larger percentage of whorls and fewer ulnar loops than the European races. But the results are doubtful because there is a large personal equation in the matter of classification. I think we must conclude with Stockis (*Revue Anthropologique*, Année 1922, p. 92) that the results reached (thirty years after Galton) are still not adequate to admit of our asserting the existence of well-defined ethnic differences in finger-prints.

† See *Biometrika*, Vol. II, p. 365, 1903. Collection made in the years 1903 to 1905.

‡ A beginning was made in the study of the organic correlation of finger-prints by Dr H. Waite, *Biometrika*, Vol. x, p. 421 *et seq.*

England, but, what in the sequel has proved more important, the use of finger-prints. The fact that such prints are now practically adopted in the Criminal Investigation Departments of all civilised countries, is striking testimony to Galton's work and to his energy. Attempts have been made to belittle his achievement in this matter. Galton's claim is not based on his being the first to *suggest* this use of finger-prints, or on being the first actually to apply them. It lies in the fact that general police adoption of finger-prints resulted from his activities. It is easy to make suggestions, it wants an additional mental quality to get them carried out by administrative bodies, always and often justly conservative in character*.

In Galton's lecture at the Royal Institution in 1888 on "Personal Identification†," he gave an account (pp. 3–5) of Bertillon's method—bertillonage as it came to be called—the basis of which lies in recording the anthropometric measurements of criminals. Galton believed in the serviceableness of this method, but he held also that its efficiency had been overrated, because its inventor much underestimated the high correlations, which Galton surmised, and which were later demonstrated to exist between the various measurements taken. He then made his first public reference, as far as I am aware, to those "most beautiful and characteristic of all superficial marks" the

"small furrows, with the intervening ridges and their pores, that are disposed in a singularly complex yet regular order on the under surfaces of the hands and the feet. I do not now speak of the large wrinkles in which chiromantists delight, and which may be compared to the creases in an old coat, or to the deep folds in the hide of a rhinoceros, but of those fine lines of which the buttered fingers of children are apt to stamp impressions on the margins of the books they handle, that leave little to be desired on the score of distinctness."

Galton then refers to the work of Purkenje in 1823, Kollmann 1883, Sir William Herschel and Dr Faulds, etc., much in the same terms as in his *Finger Prints* of 1892‡. In this lecture Galton submitted on the problem of permanence:

"a most interesting piece of evidence, which thus far is unique, through the kindness of Sir Wm. Herschel. It consists of the imprints of the first two fingers of his own hand made in 1860 and in 1888 respectively, that is, at periods separated by an interval of twenty-eight years." (pp. 12–13.)

Galton analyses the *minutiae* (see our p. 178) in an adequate, but less thorough manner than he did later by the aid of sevenfold photographic enlargements and the tracing in of the ridges. It is clear that Galton was actively interested in finger-printing, and his remarks on p. 15 show that he had been experimenting in many ways on the most advantageous and

* An executive department has not only to consider the cost of installing an innovation, and afterwards of its maintenance, but likewise whether the resulting advantages will compensate for the additional expenditure.

† See our Vol. II, p. 306; also the *Journal Royal Institution*, May 25, 1888, or *Nature*, June 28, 1888.

‡ He does not refer to the paper by Nehemiah Grew in the *Phil. Trans.* of 1684 (Vol. XIV, pp. 566–67). That paper has a very good, i.e. well engraved, illustration of the ridges on the palm of the hand and on the fingers. A rather curious representation of the pores on the ridges—not referred to in the text—appears also to belong to Grew's paper. Grew emphasises that the pores are *on* the ridges, not in the furrows, and speaks of them as little fountains for the discharge of sweat. There is no statement as to permanence or as to personal identification.

cleanliest method of taking the impressions. In his *Finger Prints* of 1892 Galton says that

"My attention was first drawn to the ridges in 1888, when preparing a lecture on Personal Identification for the Royal Institution, which had for its principal object an account of the anthropometric method of Bertillon, then newly introduced into the prison administration of France. Wishing to treat of the subject generally, and having a vague knowledge of the value sometimes assigned to finger marks, I made inquiries, and was surprised to find, both how much had been done, and how much there remained to do, before establishing their theoretical value and practical utility." (p. 2.)

in 1860.   *in 1888.*

Fig. 16. Finger Prints of Sir William J. Herschel at an interval of 28 years. From Galton's *Finger Prints*, Plate 15, Right Forefinger. Second method of marking *minutiae*.

I do not think that it can be asserted that Galton failed to recognise what work had been previously *published*, except in the case of Nehemiah Grew*, and from him he would indeed have learnt very little, had he known of him. That the pores were on the ridges, not in the furrows, Galton probably found out from his own observation†.

* Alix's paper of 1868 (see our p. 143 ftn. †) and Klaatsch's of 1888 are referred to on p. 60 of the *Finger Prints*, but more stress possibly might have been laid on the former.

† In the *Memories*, pp. 257–8, is an amusing account of Herbert Spencer's view on the relation of ridges to pores:

"I may mention a characteristic anecdote of Herbert Spencer in connection with this. He asked me to show him my Laboratory and to take his prints, which I did. Then I spoke of the failure to discover the origin of these patterns, and how the fingers of unborn children had been dissected to ascertain their earliest stages, and so forth. Spencer remarked that this was beginning in the wrong way; that I ought to consider the purposes the ridges had to fulfil, and to work backwards. Here he said, it was obvious that the delicate mouths of the sudorific glands required the protection given to them by the ridges on either side of them and therefrom he elaborated a consistent and ingenious hypothesis at great length. I replied that his arguments were beautiful and deserved to be true, but it happened that the mouths of the ducts did not run in the valleys between the crests, but along the crests of the ridges themselves. He burst into a good-humoured and uproarious laugh and told me the famous story which I had heard from each of the other two who were present at the occurrence. Huxley was one of them. Spencer, during a pause in conversation at dinner at the Athenaeum said, 'You would little think it, but I once wrote a tragedy.' Huxley answered promptly, 'I know the catastrophe.' Spencer declared it was impossible, for he had never spoken about it before then. Huxley insisted. Spencer asked what it was. Huxley replied: 'A beautiful theory, killed by a nasty, ugly little fact'."

I think, however, Galton had forgotten the date at which his attention was first drawn to finger-prints. He appears to have been collecting data before his lecture in 1888. But as early as 1880 Dr Faulds wrote a letter (February 16th) from Japan to Charles Darwin mentioning that the topic might have interest for him. The letter suggested that there were *racial* differences in finger-prints and enclosed two prints of palms of hands and of the five fingers. Darwin, strangely for him, rather overlooked the possible importance of the topic ; he was clearly busy and worried*. He forwarded the letter to Galton mentioning that it might have interest for anthropologists, and suggesting it had better be dealt with by the Anthropological Institute. Galton actually did present the letter to that Institute, and its officials appear to have then pigeon-holed it. Faulds' and Darwin's letters were unearthed many years later (April, 1894), after Galton had published his books, and returned by A. E. Peek to Galton. These letters I found in the *Galtoniana*.

Before this discovery I had no knowledge that Dr Faulds had written to Darwin in 1880, but it is clear that Galton sent the letter as suggested by Darwin to the Anthropological Institute. It cannot be said that any injustice was thus done either by Darwin or Galton. No busy scientist is bound to pay attention to the innumerable suggestions that may be made to him. Further, twenty years earlier, 1858, Sir William Herschel was using finger-prints for practical executive purposes in India, and lastly what is more to the point Dr Faulds sent much the same communication slightly later to *Nature* where it was printed on October 28, 1880†, i.e. in the year of his letter to Darwin, and called forth a response from Sir William Herschel stating what he had himself achieved‡. Galton refers to both letters not only in his Royal Institution lecture of 1888, but also in his *Finger Prints*. Before Galton issued his epoch-making papers of 1891, and his three books 1892 to 1895, no really substantial work had been published on finger-prints, since Purkenje's. A comparison of Galton's results with the two letters in *Nature* of 1880 will suffice to indicate how idle it is to attempt to belittle his claims.

* Darwin was failing in health in 1880 and correspondence with strangers had become a burden to him. See *Life and Letters*, Vol. III, p. 355 *et seq.*

† Vol. XXII, p. 605. It should be noticed that Dr Faulds states that he commenced his study of the "skin furrows of the hand" in the previous year, but he yet speaks of "the for-ever-unchangeable finger furrows of important criminals," and again in his letter to Darwin he states that photographs may grow unlike the original, but never the rugae. In other words he begs the question of permanence. At the same time he shows that he has ideas of the wide possible usefulness of the finger-print. He says that he had been studying the papillary ridges in monkeys, but appears to have overlooked the elaborate comparisons of these ridges among all kinds of primates including man in the paper by Alix: "Recherches sur la disposition des lignes papillaires de la main et du pied," *Annales des sciences naturelles* (Zoologie), T. VIII, pp. 295–362, T. IX, pp. 5–42 and corrections T. X, p. 374. The portion in T. IX contains excellent engravings of the hands and feet ridges of primates. Alix enlarges on the great variety of the finger-prints in man, and figures "double vortex," "loop" (*Amande*), "racket," "spiral," "spiral within circle" and simple "circle." In other words he was reaching and figuring a classification to enable him to compare man with other primates.

‡ *Nature*, Nov. 25, 1880, Vol. XXIII, p. 76. Herschel recorded a twenty-two years' use of finger-prints and gives some evidence of their permanence.

A serious mis-statement more or less frequently made was that Bertillon who ultimately adopted the finger-print system of identification had initiated it. This pained Galton extremely, because he actually introduced Bertillon to the method, but the latter at the time feared practical difficulties such as the want of education in his employees. Bertillon's letter of June 15, 1891, in a reply to a letter of Galton's suggesting that he should try the system, has luckily been preserved. The essential paragraph runs as follows:

"Je vous remercie de votre nouvel envoi relativement aux *impressions digitales*. Je suis fort disposé à ajouter votre procédé au signalement anthropométrique surtout pour les enfants. Mais je redoute quelques difficultés pratiques pour le nettoyage des doigts après l'impression faite, etc. Puis mes agents si peu instruits mettront-ils le zèle nécessaire pour apprendre votre méthode? Je crois que vous traversez souvent Paris, pourriez-vous, à votre prochain voyage, me consacrer une matinée au Dépôt, pour un essayage sur la vile multitude?"

The words "votre procédé" and "votre méthode" clearly indicate that Bertillon was fully aware of the originator of this process of criminal investigation. Notwithstanding, even as late as 1896, in the English translation of Bertillon's *Instructions signalétiques*, the date of the introduction of the prints of the thumb and three fingers of the right hand into the French schedule for the criminal is given as 1884, instead of 1894, and "conveys the idea that the use of finger-prints in Paris is much older than it really is, and previous instead of subsequent, to its use in England*." The 1893 edition of Bertillon's *Identification Anthropométrique, Instructions signalétiques* has no reference to finger-prints. It is still over-confident as to the infallibility of bertillonage†. Galton claimed neither finality nor infallibility for his methods; as to finger-print identification he found it a suggestion and he left it an art.

In 1905 M. Bertillon wrote in reply to a question of Dr Faulds:

"Les impressions digitales à Paris sont adjointes au signalement anthropométrique depuis l'année 1894. J'ajoute que nous nous en trouvons fort bien. Quoique nous n'ayons *jamais* fait d'identification erronée antièrement nous sommes encore mieux garantis, si possible, en ce qui regarde l'avenir." (*Guide to Finger-Print Identification*, pp. 4–5.)

---

* See *Nature*, Vol. LIV, p. 569, where there is a review by Galton of the *Signaletic Instructions*, emphasising the superiority of the English finger-print system and direct indexing of the prints to the French anthropometric system or "bertillonage." Galton therein prophesies what has since come to pass, that the former would ultimately supplant the latter completely.

† "L'absolu de nos affirmations dans les questions d'identité, et notamment dans les cas plus difficiles d'identification entre deux photographies, étonne encore les fonctionnaires de la police ou de l'ordre judiciaire auxquels une longue pratique n'a pas déjà enseigné ce qu'on appelle au Palais *notre infaillibilité*. Nous nous devions à nous-même de démontrer que le péremptoire habituel de nos réponses ne résultait pas d'un tempérament risque-tout, mais était la conséquence raisonnée de la combinaison de divers procédés dont l'application, quand elle en a été correctement faite, ne laisse pas la moindre place à l'indécision.

"Puisse le présent volume satisfaire à ce programme et contribuer ainsi à assurer la survivance de la méthode dont nous sommes à la fois et L'INVENTEUR EXCLUSIF ET PARTOUT UN PEU L'ORGANISATEUR" (pp. x–xi). Capitals in original. Galton's view was that bertillonage could not be infallible owing to the high correlations of many of its measurements which its creator neglected.

Bertillon's letter to Galton of 1894* (see above) indicates where the inspiration originated. Galton was ever ready to acknowledge others' work in any field, and not less in finger-printing. Thus in 1891 he gave a full account of Forgeot's excellent work on blurred finger-prints†, and of the latter's methods of bringing up and photographing greasy finger marks on glass or metal. We have still in the *Galtoniana* the exhibit Galton made at the Royal Society Soirée of that year of Dr Forgeot's imprints of the entire hand.

In 1905 Dr Henry Faulds published a work entitled: *Guide to Finger-Print Identification*. It would have been in some respects a useful book had it not made exaggerated claims for the author's achievements in this field, which are accompanied by remarks belittling what Sir William Herschel had practically achieved and what Galton had carried out experimentally. Dr Faulds entirely overlooks the fact that up to 1904, beyond his original letter of 1880 in *Nature*, he had himself published nothing on the subject, which could reach men of science. That letter, if suggestive, was by no means convincing; it needed the experience that Herschel provided of the *permanence* of types, and of the *practical utility* for identification to induce a man in the first rank of science to take up the subject and study it effectively. It is noteworthy that Dr Faulds in his chronological bibliography of the subject of finger-printing, starts with his own letter in *Nature* of 1880, and proceeds nearly year by year till he comes to 1890, and then passes to 1894 omitting from his bibliography all reference to Galton's memoirs and books! The whole tone of the book was distinctly unpleasing and seemed directly calculated to excite resentment in the minds of workers in the same field, who had done far more than Dr Faulds for the subject‡. In regard to the claims of Dr Faulds we must remember that his original letters to Darwin and to *Nature* were written from Japan. The letters from Kumagusu Minakata to *Nature*, Vol. LI (1894), pp. 199 and 274 prove that the use of the finger-print as a sign-manual on legal documents was familiar in Japan up to 1869; thus when a husband divorced his wife, he signed the statement of reasons with his own index-finger. The use of the finger-print as a sign-manual seems to have come

---

* A further letter of Bertillon's of July 3, 1896, indicates his view of finger-prints even two years later:

"Jusqu'à ce jour, en effet, les empreintes digitales n'ont été prises dans mon service qu'à titre de marques particulières, destinées à affirmer l'identité individuelle, et cela en dehors de toute classification au moins quant à présent."

† "Imprints of the Hand," by Dr Forgeot of the Laboratoire d'anthropologie criminale, Lyon, *Journal of the Anthropological Institute*, Vol. XXI, p. 282, November 10, 1891.

‡ "Of Sir William's mute, or at least inarticulate, musings over a period of some twenty years in India, I in Japan knew nothing" (p. 37). "Mr Galton who frequently acts as a graceful chorus to Sir William" (p. 36). "Mr Collins (now Inspector), who after some training by Mr F. Galton, who had recently begun the study, took charge of the Finger Print Department" [i.e. at Scotland Yard] (p. 5). It is needful to repeat that Galton began his researches early in 1888, that he had by 1895 accumulated from all quarters of the world a larger collection of finger-prints than any other living man, and had published more work about them of a high scientific order than any one previously and, I may add, *since*.

together with Japan's old laws from China*. Presumably these facts were unknown to Dr Faulds when he wrote his letters or he would have mentioned them. Fourteen years after his first letter, on the publication of the Parliamentary Blue Book in 1894, he wrote to *Nature* challenging the statement in the Blue Book that the Finger-Print system was "first suggested and to some extent practically applied by Sir William Herschel," and demanding that this claim should be "brought out a little more clearly than has yet been done, either by himself or Mr Galton" (*Nature*, Vol. L, p. 548, October 4, 1894). Herschel replied in a letter of November 7 (*Nature*, Vol. LI, p. 77, November 22, 1894), which must be convincing to any unprejudiced mind. A few extracts will suffice:

"To the best of my knowledge, Mr Faulds' letter of 1880 was, what he says it was, the first notice in the public papers, in your columns, of the value of finger-prints for the purpose of identification. His statement that he came upon it independently in 1879 (? 1878) commands acceptance as a matter of course. At the same time I scarcely think that such short experience as that justified his announcing that the finger-furrows were 'for-ever-unchanging.'

"How I chanced upon the thing myself in 1858 and followed it up afterwards, has been very kindly stated on my authority by Mr Galton, at whose disposal I gladly placed all my materials on his request. Those published by him were only a part of what were available (see his *Finger Prints*, p. 27, and his *Blurred Finger Prints*). To what is there stated I need now only add, at Mr Faulds' request, a copy of the demi-official letter which I addressed in 1877 to the then Inspector-General of Jails in Bengal. That the reply I received appeared to me altogether discouraging was simply the result of my very depressed state of health at the time. *The position into which the subject has now been lifted is therefore wholly due to Mr Galton through his large development of the study, and his exquisite and costly methods of demonstrating in print the many new and important conclusions he has reached.*" [Italics in the last paragraph are the biographer's.]

There follows a copy of the demi-official letter sent to the Inspector-General of Jails in Bengal on August 15, 1877. From this letter it appears that Herschel had tested the permanence of the finger-furrows for 10 or 15 years, that it had been used for years in the Registry for Deeds at Hooghly, and that Herschel had taken thousands of finger-prints in the course of the previous twenty years, and had tried it recently both in the jail and among pensioners. He recommended it strongly for similar purposes throughout India. Herschel wrote as one official to another with whom he was on friendly terms, and so, while giving the office of his correspondent did not think it needful to mention his name.

Dr Faulds did not reply at the time and one might have hoped the matter would have been considered settled. But in 1905 he returned in his book to the fray: "The letter, or report, or book, is addressed to some mysterious personality [Sir William Herschel distinctly states that it was sent to the

---

* Camel drivers in Tibet sign their contracts with a thumb smeared with ink. Even at the present day, a Japanese Professor working in my laboratory informs me, a native of his country signs a document with his personal seal, but if he has not his seal with him or has mislaid it, he is allowed legally to sign with his finger-print as *a means of identification*, the latter method being a relic of a custom established for many generations. It does not seem to me that Herschel has done full justice to Japanese claims in this matter in his *Origin of Finger Printing* (see p. 40).

PLATE V

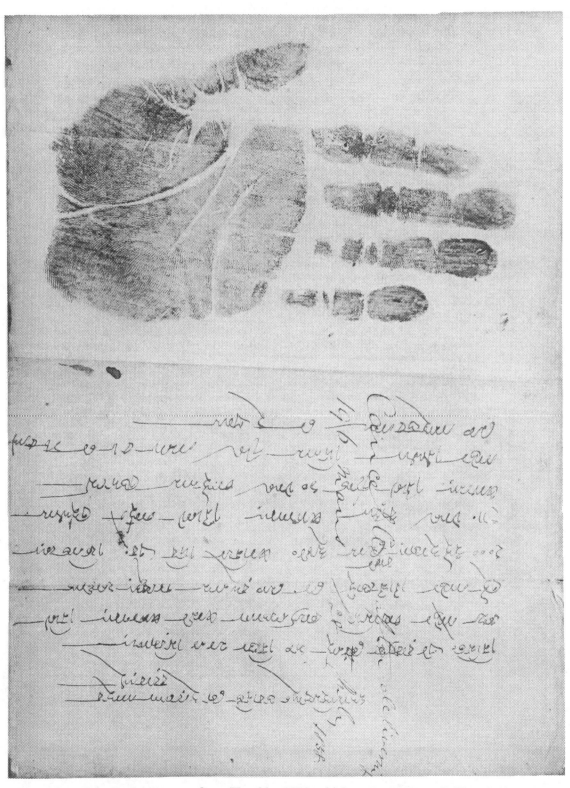

Rajyadhar Konai's Contract made at Hooghly, 1858, which at Sir William J. Herschel's request
he signed with an imprint of his right hand as an identifiable sign-manual.

Inspector-General of Jails in Bengal], known only to literature* as 'My dear B—' and is luminously certified as 'True copy of office copy,' but by whom certified is not stated." (*Guide to Finger Print Identification*, p. 36.) It was clearly impossible to deal patiently with a controversialist of this type, who first demands to see a document and when it is exhibited waits ten years before attempting to throw discredit on it! I have rarely known Galton moved. He certainly was moved on this occasion. He wrote the notice of Faulds' book which appeared in *Nature*, Vol. LXXII, Supplement, p. iv (October 19, 1905). Anyone who has read the literature on this topic up to 1905 can only agree with what Galton states. If it is severe on Dr Faulds, the severity was warranted. I cite a portion of it:

"Dr Faulds was for some years a medical officer in Japan, and a zealous and original investigator of finger-prints. He wrote an interesting letter about them in *Nature*, October 28, 1880, dwelling upon the legal purposes to which they might be applied, and he appears to be the first person who published anything, *in print*, on this subject. However his suggestions of introducing the use of finger-prints fell flat. The reason that they did not attract attention was presumably that he supported them by no convincing proofs of three elementary propositions on which the suitability of finger-prints for legal purposes depends: It was necessary to adduce strong evidence of the, long since vaguely alleged, permanence of those ridges on the bulbs of the fingers that print their distinct lineations. It was necessary to adduce better evidence than opinions based on mere inspection of the vast variety of minute details of those markings, and finally, for purposes of criminal investigation, it was necessary to prove that a large collection could be classified with sufficient precision to enable the officials in charge of it to find out speedily whether a duplicate of any set of prints that might be submitted to them did or did not exist in the collection. Dr Faulds had no part in establishing any one of these most important preliminaries †. But though his letter of 1880 was, as above mentioned, the first *printed* communication on the subject, it appeared years after the first public and *official use* of finger-prints had been made by Sir William Herschel in India, to whom the credit of originality that Dr Faulds desires to monopolise is far more justly due....

"The question of the priority of dates is placed beyond doubt, by the reprint of the office copy of Sir William's 'demi-official' letter of August 15, 1877, to the then Inspector of Prisons in Bengal. This letter covers all that is important in Dr Faulds' subsequent communication of 1880, and goes considerably further. The method introduced by Sir Wm. Herschel, tentatively at first as a safeguard against personation, had gradually been developed and tested, both in the jail and in the registering office, during a period from ten to fifteen years before 1877 as stated in the above quoted letter to the Inspector of Prisons.

"The failure of Sir Wm. Herschel's successor, and of others at that time in authority in Bengal, to continue the development of the system so happily begun, is greatly to be deplored, but it can be explained on the same grounds as those mentioned above in connection with Dr Faulds. The writer of these remarks can testify to the occasional incredulity in the early 'nineties concerning the permanence of the ridges, for it happened to himself while staying at the house of a once distinguished physiologist, who was the writer when young of an article on the skin in a first class encyclopaedia, to hear strong objections made to that opinion. His

* The *India List* for 1876–1877 would have at once informed Dr Faulds that Mr Beverley was, in August 1877, Inspector-General of Prisons in Bengal. Herschel also wrote to the Registrar-General, Sir James Bourdillon, who later expressed regret that he had allowed the suggestion to slip through his fingers. See Sir William J. Herschel, *The Origin of Finger Printing*, Oxford, 1916, p. 25.

† Actually after his letter to *Nature* of 1880, he published no scientific contribution to the subject before Galton took it up in 1888; he wasted eight years. Then Galton published his books and papers, and only in 1905 does Dr Faulds issue a work which could be even considered a scientific contribution to the subject, and then of so acrimonious a character that it is of negligible value.

theoretical grounds were that the gland, the ducts of which pierce the ridges, would multiply with the growth of the hand, and it was not until the hands of the physiologist's own children had been examined by him through a lens, that he would be convinced that the lineations on a child's hand might be the same as when he grew up, but on a smaller scale....

"Dr Faulds in his present volume recapitulates his old grievance with no less bitterness than formerly. He overstates the value of his own work, belittles that of others, and carps at evidence recently given in criminal cases. His book is not only biased and imperfect, but unfortunately it contains nothing new that is of value, so far as the writer of these remarks can judge, and much of what Dr Faulds seems to consider new has long been forestalled. It is a pity that he did not avail himself of the opportunity of writing a book up to date, for he can write well, and the photographic illustrations which his publisher has supplied are excellent."

This is a long extract and the subject is a painful one, but it has to be definitely asserted that it was to the experience of Sir Wm. Herschel and to the laborious studies of Sir Francis Galton, and not to anything Dr Faulds wrote or said, that we owe the adoption of finger-print identification for criminal investigation at first in England and since then throughout the whole civilised world*. There has been a tendency to obscure this great achievement of Galton's not only by confusing finger-printing with bertillonage, which it ultimately killed, but owing to Dr Faulds' continual attempts to monopolise all credit for both the discovery and the practical application of finger-printing. Like all arts it has developed in practice. But even as the credit for metal bridges is not due to the man who suggested that bridges might be made of metal, nor to those who changed cast iron to wrought iron or wrought iron to steel bridges, but to the man who made the first metal bridge, and induced people to walk over it†, so the credit for finger-print identification in criminal matters is due to Herschel and Galton, or even as the former has generously said—"the position into which the subject has now been lifted is therefore wholly due to Mr Galton" (see our p. 146).

On October 21, 1893, Mr Asquith appointed a Committee‡ consisting of Mr C. E. Troup of the Home Office, Chairman, Major Arthur Griffiths,

---

* My Japanese friend, referred to in the footnote on p. 146, said very definitely, that while the Japanese had resorted very early to finger-prints as personal sign-manuals, yet the Japanese criminal investigation usage did not arise from this, but was imported *de novo* from Europe.

† Thomas Paine, author of *The Rights of Man*.

‡ The origin of this Committee is fully described in a letter of Galton to the *Times*, July 7, 1893. The British Association in its Edinburgh Meeting of the previous year had listened to a paper by Manouvrier of Paris on bertillonage and another by Benedict on the modified system used in Vienna. As a result a resolution was carried by the Council in the following terms:

"Considering the need of a better system of identification than is now in use in the United Kingdom and its Dependencies, whether for detecting deserters who apply for re-enlistment, or old offenders among those accused of crime, or for the prevention of personation, more especially among the illiterate, the Council of the British Association express their opinion that the anthropometric methods in use in France and elsewhere deserve serious inquiry, as to their efficiency, the cost of their maintenance, the general utility, and the propriety of introducing them, or any modification of them, into the Criminal Department of the Home Office, into the Recruiting Departments of the Army and Navy, or into Indian or Colonial administration."

Galton was not in Edinburgh nor responsible for the resolution but he was a member of the Committee appointed in connection with it. It will be seen that the recommendation does not go beyond bertillonage. Galton, as this letter to the *Times* amply demonstrates, at once proceeded to introduce the idea of finger-printing into the proposals for a better method of identification (see also Galton's letter, *Nature*, July 6, 1893, Vol. XLVIII, p. 222), and four months later when Mr Asquith appointed his departmental committee finger-printing was *ab initio* included among the matters for examination. I know of no other reason but Galton's activities for its inclusion in Mr Asquith's programme.

Inspector of Prisons, and Mr M. L. Macnaghten, Chief Constable of the Metropolitan Police Force, to inquire (*a*) into the method of registering and identifying habitual criminals now in use in England; (*b*) into the "Anthropometric System" of classified registration and identification in use in France and other countries; (*c*) into the suggested system of identification by means of a record of finger-marks, and to report whether the anthropometric system or the finger-mark system can with advantage be adopted in England either in substitution for or to supplement the existing method. It will be seen that the inquiry resulted from Galton's work of 1892 and earlier, and if the evidence given be examined*, it will be found that the Committee were really considering whether *bertillonage,* or what we may call *galtonage* in contra-distinction, or a combination of the two should be adopted. Galton was the *only* finger-print expert examined as a witness, and the Committee visited his laboratory, saw finger-prints being taken, and the relative ease with which Galton picked out from his cabinet the finger-prints of an individual, whose prints were provided in duplicate. It is noteworthy that Galton, with a foresight for possible difficulties, gives a very simple arrangement for a drawer into which it is impossible to place a card which does not belong to that drawer. It could be easily adapted to work for a finger-print index, but Galton actually arranged it in his illustration on the basis of five bodily measurements each grouped in three categories (see p. 81 and plate). There are two other appendices by Galton, the first (p. 79) giving directions for taking finger-prints, and the second for searching a cabinet of finger-prints indexed by a simple form of bertillonage. When the Committee came to report on the Finger-Print System (pp. 25 *et seq.*) it is of Galton and his work alone that they speak. They write:

"The second system on which we are specially directed to report is that now associated with the name of Mr Francis Galton, F.R.S., though first suggested and to some extent applied practically by Sir William Herschel....A visit to Mr Galton's laboratory is indispensable in order to appreciate the accuracy and clearness with which finger-prints can be taken and the real simplicity of the method. We have during this inquiry paid several visits to Mr Galton's laboratory; he has given us every possible assistance in discussing the details of the method and in further investigating certain points which seemed to us to require elucidation. He also accompanied us with his assistant to Pentonville Prison and superintended the taking of the finger-prints of more than a hundred prisoners....The patterns and the ridges of which they [finger-prints] are composed possess two qualities which adapt them in a singular way for use in deciding questions of identity. In each individual they retain their peculiarities, as it would appear, absolutely unchangeable throughout life, and in different individuals they show an infinite variety of forms and peculiarities.

"Both these qualities have formed the subject of special investigation by Mr Galton, and having carefully examined his data, we think his conclusions may be entirely accepted." (p. 25.)

The difficulty that arose in the minds of the Committee will be a familiar one to students of the subject, namely the large classes formed by some of the loop categories. Galton was not wholly prepared to meet this difficulty of indexing, although he was already counting the ridges of loops, and differentiating them in other ways by the nature of their cores. It was not till

* *Blue Book* (C.—1763). *Identification of Habitual Criminals Report, Minutes of Evidence and Appendices.*

1895 that he was prepared with his full ideas of indexing by finger-prints alone*. He was clearly in doubt in 1893—because his own scheme of indexing was not yet fully developed—as to whether a population of 30,000 to 50,000 could be adequately indexed solely on their finger-prints, and because the Committee shared his doubts, it is a misrepresentation to assert that they condemned his work †. In his evidence he exaggerated nothing, and placed his methods and their difficulties frankly before the Committee. It is idle to say either that Galton failed to get an independent system of finger-printing carried, or that the Committee condemned his system. Finger-printing was destined to become wholly independent of bertillonage; it very soon did become so, as the study of finger-prints advanced, but in 1893 no one had published a complete system of indexing, and Galton was the only man who was able to make even suggestions in this matter. Above all to this day the all-important problem of indexing single prints seems to be unsolved‡. The Committee laid down the following three main conditions for deciding what system should be adopted:

"(1) The descriptions, measurements or marks, which are the basis of the system, must be such as can be taken readily and with sufficient accuracy by prison warders or police officers of ordinary intelligence.

"(2) The classification of the descriptions must be such that on the arrest of an old offender who gives a false name his record may be found readily and with certainty.

"(3) When the case has been found among the classified descriptions, it is desirable that convincing evidence of identity should be afforded."

Applying these conditions to galtonage, the Committee reported that:

"The 1st and 3rd of these conditions are met completely by Mr Galton's finger-print method. The taking of finger-prints is an easy mechanical process which with very short instruction could be performed by any prison warder. While in M. Bertillon's system a margin greater or less has always to be allowed for errors on the part of the operator, no such allowance has to be made in Mr Galton's. Finger-prints are an absolute impression taken directly from the body itself; if a print be taken at all it must necessarily be correct. While the working of this system would require a person of special skill and training at headquarters, it would have the enormous advantage of requiring no special skill or knowledge on the part of the operators in the prison§, who would merely forward to headquarters an actual impression taken mechanically from the hand of the prisoner. With regard to the third condition again, as we have already pointed out, Mr Galton's system affords ample material for conclusive proofs of identity....

"The Committee were so much impressed by the excellence of Mr Galton's system in completely answering these conditions that they would have been glad if, *going beyond Mr Galton's own suggestion*‖, they could have adopted his system as the sole basis of identification." (p. 29.)

* See later our account of his *Finger Print Directories*, 1895.

† Dr Faulds, *loc. cit.* p. 41, "Mr Galton's own system, afterwards expounded in a work [i.e. his *Finger-Prints* of 1892] abounding in grave errors and set forth in a way which the Blue Book of 1894 characterises." Cf. our pp. 145–147.

‡ Suppose a *single* print is found after a burglary and we need to ascertain whether the burglar was a known criminal, i.e. on the finger-print record. We may not even know of which finger it is a print, and yet the single print is perfectly individual and would identify the culprit were we able to index our single prints.

§ I have examined the finger-prints on many hundreds of practice sheets of prison warders, and can certify that this statement has been amply confirmed by experience.

‖ Italicised by biographer. The whole essence of the Report was the abandonment of Bertillon's "distinguishing marks," the use of his system as merely a method of *indexing*, and the ultimate identification by the finger-prints (see p. 20).

The result of the Committee's deliberations was the recommendation that *identification* should be made by finger-prints, but that the *indexing* of the finger-prints should be by bertillonage. After recommending the appointment of a scientific adviser to the Convict Office, the Committee remark:

"Moreover, when practical experience had been obtained of the use of finger-prints, he would be able to revise the suggestions which we have made as to the respective place of the Bertillon and the Galton methods in the system, and might possibly find it advantageous to extend the Galton method of classification further than, with the limited experience we possess of its practical application, we have ventured to propose." (p. 35.)

It appears to me that the Committee went just as far towards replacing a tried system, bertillonage, by a new system, galtonage, as it was safe at that time to do. They even foresaw that with a really scientific adviser the latter system would entirely replace the former. In 1895 Dr Garson was appointed as scientific adviser to the Convict Office, and Inspector Collins[*] was sent to Galton's Laboratory to be instructed in finger-printing, and he ultimately took charge of the Finger-Print Department. Unfortunately Dr Garson, "being a skilled craniologist and writer on human measurement, was perhaps somewhat biased towards bertillonage[†]," and little was done towards following out the Departmental Committee's suggestion of indexing by the finger-prints themselves as experience in their use increased.

Sir E. Henry, who had adopted finger-print identification in India, with as far as I can judge only small modifications of Galton's old method[‡], became Chief Commissioner of the Metropolitan Police in 1903. Of him Galton writes[§]:

"When Sir E. Henry became Chief Commissioner six years ago, full of zeal for finger-prints, well experienced in their use and master of the situation, I felt satisfied that their utilisation had become firmly established, and I ceased to do more than observe its developments from

[*] Probably the official who has risen to fame recently in less scientific activities. His teaching of the local prison warders in finger-printing certainly produced excellent results.

[†] "Identification by Finger-Prints," a letter of Galton's to the *Times*, Jan. 13, 1909.

[‡] I judge this chiefly from his letters filed in the *Galtoniana*, and notes of Francis Galton himself.

Thurs. Oct. 10/94:

"Mr Henry came today $10\frac{1}{4}$ to $12\frac{1}{2}$ to my laboratory by appointment....I showed him much about finger-prints. He had spent hours at the lab. in my absence. Agreed that my part now is to write an illustrated paper on classification. He undertakes to get me as many specimens as I want from India. I am to write to him there (he returns next week). In meantime I am to make some trials from my collection and I will talk to Macmillan." [This has reference to the proposed book, i.e. *Finger Print Directories*.]

The correspondence with regard to finger-prints continued after Mr Henry's return to India, being dated from the office of Inspector-General of Police, Calcutta. In the following year Mr Henry submitted a "Note on Finger Impressions" for the guidance of the Lieutenant-Governor. From this it appears that identification was to be by the prints and indexing by bertillonage, i.e. the system of the Report of 1893. Numerous letters, thanking Galton for communications, asking him for further information, and stating how the matter progressed in India followed in 1895, 1896 (with a further Report to the Chief Secretary in Bengal, still emphasising the doubt as to how to index the finger-prints of 20,000 persons; the letters urge the need for this indexing to replace the difficulty of exact anthropometric measurements under Indian conditions), 1899 (Henry describes his own new method of indexing) and 1900 (announcing that from April 1st the Indian Government had finally discarded anthropometry for direct finger-print indexing on Henry's system).

[§] *Ibid.*

time to time. Of course all new methods require time for development and growth, and though very much has been done under Sir E. Henry's vigorous administration, I doubt whether finality has even yet been reached ; for example, whether the power of lexiconising single prints has been developed to its utmost."

The letters of Herschel, Henry and Galton in the possession of the Galton Laboratory trace clearly the history of finger-printing. It was Galton, who by his books, memoirs and constant letters to the press got the matter ventilated and ultimately forced the subject on the attention of the police-authorities; he might not have been successful had not Herschel's practical experience and evidence of permanence* been at his service. It was Henry who during 1898–1900 in India reduced the indexing of finger-prints to a workable system, and ultimately abolished the laborious and in the hands of careless observers the even dangerous anthropometric system †.

But if a name is to be given to the system of finger-print identification in the same manner that bertillonage was attached to anthropometric measurement‡, then the right term is undoubtedly galtonage.

Of course every idler, who had not taken the trouble to investigate the subject, was up in arms against reform, as all such have ever been—"It may answer well enough on the Continent, where every one submits patiently to the inevitable, but it would not do in England, and I trust that the recommendations of the Committee—opposed as they are to the sentiments and principles of Englishmen—will not meet with the approval of the Secretary of State." (*Times*, March 23, 1894, Letter signed "Observer.") How many times have we read those words, when a powerful mind has pointed the way to a beneficial reform!

Yet even as late as 1909 misrepresentations were made as to the originator of the police system for the identification of criminals by finger-prints. The *Times* in an article on the Metropolitan Police published on January 4 of that year attributed the system of identification by finger-prints to Sir Edward R. Henry! What the writer should have said was that a

* Herschel in a letter of Oct. 28, 1896 writes: "I have just compared my own mark of *June* 1859 with that of Oct. 1896. The identity is perfectly amazing, even to me. How nature can preserve such soft tissues for 37 years, renovating them constantly, yet preserving their delineation *so precisely* is not clearly intelligible to me."

† It is only fair to Bertillon to remind the reader that the anthropometric measurements were even in Bertillon's system primarily a method of indexing, and the identification depended on the record of bodily marks and characters together with the photographs. For Galton the best bodily marks were not moles, cicatrices, etc. but the finger patterns.

‡ Those who have studied all Galton wrote on the subject of Finger-Prints before 1895 and after doing so turn to Sir E. R. Henry's *Classification and Uses of Finger Prints* (1st Edition 1901, 3rd Edition 1905) may be inclined to think that the latter work does inadequate justice to Galton's labours. Henry's system of classification follows closely on Galton's and where he departs from it by the introduction of a very heterogeneous class of "Composites," it may well be doubted, if he has really succeeded in simplifying matters. His method of "ridge-tracing" for breaking up large whorl groups is essentially that of Galton's "Basis of Classification" in the 1890 *Phil. Trans.* paper (see our pp. 163–4); while "ridge-counting" was introduced originally by Galton himself to get over the large loop aggregation difficulty. Whether Henry's numerical symbolism for the various index classes gains in brevity what it loses in perspicacity can only be determined after a wide experience of the use of both full and "shorthand" formulae.

modification of Galton's method of indexing was introduced by Sir Edward Henry*. In 1895 Galton had published his *Finger Print Directories*, which contained a great improvement on his previous method of classification; this later method was in most essential points identical with that in use in 1909 at Scotland Yard. The article in the *Times* called Sir George Darwin into the field; he concluded a letter which puts forward the simple facts of the matter with the words:

"Sir Edward Henry undoubtedly deserves great credit in recognising the merits of the system and in organising its use in a practical manner in India, the Cape and England, but it would seem that the yet greater credit is due to Mr Francis Galton."

One has to remember that identification by finger-prints was in use at Scotland Yard long before Sir Edward Henry came on the scene †, but the indexing was by bertillonage. Dr Garson, the former director, was too much of an anthropologist and had a mind of too little inventive power to give up the anthropometric index. A dozen different ways of breaking up the large loop categories would occur to an inventive mind, and as soon as one of these had been tried and found successful bertillonage was bound to disappear. The fact remains that nothing was done and no progress made in abolishing bertillonage, until Sir Edward Henry succeeded Dr Garson. This absence of progress was not Galton's fault, but lay with the Government, which selected for the post of director an old-school medical anthropologist rather than a finger-print expert.

While it is absolutely impossible for one who has really studied finger-prints to confuse A's prints with those of B, it is always possible for a clerk to make an error in extracting the dossier, which corresponds to the identified finger-prints. Such a clerical lapse occurred in a case tried at the Guildhall in 1902, and the occasion was seized upon to attack the finger-print method by certain newspapers. Galton wrote a letter on the matter to *Truth* (October 2, 1902, Vol. LII, p. 786). He pointed out that there was no doubt about the identification, but when it came to turning up the record attached to the

---

* In March 1897 Major-General Strahan and Sir Alexander Pedler reported on the system of identification by Finger-Prints as adopted in India. It was really a report on Henry's work and methods. In the course of the Report the three conditions laid down by Mr Asquith's Committee (see our p. 150) are cited and the following words occur:

"In the same report it is acknowledged that Mr Galton's finger-print method completely met the first and third conditions, but they disapproved of his method of classification."

This is a complete mis-statement of what the Committee did. Galton was not prepared at that date to provide a comprehensive method of indexing, accordingly it was impossible for the Committee to disapprove of his method of indexing. It was Galton himself who suggested indexing by bertillonage and this the Committee accepted, although both they and he looked upon it as a temporary stage. Galton's *Secondary Classification* was complete and published in 1895 (see our pp. 199 *et seq.*), and in the present writer's opinion there is little in Henry's book of 1901, which cannot be found, often better expressed, in Galton's of 1895 or in his earlier writings. The numerical notation is the chief novelty. We do not think the statement we have quoted above should have been allowed to appear without a qualifying note in Henry's *Classification and Uses of Finger Prints* (p. 112).

† I myself witnessed the rapid identification of criminals by their finger-prints in 1900.

impressions, a mistake had been made in the reference number and a wrong dossier was produced. Galton writes:

"I wish to point out the moral of this. In every system there must be some clerk-work and a consequent liability, however small, to clerical blunders. In the system by measurements at least five have to be made and recorded for each person, and they each require three figures to express them. The frequent occurrence of mistakes in this complicated process was the main motive for abolishing measurements altogether, first in India, and now in this country. In the finger-print system all the above clerk-work is done away with because the hand of the accused person prints its own impression. As regards the comparative trustworthiness of the two systems, there can be no reasonable doubt. I took, as you may be aware, great pains in testing them, with the result that it is inconceivable to me that an expert to whom the impressions have been submitted of two different persons, taken with the cleverness that is habitual in prisons, should ever mistake one set for the other."

§ II. *Popularisation of Finger-Printing.*

I propose in this section to give an account of some of the minor papers and letters to newspapers by which Galton made the idea of finger-printing familiar to his countrymen. I think this plan is better than scattering them chronologically between his more solid contributions to the science of the subject, which will be dealt with in the remaining section of this chapter.

In August 1891 Galton published in the *Nineteenth Century* (Vol. xxx, pp. 303–311) an article entitled "Identification by Finger-Tips." It contains a *résumé* of his Royal Society papers in a popular form, an account of his apparatus and a suggestion that professional photographers should take up finger-printing as part of their trade. He concludes with the prophecy:

"I look forward to a time when every convict shall have prints taken of his fingers by the prison photographer at the beginning and end of his imprisonment, and a register made of them; when recruits for either service shall go through an analogous process; when the index-number of the hands shall usually be inserted in advertisements for persons who are lost or who cannot be identified, and when every youth who is about to leave his home for a long residence abroad, shall obtain prints of his fingers at the same time that the portrait is photographed, for his friends to retain as a memento." (p. 311.)

Another matter in connection with finger-prints which excited Galton's attention and has very considerable scientific interest is the question of scars and wounds as influencing the ridges. On Plate VI are given illustrations of this matter which I have found in the *Galtoniana*. In Fig. (v) we have an enlarged print of a graft on the bulb of a thumb. In this case J. R. H., a solicitor in large practice, sliced off a piece of the flesh of his thumb; it was promptly picked up, replaced in what was thought to be its original position and the thumb tightly bandaged. The print taken thirty years later shows that the ridges had not been properly adjusted, the orientation of those on the graft being almost at right angles to their true position*!

In Fig. (i) *a–d* we have a good illustration of the effect of a burn, which occurred in the case of Sergeant Randle, Galton's assistant. In Fig. (i) *b* taken immediately after the accident, the ridges have entirely disappeared, but Fig. (i) *d* indicates that if the injury has not been too

* See *Nature*, Jan. 30, 1896 (Vol. LIII, p. 295).

PLATE VI

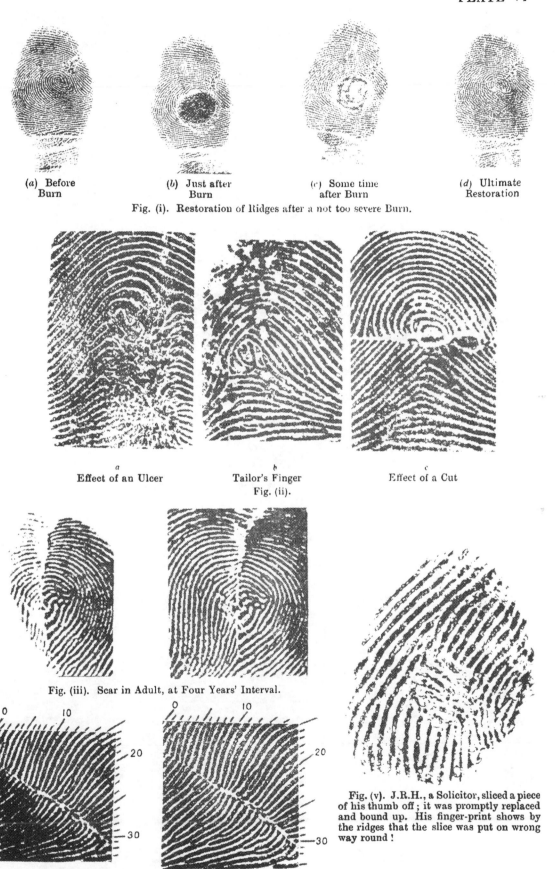

(a) Before Burn   (b) Just after Burn   (c) Some time after Burn   (d) Ultimate Restoration

Fig. (i). Restoration of Ridges after a not too severe Burn.

a — Effect of an Ulcer   b — Tailor's Finger   c — Effect of a Cut

Fig. (ii).

Fig. (iii). Scar in Adult, at Four Years' Interval.

Fig. (v). J.R.H., a Solicitor, sliced a piece of his thumb off; it was promptly replaced and bound up. His finger-print shows by the ridges that the slice was put on wrong way round !

Fig. (iv). Scar in Boy, at Two Years' Interval.

Effects of various injuries on the pattern of the Finger-Print. Enough *minutiae* are, as a rule, left for identification purposes.

great, they develop again on the old sites. Figs. (ii), (iii) and (iv)* show the effect of an ulcer in destroying the ridges, the changes produced by the occupation of tailoring, and the marks left by scars and cuts. It will be recognised at once how injuries of this kind fail to destroy completely the *minutiae* of the ridges on which identification depends. Indeed if they were in existence when the earlier print was taken, they form themselves very valuable contributory factors in the recognition of identity. It is singular how little Galton left unobserved when he came to deal with a new topic; his fruitful mind seemed to envisage all possibilities that might detract from or aid the enterprise he had in hand. He collected material from all quarters, but the subject was so vast that even he left much that would still be worth gleaning. I think a wide study of finger-prints from the standpoint of occupations might still indicate interesting possibilities. The carpenter's, the metal worker's, the shoemaker's, the seamstress's, the typist's, the laundrymaid's, and even the textile worker's finger-prints may all show individual wearings of the ridges, if they were attentively studied in large numbers; and so might well replace some of the information which Bertillon drew from the shoes and trousers, etc. of his subjects.

In an article entitled "Enlarged Finger Prints" which appears in the journal *Photographic Work*, February 10, 1893, Galton emphasises the proposal that professional photographers should master the art of finger-printing and the enlargement of prints.

"It seems not unreasonable to suppose that many persons would like to possess so curious and unchanging an evidence of their own identity, and that the wish to have prints taken of the finger might become a fashion which photographers would find it lucrative to promote."

He gives two excellent enlargements to a sixfold size, in which the sweat-glands, the "islands," ridge terminals, forkings and all the *minutiae* are very distinct. We reproduce his two figures (p. 156). Fig. 17 is the print of a well-known explorer and contains at least 39 *minutiae*, Fig. 18 some 30. As Galton says every one of these *minutiae* may be expected to persist, not only during life, but after death also, until they are effaced by decay.

Galton describes his apparatus for taking prints and refers to his original and to his later enlarging cameras.

Two further letters from Galton may here be mentioned. On October 19, 1893 he wrote to *Nature* (Vol. XLVIII, p. 595) stating that finger-prints had been adopted as a means of identification for recruits in the Native Army of India (Circular, August 25, 1893). Galton points out the necessity for clear prints and also suggests for the purpose of comparing two prints the use of a mounted watchmaker's lens†, and further of four "pointers," two for each print. One pointer is used for each print to mark an origin of reference and the others to indicate the special *minutiae* which are to be compared in the prints under comparison.

* See *Finger Prints*, 1892, Plate 4, Fig. 7 *a, b, c*, p. 59.

† Galton's own lens neatly mounted is still in use in the Galton Laboratory : see the footnote on our p. 178.

"They are T-shaped; their long arms are six or seven inches long, they are roughly made of wood as thick as the thumb, so that they are purposely not over light. Each pointer stands on three supports, viz. on the point of a bent pin, whose headless body has been thrust into the end of the long arm of the T, and on the ends of two nails or better on staples, one of which is driven under either end of the cross-arm. It is most easy to adjust the point of the bent pin upon any desired character in the finger-print. Both hands of the observer are thus left free to manipulate other pointers, when desired. The stationary pointers are a great help in steadying the eye while pursuing a step by step comparison between two finger-prints."

We may remark that, the pointer being raised from the paper, the bent pin scarcely obscures any part of the print.

The second letter appeared in the *Times* of December 27, 1893, and contains a suggestion which it was certainly undesirable that the authorities

Fig. 17.   Fig. 18.
Early Examples of Galton's Method of Finger-Print Enlargement.

should have entirely disregarded. It was that depositors in the post-office savings bank should have their fingers printed in their deposit books and that these should be used as a means of identification, when the depositor sought to draw money from a post-office where he was not known. This brings us indeed to a matter Galton had much at heart; he did not think finger-prints were useful solely as a matter of criminal identification. The art of comparing finger-prints is so easily learnt that it might well be part of the training of many minor civil servants, postmasters, Public Trustee employees, War Office and Admiralty pension-officers, and many other similar officials. Two lectures and two practical classes of a few hours each would suffice to give the necessary instruction to a group of twenty or thirty minor officials of average intelligence. It would include the taking of clear (non-blurred) prints, and the rapid identification of prints. A signature can be forged, and it changes with age and illness, or even with the nature of the pen with which it is made. The finger-print remains with all its minutiae throughout life incapable of being forged. Unless a man be a criminal there is no central office in existence even at the present day, where his finger-prints could be registered, and he could be certain of identification for legal purposes at any time during his life, and for some time after his death. For many legal purposes such a registration might be as valuable as a land-registration office, and ownership of many personal effects, securities, bonds, passports, etc. might be testified by simply finger-printing them, if the finger-prints like a trade-mark had been duly registered. It is almost a catastrophe that the process of finger-printing should have become tainted in the popular mind by a criminal atmosphere.

In 1900 Galton wrote another paper in the *Nineteenth Century* (Vol. XLVIII, pp. 118–126) under the title of "Identification Offices in India and Egypt":

"There are many Identification Offices, supported by Governments and known by various titles, in different parts of the world. Their number increases, and so does that of the purposes to which they are applied; a knowledge of them is, however, confined to a few persons. This is especially unfortunate, because a fair amount of popular interest would ensure their adequate support, and would check the common tendency of all Government institutions to slackness of management, which is particularly fatal to the efficiency of Identification Offices." (p. 118.)

He then refers to the work of Henry in India and Harvey in Egypt, where Galton had seen the working of the central office in Cairo. Speaking of Egypt he writes:

"The difficulty of identification is increased by the roaming habits of the natives, many of whom travel great distances for pilgrimages, petty commerce, or change of employment, so that witnesses may not easily be found to identify them. Again, while the natives of India and of Egypt have beautiful traits of character and some virtues in an exceptional degree, their warmest admirers would not rank veracity among them. It is not insinuated that false testimony is unknown in English courts of justice, or in England generally; indeed I find, on a rough attempt at a vocabulary (made for quite another purpose), that more than fifty English words exist which express different shades and varieties of fraud*; but if a map of the world were tinted with gradations of colour to show the percentage of false testimony in courts of law, whether in different nations or communities, England would be tinted rather lightly and both Bengal and Egypt very darkly. So, whether it be from the impossibility of identifying the mass of natives by their signatures, or from the difficulty of distinguishing them by name, or from their roving habits, or from the extraordinary prevalence of personation and false testimony among them, the need for an Identification Office has been strongly felt both in India and in Egypt."

Galton gives a list of eight ways in which finger-prints were already in use in India, namely: (1) Pensioners, civil or military; (2) Transfer of

* It may be worth while to give these words. The list is imperfect but will do: cant, cheat, chicanery, circumventing, counterfeit, chouse, connivance, cozen, crafty, cunning, deceit, defraud, delude, dishonest, dissemble, dissimulate, dodge, duplicity, fallacious, feign, flattery, fraud, furtive, hoax, humbug, hypocrisy, insinuation, intrigue, jesuitical, jobbery, knavery, lying, mendacious, peculating, perfidious, perjury, personation, rascality, roguery, scheming, scoundrel, sharper, shuffler, slanderer, slimness (a new word due to the Boers), slyness, sneaking, spying, stratagem, subterfuge, traducing, treachery, trickery, wiles [the last two added by Galton in a corrected copy of the article, which I follow. The reader will find it quite easy to add to the list, e.g. guile, imposture, fake, mislead, gerrymander, graft, etc., etc.].

property; (3) Advances by Opium Department to cultivators; (4) Receipts of employers for wages to labourers; (5) Survey of India, workers on engagement, to prevent their re-enlisting in a distant area after discharge for misbehaviour; (6) Director-General of Post-offices in similar cases; (7) Medical Department before granting certificate to examine; (8) In plague regulation and for controlling the Mussulman pilgrims to Mecca. All these cases are in addition to the matter of criminal identification. Galton suggests that finger-prints should be used in cases of life-insurance, and after registering for authenticating wills. The Indian Legislature had passed an act amending the law of evidence, by declaring relevant the testimony of those who had become proficient in deciphering finger-prints. It was clear that finger-printing had taken on in India, and most of this was due to the energy of Mr (later Sir) E. R. Henry.

Galton discusses the classification for research, alludes to the difficulty of the great preponderance of ulnar loops, which have to be distinguished mainly by lineations. Galton himself counted the number of lineations from "core" to the "V" or point of divergence of the ridges*.

"Mr Henry reckons lineations on more than one finger, with the simplification of merely noting whether their number exceeds or falls short of the average, and is thus able, as he states, to cope successfully with his far larger collection than mine. His success in this respect seems to me so surprising that I should greatly like to witness his methods tested on a really large collection, say of 100,000, in which there would probably be found no less than 6000 cases of *all-loops* of the ulnar kind, to be distinguished mainly by the method of lineations†."

Galton then speaks of the Cairo Office, of which he had seen the working. He notes several cases in which its efficiency had been proved, and this not only for criminal purposes, but for the advantage of honest men, who were given a registration and could thus demonstrate to a new employer that they were the actual men, whose merits had been testified to by former masters. Some such registration of servants would render written characters of more value than they are at present in our own country.

"Space does not permit me to go more fully into this large and interesting subject. It will be a real gain if these remarks should succeed in impressing the public with the present and future importance of Identification Offices, especially in those parts of the British Empire where for any reason the means of identification are often called for and are not infrequently absent. I think that such an institution might soon prove particularly useful at the Cape."

By such articles and frequent letters to the newspapers Galton kept the topic of finger-printing to the fore. When in 1900 he read a paper before the Khedivial Society of Geography in Cairo‡, comparing the Egypt of 1846 with that of 1900, and spoke of the influence which D'Arnaud Bey had

---

* Later termed the *delta*.

† Henry was ultimately driven to count the ridges on the first two fingers of *both* hands, and to make sixteen classes of the four numbers so obtained. But even this was not sufficient, and for his indexing he counts the ridges on the little finger of the right hand with the view of arranging in the order of that count the schedules in each of the sixteen classes. Galton started the ridge counting and had already applied it to fore and middle fingers to break up the large simple loop groups.

‡ *Bulletin de la Société Khédiviale de Géographie*, V⁰ Série, No. 7, pp. 375–380.

had on his life, converting his conception of travel from pleasure to purpose, Galton could not refrain from discussing finger-prints. It is almost impossible to overrate the energy Galton displayed in making the general public familiar with the idea of finger-print identification. We have not only a whole series of letters to such journals as the *Times* and *Nature*, but Galton did not despise more popular organs of communication. Thus there appeared a paper in the *Sketch*, entitled: "The Wonders of a Finger-Print," with a portrait of Galton (November 20, 1895), and another in *Cassell's Saturday Journal*, March 25, 1896. The latter took the form of an interview, and perhaps a few lines of it are still worth recalling:

"There are about thirty characteristic points on an average in a finger-print," Mr Galton continued. "As I have said you will find no two pairs of fingers alike; it is like comparing the ground plans of two different cities."

"But suppose an old and hardened criminal, whose finger-print was in your possession, hacked his fingers about with a knife," I asked: "would that cause you confusion on his re-capture?"

"Plenty of material for identification would still be left. He would never be able to obliterate all the ridges unless he cut off both his hands. But I don't want you to think that finger-prints are only of value for the identification of criminals. I want other people to take the finger-prints of their children for possible use in identification in after life."

"You remember what a stir there was when the rumour spread of a plot to kidnap the Duke of York's baby*. Think of all the national difficulties that would have arisen had he been lost and then professed to be found, but his identity doubted. Many people urged me at the time to propose that his finger-prints should be taken, but I hesitated to move seriously in the matter."

In the same year Galton read a paper entitled "Les empreintes digitales" at the Fourth International Congress of Criminal Anthropology†. In this paper he briefly describes the facts he had demonstrated in his *Finger Prints*, then he turns to the question of nomenclature and classification, and notes his "shorthand" method of indexing. What, however, he particularly insists upon is the need for an international concordat in the matter of nomenclature and indexing so that it would be at once feasible to telegraph the finger-print formula of a suspected person. Galton proposed:

"Qu'il soit fait des recherches dans les administrations de police des différentes nations pour déterminer la nomenclature la plus convenable et les autres détails relatifs aux empreintes digitales pour les services internationaux, c'est-à-dire pour communiquer, par lettre ou télégraphe, et en termes généralement intelligibles, le signalement par les empreintes digitales des personnes soupçonnées." (p. 37.)

The noteworthy points about this paper are:

(i) That as early as 1896 Galton had freed himself entirely from the anthropometric system; there is not a reference to bertillonage as a system of indexing, but the indexing is to depend entirely on finger-print classification.

(ii) That although the system had only been a few years at work in England and was just started in India, Galton envisages an international

---

* The present Prince of Wales.

† *Comptes-rendus, Session de Genève*, 1896, pp. 35–38.

system of finger-print Identification Offices with a common nomenclature, a common method of indexing and a common code.

Realising how such Identification Offices, depending wholly on finger-printing, now stretch from London to Tokyo, from Tokyo to San Francisco and thence to New York, we see how Galton recognised a widespread need, and how by his ceaseless energy he carried through a great reform. Whatever influence his idea of correlation may have exercised in the field of scientific investigation—and it has been indeed deep and far-reaching—the establishment throughout the world of finger-print identification is a no less astonishing mark of his power of achieving on the practical side.

On October 16, 1902, Galton has still another letter in *Nature* (Vol. LXVI, p. 606). It is entitled "Finger-Print Evidence." The problem he is concerned with here is to find the best manner of convincing a judge and jury that an accused person is really one whose finger-prints are already on the criminal register. Owing to the courtesy of Scotland Yard he had received two

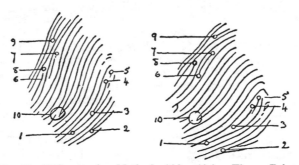

Fig. 19. Ridge-tracing Method of identifying Finger-Prints.

enlarged photographs of thumb-prints. The first is that of an impression left on the window frame of a house where a burglary had occurred, and the second that of the left thumb of a criminal who had been released and whose finger-prints were preserved and classified at Scotland Yard. Galton applies the method of his *Decipherment of Blurred Finger Prints* (see our p. 194), "believing that to be the readiest way of explaining to a judge and jury the nature of the evidence to be submitted to them....The questions of the best mode of submitting evidence and the amount of it that is reasonably required to carry conviction deserve early consideration, for we may have a great deal of it before long." In the accompanying diagrams it will be seen that Galton has selected and numbered ten *minutiae* for identification and comparison. It is scarcely conceivable that any twelve reasonably intelligent men would fail to be convinced of the identity of the two thumb-prints, although conviction would be still further strengthened were a third random thumb-print of the same type presented, which would undoubtedly lack corresponding *minutiae*.

§ III. *Scientific Papers and Books.*

A. *The Royal Society Papers.*

Galton's first important scientific paper on Finger-prints was published in 1891 in the *Philosophical Transactions**. It is entitled: "The Patterns in Thumb and Finger Marks; on their arrangement into naturally distinct classes, the permanence of the papillary ridges that make them, and the resemblance of their classes to ordinary genera." It was actually received by the Royal Society on November 3 and read November 27, 1890. It may be described as the fourth scientific contribution to the subject, the first being that of Purkenje in his *Commentatio* of 1823, the second that of Alix in his memoir of 1868, and the third the work of Kollmann in 1883 (see our pp. 141–143, and 174). While these authors endeavoured to give names to various types of finger-prints, none of them had formed a considerable collection of human prints, by aid of which it would be possible to describe anything like the variety of types and subtypes which occur, or give even the roughest measure of their relative frequencies. Galton, with his usual insight, grasped the essential point that not only a classification of types was needful, but a study of their relative frequency. He also recognised that mere assertion of their permanence must be replaced by a definite demonstration thereof†. In the present case Galton's *main* data consist of both thumb-prints of 2500 persons taken at his second Anthropometric Laboratory. I do not think Galton had fully realised at that time the amount of correlation that exists between the type of pattern and the individual finger, and that accordingly the frequencies of the thumb-prints cannot without further consideration be applied to those of finger-prints in general. Very soon after the publication of this paper Galton started to take the prints of all ten digits, and formed the large and representative collection now in the Galton Laboratory‡.

The paper first refers to Kollmann's paper (see our p. 141) for the origin of the ridges, but states that no reason has yet been given why the prominences tend to arrange themselves in continuous ridges and not to form isolated craters. Galton next describes how he takes impressions, and how it is advantageous to take duplicate impressions on tracing cloth, so that the pattern can be reversed by viewing it face downward§. (I may note that it is always an advantage to take finger-prints in duplicate, for one set can then be used for pencilling in ridges and defining the core.) If the hands be placed palms downward on the knees, so that the thumbs correspond to

---

* Vol. 182 B (1891), pp. 1–23.

† In the Royal Institution lecture of May 25, 1888, Galton had given, owing to the kindness of Sir W. J. Herschel, two illustrations of permanence, but even these had not been fully and adequately investigated (see our p. 142).

‡ He was taking prints of the fingers as well as the thumbs among his circle of friends in December 1890 and he began early in 1891 a more systematic collection.

§ It must be remembered that the finger-print is always a reversed impression of what one sees directly.

the great toes, it will be seen that the thumb is *inward* and the little finger *outward*. [Finger-prints taken in this order from left to right are in "natural order."] Inward and outward are respectively thumb-side and little finger-side, but these terms are awkward when we have to use them for the thumb and little finger themselves, and the same criticism applies to the anatomical terms radial and ulnar when applied to those bones themselves*.

Next Galton gives for the first time his explanation of the manner in which the "core" of a pattern originates; he held that it is due to the nail; the ridges, instead of going straight across the bulb of the finger, are distorted to cover the top of it; the space between the originally adjacent parallel ridges Galton terms the "core." This core or interspace is filled up

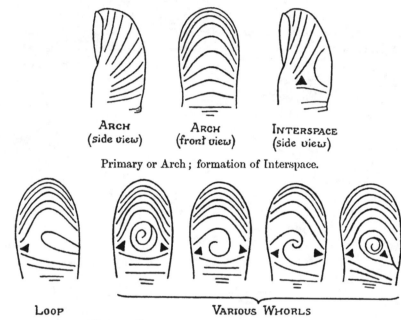

ARCH
(side view) ARCH
(front view) INTERSPACE
(side view)

Primary or Arch; formation of Interspace.

LOOP   VARIOUS WHORLS
Fig. 20. Cores in Interspace, showing "deltas."

with an additional scroll work of ridges, which in themselves form the pattern on which classification depends. When the scroll work of the core consists of a series of ridges separated on the central portion of the bulb by wider intervals than at the sides of the bulb, Galton in this memoir terms it a "primary," but later he uses the term "arch." Next the "deltas" are defined. These are the small "islands" at the points where the adjacent parallel ridges begin to diverge to form the core. In a primary there are no deltas, in a loop one and in a whorl usually two are discoverable. When there are two deltas, the line joining them and its perpendicular bisector serve as axes of reference; when there is only a single delta, in a loop, Galton

* Galton unfortunately transposed "inner" and "outer" in this memoir, calling the inside of the thumb that "nearest to the rest of the hand" (p. 4). He corrected this error in his *Finger Prints* (p. 70).

takes one axis of reference to be the "axis of the loop," i.e. a line drawn to bisect the loop "at the upper end of its innermost bend," and the other axis of reference the line through the single delta perpendicular to the loop axis. This loop axis is very important, for it is the line upon which Galton first counted the number of ridges, and much depends on two observers constructing identical loop axes before proceeding to count ridges in comparing prints*. It must be remembered that the two observers may be comparing two separate prints at a distance and, owing to the termination of ridges and to the forking of ridges, a slight difference of position in the loop axis may lead to divergent results.

In Fig. 21 W is "outside," V is "inside," the print. AH in Fig. 21 (iii) and (iv) is the loop axis on which Galton originally counted the ridges from A to H, that at A counting zero. When the loop axis exactly passed through a bifurcation Galton counted the ridges as $\frac{1}{2}(1+2)=1\frac{1}{2}$. He omits to tell us what happened when it exactly passed through a ridge terminal—presumably he counted it as $\frac{1}{2}$—or what he did in the case of an island.

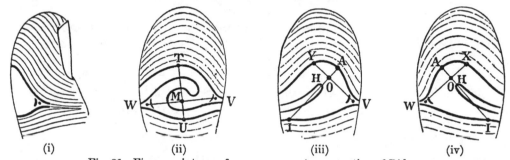

Fig. 21. Finger-print axes for measurements or counting of Ridges.

Galton next proceeds to his *first* basis of classification. It consists in paying attention solely to the deltas and the core boundaries. There may be no deltas, i.e. we have a primary or arch. If there are deltas we can trace the adjacent ridges from one or both deltas upward and downward. These ridges will either reach the other delta, or pass above or below the corresponding ridges from the other delta. There are thus three cases for the upper and three for the lower boundary of the core, or with the primary cases *ten* cases in all. Galton denotes the summit of the core on the central line of the bulb by S, and the bottom of the core on the same central line by B. Then the possible cases are

$$0 = \text{primary or arch}; \quad 1 = WSV - WBV; \quad 2 = SW - BV;$$
$$3 = SV - BW; \quad 4 = SV - BV; \quad 5 = WSV - BV;$$
$$6 = SV - WBV; \quad 7 = SW - BW; \quad 8 = WSV - BW;$$
$$9 = SW - WBV.$$

* Galton remarks: "There is usually quite enough length in a straight line of the upper most portion of the inner bend to indicate the direction of the required axis" (p. 9). I am less confident of this. I should be inclined to replace Galton's axes by drawing a tangent from the delta to the head of the loop, and taking this and the perpendicular to it through the point of contact as the axis of reference, defining this perpendicular as the "loop axis."

These are represented in the following diagram:

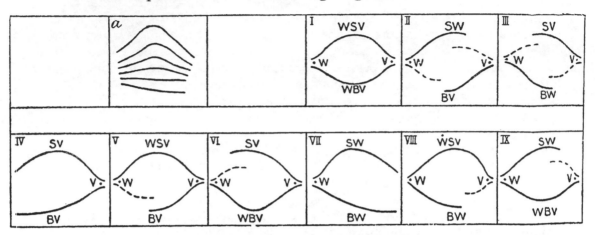

Fig. 22. Classifying by nature of Ridges from Deltas.

We can, perhaps, improve somewhat Galton's indexing in the following manner*. Consider the digits taken in order from little finger of left hand to little finger of right (as the hands are placed palms downward on the knees) to occupy the places from first to last of a ten-figure number, e.g. 32881,56490, then this would be interpreted as meaning that the little finger of the left hand was $SV - BW$, the ring finger $SW - BV$, the pointer and the forefinger $WSV - BW$ and the thumb $WSV - WBV$; the thumb of the right hand would be $WSV - BV$, and so on down to the little finger of the right which would be an arch. Thus a thousand million variations would be possible, and every individual would have his own ten-figure index number, which could be

Fig. 23. "Outlining" a rolled pattern.

recorded in numerical order in the index. The question, however, of how many of these would be "repeats" remains to be considered. Galton shows how, after outlining the pattern, it is fairly easy to classify a great variety of patterns according to his scheme (see his Fig. 9, p. 7, which contains forty

* Galton later drops without comment his classification of prints from the contours of the cores. He nowhere states why. Probably he found it not adequately discriminative for large numbers, or perhaps he discovered the personal equation involved in drawing contours.

typical forms). Unfortunately loops occur in two of Galton's classes only, i.e. as inward and outward loops, and thus the system fails to break up the large class of plain loops which is one of the difficulties of indexing. Further, to be effective, it involves pencilling in the core. However, I feel confident that the choice of a numeral place for each digit and ten classes for each pattern, *which need not necessarily be the same for each digit,* since the relative frequency of each pattern varies from digit to digit, ought to be the basis of any sound system of finger-print indexing. Here, however, it is the difficulty of breaking up the nearly 50 °/$_\circ$ of plain loops which requires ingenuity and study, and for this Galton could only suggest ridge counting. Breaking up the 25 °/$_\circ$ of "whorls" is a relatively easy matter. Even if we proceed to consider the "nuclei" of the cores (Galton, p. 8) we have five belonging to the whorls, and only two

Fig. 24. "Nuclei" of cores. *a, b, c, d, e* are "nuclei" of whorls, *f* and *g* of loops.

provided by the loops. Thus, considering "inward" and "outward" loops, we have at most four loop classes, and we require greater subdivision. This problem was left unsolved by Galton in this his first memoir (1890), unless his suggestion of ridge-counting be considered adequate. Our author next proceeds to deal with the identification of patterns. He draws attention to the fact that the patterns may become distorted, either by age or decay, if the times of taking prints are at long intervals. "They may change their shape just as the pattern on different portions of the same piece of machine-made lace may become variously stretched by wear, or shrunk by wet, or even torn" (p. 9). Exactly as we might proceed to identify the lace by counting the threads of corresponding parts of the lace pattern, so we may count the ridges on corresponding parts of two finger-prints. For this purpose Galton uses the axes of reference which I have already discussed.

Besides the counting of ridges Galton also uses the *minutiae* of which he gives examples in his Fig. 11, but he is careful to warn the reader that two

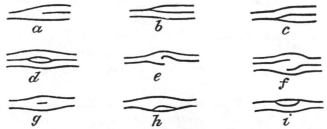

Fig. 25. Ambiguous *minutiae*.

prints of the same finger may show one a fork of ridges and the other a continuous ridge and the terminal of a new ridge, i.e. one may show *a* and the other *b* or *c* of the above figure. The reason for this is that the ridges are not all of equal height, and occasionally will escape being inked or, even if inked, fail to be pressed on the paper. Those familiar with various types of engraving by gelatine processes will have noted like divergences when comparing pulls from the same stone under a lens. In like manner, even different copies of the same plate of Albrecht Dürer's Apocalypse woodcuts show similar variations, but no connoisseur would assert on that ground that they were not pulls from the same block. Notwithstanding this it is the *minutiae* which provide the best means of identifying two prints, and these are very numerous, if the finger be rolled.

The next section in Galton's memoir deals with the *Persistence of Patterns* (pp. 10–13) and here he had the advantage of Herschel's material. We may give a brief *résumé* of his table on p. 11 :

Individual and Plate number	Age at First Print	Interval in years before Second Print	Total number of *minutiae* identified
1	7½	9	33
2	7½	9	36
3	Adult	28	27
4	Adult	28	36
5	Adult	28	55
6	Adult	31	27
7	Adult	30	50
8	Adult	31	32

Thus a total of 296 *minutiae* were identified.

"The upshot of a careful step by step study is that I have found an absolute and most extraordinary coincidence between the details of each of the two impressions of the same finger of the same person. There was, as the table shows, a grand total of no less than 296 (say roundly 300) points of comparison and not a single one of them failed, though I had much trouble in deciphering the ridges, especially about the *V*-point [inward delta] in Case 5. There was no one case found of a difference in the number of ridges between any two specified points. Never during the lapse of all these years did a new ridge arise, or an old one disappear. The pattern in all its minute details persisted unchanged, and, *a fortiori*, it remained unchanged in its general character." (p. 12.)

Galton's method of comparing *minutiae* at this time was by outlining the ridges of the two "allochronic" prints. I have arranged Galton's persistency data, outlines next prints, on our Plates VII and VIII in a manner slightly different from that of the original plates of his memoir. This outline method has distinct advantages, if it be not here as complete as in Galton's later development of it.

Galton added a line or two to the memoir on January 28, 1891, to say that he had examined a number of other pairs of impressions in the same manner, and had found only one instance of fundamental discord, where a ridge had been partly cleft in a child, but when the child had grown to a boy the cleft had disappeared. Thus Galton, with the aid of Sir W. J. Herschel's material, satisfactorily established for the first time the permanence of finger-prints.

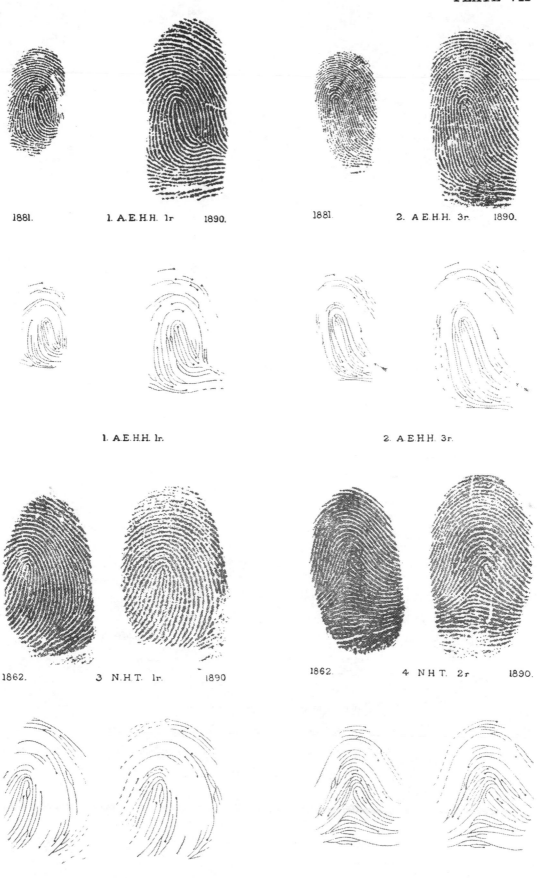

PLATE VII

1881.   1. A.E.H.H. 1r   1890.   1881.   2. A.E.H.H. 3r.   1890.

1. A.E.H.H. 1r.   2. A.E.H.H. 3r.

1862.   3. N.H.T. 1r.   1890   1862.   4. N.H.T. 2r   1890.

3. N.H.T. 1r.   4. N.H.T 2r.

Persistence of *minutiae* at intervals of nine and twenty-eight years.

After a few remarks on *scars*, which consist chiefly in noting how few he had found which destroyed the patterns to any considerable extent, and how even in these cases with "rolling" generally enough is left for sound identification (see our Plate VI, p. 154), Galton turns to another matter, which needs possibly more criticism or at least an ampler treatment. He considers that there are certain main types of finger-prints, "arches," "loops," "whorls," etc. There are also, he admits, transitional forms which create difficulty in classification, but he says the result of statistical observation shows these intermediates to be relatively few. He considers therefore the finger-print types to be analogous to ordinary genera, and in order to illustrate this he takes the case of the loop, and (*a*) counts the number of ridges in AH (see our Fig. 21 (iii) and (iv), p. 163), (*b*) measures the index VY/OI, and (*c*) the index AO/AH. Using both hands, and populations numbering 140 to 176 individuals only, he forms six frequency distributions, reducing them to percentages. For example:

Percentage Number of Ridges in 166 *Right Thumb Loop Prints.*

1	2	3	4	5	6	7	8	9	10	11	12	13	14	15	above
1	2	2	2	3	4	8	8	11	9	14	11	10	7	6	2

Percentage Value of Index VY/OI in 149 *Left Thumb Loop Prints.*

0·3—0·4	0·5—0·6	0·7—0·8	0·9—1·0	1·1—1·2	1·3—1·4	1·5—1·6	1·7—1·8	1·9—2·0	2·1—2·2	above
2	11	14	18	23	7	10	6	6	1	2

Galton does not apply an individual test for normality of distribution to these rather abnormal-looking distributions*, but reducing them to their medians and quartiles (see our Vol. II, pp. 385–6, 401) compounds them together to form a single average "ogive" curve (see our Plate II, p. 31). His final comparison is as follows:

Ordinates of Ogive Curve from Six Distributions and the computed Values.

Six Distributions	−231	−182	−117	−93	−73	−37	+1	+38	+77	+107	+139	+213	+260
Computed.........	−244	−190	−125	−100	−78	−38	0	+38	+78	+100	+125	+190	+244
Grades	5	10	20	**25**	30	40	50	60	70	**75**	80	90	**95**

Considering that we have 965 observations to start from, this does not appear on the face of it a very good agreement, and even Galton (p. 22) contents

---

* He merely places his observed values alongside the normal curve results, and says that considering the paucity of observations "there is nothing in the results that contradicts the possibility of much closer conformity when many more observations are dealt with." (p. 19.)

himself by calling it a "quasi-accordance with the theoretical law of Frequency of Error." Personally I do not see why it is needful to show accordance, quasi or otherwise, with the normal law of error. It might have served Galton's purpose to show "tailing off" in his distributions of measurements. But it does not seem to me that measurement or enumeration is really what he needs, or it must be measurement or enumeration of characters which belong alike to the various genera, not to a loop alone. The transitions from loop to whorl are qualitative rather than quantitative, and it is the frequency of these qualitative intermediates that we need to analyse. I am inclined to think that this was later recognised by Galton, for I have several times heard him say that there appeared to him nothing to be measured in finger-prints *in general*; could I not suggest a measurable character? I still know of nothing that will apply *satisfactorily* to all types, and I hold that scientific finger-print classification must be qualitative*.

As I have said, I do not see, even if Galton had proved that measurements on loop finger-prints only followed the normal law of distribution, that it would follow that types of finger-prints are genera. In order to prove this we should need to measure characters which run through the whole series of types and this is precisely what it does not appear feasible—at any rate for the present—to achieve. Undoubtedly finger-print patterns do occasionally blend, if such occasions are less frequent than the instances in which they appear to be exclusive (see our Plates XIII and XIV, p. 181). It seems therefore that the key to the matter lies in a closer study of the heredity of finger-prints than has yet been made. With this Galton certainly would have agreed. He writes (p. 21):

"There is reason to believe that the patterns are hereditary. I have no adequate amount of data, whereby to test the truth of this belief by a direct inquiry, but rest the belief partly on analogy, but more especially on the ascertained existence of a considerable tendency to symmetry. When, for instance, there is a primary pattern on one thumb, there are not far from ten chances to one in favour of its being found on the other. Again, if there is a loop in one thumb, there is a strong chance that it will be found in the other thumb also. Similarly as regards each pair of corresponding fingers. Therefore the causes of the pattern must not be looked for in purely local influences. Some of the causes why it and not another pattern is present, are common to both sides of the body and may therefore be called constitutional, and be expected to be hereditary."

Galton continuing next states that finger-prints form an "instructive instance of the effects of heredity under circumstances in which sexual selection has been neutral." He seems to think that sight could be the only sexual selective factor, for he says that finger-prints are too small to attract attention. He remarks that they appear to be uncorrelated with any desirable or repellent quality. Galton holds that they might possibly be related to sensitivity, the average breadth of a ridge-interval being possibly a measure of delicacy in the sense of touch†. But he states that this could have nothing

---

* I write this fully aware of the attempts made by Kristine Bonnevie (*Journal of Genetics*, Vol. xv, pp. 46–54) to give a common measurable characterisation to all types of finger-print pattern.

† Experiments on this point were soon after made for Galton by Titchener, who found no relation between ridge-interval and sensitivity.

to do with the *attractiveness* or otherwise of any particular pattern. I do not believe that Galton has quite plumbed the possible depths of the action of sexual selection in this matter. Touch is one of the least studied, and therefore the least understood of the sexual factors. The question is not that of sensitivity in the producer of a sensation, but of the feelings excited in the recipient. It is not, perhaps, probable, but it is still possible, considering how large a part touch plays in courtship, that the shades of feeling excited by it may be associated with finger-print pattern. Those who straight-away dismiss any slender possibility in this direction have hardly the true measure of our present scientific ignorance, and probably do not realise how much greater a part touch plays in the sensitory life of the female than of the male.

Galton holds that there must have been complete promiscuity of matings, or as it is now called, panmixia, with regard to these patterns, and that consequently they ought to have hybridised. I cannot see that this argument is any more valid than the argument that iris-colours ought to hybridise. It is true that both iris-colours and finger-prints do blend under certain rare physiological conditions that we do not yet understand, but I can see no necessity for a universal rule which anticipates that blending must follow hybridisation. The mere fact that the individual can have finger-prints of various patterns *suggests* hybridisation, and it seems to me that the question of racial differentiation in finger-print frequencies wants renewed investigation, starting very nearly from the point where Galton left it (see our pp. 140, 193–4).

We next turn to the question of natural selection and here we read:

"As regards the influence of all other kinds of natural selection, we know that they co-operate in keeping races pure by their much more frequent destruction of the individuals who depart more widely from the typical centre. *But natural selection is wholly inoperative in respect to individual varieties of patterns and unable to exercise the slightest check upon their vagaries.* Yet, for all that, the different classes of patterns are isolated from one another, through the rarity of transitional cases, just as thoroughly, and just in the same way, as are the genera of plants and animals." (p. 22.)

In the words I have italicised Galton seems to me to have departed from his usual cautious restraint in the matter of dogma, and some suspicion may be thrown on his conclusion from his own data. On p. 21 of the memoir are given measurements of the core and the number of ridges in loop finger-prints of the left and *right* thumbs. From this it appears that the *right* thumb exceeds the left thumb in these measurements and in the number of ridges. Is this relation reversed in left-handed persons? Nowadays we know that the finger-print types are not scattered at random among the digits, there is association between individual digit and individual type. Can it be that there is any reversal of this association in left-handed persons? We do not know; but if it should prove to be so, the first step would have been taken to show a relation between finger-print pattern and manual efficiency. It is never safe to dismiss all relationship of a character to natural selection because we cannot for the moment see any link between the character and fitness.

Galton having dismissed both sexual and natural selection from past or present influence on finger-print patterns, argues that natural selection has had no monopoly in producing genera.

"Not only is it impossible to substantiate a claim for natural selection that it is the sole agent in forming genera, but it seems, from the experience of artificial selection, that it is scarcely competent to do so by favouring mere *varieties*, in the sense in which I understand the term.

"My contention is that it acts by favouring small *sports*. Mere varieties from a common typical centre blend freely in the offspring, and the offspring of every race whose *statistical* characters are constant, necessarily tend, as I have often shown, to revert to their common typical centre*. Sports do not blend freely; they are fresh typical centres or sub-species, which suddenly arise, we do not yet know precisely through what uncommon concurrences of circumstance, and which observations show to be strongly transmissible by inheritance.

"A mere variety can never afford a sticking point in the forward course of evolution, but each new sport implies a new condition of internal equilibrium, and does afford one. A change of type is effected, as I conceive, by a succession of sports or small changes of typical centre, each being in its turn favoured and established by natural selection to the exclusion of its competitors. The distinction between a mere variety and a sport is real and fundamental. I argued this in a recent work [see our discussion pp. 58–62 above of Galton's *Natural Inheritance*, 1889], but had then to draw my illustrations from non-physiological experiences. I could not at that time find an appropriate physiological one. The want is now excellently supplied by observations of the patterns made by the papillary ridges on the thumbs and fingers." (pp. 22–3.)

While I am very loath to say that Galton is in error, I think that he has far from demonstrated the correctness of his views. I have cited his paper at considerable length because I want to indicate how keen a "mutationist" he was. We can claim that he was the first to assert a distinction between "mutations" (sports in his terminology) and "fluctuating variations" (varieties round a typical centre, as he would call them). If the Biometric School has been unable to follow him whole-heartedly in this path, it is because in his case the conclusion was only in a very minor degree based on observation; in the main it flowed from a misinterpretation of his own great discovery of regression†.

Finger-Print Indexing. In the year following the presentation of this memoir Galton read a second paper before the Royal Society (April 30, 1891). It was entitled: "Method of Indexing Finger-Marks," and was published in the *Roy. Soc. Proceedings*, Vol. XLIX, pp. 540–548. Our author points out that the indexing of finger-prints is not only of importance for criminal identification, but for racial and hereditary inquiries. He especially emphasises their value in the latter case:

"The patterns are usually sharp and clear and their *minutiae* are independent of age and growth. They are necessarily trustworthy, and no reluctance is shown in permitting them to be taken, which can be founded either upon personal vanity or upon an unwillingness to communicate undesirable family peculiarities." (p. 540.)

* [This is the old error of the misinterpretation of regression, which led Galton so often and so far astray; see our pp. 31, 48 and 83. K.P.]

† An additional point in this memoir (p. 20) may, perhaps, be just noted. Galton compares the index found from the ratio of means of two absolute variates, with the mean of all the indices found from the individual values of the variates. He shows that the two are nearly the same. We now know the proper corrective factor required to pass from one to the other.

It appears, possibly for reasons to which we have already referred (see p. 163), that Galton had by this time put aside his earlier method of indexing, and he remarks:

"Without caring to dwell on many of my earlier failures to index the finger-prints in a satisfactory way, my description shall be confined to that which has proved to be a success. It is based on a small variety of conspicuous differences of pattern in each of many digits, and not upon minute peculiarities of a single digit." (p. 541.)

Galton had now obtained the prints of all ten digits of 289 persons, though his indexing applies only to the first hundred of these.

He here introduces for the first time the Arch-Loop-Whorl classification*, which has formed the basis of all later attempts at indexing. If a line be drawn from the tip of the forefinger to the base of the little finger, this is roughly the usual slope of the "axes" of the finger-prints if they be not symmetrical. Galton uses the odd numerals 1, 3, 5 for symmetrical forms or for sloped forms with the usual or "normal" slope, the even numerals 2, 4, 6 for the unusual or "abnormal" slopes, in the three classes, arches, loops, whorls. There is little difficulty as a rule in allotting a print to one or other of these six classes. It is only when the rarer compounds (later termed "composites") appear that some difficulty may arise. Galton's scheme is provided in the accompanying diagram.

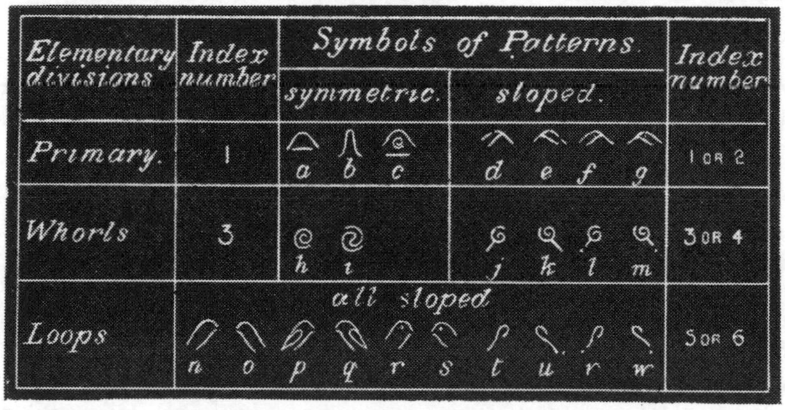

Fig. 26.

He does not arrange his numerals which denote the character of the finger-print in the natural order of the digits, i.e. from little finger left to little finger right. His reason for this is thus stated:

"The forefingers are the most variable of all the digits in respect to their patterns, their slopes being almost as frequently abnormal as not†; the third fingers rank next; the little finger ranks last, as its pattern is a loop in nine cases out of ten. I, therefore, found it convenient not to index the fingers in their natural order, but in the way that is shown at the head of the

* Galton still uses the term "primary" for arch.
† i.e. as frequently radial as ulnar.

columns of figures on the left side of Fig. 27. There, the sequence of the numerals that express the patterns on the digits is divided into two groups of three numerals and two groups of two numerals, as 355, 455, 55, 35. The first group 355 refers to the first, second and third fingers of the left hand*; the second group 455 to the first, second, and third fingers of the right hand; the

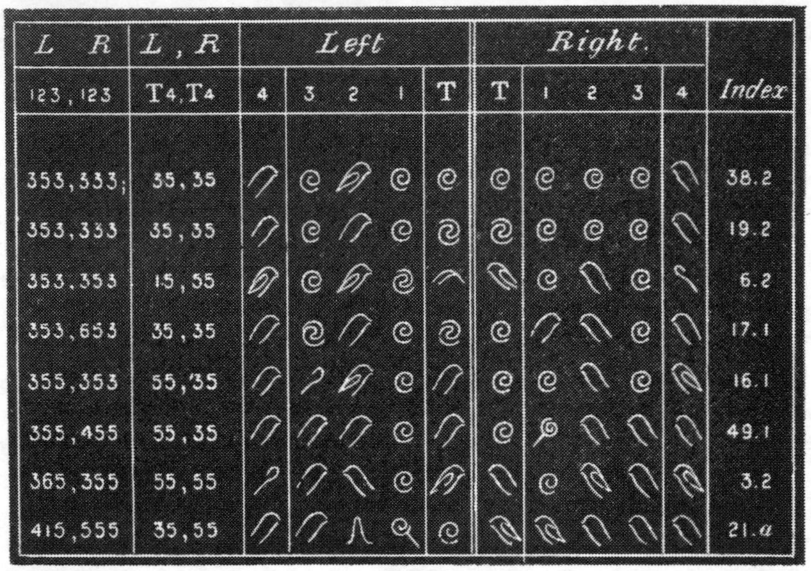

L , R	L , R		Left					Right.					
123, 123	T4, T4	4	3	2	1	T	T	1	2	3	4	Index
353, 333;	35, 35											38.2
353, 353	35, 35											19.2
353, 353	15, 55											6.2
353, 653	35, 35											17.1
355, 353	55, 35											16.1
355, 455	55, 35											49.1
365, 355	55, 55											3.2
415, 555	35, 55											21.a

Fig. 27.

third group 55 to the thumb and fourth finger of the left hand; the fourth group 35 to the thumb and fourth finger of the right hand. The index is arranged in the numerical sequence of these sets of numbers as shown in Fig. 27 †." (pp. 542–3.)

It will be seen from Fig. 27 that Galton drew a rough symbol denoting the nature of his subclasses, the *a* to *w* of Fig. 26 in his index. The symbols with dots attached mark cases in which there may be doubt as to classification. Thus the primaries *f* and *g* may have been classed by another as loops. If there has been hesitation about them, after seeking them as loops, a second reference to the index should be made, treating them as primaries. When a whorl is "crozier" shaped, as *j, k, l, m*, it lies in a loop, and may when it approaches the plain eyes *t, u* give rise to hesitation and a dot is then added, as at *l, m*. Galton says that he has not found much difficulty with transitional cases, and considers it could be well surmounted if a standard collection of doubtful forms were established to ensure that different persons would abide by a common rule.

Galton (pp. 545–6) gives an index based on the ten finger-prints of 100 persons. In this index there are nine cases of duplicated numbers and three

* Galton's first finger = forefinger, second finger = middle finger, third finger = ring finger, and fourth finger = little finger. Galton's purpose is clear, but there are distinct and greater advantages in the "natural" order.

† The last column in Galton's figure, our Fig. 27, requires explanation; it is the page reference to his records where the actual finger-prints will be found. The word "Index" at the head of the column is, perhaps, not explanatory enough.

cases of triplicated numbers. In other words the index number alone would not suffice to identify an individual in about a quarter of the indexed cases. Now if the index contains 100,000 instead of 100 individuals, it is clear that these multiple cases, instead of being counted by twos and threes, would be counted by hundreds, and the number of references required to the prints themselves would become most fatiguing. The source of this evil is fairly clear if we examine Galton's Table II (p. 548). It classifies the patterns that occur in 100 prints of left forefingers. It is obvious that we have gained very

Forefinger of Left Hand.

Pattern	Classificatory Number	Number of Occurrences
Primary, plain ... ...⎫ Primary nascent loop, slope normal ...⎭	1	26
Primary nascent loop, slope abnormal	2	4
Whorl, plain ... .⎫ Whorl, with tail, slope normal ...⎭	3	23
Whorl, with tail, slope abnormal ...	4	6
Loop, slope normal	5	21
Loop, slope abnormal	6	20
Total cases	...	100

little indeed in the case of primaries and whorls by taking the nature of the slope as a characteristic. The four groups of 26, 23, 21 and 20 still remain far too large. We need to break up the primaries into three nearly equal groups, not into two of 26 and 4; the same applies to the whorls, while each group of loops requires bisecting. This would give us ten classes, and fit in well with a ten-figure index number. The scheme of indexing Galton proposed in this paper could not be final, yet it was pioneer work*; no one but our author himself had so far published or even suggested a plan for indexing, and there still remained much spade-work to be done before an adequate scheme was evolved. Galton himself recognised the difficulty, thus he writes:

"The greatest difficulty in constructing a uniformly efficient catalogue lies in the troublesome frequency of plain loops, so that even the method of picture writing fails to analyse satisfactorily the numerous 555, 555, 55, 55 cases. When searching through a large number of similarly indexed prints for a particular specimen, it is a very expeditious method to fix on any well-marked characteristic of a minute kind such as an island, or enclosure, or a couple of adjacent bifurcations, that may present itself in any one of the fingers, and in making the search to use a lens or lenses of low power, fixed at the end of an arm, and to confine the attention solely to looking for that one characteristic. The cards on which the finger marks have been made may then be passed successively under the lens with great rapidity. I fear that the method of counting ridges (as the number of ridges in *AH* of my previous memoir [see our pp. 163, 167]) would be difficult to use by persons who are not experts. Anyhow, I have not yet been able to devise a plan for doing so that I can recommend." (p. 547.)

* The diagrammatic symbols used by Galton are the basis from which his fuller classification in *Finger Print Directories* starts (see our pp. 199 *et seq.*).

Another point dealt with by Galton in this memoir is the relative ad
vantage gained in indexing by the first two fingers of the left hand, the first
three fingers of the left hand, the first three fingers of both hands or by all
ten digits; he finds the numbers of different patterns occurring are respectively
16, 27, 65 and 83. The ten-digit indexing is now in general use, and of course
provides a greater field for identification, if the indexing be somewhat more
cumbersome.

B. *Finger Prints*, 1893.

We now reach Galton's fundamental book on finger-prints, namely *Finger
Prints** (Macmillan, 1893). Chapter I (pp. 1–21), entitled *Introduction*,
gives a brief account of the subject referring to Purkenje and the pioneer
work of Sir William Herschel; it further provides a synopsis of the contents
of the entire book †.

Chapter II (pp. 22–29) deals with *Previous Use of Finger-Prints*. It
recounts the use of nail-marks or finger-marks among barbarous or semi-
civilised people rather as a superstitious sign of personal touch than of personal

Chinese Coin, Tang Dynasty, about 618 A.D., with nàil mark of the Empress Wen-teh,
figured in relief.

Fig. 28.

identity. It notes also the frequent appearance of finger-impressions upon
ancient pottery. Here, as in the case of a Greek impression found by Sir
Charles Walston on a steatite seal at the Argive Heraeum‡, it is somewhat

* It is an interesting example of the futility of some reviewers, that the critic who wrote
the notice of Galton's *Finger Prints* in the *Athenaeum* of Dec. 24, 1892, expressed the wish that
he might devote his brilliant powers to "subjects of greater promise of practical utility"; and
again: "Whether the practical results to be derived from his researches will repay the pains
he has bestowed upon them we must take leave to doubt. It will be long before a British
jury will consent to convict a man upon the evidence of his finger-prints; and however perfect
in theory the identification may be, it will not be easy to submit it in a form that will amount
to legal evidence."

† At the end of this chapter Galton thanks Mr Howard Collins for his very material aid.
The correspondence between Galton and Collins during the progress of the work was considerable, and of some scientific value. In 1911 I issued a request in the *Times* and other journals
for letters or copies of letters written by Galton. The response was very disappointing. During
the last nine years the Galton Laboratory has had frequently to purchase letters of Galton
sold by their recipients or the assigns of the latter to booksellers or autograph dealers. Among
such purchases the Laboratory obtained from a Birmingham bookseller, whose catalogue the
Director luckily chanced to see, Galton's numerous letters to Collins on the subject of finger-
prints.

‡ *The Illustrated London News*, Feb. 7, 1925, p. 231.

doubtful if the impression was purely accidental, arising simply from touching by chance the wet clay, or was the result of moulding with the thumb the small base of an object, or was actually intended as a potter's mark. Galton next refers to Bewick's impressing his thumb-mark and a finger-mark on a

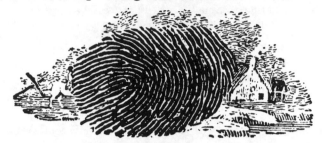

Fig. 29. Thomas Bewick, his mark.

block of wood, engraving them and afterwards using them for ornaments in his books*; this approaches the use of a finger-print for a sign-manual. Galton continues:

"Occasional instances of careful study may also be noted, such as that of Mr Faulds (*Nature*, Vol. XXII, p. 605, Oct. 28, 1880), who seems to have taken much pains, and that of Mr Tabor, the eminent photographer of San Francisco, who, noticing the lineations of a print that he had accidentally made with his own inked finger upon a blotting-paper, experimented further, and finally proposed the method of finger-prints for the registration of Chinese, whose identification has always been a difficulty, and was giving a great deal of trouble at that particular time;

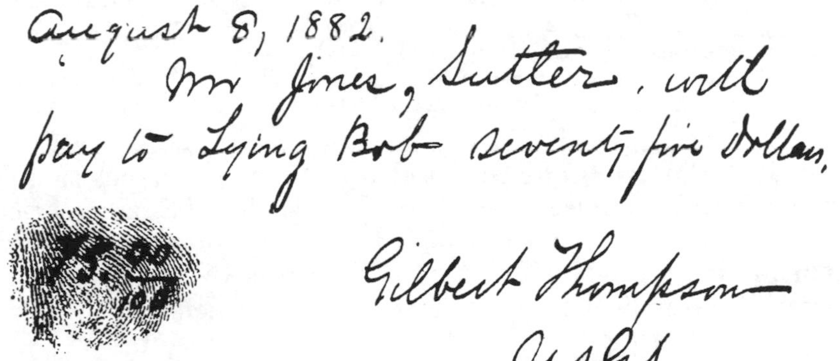

August 8, 1882.
Mr Jones, Sutler, will
pay to Lying Bob seventy five Dollars.
Gilbert Thompson
U.S.G.S.

Order on a Camp Sutler, by the officer of a surveying party in New Mexico 1882.

Fig. 30.

but his proposal dropped through. Again Mr Gilbert Thompson, an American geologist, when on Government duty in 1882 in the wild parts of New Mexico, paid the members of his party

* See for example *History of Birds*, Vol. I, p. 180, edn. 1805. It is not in my edition (1807) of the *General History of Quadrupeds*. Sir William Herschel reproduces in his book, *The Origin of Finger Printing*, 1916, p. 33, a receipt of Bewick from 1818, in 1918 in the possession of Mr Quaritch. The print is a very delicate one, and has the attached words "Thomas Bewick, his mark." Sir William thinks that these marks of Bewick, known to him as a boy, may have unwittingly led him to study such prints.

by order of [? on] the camp sutler. To guard against forgery he signed his name [? wrote the amount] across the impression made by his finger upon the order, after first pressing it on his office pad. He was good enough to send me the duplicate of one of these cheques made out in favour of a man who bore the ominous name of 'Lying Bob' [see Fig. 30 on p. 175]. The impression took the place of scroll work on an ordinary cheque; it was in violet aniline ink, and looked decidedly pretty. From time to time sporadic instances like these are met with, but none are comparable in importance to the regular and official employment made of finger-prints by Sir William Herschel, during more than a quarter of a century in Bengal. I was exceedingly obliged to him for much valuable information when first commencing this study, and have been almost wholly indebted to his kindness for the materials used in this book for proving the persistence of lineations throughout life.

"Sir William Herschel has presented me with one of the two original 'Contracts' in Bengali, dated 1858, which suggested to his mind the idea of using this method of identification*. It was so difficult to obtain credence to the signatures of the natives, that he thought he would use the signature of the hand itself, chiefly with the intention of frightening the man who made it from afterwards denying his formal act; however, the impression proved so good that Sir W. Herschel became convinced that the same method might be further utilised. He finally introduced the use of finger-prints in several departments at Hooghly in 1877, after seventeen years' experience of the value of the evidence they afforded. A too brief account of his work was given by him in *Nature* (Vol. XXIII, p. 23, Nov. 25, 1880). He mentions there that he had been taking finger marks as sign-manuals for more than twenty years, and had introduced them for practical purposes in several ways in India with marked benefit. They rendered attempts to repudiate signatures quite hopeless. Finger-prints were taken of Pensioners to prevent their personation by others after death; they were used in the office for Registration of Deeds, and at a gaol where each prisoner had to sign with his finger. By comparing the prints of persons then living, with their prints taken twenty years previously, he considered he had proved that the lapse of at least that period made no change sufficient to affect the utility of the plan. He informs me that he submitted, in 1877, a report in semi-official form to the Inspector-General of Gaols, asking to be allowed to extend the process; but no result followed. In 1881, at the request of the Governor of the gaol at Greenwich (Sydney), he sent a description of the method, but no further steps appear to have been taken there.

"If the use of finger-prints ever becomes of general importance, Sir William Herschel must be regarded as the first who devised a feasible method for regular use, and afterwards officially adopted it." (pp. 26–29.)

I have cited this long passage because I wish to give evidence that Galton did ample justice to his predecessors, more justice than has since been done to his own work†. Galton never claimed to have invented the *idea* of identification by finger-prints. What he did do was to take up the matter from the scientific standpoint to establish certain principles and the practical methods of operating them. It was his publications and his energetic demonstration of the value of finger-print identification, not occasional newspaper diatribes, which led to its adoption by the English Prison Service, and ultimately to its acceptance throughout the civilised world. Much solid

---

* One is reproduced on our Plate V, p. 146 and the other in Sir William Herschel's *The Origin of Finger Printing.*

† "In discussing the true natural history of the minute ridges upon the fingers Galton goes no further than did the first physiologist of note who drew attention to their presence. This was Nehemiah Grew." Louis Robinson in *North American Review*, May 15, 1905. Again: "Mr Galton distinctly says in his *Finger Prints*, p. 2: 'My attention was first drawn to the ridges in 1888,' etc. It is not a little remarkable to my mind that that date should so nearly coincide with the period when I was interesting Sir Wollaston Franks, of the British Museum, and other scientific authorities in the importance of this means of identification." *Birmingham Post*, May 16, 1905. Dr Faulds cites only the first words of Galton's paragraph on p. 2. For the full citation see our p. 142.

work had to be done before the mere idea of identification by finger-prints could be transformed into its full realisation as a practical criminal procedure. For that actual transformation we have to thank neither Nehemiah Grew nor Dr Faulds, but Francis Galton expanding and working on the experiences of Sir William Herschel.

Chapter III (pp. 30–53), entitled *Methods of Printing*, gives a very full description of methods for the permanent preservation of finger-marks.

Galton starts by indicating a way of getting very perfect finger-prints, which has been since used very largely for detective purposes. The reader can easily try it for himself; let him pass his finger over the hair at the back of his head, and then press the bulb of his finger on a window pane, that of a recently cleaned window if available; he will find a very perfect imprint of his finger lineation, and there it may remain decipherable for days—under post-war conditions of domestic service! If the finger be merely moistened the impression soon evaporates; the essential need is to oil the finger *very slightly*, and this is adequately achieved by the natural oiliness of the hair. Similar finger-prints may be obtained on polished steel—a razor blade—or on table plate. Now-a-days for the purposes of criminal investigation such accidental finger-prints can be reproduced and preserved. Galton next proceeds to give accounts of laboratory and also of pocket apparatus for finger-printing; the important factors are the persistent cleanliness needful in the apparatus, and the extreme thinness of the ink layer on the finger, if a good impression is to be obtained*. This chapter is replete with suggestions such as we have recorded of the younger Galton with his mechanical "dodges." A thin sheet of copper which I found in one of Galton's diaries puzzled me, till I re-read *Finger Prints*, and there noted that it was to receive soot from a candle (or even a match) to blacken fingers for their prints.

"Paste rubbed in a very thin layer over a card makes a surface that holds soot firmly, and one that will not stick to other surfaces if accidentally moistened. Glue, isinglass, size, and mucilage, are all suitable. It was my fortune as a boy to receive rudimentary lessons in drawing from a humble and rather grotesque master. He confided to me the discovery, which he claimed as his own, that pencil drawings could be fixed by licking them; and as I write these words, the image of his broad swab-like tongue performing the operation, and of his proud eyes gleaming over the drawing he was operating on, come vividly to remembrance. This reminiscence led me to try whether licking a piece of paper would give it a sufficiently adhesive surface. It did so. Nay, it led me a step further, for I took two pieces of paper and licked both. The dry side of the one was held over the candle as an equivalent to a plate for collecting soot, being saved by the moisture at the back from igniting (it had to be licked two or three times during the process), and the impression was made on the other bit of paper. An ingenious person determined to succeed in obtaining the record of a finger impression can hardly fail altogether under any ordinary circumstances." (pp. 48–9.)

I should like to have asked Galton what he would have done had there been no *paper*†; I feel sure he would have been ready with a substitute! The chapter concludes with remarks on the photography of finger-prints and on

* The Galton Laboratory, which collects finger-prints of families, finds that an operator can be easily taught to take decipherable finger-prints with a simple pocket apparatus, which it circulates for this purpose.

† Quite good impressions can be made with bird lime and candle black, specimens in *Galtoniana*.

methods of enlarging them. In the *Galtoniana* we have still his special camera for enlarging finger-prints (see our p. 215), his much enlarged series of finger-prints used for fine classification (reproduced for this work, and to be found in a pocket at the end of this volume) and the watchmaker's glass mounted on a stand for directly examining them*.

Chapter IV (pp. 54–63) deals with *The Ridges and their Use.* Galton starts with the ridges of the palm of the hand, and indicates that they are not very closely related to the "creases," so that the latter cannot be the cause of the former. He also refers to the ridges on the soles and toes, but ultimately confines his attention to those on the fingers. Here he defines two important terms: first, *Minutiae*, which are the minute peculiarities characterising an individual ridge. A ridge may divide into two or unite with another (see Fig. 31, *a* and *b*), or it may divide and almost immediately

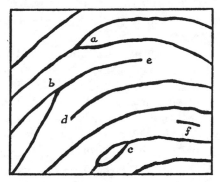

Characteristic Peculiarities in Ridges.
(about 8 times the natural size)
Fig. 31.

reunite, enclosing a small circular or elliptic space (*c*); at other times it may begin or end abruptly (*d* and *e*); or lastly the ridge may be so short as to form a small island (*f*). Secondly, *Patterns*: whenever an interspace is left between the boundaries of different systems of ridges, it is filled by a small system of its own which will have some characteristic shape. This shape is termed a *pattern* (see Figs. 20, 21 on our pp. 162, 163). The descriptions of *minutiae* and of patterns belonging to an individual *are* of special value for the purposes of identification.

On the whole there is little known of the origin and use of the ridges, beyond the fact that they carry the sweat pores. Nor is their origin or use of much importance for the purpose of identification provided we can be assured of their persistency during life. Titchener, as I have noted (p. 168), made, at the suggestion of Galton, a series of experiments with the aesthesiometer, and proved that the fineness or coarseness of the ridges in different persons had no effect whatever on the delicacy of their tactile discrimination.

* This finger-print glass appears in Furse's painting of Galton; see the Frontispiece to Vol. I. It is worth noting that Galton selected this piece of apparatus as the most characteristic of his many activities.

PLATE IX

THE STANDARD PATTERNS OF PURKENJE

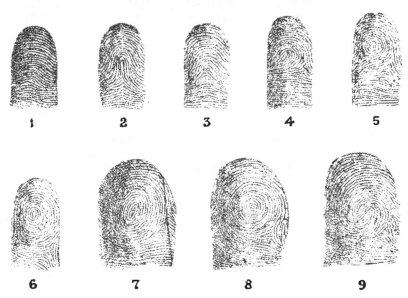

Reproduced *de novo* from the copy of Purkenje's *Commentatio* in the
Library of the Royal College of Surgeons.

THE CORES OF THE ABOVE PATTERNS.

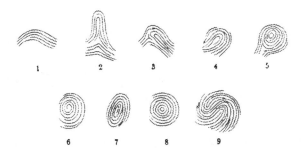

Galton's Patterns from Purkenje's Types.

Also he found it made no difference whether one or both points of the compass rested on the ridges or in the furrows. Nor again was the width of the ridge interval any test of the relative power of discrimination of the different parts of the same hand (p. 62). Galton himself suggests that the ridges may serve the purpose of enabling us to judge the relative roughness of surfaces by touch, and so to determine their nature. If a blindfold person be asked to determine an object by touch, he will be observed to rub the surface with his finger.

"The ridges engage themselves with the roughness of the surface, and greatly help in calling forth the required sensation, which is that of a thrill; usually faint, but always to be perceived when the sensation is analysed, and which becomes very distinct when the indentations are at equal distances apart as in a file or in velvet. A thrill is analogous to a musical note, and the characteristics to the sense of touch, of different surfaces when they are rubbed by the fingers, may be compared to different qualities of sound or noise. There are, however, no pure overtones in the case of touch, as there are in nearly all sounds." (p. 63.)

I should be glad to have the experience of any of my readers on this point. I wonder if this thrill is universal; personally I am unable to associate even uniform roughness of a touched surface with anything of the nature of a thrill. Two other men were like myself. Of three women tested one had no sensation of thrill, a second failed with file and a stiff brush, but was doubtful in the case of velvet; the third felt a thrill—chill in the spine—on rubbing with the finger-tip file, velvet or brush.

Chapter V (pp. 64–88) is entitled *Patterns: their Outlines and Cores.* Galton opens this chapter by referring to Purkenje's types*, and states that he had entirely failed on trial to classify prints by mere inspection and the use of Purkenje's types. He had accordingly devised his method of "outlining" the pattern in order to classify it. He took as material for his classification 504 prints of right thumbs *enlarged* two and a half times their natural size, so

---

* Galton (pp. 85–88 of this chapter) provides a translation of the portion of Purkenje's *Commentatio* which deals with types, and also a plate of Purkenje's nine types, accompanied by Galton's own outlining of the cores. Purkenje's nine types are the following: (i) *Transverse Flexures* = Galton's "primaries." In the course of his description Purkenje used the word "arch." I think this must have led Galton to replace his term "primary" of 1890 by "arch" of 1892. (ii) *Central Longitudinal Stria* = Galton's "tented arch." (iii) *Oblique Stria* = (I think) Galton's "nascent loop." (iv) *Oblique Sinus* = Galton's "plain loop." (v) *Almond* = (I think) "circlet in loop," a sub-type of Galton's "whorl" (see his Plate 8, No. 22). (vi) *Spiral* = Galton's "whorl," sub-type "spiro-whorl" (see his Plate 8, No. 26). (vii) *Ellipse*, or *Elliptic whorl* = Galton's "whorl" (ellipses). (viii) *Circle*, or *Circular Whorl* = Galton's "whorl" (circles). (ix) *Double Whorl* = Galton's "whorl" ("duplex spiral"), see his Plate 8, No. 29. The reader who attempts to classify prints by Purkenje's nine classes will soon find, if he follows Purkenje's rather elaborate descriptions, that they exclude many frequently occurring cases. His definitions are indeed not broad enough to embrace the innumerable variations which arise. It is perhaps worth noting that Purkenje under the definition (vi) of "Spiral" introduces the word "composite" but not in its modern sense to denote a compound of two patterns, but for a spiral made up not of a single line, but of two or more lines proceeding from the single focus or pole. I imagine Galton would have called Purkenje's "composite spiral" a "whorl," sub-type "twist" (see Plate 8, No. 52 and Plate 16, Nos. 36, 37, where, however, no name is provided). Purkenje does not figure his "composite." He refers to the "small triangles," Galton's "plots," "deltas" or "islands," under his definitions (iv) and (vi). This footnote will suffice to indicate the extent to which Purkenje anticipated Galton in matters of nomenclature. See also our Plate IX.

that each print was about playing card size. Galton found that on repeated trials he did not, by inspection only, deal these out into the same classes. The same failure occurred when he selected standard types and endeavoured to sort into groups by aid of these. Mere judgment by the unaided eye is liable to be influenced by the intensity of inking of some ridges; two prints will not always give the same extent of pattern. "A third cause of error is still more serious; it is that patterns, especially those of a spiral form, may be apparently similar yet fundamentally unlike, the unaided eye being frequently unable to analyse them and to discern real differences" (p. 66). Accordingly Galton introduced his system of "outlining" the pattern. To this we have already referred in discussing his *Phil. Trans.* memoir (see our p. 164). His Plate 5, here reproduced as our Plate X, shows samples of outlined patterns. Whether it is needful for an expert *always* to outline is another question, but to become an expert in classification, it is undoubtedly necessary to gain experience in grouping by outlining, even if the classification is only to be in the broadest categories. The chief reason for this is that the existing classification schemes are in truth largely artificial. There is really no generic difference between a "tented arch" and a "tented loop," or between an "eyeletted loop" and a "small spiral in loop" which Galton reckons a whorl. There are numerous such cases where the classification can only be by arbitrary standardisation. We reproduce as our Plates XI, XII and XIII Galton's Plates 7, 8 and 6 which will aid any reader desirous of learning to classify by outlines; yet even then he will undoubtedly find rare patterns, which he can only hope to thrust into a miscellaneous group of "composites." Galton's Plates 9 and 10 (see our Plates XIV and XV) give threefold enlargements of troublesome transitional patterns, the first between arches and loops and the second between loops and whorls. The beginner should attempt to classify them, and then compare his results with Galton's views on pp. 79–80.

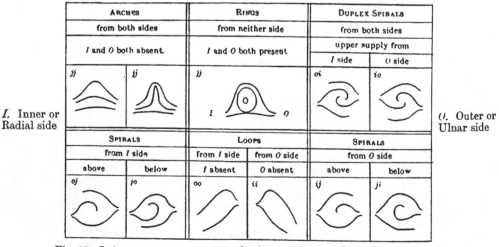

Fig. 32. It is necessary to suppose the finger-prints are from the *right* hand.

On pp. 80–81 Galton repeats the classification of his *Phil. Trans.*

PLATE X

Examples of the outlining of Patterns to assist Classification.

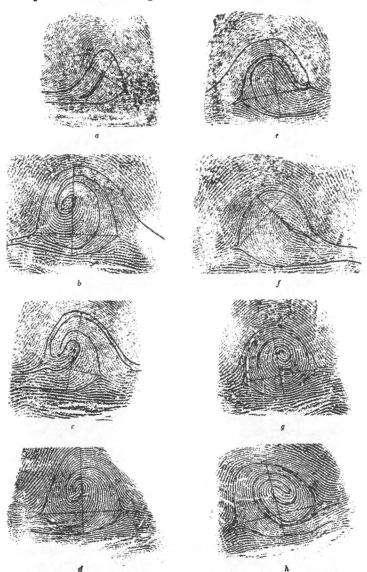

The specimens are *rolled* impressions of natural size. Galton was the first writer on the subject
to introduce "rolling." All impressions are now rolled. *a* and *f* are loops; *b, c, d, e, g* and *h*
are various types of whorls. *Finger Prints*, Plate 5.

PLATE XI

Outlines of Patterns in Arches and Loops.

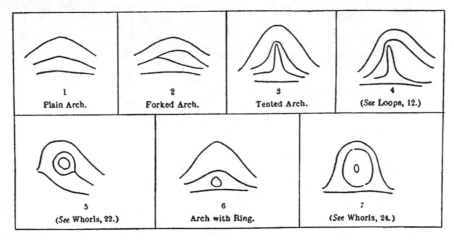

Galton's nomenclature as aids to description and classification. Arches and Loops.
From Galton's *Finger Prints*, Plate 7.

PLATE XII

Outlines to Patterns in Whorls. Types of Cores.

WHORLS.

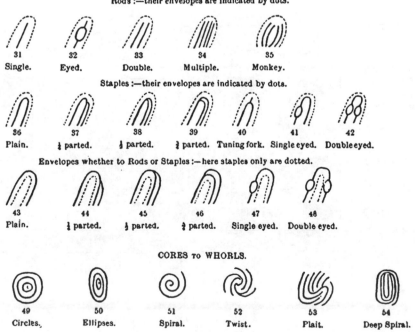

20	21	22	23
Small Spiral in Loop.	Spiral in Loop.	Circlet in Loop.	Ring in Loop.
24	25	26	
Rings.	Ellipses.	Spiro-rings.	
27	28	29	30
Simple Spiral.	Nascent Duplex Spiral.	Duplex Spiral.	Banded Duplex Spiral.

CORES to LOOPS.

Rods :—their envelopes are indicated by dots.

| 31 | 32 | 33 | 34 | 35 |
| Single. | Eyed. | Double. | Multiple. | Monkey. |

Staples :—their envelopes are indicated by dots.

| 36 | 37 | 38 | 39 | 40 | 41 | 42 |
| Plain. | ¼ parted. | ½ parted. | ¾ parted. | Tuning fork. | Single eyed. | Double eyed. |

Envelopes whether to Rods or Staples :—here staples only are dotted.

| 43 | 44 | 45 | 46 | 47 | 48 |
| Plain. | ¼ parted. | ½ parted. | ¾ parted. | Single eyed. | Double eyed. |

CORES to WHORLS.

| 49 | 50 | 51 | 52 | 53 | 54 |
| Circles. | Ellipses. | Spiral. | Twist. | Plait. | Deep Spiral. |

Galton's nomenclature as aids to description and classification.
From Galton's *Finger Prints*, Plate 8.

PLATE XIII

OUTLINES OF THE PATTERNS OF THE DIGITS OF EIGHT PERSONS, TAKEN AT RANDOM.

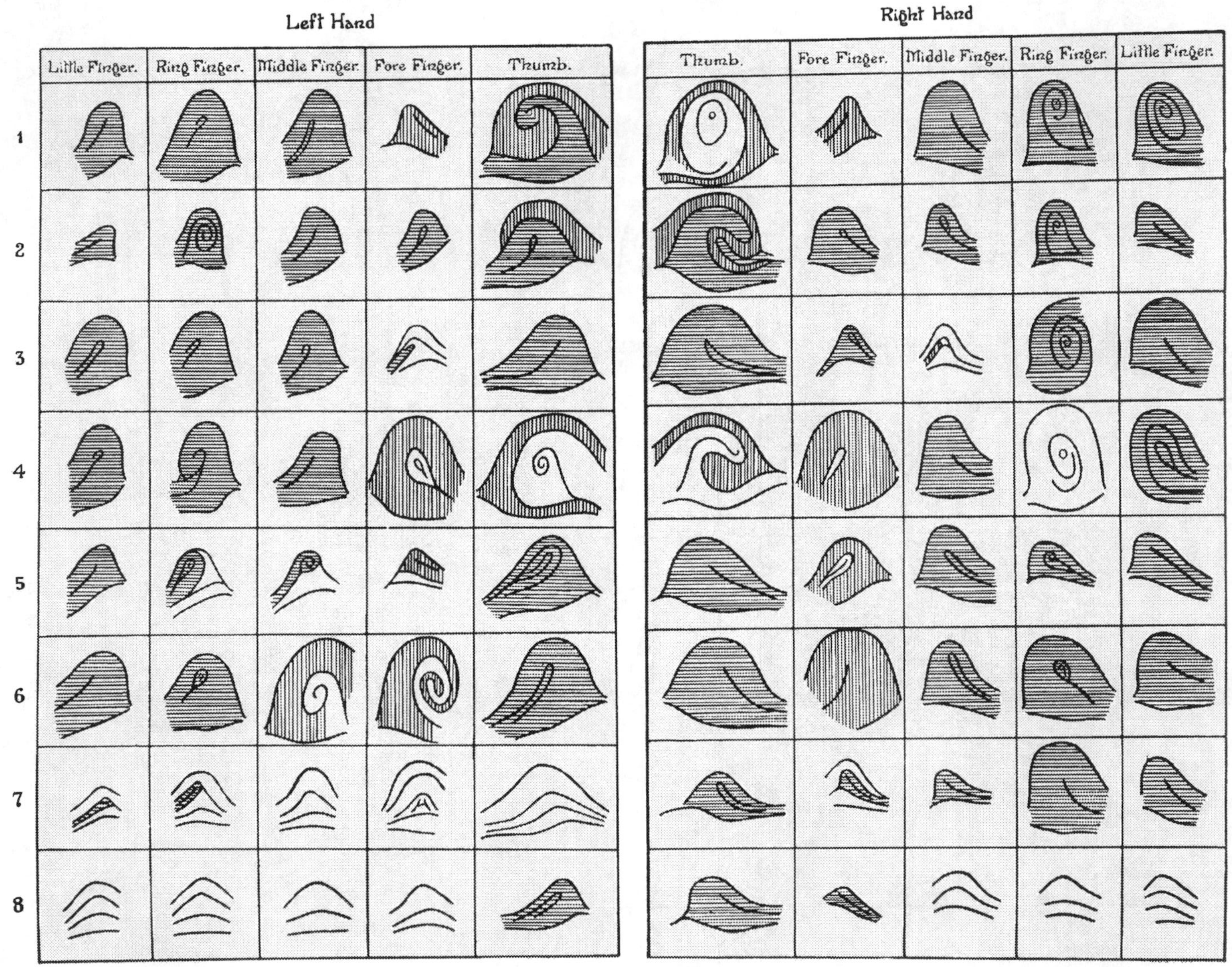

Left Hand

Right Hand

Ridges from inner (or radial) side have vertical hatching, from outer (or ulnar) side have horizontal hatching.

After Galton's *Finger Prints*, Plate 6.

PLATE XIV

Transitional Patterns—Arches and Loops.

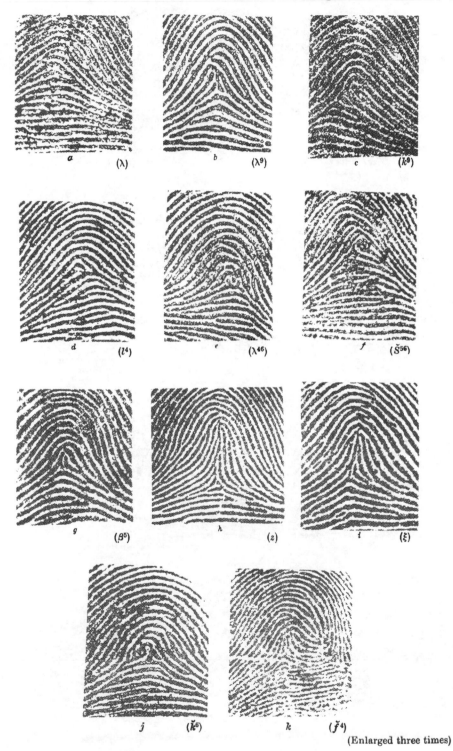

(Enlarged three times)

Transitional Patterns from Galton's *Finger Prints*, Plate 9, with suggested symbols.
The prints are supposed to be of left-hand fingers.

PLATE XV

Transitional Patterns—Loops and Whorls.

(Enlarged three times)

Transitional Patterns from Galton's *Finger Prints*, Plate 10, with suggested symbols.
The prints are supposed to be of left-hand fingers.

memoir (see our p. 164). Unfortunately he uses the letter j throughout his Plate 11 (our Fig. 32) for what he terms u in the text.

"The divergent ridges that bound any simple pattern admit of nine, and only nine, distinct variations in the first part of their course. The bounding ridge that has attained the summit of any such pattern must have arrived either from the Inner plot (I) [radial delta], the Outer plot [ulnar delta], or from both. Similarly as regards the bounding ridge that lies at the lowest point of the pattern. Any one of the three former events may occur in connection with any of the three latter events, so that they afford in all 3 × 3, or nine possible combinations. It is convenient to distinguish them by easily intelligible symbols. Thus, let i signify a bounding line which starts from the point I, whether it proceeds to the summit or to the base of the pattern; let o be a line that similarly proceeds from O, and let j be a line that unites the two plots [deltas] I and O either by summit or by base. Again let two symbols be used, of which the first shall always refer to the summit, and the second to the base of the pattern. Then the nine possible cases are jj, ji, jo; ij, ii, io; oj, oi, oo. The case of the arches is peculiar, but they may be fairly classed under the symbol jj." *Finger Prints*, pp. 80–81, with j as in figure replacing u of Galton's own text.

Galton next refers to measurements on the print and states that the average ridge interval should be taken as unit of measurement for comparative

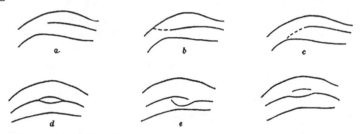

Fig. 33. Illustrations of Ambiguities in *minutiae*, a may appear as b or c, d as e or f.

purposes, especially where prints of non-adults are concerned. Plate 11 (our Fig. 33) gives illustrations of ambiguities in *minutiae* to which we have previously referred (see our p. 165).

Chapter VI (pp. 89–99) deals with *Persistence*. It is an extension of the evidence partially given in the *Phil. Trans.* memoir (see our p. 166 and Plates VII and VIII). Galton has here studied between twenty and thirty different digits and compared *minutiae* to the number of 700 (p. 96) and only found the one discrepancy to which reference has already been made (see last lines on our p. 166). We reproduce Galton's Plates 13 and 14 (our Plates XVI and XVII) as an illustration of his methods of comparing *minutiae*, and of the periods for which persistency was demonstrated. Galton again emphasises that it is in the *minutiae*, not in the measurements of the pattern, that persistency lies (p. 98). After indicating that for the four periods of life there is no change, and that we may expect in 700 *minutiae* only one to fail us, Galton continues:

"Neither can there be any change after death, up to the time when the skin perishes through decomposition; for example, the marks on the fingers of many Egyptian mummies, and on the paws of stuffed monkeys, still remain legible. Very good evidence and careful inquiry is thus seen to justify the popular idea of the persistence of finger markings, that has hitherto been too rashly jumped at, and which wrongly ascribed the persistence to the general appearance of the pattern, rather than to the *minutiae* it contains. There appear to be no external bodily characteristics, other than deep scars and tattoo marks, comparable in their persistence to these markings, whether they be on the finger, on other parts of the palmar surface of the hand, or on the sole of the foot. At the same time they are out of all proportion more numerous than any other measurable feature; about thirty-five of them are situated on the bulb of each of the ten digits, in addition to more than 100 on the ball of the thumb, which is not one-fifth of the superficies of the rest of the palmar surface. The total number of points suitable for comparison on the two hands must therefore be not less than one thousand and nearer to two; an estimate which I verified by a rough count on my own hand; similarly in respect of the feet. The dimensions of the limbs and body alter in the course of growth and decay; the colour, quantity and quality of the hair, the tint and quality of the skin, the number and set of the teeth, the expression of the features, the gestures, the handwriting, even the eye colour, change after many years. There seems no persistence in the visible parts of the body, except in these minute and hitherto too much disregarded ridges." (pp. 97–8.)

Chapter VII (pp. 100–113) is entitled *Evidential Value.* Its object is to give an approximate numerical idea of the value of finger-prints as a measure of Personal Identification. Galton's method is a somewhat elaborate one. If we take a square of one ridge interval, and place this on our finger-print, we can almost certainly draw on its surface correctly the ridge or ridges which lie behind it. When we take an opaque square of side 6-ridge intervals, and fasten this blank square to the finger-print and then reconstruct the system of ridges which lies behind it we are rather more often wrong than right in our reconstructed ridges. Galton thinks that a square of 5-ridge intervals would probably allow reconstruction as often right as wrong. He made two series of experiments of this character, with the enlargements double and sixfold. Then he made a twentyfold enlargement, and placed upon it a chequerboard arrangement of 6-ridge interval squares; he reconstructed the whole finger-print, each square from the four adjacent ones, which bordered the unseen square. There were in this case seven rightly and sixteen wrongly constructed. He now makes a rather drastic assumption

"that any one of these reconstructions represents lineations that might have occurred in Nature, in association with the conditions outside the square, just as well as the lineations of the actual finger-print (p. 107)....It therefore seems right to look upon the squares as independent variables, in the sense that when the surrounding conditions are alone taken into account, the ridges may either run in the observed way or in a different way, the chance of these two contrasted events being taken (for safety's sake) as approximately equal." (p. 108.)

There being about 24 6-ridge interval squares in any finger-print, Galton makes $1/2^{24}$ to be the chance of the actual system of ridges appearing. He now proceeds to give a rough approximation to two other chances, which he considers to be involved: the first concerns guessing correctly the general course of the ridges adjacent to each square, and the second of guessing rightly the number of ridges that enter and issue from the square. He takes these in round numbers to be $1/2^4$ and $1/2^8$, so that the whole chance of the observed system is $1/2^{36}$. Now the total number of persons in the world has been reckoned at about 16,000,000,000 and the chance of a particular observed arrangement is of the order 1/64,000,000,000, or the odds are very roughly 39 to 1 against the particular arrangement occurring on a single definite digit of any existing human being*.

* Galton in his own copy has a pencil note "repeat calculations" and corrects the total population of the world which in his text he has made ten times too great. I have corrected the figures in the last paragraph of his p. 110 accordingly.

PLATE XVI

V. H. H-D æt. 2½ in 1877,
and again as a boy in Nov. 1890.

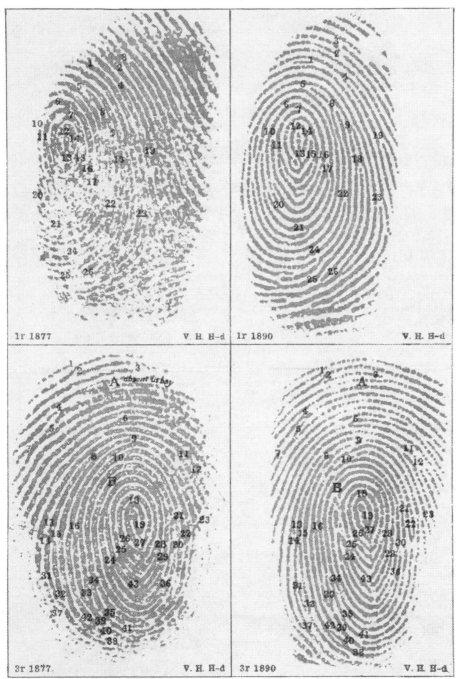

1r 1877   V. H. H-d   1r 1890   V. H. H-d

3r 1877.   V. H. H-d   3r 1890   V. H. H-d

To illustrate Persistence of Pattern in Finger Prints.
From Galton's *Finger Prints*, Plate 13.

PLATE XVII

Persistence of Finger-Print Patterns with corresponding *minutiae* like numbered.

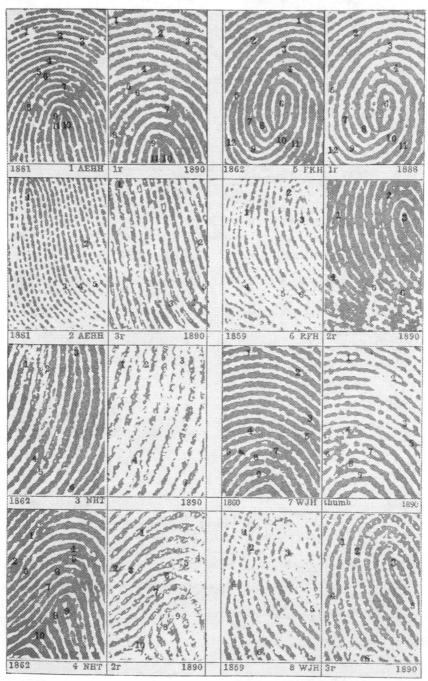

Intervals of 9, 9, 26, 28, 28, 30, 31 and 31 years. Galton's illustrations from
Herschel's material, *Finger Prints*, Plate 14.

While convinced that the chance of two individuals actually possessing the same finger-print in all its *minutiae* is infinitesimally small—as small as the chance that two woodcutters given the same topic would produce two blocks identical in every line and dot—yet one recognises that Galton's treatment, however ingenious, lacks the power of compelling conviction. Nature probably works more definitely to form a whole pattern than can be mimicked by Galton's 24 "independent variable" squares. He himself writes that

"it is hateful to blunder in calculations of adverse chances, by overlooking correlations between variables, and to falsely assume them to be independent, with the result that inflated estimates are made which require to be proportionately reduced. Here, however, there seems to be little room for such an error." (p. 109.)

It is the last sentence only we would call in question. After all it is the *minutiae*, rather than the pattern, by which identification is determined. Hence we might consider the problem as follows: These *minutiae* are not points, the ridges having a measurable thickness. Let us suppose a ridge-interval square to cover the area within which, if two such *minutiae* occurred in two prints under comparison, we should hold these *minutiae* to be identical in position. Galton's 6-ridge interval squares contain 36 little 1-ridge interval squares, and the chance of a given *minutiae* occurring in one of these is $\frac{1}{36}$, say $\frac{1}{2^5}$ roughly. Now Galton takes 24 such squares to a finger-print, and roughly there are 20 to 30 or even more *minutiae* in a print, say one to each 6-ridge interval square; then the probability that the *minutiae* will be placed each in its right compartment in its 6-ridge interval square is less than $\left(\frac{1}{2^5}\right)^{24}$, i.e. less than $\frac{1}{2^{120}}$. Actually it is considerably less than this because although the *minutiae* do not tend to cluster each one of them is not confined to its own 6-ridge interval square. Further all *minutiae* are not alike, e.g. ridge terminals. I think we may suppose a far more random, that is, less correlated, distribution of *minutiae*, than of parts of a pattern, and still conclude with Galton that it is very unlikely that two persons in the universe have the same print on any digit, as judged by its *minutiae*, still less on all ten digits.

Galton concludes this chapter characteristically as follows:

"We read of the dead body of Jezebel being devoured by the dogs of Jezreel, so that no man might say, 'This is Jezebel,' and that the dogs left only her skull, the palms of her hands, and the soles of her feet; but the palms of the hands and soles of the feet are the very remains by which a corpse might be most surely identified, if impressions of them during life were available." (p. 113.)

Chapter VIII (pp. 114–130) is entitled *Peculiarities of the Digits*. The data Galton uses in this chapter are the prints of the ten digits of 500 different persons. His objects are twofold: (i) to find the association of particular patterns with the individual digits, and (ii) to determine, if a particular digit has a given pattern, what is the chance that any other digit will have the same pattern. In discussion of these problems Galton uses only the triple classification arch, loop, whorl, and states that by including forked arches and nascent loops (see our Plate XI, p. 181) as arches, he has given a more liberal interpretation to the latter category in the tables of this chapter than he has done elsewhere. His fundamental table is the following:

Percentage Frequency of Arches, Loops and Whorls on the different Digits from Observations on 5000 Digits of 500 Persons.

Digit	Right Hand				Left Hand			
	Arch	Loop	Whorl	Total	Arch	Loop	Whorl	Total
Thumb ...	3	53	44	100	5	65	30	100
Fore Finger	17	53	30	100	17	55	28	100
Middle Finger	7	78	15	100	8	76	16	100
Ring Finger	2	53	45	100	3	66	31	100
Little Finger	1	86	13	100	2	90	8	100
Total ...	30	323	147	500	35	352	113	500
Percentage (Whole Hand)	6	65	29	100	7	70	23	100

From this table the following inferences may be drawn:

The patterns are not distributed indifferently either on the hands or on the individual digits. The right hand has a redundancy of whorls and the left of loops. The Fore Finger and to a lesser extent the Middle Finger have a redundancy of arches, the Little Finger and the Middle Finger a redundancy of loops, while the Thumb, Fore Finger and Ring Finger have the highest number of whorls. When we compare the corresponding digits of the two hands, we see little differentiation of pattern in Fore Finger, Middle Finger or Little Finger, but a more marked difference between the Thumbs and Ring Fingers of the two hands. While in the first group the percentages differ in the three fingers but are the same in the two hands, in the second group they are nearly the same in the two fingers but differ in the two hands (pp. 115–118).

Dealing with the slope of the loop Galton notes that the "inner" slope is much the more rare of the two for all the fingers but the forefingers, where the proportions of inner to outer slopes are about in the ratio of 2 to 3 (39°/$_0$ and 61°/$_0$)*.

The second problem, that of the resemblance of pattern in different digits, is divided by Galton into two sections, that of the resemblance in the same digits of the two hands, and that of the resemblance of different digits either in the same or different hands. He omits the little fingers because in 86°/$_0$ to 90°/$_0$ of cases both are loops.

---

* Purkenje appears to consider that while the inner slope is the more rare, it is actually in the forefingers in excess of the outer.

Percentage of Cases in which the same Class of Pattern occurs in the same Digits of the two Hands (500 Persons).

Couplets of Digits	Arches	Loops	Whorls	Total
Two Thumbs ...	2	48	24	74
Two Fore Fingers...	9	38	20	67
Two Middle Fingers	3	65	9	77
Two Ring Fingers	2	46	26	74
Mean of Total ...				72

This table as it stands is not very illuminating; take for example the middle fingers, and suppose there was no association of pattern between the same digits of the two hands. Then from the previous table the percentage probability of both being loops would be $100 \times \frac{78}{100} \times \frac{76}{100} = 59 \cdot 3\,^{\circ}/_{\circ}$. Similarly the percentage chances of both being arches and whorls are $0 \cdot 6\,^{\circ}/_{\circ}$ and $2 \cdot 4\,^{\circ}/_{\circ}$ respectively. Accordingly we must conclude that $62\,^{\circ}/_{\circ}$ of the observed $77\,^{\circ}/_{\circ}$ of coincidences would arise from mere chance, if the patterns were independent; it is the $15\,^{\circ}/_{\circ}$ balance which really marks the tendency to resemblance. Galton's second table (p. 120) is as follows:

Percentage of Cases in which the same Class of Pattern occurs in various Couplets of different Digits (500 Persons).

Couplets of Digits	Of Same Hands				Of Opposite Hands			
	Arches	Loops	Whorls	Total	Arches	Loops	Whorls	Total
Thumb and Fore Finger	2	35	16	53	2	33	15	50
Thumb and Middle Finger	1	48	9	58	1	47	8	56
Thumb and Ring Finger	1	40	20	61	1	38	18	57
Fore and Middle Fingers	5	48	12	65	5	46	11	62
Fore and Ring Fingers	2	35	17	54	2	35	17	54
Middle and Ring Fingers	2	50	13	65	2	50	12	64
Means of the Totals ...				59				57

The remarkable part of this table is that no marked change occurs in the percentage of resemblances whether the couplet of digits is from the same or opposite hands.

Of this result Galton writes:

"Though the unanimity of the results is wonderful, they are fairly arrived at, and leave no doubt that the relationship of any one particular digit, whether thumb, fore, middle, ring or little finger, to any other particular digit is the same, whether the two digits are on the same or opposite hands. It would be a most interesting subject of statistical inquiry to ascertain whether the distribution of malformations, or of the various forms of skin disease among the digits, corroborates this unexpected and remarkable result. I am sorry to have no means of undertaking it, being assured on good authority that no adequate collection of the necessary data has yet been published." (p. 122.)

Here again we have to remember that the amount of resemblance is not really measured by the numbers given; they might, as in the previous case, be merely the result of chance. Let us work out how much is due to chance in the case of the thumb and ring finger.

Percentage of Cases in which the same Class of Pattern occurs
in Thumb and Ring Finger.

How found	Of Same Hands				Of Opposite Hands			
	Arches·	Loops	Whorls	Total	Arches	Loops	Whorls	Total
Observed	1	40	20	61	1	38	18	57
Chance	0	36	15	51	0	35	13	48
Difference	1	4	5	10	1	3	5	9

The numbers remain very close, when we have deducted the resemblances due to chance, but perhaps do not look so impressive. Only about one-sixth of the resemblances in both cases can be attributed to the organic relationship.

Galton, on pp. 122–129, discusses a somewhat unusual method of determining the degree of association between the patterns on any two digits. To illustrate it let us take loops on the ring fingers of left and right hands. These occur in $66\,°/_°$ and $53\,°/_°$ of cases. Or, the chance of a loop is—if the results were independent—$100 \times \frac{66}{100} \times \frac{53}{100} = 35\,°/_°$ nearly. The maximum possible number of loops common to the two fingers is $53\,°/_°$ and the actually observed number is $46\,°/_°$ We have then the three numbers 35, 46 and 53. Galton takes the first as a zero relationship and the last as a perfect relationship, which is represented by him as $100°$. On the scale in which 35 represents $0°$ and 53, $100°$, we must have $46 = \frac{46-35}{53-35} 100° = \frac{11}{18} 100° = 61°$. He gives a table for these grades of association on p. 129 between digits of the same and of different hands. According to this table the highest relationship is between whorls on the middle and ring fingers ($74°$) and the lowest between loops on fore and ring fingers ($13°$). Galton is himself somewhat doubtful as to this method of measuring association, and I have not accordingly reproduced his full table (p. 129).

In Chapter IX (pp. 131–146) Galton deals with *Methods of Indexing*. It does not carry us far beyond the *Royal Society Proceedings* paper (see our pp. 170–174). In his main method Galton breaks up only the loops on the forefingers into "inner" and "outer."* He represents these by i and o. Thus five symbols are used: $a = $ arch, $l = $ loop, $w = $ whorl, and $i = $ inner loop on forefinger, $o = $ outer loop on forefinger. He breaks his ten-letter index into four groups†, i.e. R. hand, fore, middle and ring fingers; L. hand, fore, middle

* The reader must remember that the finger-print is *reversed*, and not be surprised at Galton labelling "inner" what appears to the reader, looking at his hand, as an "outer" slope.

† The reason for this has already been referred to (see our p. 184), namely, the greater variety in the types of forefinger prints.

and ring fingers; R. hand, thumb and little finger; and finally L. hand, thumb and little finger. Thus Galton's own index formula (see his prints on our p. 138) is *wlw, oll, wl, wl*. He indexes 100 individuals in this manner. On the basis of 500 sets of digits he gives the frequency per cent. of all index-headings which occur more often than $1°/_{o}$. The worst of these is *oll, oll, ll, ll*, which occurs in $4°/_{o}$ of occurrences. Thus, if we were dealing with 100,000 cases, we might have to search among 4000 individuals with this index-heading. The rapid fall in the number of entries having only a single individual is evidenced by the following returns which Galton gives on his p. 141:

Total Number of Entries ...	100	300	500
Percentage of Entries which are the sole members of their class	63·0	49·0	39·8

When we come therefore to indices which embrace 50,000 to 100,000 individuals, it will be seen that it may be needful to go through a large number of the cards on the *o, i, a, l, w* system of indexing before we identify a given individual. Thus even with the use of inner and outer loops on the forefingers, the great frequency of loops renders this system cumbersome for large finger-print collections. I do not think that Galton in 1892, although he suggests (p. 145) counting approximately the ridges, saw his way clearly out of this difficulty of loop redundancy. Possibly he did not fully realise the difference between his small collections and those of a national index of criminals.

In this chapter Galton describes the form of card he used for printing and his manner of storing such cards (p. 145).

Chapter X is entitled *Personal Identification*. This chapter contains much of general interest, which, however, we can only afford space to summarise briefly here. After referring to the ease with which any printer could take finger-impressions Galton again emphasises the suitability of the photographer for this work (see our p. 155), as not only can he easily enlarge prints, but he keeps an index to his negatives. Galton then passes to the many purposes for which identification is not only desirable but necessary. He cites some very interesting remarks (pp. 150–152) of Major Ferris, of the Indian Staff Corps, who, ignorant of Herschel's work, had found the same series of difficulties in identification and who had seen with much appreciation the finger-print method of identification at work in Galton's Laboratory— even as Sir E. R. Henry did later.

In the next place Galton gives on the whole a favourable account of bertillonage (pp. 154–158), questioning, however, the statements made as to the *independence* of the characters measured; Bertillon had asserted without demonstrating this *independence*. Galton shows from data of a similar kind drawn from his own Anthropometric Laboratory that such variables are *not* independent. Starting with five characters, head length, head breadth, span, sitting height, and middle-finger length, he shows that 167 out of 500 persons

24—2

measured fall into classes in which there are 7 to 24 repetitions*. But even the group of 24 individuals could be separated out by taking finer divisions of the head measurements than the three classes and introducing seven eye-colour classes. I think Galton was not unnaturally critical of bertillonage, because it started by theoretically asserting the independence of measurements which he knew to be *correlated*†; it did in fact overlook one of his greatest discoveries, the quantitative measurement of the correlation of bodily measurements. Nevertheless Galton is fair to the *results* of the system:

> "It would appear from these and other data, that a purely anthropometric classification, irrespective of bodily marks and photographs, would enable an expert to deal with registers of considerable size...it seems probable that with comparatively few exceptions, *at least* two thousand adults of the same sex might be individualised, merely by means of twelve careful measures, on the Bertillon system, making reasonable allowances for that small change of proportions that occurs after a lapse of a few years, and for inaccuracies of measurement. This estimate may be far below the truth, but more cannot be safely inferred from the above very limited experiment." (p. 163.)

It may be remarked that Bertillon does not appear to have made even such a limited experiment before he started his vast collection on the basis of his "independence" dogma!

Some account is then given of an American system of identification in the case of recruits and deserters. It seems to be based on height, age (how judged?), hair and eye colours for indexing purposes and then on a careful record of the body-marks placed on outline figures. Body-marks form of course an important factor of bertillonage (pp. 164–5). Galton remarks that no system he knows of appears to take account of the teeth. If teeth are absent when a man is first examined, they will be absent when he is examined a second time. He may have lost others in addition, but the fact of his having lost certain specified teeth prevents his being mistaken for a man who still possesses them (p. 166).

M. Herbette, speaking at the International Prison Congress in Rome, remarked of bertillonage:

> "In one word, to fix the human personality, to give to each human being an identity, an individuality which can be depended upon with certainty, lasting, unchangeable, always recognisable and easily adduced, this appears to be in the largest sense the aim of the new method."

Galton fitly remarks that these perspicacious words are even more applicable to the method of finger-prints than to that of anthropometry. Bertillonage can rarely supply more than grounds for very strong suspicion, finger-prints alone are amply sufficient to produce absolute conviction of identity.

* Number of Repetitions	7	8	9	10	11	14	19	24
Number of Individuals	28	8	18	20	22	28	19	24

† Some of the Bertillon measurements are indeed *highly* correlated. See Macdonell, *Biometrika*, Vol. I, pp 202, 212.

Chapter XI (pp. 170–191) discusses the subject of *Heredity* in finger-prints. This is a most difficult problem; it is not only that certain fingers favour certain classes of patterns, but that certain patterns classed in different broad groups are closely associated with each other. If we classify merely in arches, loops and whorls, we may find two kinsmen who have really kindred patterns, e.g. one having a plain arch and the other a nascent loop, classified as being as widely apart, as if the one had shown a tented arch and the other a twined loop. Again, supposing an extremely rare pattern occurs on the ring finger of one kinsman, and on the forefinger of the second, are we to dismiss this from our consideration of hereditary resemblance? It is almost inconceivable that a mere Arch-Loop-Whorl classification, especially if confined to a few fingers, can provide a true measure of inheritance although it may demonstrate that heredity is a factor of finger-print determination. Galton, in his first series of observations, confines himself to the fraternal relationship (boys and girls) of 105 pairs, dealing with right hand forefinger only and using the simple Arch-Loop-Whorl system. As we have remarked, this may show the existence of heredity, but it cannot really measure its intensity. He obtains the following table:

Observed Fraternal Couplets.

First Child

Second Child	Arch	Loop	Whorl	Total
Arch	5 (1·7) [10]	12	2	19
Loop	4	42 (37·6) [61]	15	61
Whorl	1	14	10 (6·2) [25]	25
Total	10	68	27	105

Galton then pays attention only to the numbers occurring in the diagonal column, i.e. identical prints in the fraternal couplet with the Arch-Loop-Whorl classification. The numbers in round brackets are what are to be randomly expected, the numbers in square brackets, the highest values attainable for resemblance, on the hypothesis of independence of the marginal totals. In every case the observed values lie between the random and the highest values and Galton takes this as evidence of heredity.

It will be seen that Galton is here aiming if rather ineffectually at some process like the modern method of contingency. If we apply now his method of centesimal grades we find for the degree of resemblance:

Arches: 39·8°; Loops: 18·8°; Whorls: 20·2°.

Of these the last two are probably equal within the error of random sampling. The first shows about double the relationship of the other two. I do not believe this is due to a greater intensity of the force of heredity in arches than loops, but solely to the fact that arches form a relatively rare and homogeneous group, while loops and whorls are conglomerates and the use of these terms tends to obscure finer resemblances. This peculiarity of the loops recurs in further investigations made by Galton with the aid of Howard Collins, and the former writes:

"I am unable to account for this curious behaviour of the loops, which can hardly be due to statistical accident, in the face of so much concurrent evidence." (p. 185.)

But I think the explanation lies in the fact that resemblance is lost when a very broad category such as "loops" is taken.

Galton, however, did see the difficulties of the Arch-Loop-Whorl classification, though not as far as I can judge of the limitation due to "corresponding finger." He accordingly prepared a set of 53 standard patterns, of which 46 are in pairs for "inner" and "outer," i.e. each pair is a mirror reversal. They are for the right hand, and the numbers of each pair of the last 46 must be reversed when we deal with the left hand. He calls this the "*C*-set of Standard Patterns," as Mr Howard Collins performed most of the tabulation under the *C*-set of patterns. The data consisted of right fore, middle and ring fingers in 150 couplets of siblings*, 900 digits in all. Unluckily the "*C*-set of Standard Patterns" is in one, the most important, respect almost as defective as the Arch-Loop-Whorl classification. While in the former treatment 129 out of 210 finger-prints fell into the loop category here 291 out of 900 finger-prints fall under the pattern No. 42, which is practically the simple loop; it is clear that this standard set of 53 patterns has failed to meet the inherent difficulty of breaking up the bulk of the loops.

Our author proceeds in the same way to deal only with complete re semblances, i.e. he deals only with the diagonal of his contingency table, disregarding the possibility that a deficiency below the random value may be as important in measuring association as an excess above that value. Comparing in this way random values, observed values and maximum possible values, and applying his method of the centesimal scale, Galton obtains the following results:

Resemblance of Siblings, 150 Couplets: forefinger, 9°; middle finger, 10°; ring finger, 12°. We have no probable error given for this method of computing association, but it may be to some extent estimated by the fact that an additional 50 couplets, worked out for middle finger only, gave a value of 21°. For loops on the middle finger only, the 150 couplets gave 1·25°, and the 50 couplets 8°, indicating little if any association. In nearly all cases the random values were below the observed; in the few cases where they are not so they were only slightly in excess. I think there is enough to show that fraternal resemblance exists, but I personally hold that the classification is rather inadequate, and the statistical method of reduction is unsound.

Galton next turns (p. 185) to the degree of resemblance in twins. Here he has two series, each of 17 sets of twins for the fore, middle and ring fingers of the right hand. In the first series 19 of the 51 finger-print pairs gave the same pattern for the same fingers of both twins, 13 gave partial resemblance and 19 disagreement. Or, as he puts it, of 17 sets of three fingers, two

* Pairs of children with the same parents without regard to sex.

PLATE XVIII

LEFT HAND

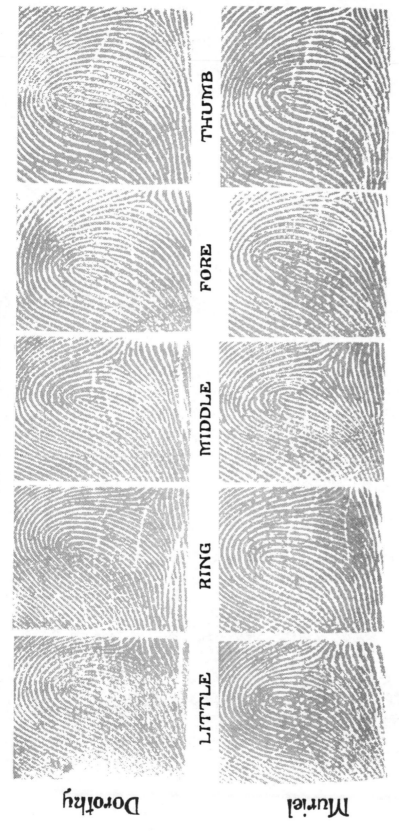

THUMB FORE MIDDLE RING LITTLE

Dorothy

Muriel

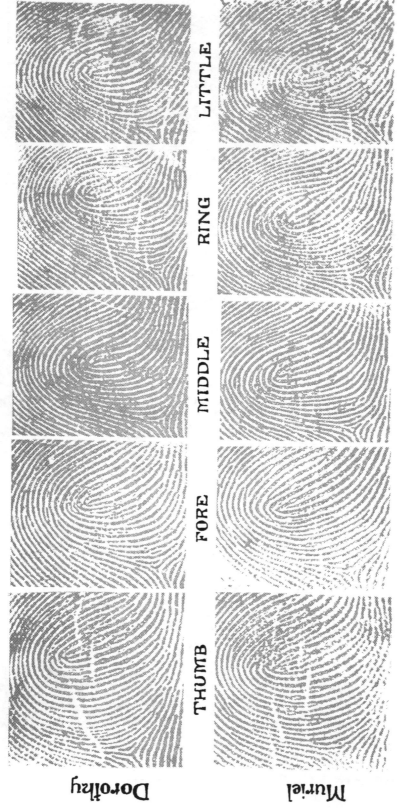

RIGHT HAND

THUMB   FORE   MIDDLE   RING   LITTLE

Dorothy

Muriel

FINGERPRINTS OF LIKE TWINS

From the Collection in the *Galtoniana.*

sets agreed in all their three couplets of fingers; four sets in two, and five sets in one of their couplets. There are instances of partial agreement in five others, and only complete disagreement in one. Of the second series of 17 twins Galton contents himself by saying that two sets agreed in two of their couplets and five agreed in one, without giving details. He concludes that ·

"there cannot be the slightest doubt as to the strong tendency to resemblance in the finger patterns of twins." (p. 186.)

Unfortunately Galton gives no measure of the probability of the random occurrence of similar resemblances, and we are unable to compare what is the relative degree of resemblance of twins and ordinary siblings.

Perhaps the best appreciation the reader can rapidly form of the degree of resemblance in the finger-prints of like twins can be obtained by carefully examining our Plate XVIII which gives the finger-prints of a pair of like twins from the *Galtoniana*.

The last problem Galton touches on is that of parental heredity. Here he has only 27 pairs of parents, whom he chooses because on one of the three fingers, fore, middle, or ring, they have the same pattern. He has 4 cases of the forefinger, 14 of the middle finger and 9 of the ring finger. These 27 pairs of parents have 44 sons and 65 daughters; 22 out of the 44 sons, 37 out of the 65 daughters have the same pattern on the same finger as their parents. In 19 cases out of the 27 both parents had loops of type No 42, and in 48 cases out of their 75 children there was also a loop on the same finger; that is to say, in about 64 $°/_o$ of cases, while the normal percentage is about 33 $°/_o$. Thus, according to Galton's method, the resemblance is about 48°. This seems to show a much greater value for filial resemblance in looping than had been found for fraternal resemblance. Yet in analysing these parental sets, Galton is rather apt to desert the method he adopted for fraternal resemblances, namely, of terming two points like or unlike according as they are of the same or not the same pattern in his *C*-set of 53 patterns. Thus he has 3 parental sets with No. 14 tendrilled loops; they have 17 children of whom only 3 have No. 14 pattern; he says, however, that No. 14 counts as a whorl, and that the 17 have 11 whorls and only 6 loops. Few, however, of the remaining 8 whorls bear close resemblance to No. 14. Galton gives no general measurement of parental heredity.

This raises, indeed, the broad question whether it is really the pattern which is inherited, or merely a tendency to arch, to loop, or to whorl without regard to the individual character of the pattern. Galton remarks (p. 187) that the finger-prints of twins while tending to be of the same pattern, cannot be mistaken one for the other; in other words, the number of ridges and the *minutiae* differ*. Thence he leads us to a very fertile suggestion, which neither he nor anyone else later, so far as I know, has ever worked out:

"It may be mentioned that I have an inquiry in view, which has not yet been fairly begun, owing to the want of sufficient data, namely to determine the minutest biological unit that may be hereditarily transmissible. The *minutiae* in the finger-prints of twins seem suitable objects for the purpose." (p. 187.)

* Our Plate XVIII suggests that Galton in this statement has somewhat over-emphasised the divergence between the finger-prints of twins.

The last section of this chapter is entitled the *Relative Influence of the Father and the Mother*. The fore, middle, and ring fingers of the right hand of the father and mother of 136 sons and 219 daughters were tabled under the 53 standard patterns, and I present Galton and Collins' results in the form of percentages of likenesses found in the case of the three fingers. It will be seen that for the fore and ring fingers there is no difference.

Percentages of Same Finger-prints in Parents and Offspring on the basis of 136 Sons and 219 Daughters.

Relationship	Fore Finger	Middle Finger	Ring Finger	Total Percentage of Sameness	
Father and Son ...	12·5 °/₀	25·7 °/₀	20·6 °/₀	58·8 °/₀ }	54·7 °/₀
Father and Daughter	13·2 °/₀	23·5 °/₀	14·0 °/₀	50·7 °/₀	
Mother and Son ...	13·2 °/₀	36·8 °/₀	19·1 °/₀	69·1 °/₀ }	68·4 °/₀
Mother and Daughter	17·4 °/₀	34·3 °/₀	16·0 °/₀	67·7 °/₀	

I think it may be safely inferred from these percentages:

(i) that the Son has no greater degree of resemblance to the Father than the Daughter has;

(ii) that the Son has no greater degree of resemblance to the Mother than the Daughter has;

(iii) that there is no sensible degree of difference between the resemblances of Father and Mother to their offspring in the fore and ring fingers;

(iv) that there does appear to be a difference in the middle finger, and this alone causes the Mother's total of resemblances to be greater than the Father's.

Are we to assert as a result of these conclusions (*a*) that the heredity factor has greater influence in the case of the middle finger, and (*b*) that the mother has more influence than the father on the finger-prints of the offspring?

Galton does not pledge himself to (*b*), but merely throws it out as a suggestion. We must, however, note that the resemblances here given include not only the hereditary but the organic factor, and the values of the percentages given if they were corrected for random agreement might show very different results. The middle finger has a far higher percentage of loops (see the table on our p. 184) than the fore or ring fingers, hence there will be a far larger number of random coincidences to be corrected for. Until that is done we cannot accept (*a*) as true on the basis of the above table. Further, Galton has not given the digital distribution of patterns for the *two* sexes, and if these be not the same we cannot straightaway assume that (*b*) holds, or indeed that either parent has the like influence on son and daughter.

I have discussed this chapter at length, primarily because Galton was undoubtedly the first to take up the subject of the inheritance of finger-print patterns, and it is desirable that later workers should see how he approached the problem, and so try to avoid the difficulties he encountered. Our statistical tools are better now than such tools were in 1892, but still the problem remains of transcendent difficulty. Secondly, I have done so because Galton provides as usual many suggestions for further inquiry. Here as elsewhere we come across the urgent problem of a standard set of patterns, which will subdivide plain loops into small approximately equal subclasses. Galton's set of 53 standard patterns provides at once too many and too few. There is no great advantage gained by dividing whorls into "inner" and "outer," and the division of loops into "inner" and "outer" is not division enough.

Chapter XII (pp. 192–197) deals with *Races and Classes*. Galton obtained finger-print series for the English, Pure Welsh, Hebrew, Negro and Basque races. These were dealt with in a variety of ways and he concluded that there was no *peculiar* pattern which characterises persons of the above races. Many tabulations to discover racial differentiations appear to have been made without any great success. As an illustration Galton gives the following table:

Percentages of Arches in the Right Forefinger.

Number of Persons	Race	Percentage
250	English	13·6
250	Welsh	10·8
1332	Hebrew	7·9
250	Negro	11·3

Galton considers that there may be a significant difference between the percentages of arches in the English and Hebrew races. Now the probable error of his percentage value for English is 1·5 with a slightly greater value for the Welsh and Negro. Accordingly we see that the three series of 250 are too small to show significant differences if they really exist between these three races. The difference between Hebrew and English is 3 to 4 times its probable error and may be significant. The point needs further inquiry on longer series. Although no statistical differentiation of the Negro was found, Galton remarks:

"Still, whether it be from pure fancy on my part, or from some real peculiarity, the general aspect of the Negro print strikes me as characteristic. The width of the ridges seems more uniform, their intervals more regular, and their courses more parallel than with us. In short, they give an idea of greater simplicity, due to causes that I have not yet succeeded in submitting to the test of measurement." (p. 196.)

Galton considers that this matter should be pursued further, especially "among the Hill tribes of India, Australian blacks and other diverse and so-called aboriginal races." I would venture to add the amplest study of the

Oriental Races, Japanese, Chinese, Aino and Tibetans, whose anthropological characters are so distinctive*. Further, an investigation should be made of the finger-prints of prehistoric man, especially of palaeolithic man in the caves†. Nay, we may go back further and ask what are the finger-prints of *Tarsius*, to whom some anatomists, at any rate in the matter of the hand, believe man to be more closely linked than to the anthropoids. The ancestry of man might possibly be illuminated by still further study of the primates' finger-prints. It is almost impossible to believe that the *Urmensch* had all men's present finger-print patterns scattered in a roughly promiscuous way over his digits! If he had, then it forms a huge stumbling-block in the evolution of man from a primate form.

Galton concludes his chapter by stating that he has studied the finger-prints of men of much culture and of scientific achievement, of labourers and artists and of the worst idiots.

"I have prints of eminent thinkers, and of eminent statesmen that can be matched by those of congenital idiots. No indications of temperament, character, or ability can be found in finger marks, so far as I have been able to discover." (p. 197.)

Chapter XIII (pp. 198–212), the final chapter, is entitled *Genera*, and as it is substantially a reproduction of the matter on this topic in the *Philosophical Transactions* (see our pp. 167–169), it seems unnecessary to analyse its contents or repeat the criticisms already made on it by the present writer.

Taking Galton's work as a whole we have to remember that it is the first treatise on finger-printing and none has been published since. That it is full of novel matter and teems with suggestions. That from the time of Purkenje (1823) to Alix (1868) there had been no scientific contribution to the subject, nor anything published which could provide Galton with material for study, until his own Royal Society memoirs were issued. The whole of the scientific treatment of finger-prints and the art of identification by means of them, now spread over the civilised world, arose from Galton's labours, especially those in this book. If anyone doubts this let him point to a single scientific memoir on identification by finger-prints which antedates Galton's publications, or his campaign for finger-printing as an expert art. No one can realise how insignificant were the results before Galton, who has not read his *Finger Prints*.

Decipherment of Blurred Finger Prints, 1893. In the following year Galton issued a booklet of the above title, with the subtitle *Supplementary Chapter to "Finger Prints."* Slender as is this volume (18 pp.), the important part of which consists of sixteen plates, it is again a pioneer work. It shows for the first time in numerous instances how evidence should be prepared which might convince a jury of the identity of two finger-prints, even if one or both those prints are badly impressed, or, as Galton puts it, "blurred."

* I have already indicated why I do not think the researches of Kubo or Collins conclusive as to racial differences. See the footnote p. 140 above.

† See E. Stockis, "Le dessin papillaire digital dans l'art préhistorique," *Revue Anthropologique*, année 30, 1920, p. xliii *et seq.*

" The registration of finger-prints of criminals, as a means of future identification, has been thought by some to be of questionable value on two grounds—first, that ordinary officials would fail to take them with sufficient sharpness to be of use; secondly, that no jury would convict on finger-print evidence. These objections deserve discussion, and would perhaps by themselves have justified a supplementary chapter to my book. It happens, however, that there are strong concurrent reasons for writing it. I have lately come into possession of the impressions of the fore and middle fingers of the right hand of eight different persons made by ordinary officials, in the first instance in the year 1878 and secondly in 1892. They not only supply a text for discussing both of the above objections, but they also afford new evidence of the persistence of the *minutiae*, that is of the forks, islands and enclosures, found in the capillary ridges." (p. 1.)

The reader will remember (see our p. 176) that Sir W. J. Herschel in 1877 had taken finger-prints for the registration of deeds at Hooghly. Galton in his *Finger Prints* (p. 89) had suggested that it might be well worth while to hunt up such of these Hindoos as were still alive and retake their finger-prints. Through the mediation of Sir William it was possible to obtain from the magistrate and sub-registrar of Hooghly not only fresh prints of the fore and middle finger-prints of eight persons, who had impressed their finger-prints in the Register of Deeds of 1878, but also these earlier prints themselves. In all cases the range of interval was about 14 years, so that Galton got evidence of persistence roughly between the following ages: I, 51 to 65; II, 50 to 64; III, 38 to 52; IV, 28 to 42; V, 48 to 62; VI, 38 to 52; VII, 40 to 54; VIII, 32 to 46 (p. 4). But his task was not an easy one; not only were the paper* and the inking on both earlier and later prints very defective, but the prints were not *rolled* prints and in a number of cases only a portion of the bulb had been impressed. Thus some of the *minutiae* were lost on each separate print and this in itself caused a double loss on comparison. Galton contented himself with a full discussion of eight out of the sixteen finger-prints and found the following results:

Personal Number	Finger	Number of Agreements	Number of Disagreements	Patterns
I	Fore	9	0	Loop
II	Middle	5	0	Loop
III	Middle	21	0	Whorl
IV	Fore	19	0	Whorl
V	Fore	7	0	Loop
VI	Fore	19	0	Loop
VII	Middle	15	0	Loop
VIII	Fore	30	0	Whorl
Average	—	15·6	0	—

Galton discusses each finger in detail (pp. 11–15), commenting on various peculiarities and difficulties. He remarks that his evidence for correspondence

* They were on a common kind of native-made paper, worm eaten, with many holes Several of the Hindoos were old for their race and showed signs of much manual labour wearing down the sharpness of the ridges.

is drawn from the *minutiae* and not from the general pattern; for though no one can mistake a decided whorl for a decided loop, lesser differences are often deceptive to the untrained eye, especially when only a portion of the pattern has been impressed.

But the chief interest of Galton's present work lies not in the identification of poor impressions at fourteen years' interval by aid of their *minutiae* but in his manner of presenting the evidence. His aim is to show that rough impressions such as may be taken by ordinary officials (or left behind by the burglar) can be made to afford evidence strong enough to convince a jury that two finger-prints had been made by the same person. "It is of course supposed that the cogency of the finger-print argument will be presented to the jury in that lucid and complete form in which it is the business of barristers to state and support their case, when they are satisfied of the integrity of the evidence on which it is based" (p. 2). Galton's method is best grasped from his plates rather than from a verbal description. He first enlarges his prints 2½ times photographically. The enlargements, eight to the page, occupy Plates I–IV. These give him a general impression of the patterns, and the particular cases and the *parts of the particular cases* he considers it desirable to study further. These selected parts of particular cases are now photographically enlarged to seven times natural size. These enlargements occupy Plates V–VIII, and are printed in black. Thus far the work, except for the choice of parts, has been largely mechanical. Now comes the labour of the expert: the outlining of the ridges on these blurred prints. In doing this tracing paper may be used by the draughtsman, but Galton thinks a better plan is to do the outlining on the back of the print placed against a pane of the window or on a photographic retouching frame.

"The axes should be drawn with a finely-pointed pencil, and with care, down the middle of the ridges. Slap-dash attempts are almost sure to be failures. It is advisable to take pains to determine a common starting point, before proceeding to draw any lines at all; then to proceed from point to point in the two prints alternately, at first with wariness but afterwards much more freely....The continuous course of every line has to be made out from beginning to end, and the lines must nowhere be too crowded or too wide apart, and they must all flow in easy and appropriate curves; also as much regard must be paid to such blanks as are not obviously due to bad printing as to the markings. The general effect of these conditions is that a mistake in deciphering any one part of the impression nearly always introduces confusion at some other part, where the lines refuse to fit in." (pp. 10–11.)

On Plates IX–XII Galton gives his outlinings, the blurred ridges being now printed in orange with the outlining in black, still on a sevenfold scale. Tiny circles mark the ends or beginnings of ridges, but as Galton warns his readers some of these may well be forks (see his p. 8 and our pp. 165, 181).

Lastly Galton provides on the same sevenfold scale the outlinings of the ridges without the blurred ridges at all. Here in the juxtaposed prints corresponding *minutiae* are given the same small numbers, so that it is perfectly easy to refer to one after another of the correspondences. The whole series of plates forms a singularly lucid illustration of what it is possible to do even with badly printed and partial impressions. No reasonably thoughtful counsel ought with such evidence to fail to convince a jury that Dwārikā

PLATE XIX

Illustrations of Galton's Treatment of Blurred Finger-Prints (Data obtained through Sir William J. Herschel).

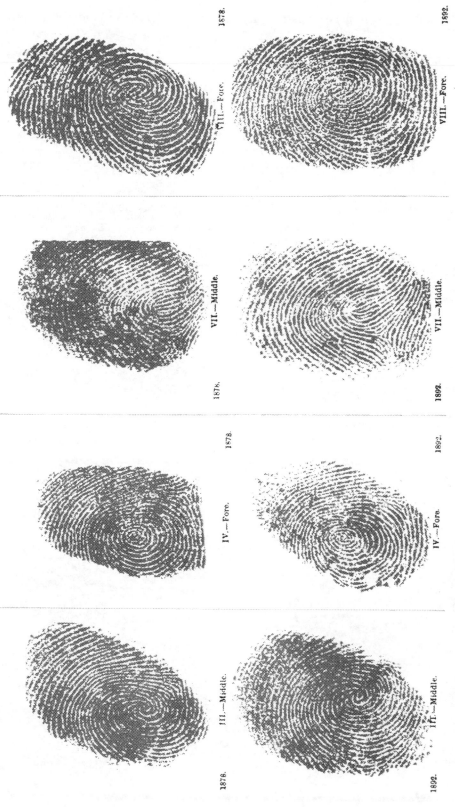

1878.

1892.

VIII.—Fore.

VIII.—Fore.

VII.—Middle.

VII.—Middle.

1878.

1892.

1878.

1892.

IV.—Fore.

IV.—Fore.

III.—Middle.

III.—Middle.

1878.

1892.

Fore and Middle Fingers of Persons at Hooghly, Bengal, taken first in 1878 and afterwards in 1892. From Plates II, III and IV of Galton's *Decipherment of Blurred Finger Prints*, 1893. Non-rolled Prints, enlarged 2½ times.

PLATE XX

Selected Corresponding Portions of the Hooghly Doublets (1878 and 1892) from **Plate XIX.**

Enlarged seven times preparatory to drawing central lines of ridges.

PLATE XXI

Skeleton Charts of the Central Lines of the Ridges of the Hooghly Doublets 1878 and 1892,
drawn by aid of tracing paper from the prints on Plate XX.

Corresponding numbers in upper and lower prints indicate persistence of *minutiae*.

PLATE XXII

Superposition f Central Lines of Ridges on enlarged Finger-Prints, i.e. Plate **XXI overprinted** on Plate XX, reproduced in fainter ink.

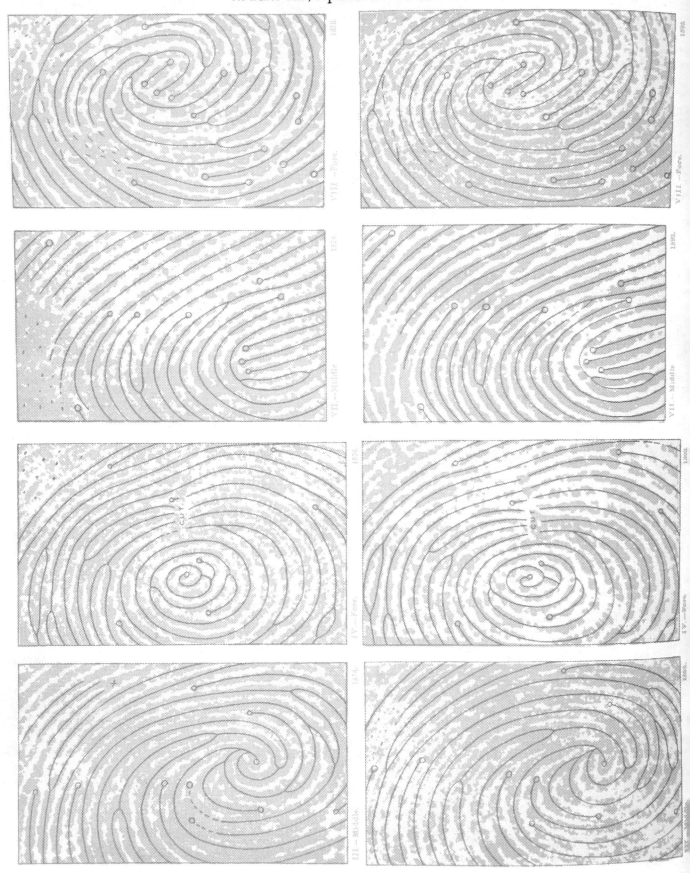

Nath Banerji, who had impressed his fingers in 1892 afresh, was the same man who had impressed them on Deed No. 28 in 1878!

We reproduce Galton's:

Plate II, Plate III left-hand side, Plate IV left-hand side (see our Plate XIX);

Plate VI and Plate VIII (see our Plate XX);

Plate X and Plate XII (see our Plate XXI);

Plate XIV and Plate XVI (see our Plate XXII).

These plates form the best—a graphical—illustration of Galton's methods.

On pp. 17–18, we have some useful suggestions as to enlarging finger-prints, but such work is now much more generally understood and accurately done than in 1892. Galton's two enlarging cameras are in the possession of the Galton Laboratory (see our p. 215). Our Author concludes with the following remarks:

"Photographic enlargements save a great deal of petty trouble. It is far easier to deal exhaustively with them than it is with actual impressions viewed under a magnifying glass. In the latter case, a few marked correspondences, or the reverse, can readily be picked out, and perhaps noted by the prick of a fine needle, the point of a pin being much too coarse. It is thus easy to make out whether a suspicious print deserves the trouble of photographic enlargement, but without previous enlargement a *thorough* comparison between two prints is difficult even to an expert, and no average juryman could be expected to make it." (p. 18.)

The Second Attempt at Indexing Finger-Prints. Galton provided another Finger-Print Index to 100 persons in July 1894. It is entitled "PHYSICAL INDEX to 100 persons on their measures and finger-prints (*set up in two parts as an experiment*)." Here the two parts consist: first, of an index based primarily on five measurements as in bertillonage, and secondarily on finger-prints; and again of an index based primarily on finger-prints, and secondarily on the five measurements. I cannot find that this index was ever published although it appears to have been printed, stereotyped and circulated among Galton's friends and correspondents. It possesses in arrangement greater brevity than that of the *Finger Print Directories* of the following year, and yet gives more information since the anthropometric measurements and certain other data are included. The whole space occupied by any entry is 36 × 17 mm., and Galton considers that, if the entries were cut up and pasted on to cards,

"a cabinet of 27 broad and shallow drawers measuring, over all, less than 12 inches in height and 4½ feet in width, would contain more than 100,000 of these small cards arranged as a catalogue."

Each entry or label consists of four lines (see table on p. 198). In the first line on the left is the anthropometric formula, on the right the finger-print formula. These are the bases on which the indices of Part I and Part II respectively are formed, the entries being made in order of letters and numbers in the formulae taken in consecutive order.

The second line gives the five anthropometric measurements in the order from left to right of (i) head length, (ii) head breadth, (iii) extreme breadth between cheek bones, (iv) length of left cubit, (v) length of left middle finger. To obtain the anthropometric formula, these are divided into a, l, w, which signify short, medium, long. The medium limits are for (i) 191 to 196,

(ii) 150 to 156, (iii) 129 to 136, (iv) 450 to 464, (v) 113 to 116, all inclusive. The danger of the anthropometric formula will arise when we have one or more measurements in the neighbourhood of these limits. Galton uses the five-symbol classification for his finger-print formula, namely A = arch, L = loop, W = whorl, U and R being used for ulnar and radial loops on forefingers only. He adopts the numerical abbreviations of his later work, i.e.

$$1 = aa \text{ or } AA, \qquad 2 = al \text{ or } AL, \qquad 3 = aw \text{ or } AW,$$
$$4 = la \text{ or } LA, \qquad 5 = ll \text{ or } LL, \qquad 6 = lw \text{ or } LW,$$
$$7 = wa \text{ or } WA, \qquad 8 = wl \text{ or } WL, \qquad 9 = ww \text{ or } WW.$$

The third line is the secondary classification of the finger-prints, but he takes only the following six fingers in the order: fore, middle and ring fingers of the right hand, and then fore, middle and ring fingers of the left hand. In the secondary classification the symbols Galton uses are those of his *Finger Print Directories* with two additions, i.e. b = partially burnt by fire or chemicals, or so spoilt by work as to leave granulations in place of ridges, and m = the pattern is minute, so small that two specimens of the characteristic portion would occupy less space than that covered by a single dabbed print. As there is no secondary classification for thumb or little finger, the description is not so full as in the later work. Ridges are counted in the same manner as we describe on pp. 201–2; and are given for the forefingers when they are loops, and for the middle finger when it is needful to distinguish between individuals having the same primary classifications. The fourth line gives the initials of the subject, the year of birth, the year of measurement, and the registered number of the subject, so that his finger-prints may be found. The following individual cases will illustrate the compactness of the arrangement and explain its interpretation:

58a				*A*6 *R*5, 88	89w				*U*5 *R*5, 55	47a				*W*6 *W*6, 88
196	153	138	454	111	203	151	137	486	121	195	148	140	446	111
y *v* *y*	‖	—	*k*	—	6 16	—	‖	2a	— *y*	*o* *v*	—	‖	*o*	*wv* —
G. K. 1862–94				6590	C. J. E. 1870–94				6547	G. A. 1839–94				6578

They are taken from the finger-print index. $58a = llwla$; thus G. K., whose finger-prints were registered as No. 6590 and who was born in 1862 and measured when he was 32 years old, was in the medium classes for length and breadth of head and for left cubit; he was wide in bizygomatic or cheek bone width and had a short left middle finger; the second line gives his actual measurements. His finger-print formula was ALW, RLL, WL, WL. Both his thumbs were whorls, and his little fingers loops; no further information is given. His right forefinger was an arch, there being a needle or racquet-shaped ridge therein; his right middle finger was a loop invaded by a blunt system of ridges; his right ring finger was a whorl with a racquet-shaped core. On the left hand the forefinger was a radial loop, and the ring finger a non-radial loop, the middle finger was a non-radial loop with the inner part of the pattern more or less hooked. C. J. E. was born in 1870 and measured when 24 years old. His finger-prints will be found under register number 6547. His anthropometric formula is $89w = wlwww$, or he is of medium head breadth, but large in all his other measurements. His finger-print formula is

$$U5\ R5,\ 55 = ULL,\ RLL,\ LL,\ LL,$$

or he belongs to the class of which all the ten prints are loops. We are only told that the right forefinger has an ulnar and the left a radial loop. The number of ridges on the right forefinger is 6, and on the right middle finger 16. The left forefinger with its radial loop has only two ridges and might also be called an arch (a); the left ring finger loop has a racquet-shaped core.

Finally G. A., born in 1839 and measured at 55 years of age, has for register number 6578. His anthropometric formula is $47a = lawaa$, or he is small in head breadth, left cubit and left middle finger, medium in head length and large in facial breadth (bizygomatic). His finger-print formula is

$$W6\ W6,\ 88 = WLW,\ WLW,\ WL,\ WL.$$

Thus his thumbs are whorls and both his little fingers loops; both his fore-fingers are whorls with well-defined rings round the core; his right middle finger is a loop invaded by a blunt system of ridges and the same is true for the left middle finger, the print of which might, however, be mistaken for a whorl; there is no characterisation for either ring finger beyond the statement that both are whorls.

It is clear that Galton was at this date feeling his way up to a more complete secondary classification. Dropping the anthropometric data—although be it remembered they are useful when the police need to give the public some rough particulars of a criminal—there is ample space for a full 10-digit print formula in the first line, which would get much more differentiation into the uncharacterised L's and W's. Something of this was introduced by Galton into his *Finger Print Directories* of the following year, and we shall see that it can be easily extended. We note that for the all-loops formulae he introduces ridge counting on fore and middle fingers, and this was the method adopted by Henry from Galton, although he then proceeded for ridge frequency to follow Bertillon in using only broad categories. Galton admits that this index was only experimental, but its arrangement is suggestive especially in the cases where anthropometric measurements are also desirable. It has the advantage that as the frequency under any formula increases, it is always feasible to add more detailed secondary classification in the third line. For example, it would be at once feasible in the last illustration to break up the six whorls into those fed radially, ulnarly or from both sides, and again into right-handed and left-handed screw classes.

The Final Work on Indexing Finger-Prints. Galton's third volume on the subject of finger prints appeared in 1895; it is entitled *Finger Print Directories*, and is gracefully dedicated to Sir William J. Herschel*. The main purpose

---

* " I do myself the pleasure of dedicating this book to you, in recognition of your initiative in employing finger-prints in official signatures, nearly forty years ago, and in grateful remembrance of the invaluable help you freely gave me when I began to study them." Here, as elsewhere, Galton very fully acknowledges his indebtedness to Herschel's aid.

of the book is to provide a method of indexing 200,000 to 300,000 individuals. Galton assumes that five anthropometric characters will each be divided into three classes as in bertillonage, and accordingly, if this provides for $3^5 = 243$ classes, we need only to secure some method of finger-print indexing which will leave very few multiple entries in 1000 cases. This is the problem Galton sets himself; it will be seen that in 1895 he still thought it desirable to use a small dose of bertillonage to aid his index, if it was to provide rapid references to more than 1000 to 3000 individuals.

Galton here starts from the old Arch-Loop-Whorl classification with the addition of the inner and outer slope of loops on the forefinger, only now, I think unfortunately, he changes many of his symbols and some of his previous terminology. Having preferred in his earlier works "inner" for the thumb side and "outer" for the little finger side, he now adopts radial and ulnar formerly rejected; thus the symbols i and o are replaced by r and u. He still works in this index with the 10 digits arranged thus*: Right, fore, middle, ring fingers; Left, ditto. Right, thumb, little finger; Left, ditto—which in his old treatment gave 10 letters. He reduces them, however, to eight, by noting that a, l, w can only occur pair by pair in nine ways, and he gives the first nine figures to these, so that it is possible to represent thumb and little finger prints by a single figure. Thus far it is difficult to see that much has been gained on his earlier classification. Indeed with slight changes of notation Galton's present *Primary Classification* is his old a, l (i, o), w system. Now the defects of this as the sole classification are well exhibited in the following table which he gives (p. 77):

Formulae with Frequencies 10 and over in 1000 Tests.

Order of Frequency	Formula	Frequency of Occurrence
1	*ull, ull, ll, ll*	59
2	*rll, rll, ll, ll*	35
3	*ull, rll, ll, ll*	24
4	*www, www, ww, ww*	19
5	*rll, ull, ll, ll*	17
6	*ulw, ull, ll, ll*	14
7	*ulw, ulw, ll, ll*	12
8	*www, www, wl, wl*	11
9	*rll, ull, ll, ll*	10
10	*ull, ull, ll, ll*	10
		Total 211

* He states, however (pp. 72 and 111), that he has modified this view for the purpose of indexing and now prefers to take his finger-prints in order from left little finger to right little finger. There is little doubt that the latter, the "natural" order, and also the one in which the impressions are collectively dabbed and usually rolled, is less liable to errors of reading. At the same time as it starts with the little finger it gives far less variety to the *initial* letters of the index.

In other words, between a fifth and a quarter of the sets fall into groups which are far too unwieldy for rapid index searching. It is clear that the loops and whorls are the chief source of this trouble (see our pp. 149, 165 and 173) and Galton proceeds to break them up by what he terms a *Secondary Classification*, or a system of adding subscripts to the letters of his primary classification. The subscripts or suffixes as Galton calls them are very numerous, although some can only be attached to certain patterns. For example, what would have appeared in his old (his present primary) classification as

$$oww, \; oll, \; ww, \; ll,$$

now becomes

$$uw_yw, \; ul \dagger l_{vy}, \; w_{lvy}w, \; ll_v,$$

where subscript y means that the core of the corresponding whorl is pear- or racquet-shaped; $\dagger$ denotes that there was a scar on the middle finger of the left hand; l_{vy} denotes a loop with invasion of ridges from the side and with a racquet core; w_{lvy} means a whorl which might be mistaken for a loop, has an invasion of ridges from the side and a racquet core, and l_v denotes a loop with a like invasion only. Thus 18 symbols are used to index the set. Galton defines and discusses 28 letters and symbols which may be used as suffixes. Obviously the above system of subscripts is one liable to error either in writing or printing, and Galton, although he suggests its use, does not actually adopt it in the *Directory* he publishes of 300 sets of prints of the 10 digits. Here he gives the primary classification symbols on the left of his page, and then on the right in 10 columns the suffixes to be attached to each of these symbols. For example, the above formula appears as

$$| \, Uww \, | \, ull \, | \, 9, 5 \, \| - , \, y, \, - \, | - , \, \dagger, \, vy \, | \, lvy, \, - \, | - , \, v \, |,$$

where the last 10 columns correspond to the digits in order of the primary formula ($9 = ww$, $5 = ll$, the thumb and little finger formulae of right and left hands: see our p. 198).

Besides the 28 symbols which are chiefly devoted to breaking up the large loop and whorl groups, Galton introduces for the troublesome all-loops group the counting of the ridges on the forefingers. This counting he now does in a different manner from that of his earlier papers, and one which seems less liable to misinterpretation. He first determines a better line for counting the ridges on (see his pp. 78–80) than he had previously selected (see our pp. 163 and 165). The following are his rules (see Fig. 34, p. 202):

"The terminus from which the count begins is reckoned as 0; it proceeds thence up to, and *including*, the other terminus.

"The inner terminus lies at the top of the core of the loop, the outer terminus at the delta, but it is necessary to define their positions more exactly, as follows:

"*Inner terminus.* There are two cases:

"(a) The core of the loop may consist of an uneven number of ridges, as in each of the two figures, a¹ and a²; then the top of the central ridge is the inner terminus*.

---

* I think there is a risk of confusion here to which Galton does not refer. The ridge or ridges within the "staple" may or may not meet the latter. In Figs. a¹ and a² the inner ridges are made to meet the staple, and the inner terminus is not put at the *top* of the

"(*b*) The core may be a circumflex or 'staple'; then, the shoulder* of the staple that is farthest from the delta is taken for the inner terminus, the nearer shoulder counting as a separate ridge (Fig. b).

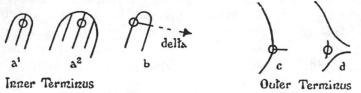

Inner Terminus   Outer Terminus

Fig. 34. Inner and Outer Termini for Ridge Counting.

"*Outer terminus.* Here also are two cases

"(*c*) Where the upper and lower sides of the delta are formed by the bifurcation of a single ridge. Here the point of bifurcation forms the outer terminus. It sometimes happens that successive forks or branches are thrown off from the same ridge first at an acute angle and progressively becoming more obtuse. In this case the branch to be considered as forming one side of the delta is the first that makes not less than a right angle with the stem (Fig. c).

"(*d*) Where the upper and lower sides of the delta are formed by two ridges that had previously run side by side, and then suddenly diverge. Here the base of the delta is the outer terminus. The nearest ridge in front of the place where the divergence begins, even if it be a mere dot, and whether or no it is independent of, or springs from one of the divergent ridges, is considered to form the base of the delta, and the outer terminus.

"If scrupulous care is taken by the beginner, first in selecting the termini that best fulfil the above conditions, and afterwards in counting the ridges, his eye will soon become accustomed to the work, and the process may then be effected both quickly and trustworthily†. It is usually easy to determine narrow limits within which the number of ridges will always be held to lie."

Galton tells us that the 156 (*ull, ull, ll, ll*)'s of his collection of 2632 sets showed, counting as above, all numbers of ridges from 3 to 16 with fairly equal frequency. He had also a few "under 3" and eight cases above 16; roughly these 15 groups would reduce the 156 to groups of about 10 sets. But Galton considers we must search not only the observed count-number, but two count-numbers on either side of it, or practically (having regard to

central ridge. If it had been, then, I think, it is clear that with the delta in relatively the same position as in Fig. b one less ridge would be counted in Fig. a¹ and two less in Fig. a². It is possible that the engraver erred in carrying the ridges quite up to the staple. Or, it may be, remembering what Galton has said about *cols*, i.e. that we cannot be certain whether a ridge terminates or forks, we ought always to put the inner terminal, as in Figs. a¹ and a² above, not where the central ridge meets the staple, but at about a ridge interval from the meet.

* The term "shoulder" is somewhat vague; the ridge-counts might well differ according to the choice of "shoulder." If the word means where the sides of the staple become parallel, then the engraver of Fig. b has hardly hit this off. I believe it would be preferable to define the shoulder as about a ridge interval below the summit. Galton's Plate 4 (our Plate XXVI), entitled "Counting Ridges," hardly seems to meet my difficulties in this and in the previous footnote. If Fig. 82 be a case of Fig. a¹, then Galton does not appear to put the inner terminus at the top of the central ridge; had he done so, I think the ridges would be 12 instead of 13.

† Galton's illustrations of ridge-counting are given on our Plate XXVI and would have been more helpful with a finer counting line. A thick line runs into the stem and occasionally obscures the finer parts of the delta.

terminal groups) about a group containing $4\frac{1}{2}$ ridges on the average. Each of these groups would contain 40 to 50 individuals of the 156, or less than $\frac{1}{3}$ and more than $\frac{1}{4}$ of the whole. Hence to count ridges in the first finger presenting a loop would reduce to less than a frequency of 10 all the groups of large frequencies except those under the formulae *ull, ull, ll, ll* and *www, www, ww, ww* (see the table on our p. 200). For the former group Galton suggests in addition counting ridges on the middle finger, and is thus able to break up his material into groups of less than 10 sets*. Here he introduces an interesting point; he gives a partial table (p. 82) for the number of ridges which occur in right middle and ring fingers for certain values of the count on the right forefinger. If the means of the former be found we have:

Number of Ridges in Fore Finger	Mean number of Ridges in	
	Middle Finger	Ring Finger
4	9·8	14·4
8	10·4	14·9
12	11·8	14·2
16	13·7	15·3

This suggests that there is correlation between the number of ridges at any rate in the fore and middle fingers of the same hand, and indicates a possible line of inquiry for the inheritance of ridge-numbers, when loops are available in both relatives.

We have next to consider how Galton meets the difficulty of the *www, www, ww, ww* class of pattern and others with numerous whorls. The main idea he uses is that if the tail of a whorl or the ridges which form it come from the radial side, the subscript or suffix *r* is used. If they come from the ulnar side the suffix *u* might be used, but Galton says this is so frequent that he does not use it. Hence *w* standing alone might mean fed from both sides, from neither side, or from the ulnar side. The suffix *s* is, however, used for whorls fed from both sides, but this may occur in three different ways:

(i) The ridges from either side may double back upon themselves, so that the contributory portions have blunt ends $= sb$.

(ii) The ridges from the two sides may be twisted together almost to a point $= sq$.

(iii) One set of contributory ridges may spring normally from one side of the finger, the other from one side of the tail of a tailed whorl $= sv$.

There are other symbols used by Galton in relation to whorls, namely *g*,

---

* The reader must remember that these numbers are based on a standard of 1000 sets. For 100,000 sets some of the groups might still be too large.

when the whorl has a great core, o, when there is at least one complete and detached ring in the whorl*.

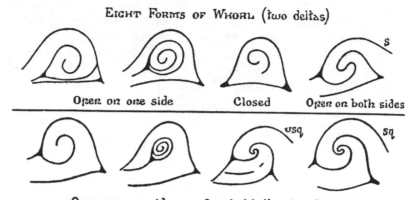

EIGHT FORMS OF WHORL (two deltas)

Open on one side   Closed   Open on both sides

Open on one side   Supplied both sides   Open on both sides

Fig. 35. Types of whorls from Galton's *Finger Print Directories*.

Galton remarks that it is best to leave a whorl ambiguous rather than attach a v or a q to it which it does not clearly and distinctly demand. "The omission of a suffix is of little harm; the insertion of a wrong one is. Cases should be dealt with merely as ambiguous, no suffix being attached to them, when the outline followed from the inner delta to a point above the outer delta or below it, as the case may be, does not suggest the same suffix as it does when the outline is followed in the opposite direction. The test in question is rapidly made and effective" (p. 94). It is, however, on the r and s sub-classification that Galton chiefly depends for breaking up the all or many whorl groups. Thus he writes:

"It is mainly through the help of the r and s suffixes that it is possible to discriminate between the all-whorls which occur 19 times in every 1000 cases [see our p. 200]. The whorls

* According to Galton's nomenclature, when in tracing any part of a pattern the direction changes so as to have pointed to all parts of the compass, that pattern is to be called a whorl.

PARTIAL AND COMPLETE CIRCUITS

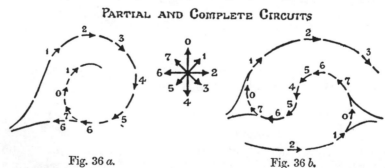

Fig. 36 a.   Fig. 36 b.

Illustration of complete circuits needed to classify a pattern as a whorl.

Hence arches with elliptic or circular rings between their arched ridges are classed as whorls. See Plates 7 and 8 of *Finger Prints* (our Plates XI and XII) and the accompanying cut, Figs. 36 a and 36 b, where, however, a print like Fig. 36 b, for which the compass point 4 might easily be non-existent, is still counted a whorl.

in that particular group are curiously monotonous in their general aspect and size, the conspicuous characteristics of *b, q* and *v* appearing rarely, and being therefore of little service in differentiation; neither is any method of counting ridges of value, for their numbers are much alike. But when the whorls are looked at carefully, and their contours followed a short way with a pointer, the variety in their *r* and *s* characteristics becomes distinctive. It may be pressed into the service of sub-classification, the sets admitting of being arranged in the order of the number of *r*'s that they severally contain, irrespective of the fingers on which those *r*'s appear." (pp. 95–6.)

A point which I think would be of value, but has not, I think, been noted by Galton, is the character of the whorl or spiral. Starting from the pole of the spiral does it correspond to a right or left-handed screw motion, i.e. is the rotation clockwise or counter-clockwise? It appears to me that these two types occur in not such unequal numbers, and at once divide whorls into two classes. Of course a clockwise or right-handed screw whorl on the actual finger is reversed on the imprint, but we may confine our classification to the imprints.

A further classification which might also be made in the case of simple spirals—and which easily admits of four classes—is the direction of the whorl or spiral at its pole or terminal. Is this direction generally upwards or downwards, generally radial or ulnar? There would be some doubt as to the 45° slopes, but as a rule the general polar slope is fairly obvious. I think there is thus actually small difficulty in breaking up the whorls for the purposes of indexing.

Galton makes only one division of arches in his Primary Classification, namely into Plain and Tented Arches (see our Fig. 37). The symbols *k, r, u,* or *v* may, however, be attached in the Secondary Classification.

We have already seen that Galton uses counting of ridges on the fore-finger and if necessary on the middle finger in order to break up the loop groups. But he admits that this is scarcely adequate in itself to deal with an index of 3000 sets or persons. Accordingly he uses other suffixes to differentiate loops by their cores. He considers the following three types will suffice:

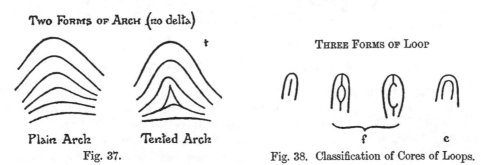

TWO FORMS OF ARCH (no delta)

Plain Arch   Tented Arch
Fig. 37.

THREE FORMS OF LOOP

Fig. 38. Classification of Cores of Loops.

i is a central rod, whose head stands *quite* distinct and separate from the ridge curving round it. Galton says there is no need to fear a *col*, if there be the distance of a furrow between central rod and staple. *f* covers the cases in which the central rod forks whether it reaches the staple or not; it may reunite forming the eye of a needle, or there may be an imperfect eye. The main point is that the core is not a simple rod; the several conditions do not need symbolising severally, they are all expressed by *f*.

c represents the case when the core within the loop is a second staple wholly detached from the outer staple which curves round it.

Galton uses still further symbols in his secondary classification—*k, v, x, y,* and three others to denote conditions of the print itself, namely: *d*, † and *. *d* marks a damaged print, either owing to the condition of the finger, or to the printing. If the print be wholly unreadable, then *d* is inserted in its proper place in the primary 10 symbols; if the print be only partially damaged, then *d* is to be used as a suffix. † denotes the scar of a cut, and should be used, however small the scar may be, as it is a valuable means of identification. * denotes that a portion of the finger has been more or less smashed, and should be combined with *d*.

Of the other four symbols *x* denotes that there is something very peculiar or questionable about the pattern.

v indicates what Galton terms an invaded loop. Usually the ridges enter through the open mouths of the loop, curve round and take their exits

Fig. 39. Secondary Classification of Loops, Galton's *y* and *v*.

FOUR FORMS OF LOOP (one delta)

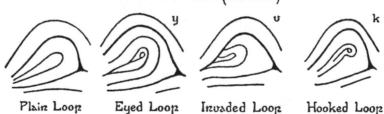

Plain Loop   Eyed Loop   Invaded Loop   Hooked Loop

Fig. 40. Secondary Classification of Loops, Galton's *y, v* and *k*.

parallel to their entrances. Sometimes, however, a system of ridges instead of entering from the mouth, springs out from one of the sides and destroys the symmetry of the pattern. Such a loop is an "invaded loop" and symbolised by *v*. Galton holds *y* to be one of the most generally useful of suffixes; it is the formation in the inner part of the loop of an eyed form. In the ordinary loop the ridges after turning back run parallel; in the eyed loop they reunite after recurving and enclose a minute plot. *y* must be distinguished from *f*, which latter is an island or approximate island in a central rod.

Finally *k* denotes a curvature sometimes affecting the whole of a loop, turning it into more or less of a solid hook, i.e. not a hook formed by a single terminated ridge. It may be applied not only to loops, but to whorls and even to arches to signify that they have an inner curl or hook.

It must at once be admitted that Galton requires a most imposing battery of additional suffixes and symbols to obtain his *Secondary Classification*, and further that when this has been accepted and we are able to classify some 3000 sets, so that only the slightest difficulty arises in entering or leaving an index, there still remains the fact that difficulties will steadily increase as we mount up from 3000 to 100,000 entries. There may, as Galton himself thought, still be need for three or four anthropometric characters—not for the purpose of identification but for classification. Again, this heavy array of symbols involves much additional work at first in indexing new sets of prints and in reading sets for identification‡. When I first read Galton's *Secondary Classification* and became acquainted with his battery of suffixes, it seemed too unwieldy to be practically applicable, but a little examination of impressions under a lens convinced me that it was reasonably easy for a moderate expert to get a grip upon it. Such experts would be in every "Identification Bureau," and for the mere trained impressor of fingers, such as the prison warder, it is rarely that any necessity arises for reading the prints themselves.

More serious defects of Galton's classification are its cumbrous character, and the fact that the letters he uses as subscripts do not convey any hint of their significations. I doubt whether the latter difficulty can possibly be met; *characteristic* symbols cannot be found for 20 to 30 subclasses, and if we once realise this, then it does not much matter whether we use numbers or letters provided we use a single one only. If we exclude numbers of ridges in loops, which might be placed as Roman figures in brackets after the index number itself, I believe that ten symbols with powers ought to describe any set of prints. The particular finger—supposing these taken in natural order from left to right—is indicated by the corresponding symbol taken in order from left to right. Now what are the symbols to be? They may be either letters or numbers. At first one might prefer the latter, because if we choose three forms of alphabet, say Greek, Roman and Italic letters, although we can go beyond ten corresponding letters of each, the printed mixture looks clumsy and can only be read out letter by letter. On the other hand, if we use numerals of three types—say, Roman, Italic and Block—the printed number, while still looking clumsy, if less so, is capable of being read aloud as so many millions, so many thousands, hundreds, tens, etc. The grave disadvantage of the numerical scheme is that it is far less readily adaptable to a *written* index, where it is not easy to distinguish between Roman, Italic and Block numerals. We shall probably do better therefore to adopt three alphabets, say, the Greek, Italic small and Italic capital. Let us see how this will work. We will first get Galton's symbols d, x,* and † into slightly simpler form. A simple note of interrogation (?) denotes that the print is missing, cannot be taken or is unreadable. A short rule over a letter denotes the print is

‡ Of course in about three-quarters of the inquiries it would not be needful to examine the secondary classification at all.

damaged ($\bar{c}$) = Galton's d. A short quantity over a letter ($\breve{e}$) denotes a questionable pattern = Galton's x. A single dot (sign of fluxion), as $\dot{m}$, denotes the scar of a cut = Galton's †; two dots (second fluxion or "Umlaut"), as $\ddot{c}$, denotes a smashed finger = Galton's *. Thus we replace these four subscripts by symbols already familiar to the printer. We then propose to adopt the Greek alphabet to represent arches, small italic letters to represent loops, and capitals to represent whorls. It is thus at once feasible to disregard all individual letters and write down the common Arch-Loop-Whorl formula by regarding alphabets only. The individual subspecies are represented by the individual letters. But we soon find that if we are to have only as many subspecies as Galton deals with, we shall need more letters than exist in any of the three alphabets! We are thus driven back to suffixes, but here we find it easier to write numerical powers than to use subscript letters. Further, as we only want 10 characteristics, the 10 numerals will suffice. They are as follows:

0 = Galton's o, or the core of the whorl has a detached ring.

1 = Galton's b, or the end of a single spiral or the two ends of a double spiral are blunted.

2 = Galton's q, or the core of the spiral is made of ridges twisted up into a point.

3 = Galton's g, or the core of the whorl is very large.

4 = Galton's k, or the body of the loop or whorl is curved like a hook, or some of the inner ridges are hooked.

5 = Galton's v, or there is an invasion of ridges from the side of loop or whorl.

6 = Galton's y, or the core of a loop or whorl, or even sometimes of an arch, has an eye shaped like a pear or racquet.

7 = Galton's c, or the upper part or innermost core of the loop is shaped like a staple detached from the enveloping ridge.

8 = Galton's f, or the innermost core of the loop forks like a tuning fork; it may afterwards reunite, enclosing a space like the eye of a needle (or like a broken eye).

9 = Galton's i, or the innermost core of the loop is a rod whose head is separate from the enveloping ridge. Multiple rods may also be included under 9.

It will be seen that the first four numerals (0, 1, 2, 3) apply only to whorls; the last three (7, 8, 9) only to loops; the remaining three (4, 5, 6) to any species of print. A little practice soon causes one to remember the significance of these numerals as easily as Galton's letters. Any combination of these numerals may appear as a power. Thus k^{54} we shall see denotes a radial loop with some resemblance to an arch, with an invasion of ridges from the side, and one or more hooked ridges; again A^{40} denotes a simple right-handed screw radial whorl with a completed circle and a ridge hooked round. Galton would represent this as $w(r, ko)$, where w denotes the whorl, r that it is radial, and ko that there is a coil of ridges enclosed in a complete or nearly complete ring. So much for the power suffixes.

It should be noted that the order of the numerals in the power is indifferent. We may now turn to the subspecies of the main species indicated by different letters of their special alphabets.

Arches:

a = simple arch; β = tented arch; γ = arch with a central dot or very small circle;
κ = arch approaching radial loop;
λ = arch approaching ulnar loop;
μ = arch which might equally well be classed as a radial loop;
ν = arch which might equally well be classed as an ulnar loop;
π = arch approaching a radial whorl;
ρ = arch approaching an ulnar whorl;
σ = arch which might equally well be classed as a radial whorl;
τ = arch which might equally well be classed as an ulnar whorl;
ζ = tented arch which might be confused with a loop fed from both sides.

It will be seen that κ, λ are nascent loops, π, ρ nascent whorls, and μ, ν, σ, τ quite ambiguous forms, which it may be needful to look out under other headings when searching the index.

Loops:

a = radial loop; b = ulnar loop; c = loop fed from both sides;
d = loop which cannot be clearly classed under a, b or c;
e = double adjacent loops; f = double superimposed loops;
g = loop resembling a tented arch;
h = loop which somewhat exceeds the limit at which it could be classed as an arch (or nascent loop);
k = radial loop which has some likeness to an arch;
l = ulnar loop which has some likeness to an arch;
m = radial loop which might equally well be classed as an arch;
n = ulnar loop which might equally well be classed as an arch;
u = radial loop which has some likeness to a whorl;
v = ulnar loop which has some likeness to a whorl;
x = radial loop which might equally well be classed as a whorl;
y = ulnar loop which might equally well be classed as a whorl;
z = loop fed from both sides which might be classed as a tented arch.

As before it will be seen that m, n and z are ambiguous cases interchangeable with μ, ν and ζ; k and l ought not to be, but may sometimes be confused with κ and λ.

Whorls:

Thus far our symbolism has only been an attempt to abbreviate Galton's. In the case of whorls we think it desirable to introduce certain additional broad classes, besides Galton's radial (r), ulnar (u) and fed from both sides (s). In the first place we distinguish between a simple spiral and a compound spiral with several whorling ridges linked at the pole. In the next place we distinguish starting from the pole between clockwise and counter-clockwise, or right handed and left-handed screw motion. We have thus twelve primary classes:

$A=$ simple radial right-handed screw whorl;
$B=$ „   ulnar   „   „   „
$C=$ „   fed from both sides right-handed screw whorl;
$D=$ compound radial right-handed screw whorl;
$E=$ „   ulnar   „   „   „
$F=$ „   fed from both sides right-handed screw whorl;
$G=$ simple radial left-handed screw whorl;
$H=$ „   ulnar   „   „   „
$I=$ „   fed from both sides left-handed screw whorl;
$J=$ compound radial left-handed screw whorl;
$K=$ „   ulnar   „   „   „
$L=$ „   fed from both sides left-handed screw whorl.

For the resembling and the ambiguous cases we have:

$P=$ radial whorl approaching arch;
$R=$ ulnar whorl approaching arch;
$S=$ radial whorl which might equally well be classed as arch;
$T=$ ulnar whorl which might equally well be classed as arch;
$U=$ radial whorl approaching loop;
$V=$ ulnar whorl approaching loop;
$X=$ radial whorl which might equally well be classed as loop;
$Y=$ ulnar whorl which might equally well be classed as loop.

Clearly X, Y are interchangeable with x and y, and if the index shows no U or V, then u or v should be sought for.

Unfortunately Galton's index does not record directly whether his whorls were simple or compound, or whether they were right or left-handed screws. Accordingly, in writing down his symbolism and that above for a few cases, we shall assume, where there is nothing to guide us, that his whorls were simple spirals and right-handed screws. I have chosen ten cases nearly at random from Galton's index of 300 sets of prints, only taking care that the selected individuals had very ample secondary classifications.

The table below gives the two notations.

In the condensed system, the indexing should be by order of letters, but for the same letter the Greek should stand before the small italic letter and the small italic before the capital, e.g. β before b and b before B.

It will be seen that it is possible to put an even finer classification based on Galton's into a very concentrated form. Therein alphabets indicate the genera, or primary classification, letters the species or subclasses, and powers the individual peculiarities. In this way many thousand finger-print sets may be indexed without reference to anthropometric characters. But we have always to remember that to avoid multiple entries more and more symbols must inevitably be used. A very little practice, however, teaches anyone the meaning of the symbols employed. It does not seem possible to adopt any system in which the symbols will be self-explanatory, and neither in Galton's original, nor in the present condensed system has this been attempted. The problem of the Identification Bureau is to balance the time lost in writing down and in reading a complicated system, against the time lost in examining the multiple entries of a more simple classification.

Table illustrating how Galton's System of Finger-Print indexing may be condensed and at the same time further developed.

Directory

1	2	3*		4			5			6		7		8
Right Hand	Left Hand	Th. L. Right	Th. L. Left	Right Hand			Left Hand			Right Hand		Left Hand		Number in Register
F.M.R.	F.M.R.			F.	M.	R.	F.	M.	R.	Th.	L.	Th.	L.	
Aal	*all*	5	5	—	*l*	—	—	—	*yw*	*v*	—	*vw*	—	3550
Rll	*rll*	5	5	12	—	*vy*	—	*ky*	*y*	*v*	—	—	—	3531
Rll	*rww*	8	8	*k*	—	*v*	*vk*	*sb*	*sb*	—	—	*sb*	—	2351
Rlw	*ull*	5	5	*kvw*	*r*	*by*	*3ay*	—	—	*v*	—	—	—	3660
Rlw	*www*	9	5	†*k*	—	—	*rko*	*s*	*s*	*sb*	—	—	—	1985
Ull	*ull*	5	5	10*c*	—	*vyc*	*i*	*c*	—	—	—	—	—	3617
Ull	*www*	5	8	*v*	*v*	*v*	*r*	*ko*	—	*vw*	—	*sq*	—	3560
Ulw	*rlw*	8	8	—	—	—	*avk*	—	*ryl*	†	—	*s*	—	3498
Wll	*rll*	8	8	*kvr*	—	—	2	†	—	*g*	—	*gs*	—	3554
Wlw	*www*	8	8	*r*†	—	—	*ly*	—	†	*o*	*v*	*vs*	—	738

The following are the values on our present condensed system:

	b	u^6	b	a	u^5	b^5 a κ b b			3550
	b	b^6	b^{46}	a	b	b^5 a b b^{56} b {xii}			3531
	b	C^1	C^1	a^{54}	C^1	B a^4 b b^5 b			2351
{3}	b	b	b	l^6	b	b^5 u^{45} a B^{16} b			3660
	b	C	C	A^{40}	b	C^1 $\dot{a}^4$ b B B			1985
	b	b	b^7	b^9	b	b b^7 b b^{567} b {x}			3617
	b	B	B^{40}	A	C^2	v^5 b^5 b^5 b^5 b			3560
	b	U^6	b	k^{54}	C	$\dot{B}$ b b B b			3498
{2}	b	b	$\dot{b}$	a	C^3	B^3 A^{45} b b b			3554
	b	$\dot{B}$	B	V^6	C^5	B^0 $\acute{A}$ b B b^5			738

With Galton: Th. = Thumb; F. = Forefinger; M. = Middle finger; R. = Ring finger; L. = Little finger. In the condensed system, the fingers are in "natural order from left little to right little finger." The vertical is placed between the two thumbs.·

We have now to consider briefly the remainder of this last finger-print work of Francis Galton.

In the *Introductory Chapter* Galton clearly defines his aim. Scotland Yard was beginning to form a vast collection of finger-prints, but these were to be primarily classified by four or five anthropometric measurements, so that the number of finger-prints in a group would not amount to more than a few hundreds or at most to, perhaps, 3000. It was the large groups in these subindices which Galton desired to break up. He was not describing how to deal with indices of 100,000 to 200,000 sets; that is a more modern problem.

* See our p. 198.

Yet, I believe, the extension I have given above of Galton's classification would readily admit of dealing with far larger numbers than he was considering. The important feature of Galton's present work is that he does not give merely definitions of his symbols for secondary classification, but he provides also illustrations of the various finger-print anomalies and characteristics he has symbolised. It is a misfortune that of the nine plates of finger-prints which accompany the memoir, all but two have the impressions natural size, and very often, to detect Galton's point, it is needful to use a magnifying glass. As the work has been long out of print and as there is, so far as I am aware, no published series of typical prints available, these plates are reproduced here on an enlarged scale to indicate Galton's ideas*. He is very modest about what he has achieved. The work, I think, shows some signs of haste, not in the studies on which it is based, but in the manner in which it is put together as if to supply some pressing need. He writes:

"The methods I have used undoubtedly admit of many improvements, and I shall myself suggest important ones; still they are the result of prolonged trials and much painstaking. They are therefore more likely to fulfil their purpose than any one alternative scheme that has not been worked out under similar conditions. In short, those who will consent to stand on my shoulders, are likely to see their way to improvements more surely than if they do not accept that aid.

"It must not be supposed that the classification of sets of finger-prints for the purpose of a directory is especially difficult. The art of classifying rapidly and correctly, like every other art, requires instruction and practice, but it does so in no exceptional degree. I can speak with much more assurance on this point than was possible three years ago, when I wrote my first book on *Finger Prints*, or even than was possible one year ago, at the time when that committee was sitting [see our pp. 148 and 174]....Having studied and during the last few months having re-studied many thousands of sets of finger-prints, and therefore many tens of thousands of individual ones, I can say with confidence that it is rare to find a pattern whose peculiarities are not due to a few easily recognisable characteristics, occurring singly or in combinations of two or three. It is true that patterns occasionally fall between two of my primary headings, and that a double reference may be needed; but these ambiguous patterns are recognised at a glance, and the alternative references that have to be made are obvious."

Chapter II (pp. 7–47) largely reproduces the *Report of the Departmental Committee*. With this I have dealt very fully on our pp. 148–151. Chapter III (pp. 48–59) contains *Conditions and Requirements* which to the reader of our present chapter will be already familiar; they concern the breaking up of the larger groups which arise in the $ALW(+UR)$ primary classification. On pp. 58–9 are some interesting observations on the amount of work which would be needful in order to register the 35,000 annual recruits to the British Army by their finger-prints, and so to stop desertion followed by re-enlistment.

Chapter IV (pp. 60–77) *Primary Classification*, and Chapter V (pp. 78–107) *Secondary Classification*, we have already summarised (see our pp. 203–205 above). Together with the plates (our Plates XXIII—XXX), they form by far the best account a novice in finger-printing can study even to-day. Chapter VI, a brief one of only three pages, deals with *Ambiguous Patterns*. This is a most valuable chapter as indicating how we must treat intermediate

* Some further understanding of his classificatory system may be obtained from the much reduced set of standard types, which will be found in the pocket at the end of this volume. The originals in three large frames are in the Anthropometric Laboratory at University College, London.

PLATE XXIII

Types treated by Galton as Arcnes.

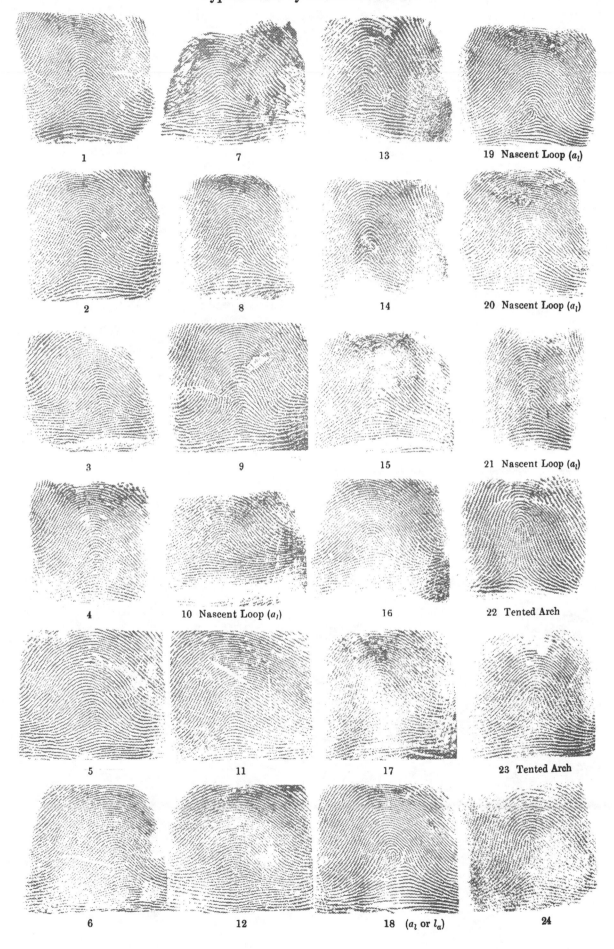

1

7

13

19 Nascent Loop (a_l)

2

8

14

20 Nascent Loop (a_l)

3

9

15

21 Nascent Loop (a_l)

4

10 Nascent Loop (a_l)

16

22 Tented Arch

5

11

17

23 Tented Arch

6

12

18 (a_l or l_a)

24

PLATE XXIV

Types treated by Galton as Loops.

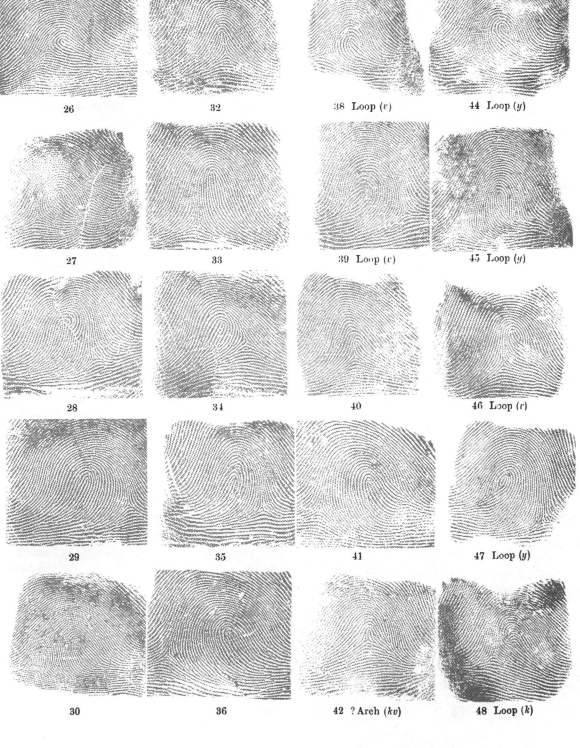

25

31

37 Loop (r)

43 Loop (y) ·

26

32

38 Loop (v)

44 Loop (y)

27

33

39 Loop (v)

45 Loop (y)

28

34

40

46 Loop (r)

29

35

41

47 Loop (y)

30

36

42 ? Arch (kv)

48 Loop (k)

PLATE XXV

Types treated by Galton as Whorls.

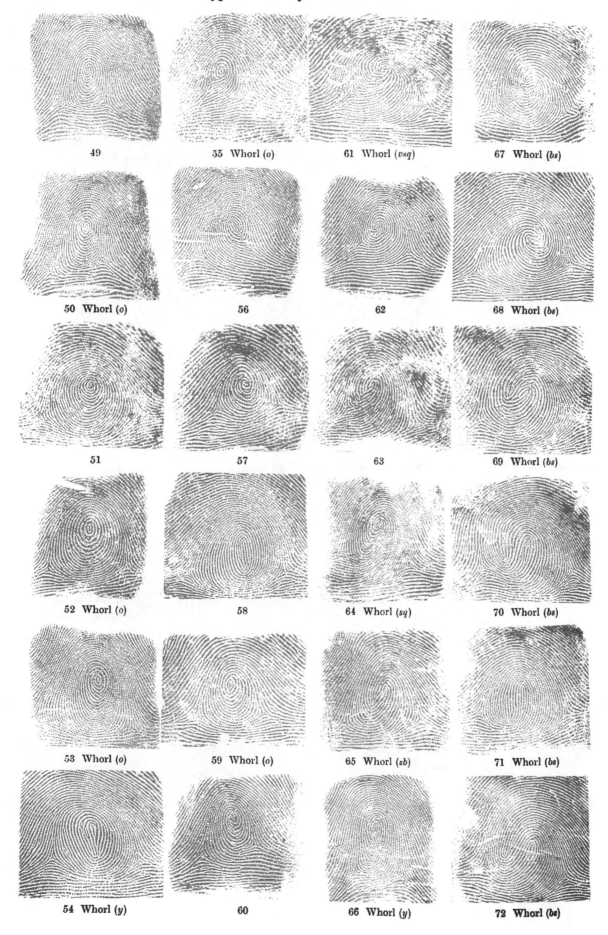

49	55 Whorl (o)	61 Whorl (vsq)	67 Whorl (bs)
50 Whorl (o)	56	62	68 Whorl (bs)
51	57	63	69 Whorl (bs)
52 Whorl (o)	58	64 Whorl (sq)	70 Whorl (bs)
53 Whorl (o)	59 Whorl (o)	65 Whorl (sb)	71 Whorl (bs)
54 Whorl (y)	60	66 Whorl (y)	72 Whorl (bs)

PLATE XXVI

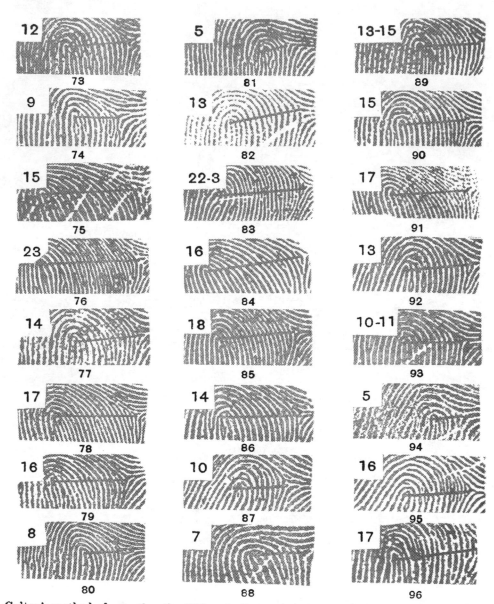

Galton's method of counting the Ridges in Loops. The number of ridges as determined by him are given in the left-hand top corner of each print: see our pp. 201–202.

A dabbed and a rolled print of the same finger to indicate how the former may lead one to classify as a loop, what the latter shows to be really a whorl: see our p. 213.

PLATE XXVII

Illustrations of Galton's Symbols i, f and c. (See our p. 205.)

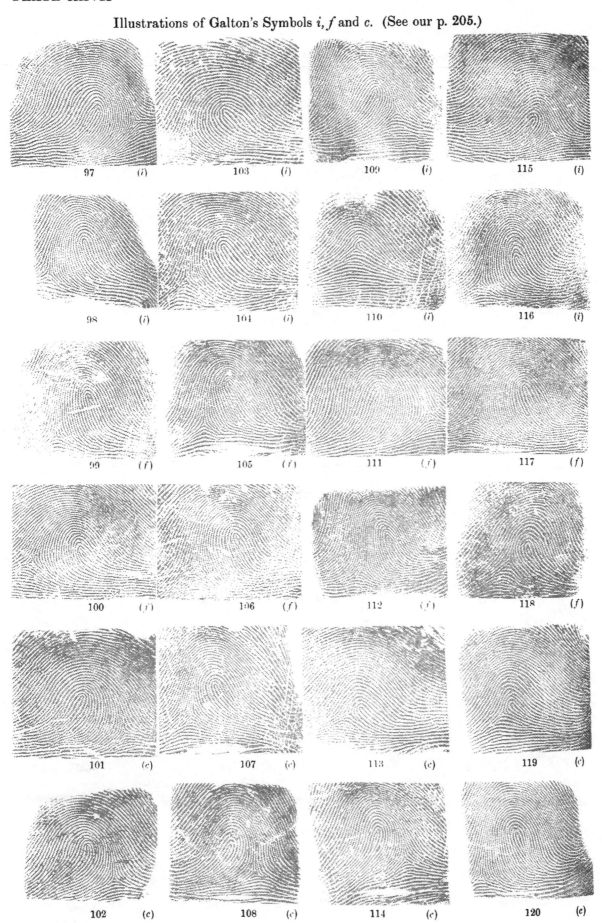

97   (i)   103   (i)   109   (i)   115   (i)

98   (i)   104   (i)   110   (i)   116   (i)

99   (f)   105   (i)   111   (i)   117   (f)

100   (i)   106   (f)   112   (i)   118   (f)

101   (c)   107   (c)   113   (c)   119   (c)

102   (c)   108   (c)   114   (c)   120   (c)

PLATE XXVIII

Illustrations of Galton's Symbols *y*, *v*, *vy*. (See our p. 206.)

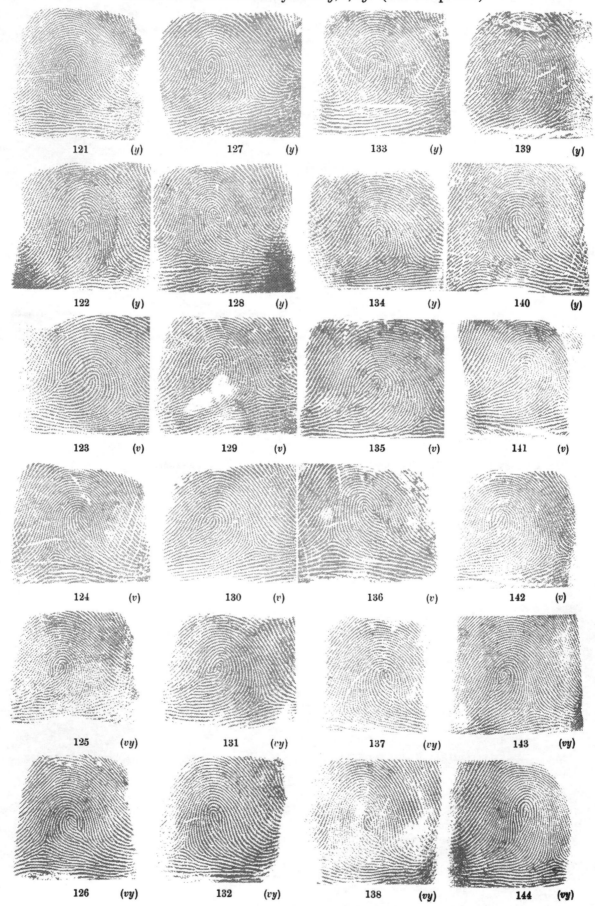

121   (*y*)   127   (*y*)   133   (*y*)   139   (*y*)

122   (*y*)   128   (*y*)   134   (*y*)   140   (*y*)

123   (*v*)   129   (*v*)   135   (*v*)   141   (*v*)

124   (*v*)   130   (*v*)   136   (*v*)   142   (*v*)

125   (*vy*)   131   (*ry*)   137   (*vy*)   143   (*vy*)

126   (*vy*)   132   (*vy*)   138   (*vy*)   144   (*vy*)

PLATE XXIX

Galton's Symbols applied to Noteworthy Peculiarities. (See our p. 208.)

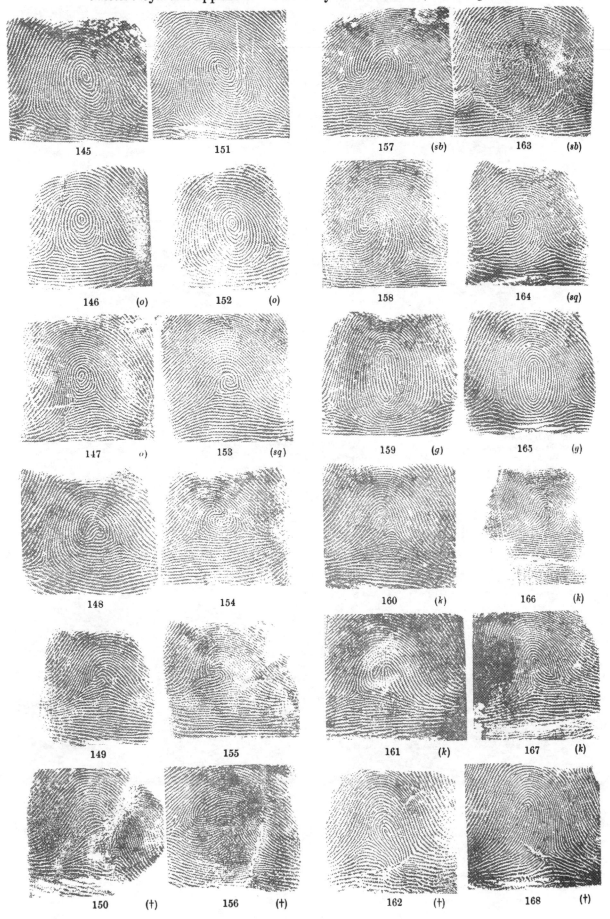

145   151   157   (sb)   163   (sb)

146   (o)   152   (o)   158   164   (sq)

147   o)   153   (sq)   159   (g)   165   (g)

148   154   160   (k)   166   (k)

149   155   161   (k)   167   (k)

150   (†)   156   (†)   162   (†)   168   (†)

PLATE XXX

Various Prints with Galton's Classifications.

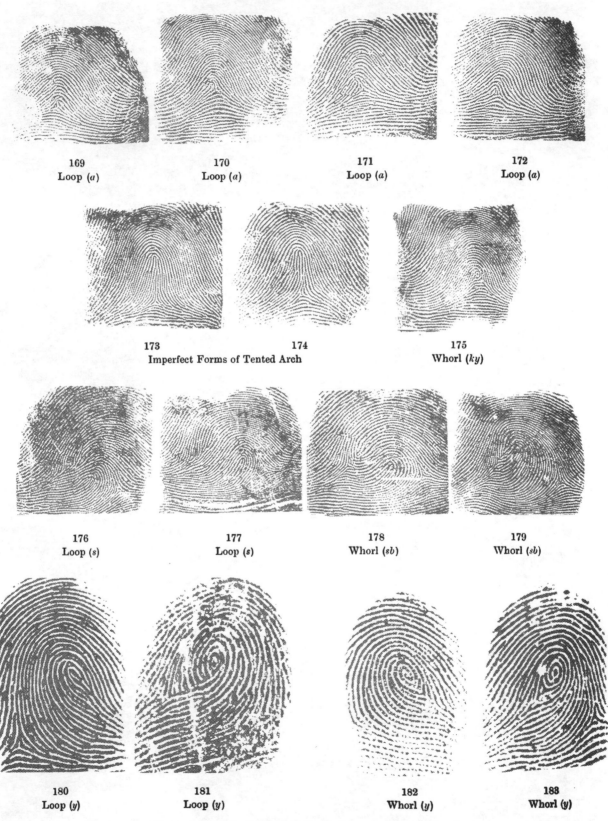

169
Loop (*a*)

170
Loop (*a*)

171
Loop (*a*)

172
Loop (*a*)

173

174

Imperfect Forms of Tented Arch

175
Whorl (*ky*)

176
Loop (*s*)

177
Loop (*s*)

178
Whorl (*sb*)

179
Whorl (*sb*)

180
Loop (*y*)

181
Loop (*y*)

182
Whorl (*y*)

183
Whorl (*y*)

To illustrate the Symbols of Galton's Secondary Classification.

patterns. I transcribe two-thirds of it for the benefit of those who can no longer obtain Galton's original work*.

'The chief peculiarities of individual Arches, Loops and Whorls having now been described, it becomes easy to discuss the frontiers of the primary classes and the debatable country between them.

"*A* to *L* [i.e. Arch to Loop]. The frontier between *A* and *L* ceases to be distinct at the point where *A* is just short of developing into a nascent loop. In the Figures 169 to 172 that point is just, but only just passed, so all those figures should count as loops with an *a* suffixed. The debatable ground lies between these and unmistakable arches, and in that debatable ground, *A* is held to predominate over *L* under any one of the following conditions:

"1. When the loop is formed by no more than one complete bend or staple, which may, however, be perfectly distinct, and may also enclose a rod (Fig. 21).

"2. When it consists of two or even three imperfect bends (Figs. 19, 20), especially if they converge and unite.

"3. Offsets at acute angles (Fig. 10) from the same ridge or from the same furrow do not rank as heads to loops.

"4. When two symmetrically disposed loops are enclosed in the same curved ridge (Figs. 173, 174) they are counted as an imperfect form of tented arch, being noted as *A* with the suffix *t* or *tur*.

"Generally speaking *A* is held to predominate whenever the pattern has no continuous contour, even though there may be a fairly distinct delta (Fig. 20), but it would be proper to unite the suffix *l* to this." (pp. 108–9.)

Clearly since Arches form a relatively small group, it would be to the advantage of the indexer, if frontier cases were allotted as far as possible to Arches.

"*A* to *W* [i.e. Arch to Whorl]. Between *A* and *W* a very small, or else an imperfect circle, or dot sometimes appears between two ridges of a pattern which is an arch in all other respects (Figs. 15, 17 and perhaps 18, which is ambiguous, and might be called a loop). If the diameter of the whorl does not exceed the width of one of the adjacent ridge intervals, the pattern does not lose the right to be called an *A*, but should for distinction's sake have a *y* suffixed to it. *W* is certainly reached when the little circle contains a central dot as in Fig. 175 which I should call *Wky*.

"*L* to *W* [i.e. Loop to Whorl]. Between *L* and *W* a large class of transitional cases have been sufficiently discussed in speaking of complete and incomplete circuits †. See Figs. 180–183.

"The specimens Figs. 176 to 179 show the relationships between whorls to which the suffix *sb* is applied (Fig. 178), with loops. In Fig. 176 we see a loop that throws off a curious crest from the upper part of its outline, and which is here and elsewhere a striking appearance; but in Fig. 177 the same peculiarity is much less distinct, while the number of cases that exist between extreme distinctness and extreme indistinctness is so great that crests are not allowed to have a suffix. Their conspicuousness in individual cases certainly depends to a considerable degree on the printing, whether more or less ink and pressure are used. When, however, the ridges cease to be given off from the outside of the contour of the loop, and recurve upon themselves as in Fig. 178, forming a blunted end to that part of the pattern, the result is a well-defined whorl. Another intermediate form between a loop and a whorl is produced in another way, and is recorded by *vy* as already explained." (pp. 109–10.) [See our p. 206.]

Lastly Galton refers to the case in which a real whorl may be mistaken for a loop because enough of the finger ridges have not been imprinted by rolling. This is especially a danger with "dabbed" prints. See our Plate XXVI.

Chapter VII (pp. 111–115) is entitled *Suggested Improvements*. Here, as I have said, Galton gives up his special finger arrangement in favour of the

---

* I have retained Galton's figure numbers, and the figures to which he refers will be found on our Plates XXIII—XXX.

† See our p. 204 footnote.

"natural order" (see our p. 200). We have seen that in the earlier publications Galton used *o* and *i*, "outer" and "inner," to mark his directions; in this work, to begin with, he uses "ulnar" and "radial" and the symbols *U* and *R* (or *u* and *r*) instead of *o* and *i*. He now appears to discard *U* and *R*, writing as follows:

"As regards the *U* and *R* notation, I am now decidedly in favour of the plan tentatively suggested in my answer to Question 207 [*Departmental Committee Report* (Evidence)], namely that it would be far better, on the grounds of diminishing error and fatigue, to regard the slope of the print relatively to the paper on which it is made, and not relatively to the Radial or Ulnar direction in the hand that made it. The slope relatively to the paper admits of uniform interpretation; the slope relatively to the hand does not, for what is *R* in the one hand is *U* in the other*." (p. 112.)

Galton next suggests a symbolic notation for the arch, whorl and two kinds of loops, i.e.

⌒	╲	╱	◯
Arch	*Loop*	*Loop*	*Whorl*

He says that the relief to eye and brain by this simple notation is very great. The pencil seems inclined to gallop over the cards automatically, because the attention is no longer strained by an endeavour to interpret the prints into alien symbols. The hand has merely to make abbreviated copies of what the eye sees, and thought is almost passive while doing so (p. 112). Galton does not, however, suggest how with such symbols the secondary classification is to be worked out.

This chapter concludes with an account of Galton's finger-print enlarging camera, which will magnify up to sixfold. We have already referred to this instrument (see our p. 197). Chapter VIII (pp. 116–123) contains the *Specimen Directory* of 300 *Sets*. At first the variety of symbols in the Secondary Classification is somewhat trying, but after a little becomes easily interpretable. Besides the numerals which are provided for the forefinger in the case of the ridge-counts in the formula *lll, lll, ll, ll,* other numerals occur in the index; they never exceed 4, and they may stand alone or be associated with *a* or *l*. They are in the *Secondary Classification*, and I cannot find that Galton has anywhere explained their meaning. This I am unable to supply. As I have said, I think the secondary classification needs condensation. It is also desirable that the method should be applied to several thousand sets of prints to ascertain, by an actual statistical experience, where the grading is still too coarse, or where it is over fine. If a student of finger-prints should, however, question me as to where he could learn how to index several thousand sets of finger-prints, I still could not refer him to anything better than Galton's *Finger Print Directories* of more than thirty years ago!

* Galton does not say how he proposes to symbolise the particular slope. As far as I can see, the result would be that two radial whorls on homologous fingers, say, which might be practically identical instead of being represented by the same symbol, would be represented by different symbols, which for any scientific purpose (e.g. inheritance) would be disastrous. If the finger-prints are taken in natural order, I see no difficulty in inscribing the letter *U* outside both little fingers and the letter *R* in the middle of the set of prints, between the adjacent thumbs. They might even be printed in these positions on the blanks which serve for the finger impressions. If the slope is then *downwards* from forefinger towards the little finger it is ulnar, otherwise radial.

The reader who has had the courage to follow Galton's biographer through the intricacies of this chapter will, I am sure, be convinced not only of the labour Galton devoted to his finger-print studies but also of the amazing energy he exhibited in acquainting not only administrative bodies but the public at large with the possibilities which then lay hidden in finger-printing, and this not solely for scientific but also for practical purposes. If the reader can find anyone who before 1895 had published a tithe of what Galton had

Fig. 41. Galton's Finger-Print Enlarging Camera.

issued on this topic, then I will admit him also to be a pioneer; if he can find anyone who has since 1895 done more than amplify in minor, often in very minor points Galton's work, then I will admit him a worthy successor to Galton.

Finger-printing as a science and finger-printing as an art are both alike the product of Galton's insight, ingenuity and tireless activity; the attempts to belittle the credit due to him can only spring from those who for their own purposes choose to ignore the literature of the subject.

Note to Chapter XV.

Finger-Prints as Reminiscences. As some collect autographs and others photographs, so we may collect finger-prints as mementoes of friends or of great men. Such a collection was formed by Francis Galton, and, the circumstances not always being favourable for a printer's ink impression, he not infrequently fell back on sealing-wax. In the *Galtoniana* are many sealing-wax impressions of Galton's friends. Thus we have Herbert Spencer's and quite a number of Sir W. R. Grove's* prints. The process of pressing the finger on hot wax was not always without pain, as is indicated in the accompanying 1893 Christmas greeting of Addington Symonds' daughter Katherine to Francis Galton.

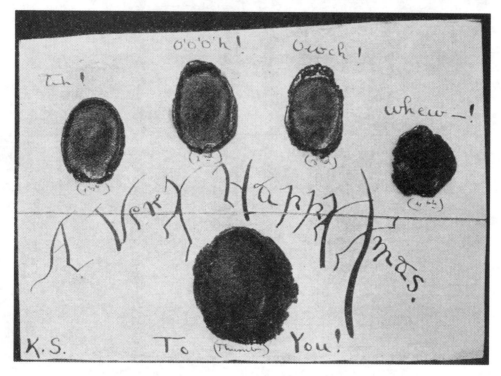

Fig. 42. A Christmas Greeting to Francis Galton "from an affectionate and admiring friend."

Among the prints of famous men to be found in Galton's Album of Prints are those of Gladstone, Zola, Wallace, Herbert Spencer, etc.; the Darwins, the Vernon Harcourts, the Garrods and many other families also appear. Galton himself had a seal cut from his right ring finger print, and this is still used on the name cards at the Annual Galton Laboratory Dinner. There are many other relics of Galton's early finger-print collecting days, e.g. prints of idiots, of farm labourers, of the Herschels at different ages, and occasionally foot and hand prints, as well as some finger-prints of apes. For some years Galton must have always had a finger-printing apparatus in his pocket, and possibly, like all men with a dominating hobby, have been somewhat of a trial to his acquaintances.

* Of combined legal and scientific fame!

PLATE XXXI

Francis Galton, the Founder of the Science of Eugenics, from a photograph of 1902, by the late Mr Dew-Smith. (By kind permission of Mrs Dew-Smith.)

CHAPTER XVI

EUGENICS AS A CREED AND THE LAST DECADE OF GALTON'S LIFE

"No custom can be considered seriously repugnant to human feelings that has ever prevailed extensively in a contented nation, whether barbarous or civilised. Any custom established by a powerful authority soon becomes looked upon as a duty, and before long as an axiom of conduct which is rarely questioned." Francis Galton, 1894.

(1) *Introductory.* The careful reader of this work will have realised how deeply impressed Galton was by the idea that with man himself lies the possibility of improving his race ; and this impression existed long before Galton initiated active propaganda for Eugenics as a social and political creed. Indeed, although Galton's earlier writings reached a limited and partly prepared audience, it was not till the beginning of the present century that he considered the time ripe for a more general public appeal, or sought proselytes to the new faith. There are some creeds, and more sciences, of which it is nearly impossible to name a single individual as the creator. When we speak of Christianity we forget, or wilfully disregard, Paul ; Einstein was not the first to see material phenomena in the curvature of space ; nor did Darwin stand alone when he propounded evolution through natural selection. But what student of evolution before Galton, realising the past ascent of man, grasped that his future lies with himself, if he be willing to study and control his own breeding ? It is given to few men to name a new branch of science and lay down the broad lines of its development ; it is the lot of fewer still to forecast its future as a creed of social conduct. In the thirty years which have elapsed, since Galton started his public teaching, what gratifying progress has been made, not only in establishing institutes and laboratories for research in Eugenics*, but also in familiarising the people at large with the code of conduct which an acceptance of eugenic principles involves ! It is as if the Great War had so thoroughly demonstrated the pitiable failure of humanity, that its thinkers and leaders felt that the old man must be replaced by a new-born Apollo†, the worn-out creed which had failed him by a more adequate

---

* Institutes primarily for Eugenics research exist to my knowledge in England, America, Sweden, Norway, Russia, Switzerland, Germany, Poland and probably elsewhere. Popular Journals or Eugenics Societies have been started in England, America, Germany, France, Italy and Russia.

† ...."Grief overcame,
"And I was stopping up my frantic ears,
"When, past all hindrance of my trembling hands,
"A voice came, sweeter, sweeter than all tune,
"And still it cried, 'Apollo! young Apollo!
"The morning-bright Apollo! young Apollo!'
"I fled, it follow'd me, and cried 'Apollo!'" Keats, *Hyperion.*

faith. We know little of how it came about that Aurignacian man replaced Mousterian man ; but the ascent was a steep one, and man needs once more some such rapid elevating. With our present acquaintance with the laws of heredity, with our present knowledge of how customs and creeds have changed, can we not hasten the evolutionary process of fitting man to the needs of his present environment ? It is indeed a great task because it involves control of the most imperious instinct of living beings, so imperious that Nature's method of improvement has been to provide quantity and seek therein for quality. The new creed bids us seek quality and restrict quantity; separate, where race demands it, the scarce controllable instinct of mating from the parental instinct, and teach nations to pride themselves on the superior type of their citizens, rather than on their material resources. The eugenic dreamer sees in the distant future a rivalry of nations in the task of bringing to greater perfection their human stocks, and this by an intensive study of biological law applied to man, and its incorporation, it may be gradually, but surely, in a revised moral or social code.

(2) *Address to the Demographers.* A paper which bridges the gulf between the *Inquiry into Human Faculty* of 1883 and the Huxley Lecture of 1901 is Galton's "Presidential Address" of August 11, 1891, to the Division of Demography of the *Seventh International Congress of Hygiene and Demography**. The word "Eugenics" does not occur in the address, it has no topical title, and yet it is an insistent demand for the study of eugenic problems. The paper has escaped and is likely to escape attention, it is not as far as I am aware included in any list of Galton's published papers, nor are copies of it among his offprints or in the bound volumes of his memoirs. Yet the address is of very great interest, not only for its intrinsic suggestiveness, but because it shows how during twenty years Eugenics had retained a foremost place in Galton's mind. His appeal, however, produced as little effect on the demographers as it did later on the anthropologists.

The topics with which the address deals are the relative fertility of various classes within a nation, and the relative fertility of nations among themselves—intranational and international fertilities—whereby tendencies arise for one class or one race to supplant another. Referring to the hypothesis of Malthus, Galton asks :

" Is it true that misery, in any justifiable sense of that word, provides the only check which acts automatically, or are other causes in existence, active, though as yet obscure, that assist in restraining the overgrowth of population ? It is certain that the productiveness of different marriages differs greatly in consequence of unexplained conditions....One of the many evidences of our great ignorance of the laws that govern fertility, is seen in the behaviour of bees, who have somehow discovered that by merely modifying the diet and the size of the nursery of any female grub, they can at will cause it to develop, either into a naturally sterile worker, or into the potential mother of a huge hive." (p. 8.)

Galton is here foreshadowing the sterilisation of those sections of the community of small civic worth, which has since become a pressing question of practical politics. He suggests that if persons are graded in a nation on

* *Transactions* of that Congress, pp. 7–12, London, 1892.

physical, intellectual and moral grounds, there must essentially be a least efficient as there will be a most efficient class. If inheritance holds for these characteristics then the relative fertility of these classes is of the utmost national importance. The same is true of the relative fertility of races and nations :

"The frequency in history with which one race has supplanted another over wide geographical areas is one of the most striking [incidents] in the evolution of mankind. The denizens of the world at the present day form a very different human stock from that which inhabited it a dozen generations ago, and to all appearance a no less difference will be found in our successors a dozen generations hence." (p. 10.)

Galton notes the Europeans who have swarmed over all the temperate regions of the globe, forming the nuclei of many future nations, the disappearance of the American Indian and the appearance of 8,000,000 negroes in America. He might have added many other instances even within Europe itself. It is indeed true that we hardly allow our thoughts to rest on the startling racial changes which have occurred in Europe in the last three or four thousand years, and on the still more significant changes in dominant races all over the world during the last few hundred years. Those who fully realise the marvellous evolution of certain types of humanity at the expense of others will smile—sadly, it may be—and wonder whether it is feasible for any League of Nations, however strong, to fix and maintain national and racial boundaries, unless it shall have first fixed the relative fertility of all the tribes of man and, what is more, internationalised all the world's resources! As interclass struggle finds its hope of solution only in the socialism which teaches the nationalisation of the materials and means of production, so international struggle can only reach its conclusion by the universalism which demands internationalisation of the world's wealth. In the first case, national eugenics is the only means left to provide any nation with men strong in mind and body; in the second case, international eugenics is the sole possibility of producing finer races of mankind. The men or group of men who can say to a nation large or small : "This is your frontier and you must keep to it," will be forced ultimately and logically to the point, not only of internationalising the world's wealth and its means of transport, but also of saying : "This is your appropriate fertility and you must keep to it." New modes of transport are rapidly making the world too small for mankind. Any plant or animal that overcrowds its proper region ends by destroying its fellows. The domesticated herd can alone thrive and progress on a limited pasture because the breeder stringently restricts its numbers, and picks from them those best fitted to their environment. Man, if he is to be freed from class struggle and from racial contests—that is to say, if he is to become thoroughly domesticated—can only thrive and progress if he breeds himself; in other words he must replace the harsh processes of Nature, which in the long run grant survival solely to the physically and mentally strong—to brain and muscle—by the milder practice of eugenics studied from the national and even the international standpoint. In the dimmest of distant futures we may see man fitting man to each region of his earth, and not

28—2

Nature very slowly developing man, or man hoping to mould Nature to his present self. But such knowledge is far from us at present. As Galton puts it:

"Much more care is taken to select appropriate varieties of plants and animals for plantation in foreign settlements, than to select appropriate types of men. Discrimination and foresight are shown in the one case, an indifference born of ignorance is shown in the other." (p. 11.)

But Galton was not pressing for immediate action, only for early study, because these great questions of civic and racial worth

"may unexpectedly acquire importance as falling within the sphere of practical politics, and if so, many demographic data that require forethought and time to collect, and a dispassionate and leisurely judgment to discuss, will be hurriedly and sorely needed." (p. 7.)

In conclusion he emphasised the fact that

"the improvement of the natural gifts of future generations of the human race is largely, though indirectly, under our control. We may not be able to originate, but we can guide. The processes of evolution are in constant and spontaneous activity, some pushing towards the bad, some towards the good. Our part is to watch for opportunities to intervene by checking the former and giving free play to the latter. I wish to distinguish clearly between our power in this fundamental respect and that which we also possess of ameliorating education and hygiene. It is earnestly to be hoped that demographers will increasingly direct their inquiries into historical facts, with the view of estimating the possible effects of reasonable political action in the future, in gradually raising the present miserably low standard of the human race to one in which the Utopias* in the dreamland of philanthropists may become practical possibilities." (p. 12.)

The garden of humanity is very full of weeds, nurture will never transform them into flowers; the eugenist calls upon the rulers of mankind to see that there shall be space in the garden, freed of weeds, for individuals and races of finer growth to develop with the full bloom possible to their species. I believe I am justified in the interpretation I have placed on Galton's address, and if there be a "national" eugenics, those words in themselves connote—as he himself indicates in his discussion of relative racial values—that there is also a science of "international" eugenics. This, if as we all trust the League of Nations survives, is bound to be the League's helpmate in the treatment of the most difficult problem with which its future is threatened.

I may indicate here what I think Galton planned as the course to be run by his new science. Laboratories were to be created where man should be studied from the standpoints of heredity, anthropology and medicine; journals and lectures were to be provided whereby the results reached should be popularised and a new morality inculcated. He had in view Eugenics not only as a science, not only as an art, but also as a national creed, amounting, indeed, to a religious faith. He never to my knowledge underestimated the difficulties, nor the slowness of its probable progress. A letter to William Bateson written in 1904 will indicate how Galton at that date visualised eugenic progress† :

42, RUTLAND GATE, S.W. *June* 12, 1904.

DEAR MR BATESON, Your letter of May 28 should have been answered earlier, had I not delayed in hope of receiving your promised answers to my "Ability in Families" circular‡, and replying to both at once.

* For Galton's own "Utopia in the dreamland of philanthropists" see later in this chapter.
† I am permitted to publish this letter by the courtesy of Mrs William Bateson.
‡ See p. 121 of this volume.

I quite understand now (I think) your point, and to a great extent agree with it. But what are we humans to do, if any "eugenic" progress is attempted? We can't mate men and women as we please, like cocks and hens, but we could I think gradually evolve some plan by which there would be a steady though slow amelioration of the human breed; the aim being to increase the contributions of the more valuable classes of the population and to diminish the converse. We now want better criteria than we have of which is which.

Do what we can (within reasonable limits as regards mankind), fraternal *variability* will never be much lessened; but I do think that the fraternal *means* might on the whole be raised.

That is the problem, as it seems to me, to be held in view; also that an exact knowledge of the true principles of heredity would hardly help us in its practical solution.

I do indeed fervently hope that exact knowledge may be gradually attained and established beyond question, and I wish you and your collaborators all success in your attempts to obtain it.

Very faithfully yours, FRANCIS GALTON.

Do you want your cobs of maize back?

This letter is of great importance; it indicates that Galton had in view only a "steady though slow amelioration of the human breed"; but it further shows that in his opinion the exact mechanism of heredity, even if we could find it out, was not of the highest importance. As an evolutionist he saw mass-changes taking place, and he recognised that the statistical solution is the one that has most importance for the eugenist. His statement that fraternal variability—by which he certainly meant heritable variability—will never be much lessened, is one with which I should personally agree, but the reader must remember that it cuts at the root of the "pure line" hypothesis*, and must not pass over its significance for Galton's own views. His remark also that the fraternal means might on the whole be raised suggests that the work of the biometricians had convinced him before 1904 that there was not a continuous regression of a selected group to the population mean; and that sports were not essential to progress.

(3) *Definition of Eugenics.* We have already seen that the term "Eugenics" was introduced by Galton in 1883 into his *Inquiry into Human Faculty.* See our Vol. II, pp. 249 ftn., 251, 252. Romanes in a review in *Nature*† of Galton's *Record of Family Faculties* and *Life History Album* in the following year (1884) uses the term "Eugenics" thrice and in one case speaks of the "science of Eugenics." "Mr Galton," he also tells us, "is indefatigable in his zeal to promote the cause of Eugenics." Thus born in 1883, the term had come into an accepted use in 1884.

Before we turn to Galton's propagandist lectures it is well to consider the definition of Eugenics. In 1883 Galton had defined Eugenics as the science of improving stock, not only by judicious mating, but by all the influences which give the more suitable strains a better chance. In 1904 Galton determined to take a step forward in his purpose by founding a research fellowship in National Eugenics, and addressed the following letter to the Principal of the University of London, Sir Arthur Rücker. This letter

* The reader may consult "A New Theory of Progressive Evolution" by the present biographer in the recently issued Vol. IV, Part I, of the *Annals of Eugenics*, published by the Galton Laboratory; it contains a discussion of the present position of the "pure line" hypothesis.

† Vol. XXIX, p. 257, January 17, 1884.

contains his own first definition of Eugenics, and whereas in the *Inquiry* we find the term may be applied to animals as well as man, it is now implicitly limited to mankind:

UNIVERSITY OF LONDON. *October* 10, 1904*.

DEAR SIR ARTHUR, I desire to forward the *exact* study of what may be called *National Eugenics*, by which I mean the influences that are socially controllable, on which the *status* of the nation depends. These are of two classes: (1) those which affect the race itself and (2) those which affect its health. It is the numerous influences comprised in (1), whose several strengths are as yet only vaguely surmised, that I especially want to have submitted to exact study. Class (2) is already the subject of much research, but I fear that here also the results arrived at require much more exact analysis by the higher methods of statistics than they have yet received.

If a scheme can be worked out that, on the one hand, fits in with the arrangements of the University of London and, on the other hand, is satisfactory to myself, I am prepared as a first instalment to give £1500 to serve for three years to carry out my purposes. If, but only if, the working of the proposed plan proves as satisfactory as I hope, I will reconsider the question with the view of making the endowment permanent of about £500 a year.

I presume that the University will supply accommodation for the person appointed at, say, £200 to £250 a year, and for a clerk, say, at £80 to £100 a year, leaving £150 to £200 for expenses. Also that the stamped official writing paper of the University may be used.

One part of his [the Fellow's] duties would be to establish a collection of records relating to those families of England who are remarkable for the number of near kinsfolk whose deeds have been noteworthy.

I feel some hesitation in drafting a statement of proposed duties for the "Research Fellow," or whatever his title may be, as they ought to fit into, and not overlap, what is already well done. Be that what it may, I think that "National Eugenics" would be good, as it is an exact title for what I wish to see done.   Yours very faithfully, FRANCIS GALTON.

This letter is important with regard to the definition of Eugenics, as it clearly indicates when and why the term "National" was introduced. The University appointed a committee to consider the offer and draft a scheme for the Research Fellowship in National Eugenics. It consisted of Sir Edward Busk (Chairman of Convocation), Francis Galton, the Principal of the University, Mr Mackinder and myself. This committee met on Oct. 14th and drew up a scheme for the Fellowship. My recollection of the meeting is that most of the time was spent in drafting a definition, which ultimately differed somewhat widely from that of Galton's letter of Oct. 10th, but which he finally approved. It heads the Draft Scheme and runs:

"The term National Eugenics is here defined as the study of the agencies under social control that may improve or impair the racial qualities of future generations either physically or mentally."

The scheme itself contains the usual regulations as to manner of appointment, the constitution of a special recommending Committee, Galton reserving a right of veto on the first nomination, the salary of the Fellow and his assistant, who if suitable was to be termed the Francis Galton Scholar. The duties of the Fellow are of more permanent interest: he was to devote all his time to Eugenics, in particular he was required:

"(*a*) To acquaint himself with statistical methods of inquiry, and with the principal researches that have been made in Eugenics, and to plan and carry out further investigations thereon.

* I do not know whether this is a clerk's error in printing Galton's letter or whether he actually wrote it in the precincts of the University.

"(b) To institute and carry on such investigations into the history of classes and families as may be calculated to promote the knowledge of Eugenics.

"(c) To prepare and present to the Committee, though not necessarily for publication, an annual Report on his work [to be done under general direction of the Committee]. To give from time to time, if required or approved by the Committee, short Courses of Lectures on Eugenics, and in particular on his own investigations thereon.

"(d) To prepare for publication at such times and in such manner as may be approved by the Committee (and at least at the end of his tenure of the Fellowship), a Memoir or Memoirs on the investigations which he has carried out."

The origin of the trend on which the Galton Laboratory of National Eugenics was developed later will be found in this Draft Scheme.

The University Senate on October 17th accepted the Draft Scheme without emendation, voted its cordial thanks to Francis Galton for his gift, and appointed as a Special Committee to recommend a Fellow and afterwards direct him*, Sir Edward Busk, Mr Mackinder, Francis Galton and myself. It also directed the Principal to issue an advertisement of the Fellowship and its conditions. This Sir Arthur Rücker did, but either out of sheer perversity, or through some clerical error, the word "morally" was substituted for "mentally" in the definition, and National Eugenics appeared in the advertisement as "the study of agencies under social control that may improve or impair the racial qualities of future generations either physically or *morally*." Quite recently this absurd definition was communicated to me by a member of the executive of the University as the work of the Special Committee! It has, I believe, no standing whatever, except that of an advertisement issued by the executive†, for which neither Galton, nor the Special Committee, had any responsibility. Galton, in his Herbert Spencer Lecture at Oxford in 1907, cites the definition correctly, and in his *Memories of my Life*, 1908 (p. 321), he writes that Eugenics is officially defined in the Minutes of the University of London as "the study of agencies under social control that may improve or impair the racial qualities of future generations either physically or mentally." I do not know whether this definition fully covered his original views or not. I only know of one occasion on which during his life he departed in public from it. This was during a talk with an interviewer from the *Jewish Chronicle*, July 20, 1910. He then defined Eugenics with a slight difference as "the study of the conditions under human control which improve or impair the inborn characteristics of the race‡." It

* There was too much "direction" about the scheme as originally planned. Galton, as I have previously remarked (see p. 135 above), was in my judgment too fond of working through committees. Beside the University Special Committee, which on the whole did little more than leave the first Research Fellow and Galton to their own devices, there was an "Advisory Committee" nominated by Galton, which met at the Eugenics Record Office and achieved little beyond hampering the Fellow. On this point the reader will find further remarks later.

† It is to be noted that in an announcement of the Fellowship in *The Times* of Oct. 26, 1904, the word "mentally" occurs in its proper place.

‡ In this interview Galton stated that it is one part of Eugenics to encourage the idea of parental responsibility, the other part is to see that the children born are well born. Galton considered that the Mosaic code had enjoined the multiplication of the human species, but it was really more important to prescribe that the children should be born from the fit and not

is clear from this wording that Francis Galton was not wholly satisfied with the term "qualities." When did he change it? In the Codicil, dated May 25, 1909, of his Will of October 20, 1908, and in the cancelled clause, Galton defined the purpose of his foundation to be:

" to pursue the study and further the knowledge of National Eugenics, that is of the agencies under social control that may improve or impair the racial faculties of future generations physically or mentally."

He thus cast his vote for "mentally." And this was undoubtedly well, for the term "mental" is wider than "moral," and the latter does not include the former, while at least many will be content to consider morality a mental characteristic. Galton was less fortunate, I think, in replacing "qualities" by "faculties." There seem to me many characteristics or qualities of the mind or body which it is desirable for the Eugenist to study, and which it is difficult to force into the category of "faculties." Perhaps they may be admitted to our studies as often associated with the faculties of mind or body to which the definition appears to limit eugenic research. It is worth noting that Galton's *Memories* citing the Committee's definition of Eugenics appeared in October, 1908—I got my copy on the 9th—and that on October 20th Galton signed a will in which "qualities" is replaced by "faculties." It might be thought that " faculties " was a word handed down from an earlier will, but this is not so. It was in the autumn of 1906 that Galton first told me of his plan to found a professorship of Eugenics in the University of London. I find that his letters to me of November and December, 1906, deal largely with the wording of the clauses in his Will as to his foundation for the study of Eugenics; they also deal with the proposed Weldon memorial and of his own desire to free himself from the direction of the University Eugenics Record Office, which was becoming too much for his strength. To these matters I shall refer later, but I think the reader will pardon me for taking one letter here out of its natural order in the history of Galton's plans for Eugenics; it demonstrates that even in his testamentary deposition of 1906 he fully accepted the definition of 1904. The letter runs as follows:

7, WINDSOR TERRACE, THE HOE, PLYMOUTH. *Nov.* 15, 1906.

MY DEAR KARL PEARSON, Enclosed is Mr Hartog's reply (1695. 11) to my "semi-private" letter. Please ultimately return it to me. It is quite satisfactory from my point of view, how would it be from yours?—Could you be persuaded to take control of the Eugenics Office as a branch of the Biometric Laboratory, working it in your way on " secular " biometric problems that have a distinct bearing on Eugenics? It cannot be under two heads or guidances so I willingly resign mine, perhaps keeping a nominal connection with it as "consultative*." It

the unfit. He did not allow that this latter principle was inculcated by the Jewish code. The *Jewish Chronicle* in a leader on the interview endeavoured to magnify the eugenic influence of the Mosaic code, in particular quoting the warning words spoken from Sinai about "visiting the sins of the fathers upon the children unto the third and fourth generations." But surely these words had no relation to physically or mentally feeble parents refraining from parenthood, but were a threat of the law-giver to induce his race to be faithful to their tribal deity, and prevent them worshipping (should their god fail them) at the altars of other gods! It is only in modern days that we have adopted them as appropriate to heredity in disease.

* Galton was a "consultative" editor of *Biometrika*, see below, p. 245.

would enlarge your means of work and from that point of view would be agreeable, I think and hope.

It is, perhaps, well that I should copy out the paragraph in my Will, which refers to the residue after paying various legacies, the amount of which residue will be fully what I told you and somewhat more.

"I devise and bequeath all the residue of my estate and effects both real and personal unto the University of London to be held, assigned and disposed of by the Senate of that University in the furtherance of the study of National Eugenics, that is of the agencies under social control that may improve the racial qualities of future generations either physically or mentally. Provided always that it shall be lawful for the Senate by a majority of not less than two-thirds of all its members at any time after ten years shall have elapsed from the date [1906] of this my Will to divert part or the whole of the then remaining sum to the study of such other branch or branches of Biometry, Statistics or of Sociology as they may then think more worthy of support."

If you think this could be amended by a Codicil, pray tell me.

Mr Heron comes to see me tomorrow till, I believe, Monday morning; I will write the results of what I may learn from him, etc.

I *hope* that your reply to this may justify my telling Hartog that all the arrangements for filling up the Eugenics vacancy and its future control will be in *your* hands, and no longer in mine, that I wish to retire wholly and that in all matters concerning its management, except financial, he must henceforth communicate with you—May it be so!

Affectionately yours, FRANCIS GALTON.

I shall think of you on the 24th*.

Plymouth is a success in all essentials as warmth, cooking and comfort, but the sky and air are somewhat depressing.

This letter shows that in 1906 Galton preferred "qualities" to "faculties" in his definition of Eugenics. In the wording of both the Will of 1908 and the Codicil of May 25, 1909, the latter term replaces the former. I find from letters that passed between Galton and myself between May 4 and May 18, 1909, that he consulted me as to the drafting of this later Codicil, actually putting a copy (returned to him) before me for my suggestions, which turned solely on the desirability of granting power to the University to delay the appointment of a Galton professor, if no suitable man was at once available. If the word "faculties" replaced "qualities" in this draft, probably Galton, and certainly I, overlooked its introduction.

Historically the origin of the definition of Eugenics is of interest; its three forms, that in the Minutes of the University as to the duties of the first Galton Research Fellow, which has been invariably used by the Galton Laboratory; the unsanctioned change in Sir Arthur Rücker's advertisement; and finally that of the Codicil defining the bequest to the University, have already been the subject of inquiry from America. If the University were ever to insist in practice on a rigid interpretation of the phrasing of the bequest, the word "faculties" might hamper a future† occupant of the Galton Chair. It would be most undesirable that he should be precluded from studying any characteristic quality—iris pigmentation, constitution

---

* I was probably giving a public lecture on that date, but do not remember topic or place.

† Hardly in the case of the present Galton professor, as the Will permits him to associate the Biometric Laboratory with the Galton Laboratory, and biometry at least covers the "qualities" as well as the "faculties" of man!

of blood or size of thyroid gland, etc.—which, without being a "faculty," might tend to throw light on hereditary processes in man. I have therefore ventured to place on record here that to the best of my knowledge and belief Galton, by the use of the term "faculties" in the Codicil of 1909, in no wise wished to set any limitation on the definition of Eugenics which he fully accepted in his *Memories* of 1908 (p. 321).

(4) *The Huxley Lecture of* 1901, *and Allied Matters.* Before entering into more detail as to the steps Galton took to develop the research side and the popular side of Eugenics, it may be convenient to pass under review the publications which he issued in this last period of his life. It is true that they were written more from the popular standpoint than his earlier papers on statistics and heredity, but they lacked little of the old fire, and were eminently suited to his purpose, viz. that of creating a national movement in favour of a eugenic policy. His work may best be reviewed in chronological order, thus forming a history of the last eleven years of his life, 1901 to 1911, from his 79th to 89th year. We have seen* that in the winter of 1900 Galton was in Egypt and spoke before the Khedivial Society for Geography on the Egypt of 1846 † and of 1900. On his return in 1901, he was invited to give the Huxley Lecture and receive the Huxley medal of the Royal Anthropological Institute. These events took place on October 29th ‡, and the lecture, entitled "The possible Improvement of the Human Breed under the existing Conditions of Law and Sentiment," was published in *Nature*, Nov. 1, 1901, and again in the *Report of the Smithsonian Institution*, pp. 523–538. It seems to have been published only in abstract by the Anthropological Institute. It is noteworthy that Galton in his early days tried to induce the physical anthropologists of that Institute to adopt a scientific technique. In his old age he endeavoured to prove to them that a study of racial characters finds its practical outcome in the art of Eugenics. In neither case was he really successful. It is the Eugenics Laboratories springing up over Europe which are adopting anthropology as an auxiliary science and revivifying its technique and aims; it is the older institutes of anthropology which have not grasped that their study of the evolution of man's past has for its main purpose the direction of man's future—therein alone it finds its full justification.

Galton opened his Huxley Lecture by stating that he proposed to treat broadly a new topic belonging to a class in which Huxley himself would have felt a keen interest. He had accordingly selected a topic, which had occupied his thoughts for many years, and to which a large part of his published inquiries had borne a direct though silent reference. His remarks would serve as an additional chapter to his books on *Hereditary Genius* and *Natural Inheritance*, and we may add also to his *Inquiry into Human Faculty*, wherein he first defined and used the term "Eugenics," and talked of the possible purposeful improvement of the human breed§.

* See the present volume, p. 158.
† Actually 1845–6: see our Vol. I, pp. 198–205.
‡ With Lord Avebury (formerly Sir John Lubbock) in the chair, a very fit choice.
§ See our Vol. II, pp. 252, 264 *et seq.*

The topic, he stated, had not hitherto been approached along the path that recent knowledge has laid open, and as a result the subject had not held as dignified a position in scientific estimation as it ought to do. "It is smiled at as most desirable in itself and possibly worthy of academic discussion, but absolutely out of the question as a practical problem" (p. 523*). The object of the lecture was to show cause for a different opinion.

> "Indeed I hope to induce anthropologists to regard human improvement as a subject that should be kept openly and squarely in view, not only on account of its transcendent importance, but also because it affords excellent but neglected fields for investigation. I shall show that our knowledge is already sufficient to justify the pursuit of this, perhaps the grandest of all objects, but that we know less of the conditions upon which success depends than we might and ought to ascertain. The limits of our knowledge and of our ignorance will become clearer as we proceed." (p. 523.)

Thus Galton attempted to introduce the science of Eugenics to anthropologists, cautiously screening the label on his draught!

He first pointed out that the natural characters and faculties of human beings differ at least as widely as those of domesticated animals, such as dogs and horses :

> "In disposition some are gentle and good-tempered, others surly and vicious; some are courageous, others timid; some are eager, others sluggish; some have large powers of endurance, others are quickly fatigued; some are muscular and powerful, others are weak; some are intelligent, others stupid; some have tenacious memories of places and persons, others frequently stray and are slow at recognizing. The number and variety of aptitudes, especially in dogs, is truly remarkable; among the most notable being the tendency to herd sheep, to point and to retrieve. So it is with the various natural qualities which go towards the making of the civic worth in man. Whether it be in character, disposition, energy, intellect or physical power, we each receive at our birth a definite endowment, allegorized by the parable related in St Matthew, some receiving many talents, others few." (p. 524.)

It is to be noted how artfully Galton chose the very characteristics of the dog which correspond to those of man, and led up his artless listeners without direct statement to the inference that what you can certainly breed for in the dog, you might equally well breed for in man! Galton realised to the full that the best method of making converts is to allow the average man an opportunity of independently discovering your truth. In the pride of himself finding a nugget (conveniently placed), he is far less inclined to assert without examination that the whole field is non-auriferous.

Pushing the parable of the talents further, Galton, rather quaintly, proceeds to put it into numbers, taking the quartile deviation ("probable error") to represent one talent, and using the normal frequency distribution to express the frequency of the various grades of qualities in a nation. He justifies the use of the normal distribution on the ground that experience has shown that it is a fair approximation in the case of a number of qualities.

---

* My references are to the pages of the *Smithsonian Report.*

He thus obtains the following distribution for 10,000 individuals of any character in a nation:

Defect talents						Excess talents					
Under −4	−4	−3	−2	−1	Mean	1	2	3	4	Over 4	Total
35	180	672	1613	2500	2500	1613	672	180	35		10,000
v	u	t	s	r	R	S	T	U	V		

The letters below mark the particular classes for purposes of reference, the small letters denoting classes with the corresponding range of defect of talents below mediocrity and the capital letters the classes with excess of talents above mediocrity. The reader will note that with a different nomenclature the distribution is one very familiar to statisticians. Beyond V and v Galton supposes classes W, X, etc., w, x, etc., each corresponding to a range of one talent. He illustrates this scheme from his own data for male stature where the mean was 5′ 8″, the "talent" $1\frac{3}{4}''$ nearly, and where accordingly class U would contain men over 6′ $1\frac{1}{4}''$, "quite tall enough to overlook a hatless mob." Then he continues:

"So the civic worth (however the term may be defined) of U-class men, and still more of V-class, are notably superior to the crowd; though they are far below the heroic order." (p. 526.)

In round numbers about one man in 300 belongs to the V-class.

In the next place Galton proceeds to compare his normal distribution scale with the classes A, B,...H, into which Mr Charles Booth divided the population of London in his noteworthy survey. He concludes that Mr Booth's class H corresponds to his own T, U, V and above. Further, his own t, u, v and below correspond to Mr Booth's class A, criminals, semi-criminals, loafers, and a few others, and to his class B, very poor persons who subsist on casual earnings, many of whom are inevitably poor from shiftlessness, idleness or drink. Galton rightly considers that, from the standpoint of civic worth, classes t, u, v and below are undesirables.

The next section of the lecture is entitled *Worth of Children*. The lecturer points out that the brains of the nation lie in the W and X-classes, and if the people, who would be placed in them as adults, could be distinguished as children, were procurable by money, and could be reared as Englishmen, it would be a cheap bargain for the nation to buy them at the rate of several hundreds or even thousands of pounds per head. He refers to Dr Farr's estimate of the value of the baby of an Essex* labourer's wife at £5 and says he believes that on the same actuarial principles an X-class baby might be reckoned in thousands of pounds. While some such "talented" folk fail, most succeed and many succeed greatly:

* Dr Farr's analysis seems based on the wages of agricultural labourers in Norfolk, not Essex: see *Journal of R. Statistical Society*, Vol. XVI, pp. 38–44.

"They found great industries, establish vast undertakings, increase the wealth of multitudes, and amass large fortunes for themselves. Others, whether they be rich or poor, are the guides and light of the nation, raising its tone, enlightening its difficulties, and imposing its ideals. The great gain that England received through the immigration of the Huguenots* would be insignificant to what she would derive from an annual addition of a few hundred children of the classes *W* and *X*. I have tried but not yet succeeded to my satisfaction, to make an approximate estimate of the worth of a child at birth according to the class he is destined to occupy when adult. It is an eminently important subject for future investigators, for the amount of care and cost that might profitably be expended in improving the race clearly depends on its result." (p. 528.)

Thus far it will be clear to the reader that all that Galton does is to assert and assert with truth that in any scale of civic worth, whether it be one of brains or energy, artistic power or skill, the classes *W* and *X* are of the highest value to a nation, and should be multiplied if possible, the classes *t, u, v* and below are undesirable, and should be decreased if feasible. It is difficult to see how anyone can deny this, for by the very definitions of those classes they are the best and the worst in the community.

Galton now passes to "the descent of qualities in a population." Here he makes use of the conception of regression as he has discussed it in his *Natural Inheritance*, and makes the parental correlation one-third. As in that work he indicates with a diagram how a population reproduces itself. The same criticism may be made here as earlier on our pp. 18, 23 and 65, namely in the first place the parental correlation is actually much higher than he assumes it, and secondly he supposes the ancestors of the parents in all cases to be mediocre, whereas these ancestors will most probably deviate from mediocrity in the same direction as the parents themselves do. Luckily these slips do not invalidate his conclusions, for, if corrected, his case for obtaining *V*-class offspring most economically by encouraging parentage in *V-, U-,* or *T*-class individuals is greatly strengthened. If the reader will bear in mind that Galton's statements owing to the above reasons give results far less favourable than they should be to *V*-class parents, we need not hesitate to cite his sentences on p. 531:

"Of its [the *V*-class in new generations] 34 or 35 sons, 6 come from *V* parentage, 10 from *U*, 10 from *T*, 5 from *S*, 3 from *R*, and none from any class below *R*; but the number of the contributing parentages has also to be taken into account. When this is done, we see that the lower classes make their scores owing to their quantity not to their quality, for while 35 *V*-class parents suffice to produce 6 sons of the *V*-class, it takes 2500 *R*-class fathers to produce 3 of them. Consequently, the richness in produce of *V*-class parentages is to that of the *R*-class in an inverse ratio, or as 143 to 1. Similarly the richness in produce of *V*-class children from parentages of the classes *U, T, S*, respectively, is as 3, 11·5 and 55 to 1. Moreover nearly one-half of the produce of *V*-class parentages are *V* or *U* taken together, and nearly three-quarters of them are either *V, U,* or *T*. If, then, we desire to increase the output of *V*-class offspring, by far the most profitable parents to work upon would be those of the *V*-class, and in a three-fold less degree those of the *U*-class." (p. 531.)

Here we see Galton fully cognizant of the solution of the paradox which nearly thirty years later was still troubling the non-statistical mind of Professor Leonard Hill†.

* This is an illustration often used by Galton, e.g. in his Presidential Address to the Demographic Congress, 1891, and in the *Jewish Chronicle*, July 30, 1910.
† See this volume, p. 27.

STANDARD SCHEME OF DESCENT

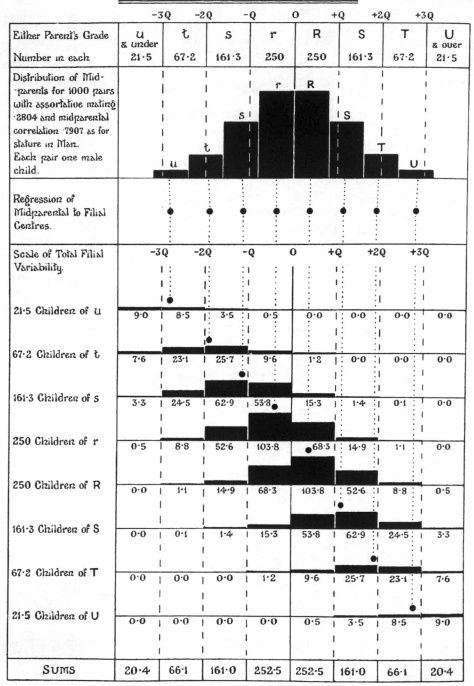

	−3Q	−2Q	−Q	O	+Q	+2Q	+3Q	
Either Parent's Grade	u & under	t	s	r	R	S	T	U & over
Number in each	21·5	67·2	161·3	250	250	161·3	67·2	21·5

Distribution of Mid-parents for 1000 pairs with assortative mating ·2804 and midparental correlation ·7907 as for stature in Man. Each pair one male child.

Regression of Midparental to Filial Centres.

Scale of Total Filial Variability.	−3Q	−2Q	−Q	O	+Q	+2Q	+3Q	
21·5 Children of u	9·0	8·5	3·5	0·5	0·0	0·0	0·0	0·0
67·2 Children of t	7·6	23·1	25·7	9·6	1·2	0·0	0·0	0·0
161·3 Children of s	3·3	24·5	62·9	53·8	15·3	1·4	0·1	0·0
250 Children of r	0·5	8·8	52·6	103·8	68·3	14·9	1·1	0·0
250 Children of R	0·0	1·1	14·9	68·3	103·8	52·6	8·8	0·5
161·3 Children of S	0·0	0·1	1·4	15·3	53·8	62·9	24·5	3·3
67·2 Children of T	0·0	0·0	0·0	1·2	9·6	25·7	23·1	7·6
21·5 Children of U	0·0	0·0	0·0	0·0	0·5	3·5	8·5	9·0
SUMS	20·4	66·1	161·0	252·5	252·5	161·0	66·1	20·4

Modified from Galton's original scheme by taking better numerical values for stature in Man, and the assortative mating not perfect.

Fig. 43.

Next Galton refers to the important fact that in each class of a community there is a strong tendency to intermarriage; this not only produces a "marked effect in the richness of brain-power of the more cultured families" but further an effect of another kind in the lowest stratum of civic worth. After citing Charles Booth on this "handful of barbarians*," Galton proceeds as follows:

"Many who are familiar with the habits of these people do not hesitate to say that it would be an economy and a great benefit to the country if all habitual criminals were resolutely segregated under merciful surveillance and peremptorily denied opportunities for producing offspring. It would abolish a source of suffering and misery to a future generation, and would cause no unwarrantable hardship in this." (p. 532.)

Galton, in his scheme of Standard Descent on p. 529, makes the assortative mating coefficient perfect. I have replaced it by one [see the opposite page] in which that coefficient has the observed value for stature. He has also supposed his filial arrays to regress from the midpoints of the parental blocks instead of from their means, and used a value lower than I have adopted for the filial regression. I think my diagram emphasises the conclusions he has drawn above. The fact that the population does not reproduce itself absolutely is due to grouping into blocks instead of dealing with a continuous distribution.

The following section is headed *Diplomas*†. Galton considers, and probably correctly, that there would not be a serious difficulty, if a strong enough desire were felt, in picking out young men whose grade was of the *V, W* or *X* order. He points out that at any great university the students are in continual competition in studies, in athletics and in public meetings, and that thus their faculties are well known to their tutors and associates; he remarks that civic worth may take various forms, and a considerably high level both intellectually and *physically* should be required as a qualification for candidature. Galton considers that when a limited number had thus been selected they "might be submitted in some way to the independent votes of fellow students on the one hand and tutors on the other whose ideals of character and merit necessarily differ." Finally he would have an independent committee, who would examine the candidates personally and consider the favourable points of their family histories, making less of the unfavourable points, unless they were "notorious and flagrant," because of the difficulty of ascertaining the real truth about them—a view which is perhaps not wholly to be commended. As examples of successful working of such committees Galton cites the selections made by scientific societies, including, perhaps, the award of their medals, "which the fortunate recipients at least are tempted to consider judicious‡" (p. 533).

---

* Of this *A*-class Charles Booth wrote very curiously: "It is much to be desired and it is to be hoped that this class may become less hereditary in its character; there appears to be no doubt that it is now hereditary to a very considerable extent." This seems to be a misuse of the word "hereditary."

† The proposal for diplomas or certificates for eugenically fit young people was first made by Galton in 1873; see our Vol. II, pp. 120–1.

‡ The reader may be reminded that Galton was to receive the Huxley medal at the conclusion of this lecture before the Institute.

Galton next turns to the selection of women which he apparently considers harder than that of men students, because they are fewer. He would lay stress on their athletic proficiency and on their capacity to pass a careful medical examination, and he would pay more attention to their hereditary family qualities, under which he includes those of fertility and prepotency.

This idea of diplomas may raise a smile, but experience has shown the present writer its feasibility, when public opinion is ripe for it. In any university the anthropometric laboratory which tests some 25 or 30 physiological, mental and physical characters, the eugenics laboratory which studies family pedigrees, the academic examinations and the numerous athletic competitions could in combination, if guided wisely, place university students into classes graded sufficiently finely for Galton's aims. I believe there would be no greater difficulty and considerably more accuracy than was reached during the Great War in grading conscripts into *A*, *B* and *C* classes and their subdivisions. But having admitted the possibility of at least approximately selecting our promising youths* can we be certain of their subsequent performance? This is the subject of Galton's next section.

He remarks on the real difficulty of the problem whether a classification in youth would be a trustworthy forecast of qualities in later life, but states that for eugenic purposes this classification of the relatively young is essential:

"The accidents that make or mar a career do not enter into the scope of this difficulty. It resides entirely in the fact that the development does not cease at the time of youth, especially in the higher natures, but that faculties and capabilities which were then latent subsequently unfold and become prominent. Putting aside the effect of serious illness, I do not suppose there is any risk of retrogression in capacity before old age comes on. The mental powers that a youth possesses continue with him as a man, but other faculties and new dispositions may arise and alter the balance of his character. He may cease to be efficient in the way of which he gave promise, and he may perhaps become efficient in unexpected directions.

The correlation between youthful promise and performance in mature life has never been properly investigated†. Its measurement presents no greater difficulty, so far as I can foresee, than in other problems which have been successfully attacked....Let me add that I think its neglect by the vast army of highly educated persons who are connected with the present huge system of competitive examinations to be gross and unpardonable. Neither schoolmasters, tutors, officials of the universities, nor of the State department of education‡, have ever to

* It will be seen that the lecturer does not deal with the equally, perhaps more, important classification of other social grades, for example craftsmen and factory workers.

† E. Schuster, the first Galton Research Fellow, broke ground in this direction in his paper in the *Eugenics Laboratory Memoirs*, No. III, "The Promise of Youth and the Performance of Manhood," but the subject demands the treatment of still ampler material.

‡ Some years ago our Civil Service Examinations—the most elaborate system of State marking—were analysed in the Biometric Laboratory, not only with a view to testing the very empirical system of marking therein adopted, but also of ascertaining whether the marks thus settled were a real criterion of relative ability. The sole additional data needed were appreciations of success in State service after a period of 20 or 25 years. At first one believed salary might be such a test, but it was soon clear that other factors than ability were liable to determine salary. A control which I proposed, namely a classification in five classes of success, the judgment to be made by those acquainted with the inner working of the several offices (and to be treated as strictly confidential as to the individual), was at first accepted, but later rejected. Meanwhile the Government appears to have no proof—which must of course be statistical—either that its system of marking is a real measure of relative ability, or that the individuals thus selected fulfil in manhood the promise of their youth.

my knowledge taken any serious step to solve this important problem, though the value of the present elaborate system of examinations cannot be rightly estimated until it is solved." (pp. 533–4.)

Here Galton's judgment must appeal to every thoughtful man. Educational methods both in teaching and examination are put into practice on the balance of opinion in committees, or even by the arbitrary will of particular headmasters, and when the system is developed no attempt is made to determine statistically whether it really achieves what it professes to do. The preparatory schools prepare for the public schools' examinations, the public schools are again in their teaching controlled by the examinations on which the universities distribute their prizes, and finally distinction in the academic graduation examinations is an all-important factor in many lucrative appointments. Our educational system may be the very best available, as apparently its administrators believe it to be; but public confidence in it would be based on a firmer footing if those administrators would occasionally take stock and prove to us that the promise of youth has been fulfilled in adult performance. We debate and we legislate, we educate and we examine— and never take the trouble to inquire after a few years whether the results we aimed at have been achieved!

Galton next turns to the question of the augmentation of favoured stock. It is clear that the improvement of the stock of a nation depends on our power of increasing the productivity of its best members. He considers this of more importance than repressing the productivity of the worst stock; he does not give his reasons for this view, possibly he holds the production of one superman to be in the long run more profitable to a nation than the repression of fifty subhumans; it is better to spend all available funds in the production of men of outstanding civic worth, rather than in the reduction of the number of undesirables. Galton's main proposal certainly would involve considerable expense; it is that his youths and maidens, selected for all types of outstanding civic worth, should be put under conditions where early marriage is feasible and large families are not detrimental to success. He holds that with able and cultured women in particular there might be a reduction in the age at marriage from 28 or 29 to 21 or 22, thus prolonging marriage by seven years. This would not only save from barrenness the earlier part of the childbearing period of these women, but would shorten each generation by some seven years. Galton considers that it is no absurd idea that outside influences should hasten the age of marriage or lead the best to marry the best. "A superficial objection is sure to be urged that the fancies of young people are so incalculable and so irresistible that they cannot be guided." So they are—in the exceptional case which only proves the contrary rule*. But the anthropologist is only too familiar with the fact that marriage is the most custom-ridden institution of humanity, and the variations in its customs are as wide as the races of mankind. At least 95 °/$_\circ$ of men and women marry not only according to the custom of their nation, but according to the habits of

---

* Galton cites as such the lady who scandalised her domestic circle by falling in love with the undertaker at her father's funeral and insisting on marrying him!

the small section of it to which they belong; the agricultural lad and lass early and within their district; the cultured man and woman late and yet within their own circle.

"An enthusiasm to improve the race would probably express itself by granting diplomas to a select class of young men and women, by encouraging their intermarriages, by hastening the time of marriage of women of that high class, and by provision for rearing children healthily. The means that might be employed to compass these ends are dowries, especially for those to whom moderate sums are important, assured help in emergencies during the early years of married life, healthy homes, the pressure of public opinion, honours, and above all the introduction of motives of a religious or quasi-religious character.

"Indeed an enthusiasm to improve the race is so noble in its aim that it might well rise to the sense of a religious obligation. In other lands there are abundant instances in which religious motives make early marriages a matter of custom and continued celibacy to be regarded as a disgrace, if not a crime. The customs of the Hindoos, also of the Jews, especially in ancient times, bear this out. In all costly civilizations there is a tendency to shrink from marriage on prudential grounds. It would, however, be possible so to alter the conditions of life that the most prudent course for an X-class person should lie exactly opposite to its present direction, for he or she might find that there were advantages, and not disadvantages in early marriage, and that the most prudent course was to follow their natural instincts."

When Galton comes to the consideration of "Existing Agencies," we are bound to admit how few endowments of real eugenic value exist at present. Galton suggests what might be done rather than what is already available. With an annual expenditure of £14,000,000 on charities might not more be achieved in producing the healthy fit than in tending the unhealthy weak? How much of this huge charitable expenditure may not really be opposed to eugenic doctrine in its effects? Galton refers to endowments by scholarships and fellowships, but does not say that their present length of tenure is inadequate for his purpose; he thinks that wealthy men might be proud to befriend poor but promising lads without the patron being "a wretch who supports with insolence and is repaid by flattery." He commends the wise landlord of a large estate who builds healthy cottages and prides himself upon having them occupied by a class of men markedly superior to those in similar positions elsewhere.

"It might well become a point of honor, and as much an avowed object, for noble families to gather fine specimens of humanity around them as it is to procure and maintain fine breeds of cattle, etc., which are costly but repay in satisfaction." (p. 537.)

Our author has his Utopias, as many men have had with less scientific insight behind them. He dreams of settlements or colleges where promising young couples might be provided with healthy and convenient quarters. " The tone of the place would be higher than elsewhere on account of the high quality of the inmates, and it would be distinguished by an air of energy, intelligence, health and self-respect, and by mutual helpfulness." He dreams again his dream of 1873 * of a great society with ample funds recording the abler of every social class, seeing to their intermarriage, and establishing personal relations between them.

But while he dreams he realises that the first thing is to justify a crusade in favour of race-improvement; to show step by step that it is both from the

* See our Vol. II, pp. 119–122. He dreamt it again in the Utopia he described in the last few months of his life: see the letters of the autumn of 1910 below.

scientific and the practical standpoint possible; to fill up by research the gaps in our ignorance and make every stepping-stone safe and secure. He would be content if his lecture justified men "in following every path in a resolute and hopeful spirit that seems to lead towards that end." And he concludes :

"The magnitude of the inquiry is enormous, but its object is one of the highest man can accomplish....We cannot doubt the existence of a great power ready to hand and capable of being directed with vast benefit as soon as we shall have learned to understand and apply it. To no nation is a high human breed more necessary than to our own, for we plant our stock all over the world and lay the foundation of the dispositions and capacities of future millions of the human race." (p. 538.)

Thus Galton concludes the second Huxley Lecture of the Anthropological Institute; it is possibly the only one of the series which is destined to live, for it founded a new science, which in truth carried with it the germs of a great future social movement. But the seed fell on barren soil, it found no echo in the researches of British anthropologists, and the lecture, perhaps the most weighty paper their Institute had heard, was never fully published in their Journal. It attracted more attention and bore ampler fruit in America than in this country.

Nothing daunted Galton determined to appeal to a wider public and another class of mind. From now on he made it his chief purpose to spend his remaining years and energies in teaching the public that they had to take Eugenics as seriously as any other branch of science with practical applications.

It must not be supposed, however, that Galton's devotion of his remaining years to Eugenics cut him off entirely from other interests and from his habitual helpfulness to other allied causes. I find that the letters interchanged between us during the years 1900 to 1902 turn largely on the foundation of *Biometrika*, and it is pleasing to recall the sympathy expressed and the help which the Master's letters in those days of stress were to Weldon and myself, his disciples. Unfortunately it is not possible to understand the setting of Galton's letters or the frank and generous relationship between the older man and his lieutenants without publishing certain letters of the latter, which maintain the thread of the narrative. My own correspondence with Francis Galton is scattered over nineteen years, and only small portions of it can be published in this chapter. I shall select here a portion from the correspondence for the years 1900–1902, which, we must remember, were marked for Galton by (i) the foundation of *Biometrika*, (ii) the delivery of the Huxley Lecture, (iii) the award of the Darwin Medal, and (iv) the election to an Honorary Fellowship at Trinity College, Cambridge.

The following letters bearing on these points may first be cited as throwing light on parts of that correspondence :

INNISFAIL, HILLS ROAD, CAMBRIDGE. 24 *June* 1901.

MY DEAR MR GALTON,   I have been commissioned by the Council of the Anthropological Institute to ask whether you would do us the honour to deliver the Huxley Lecture this autumn or early winter, and at the same time to receive the Huxley Medal.

We would like in this way to emphasise our appreciation of the value of your researches, which have placed biological data on a prime mathematical basis. You have been the pioneer in the Mathematical School of Evolution, and Anthropology has benefitted enormously, not only by your investigations, but by those which you have directly or indirectly instigated and inspired. Who then is better fitted to discourse to us than a Pioneer Investigator in one corner of that field of which in other departments Huxley was a brilliant exponent?

We sincerely trust that you will add another self-denying good deed for the sake of Anthropology, and will favour the Institute, and benefit our Science, by acceding to our urgent request.　　　　Believe me, my dear Mr Galton,

Yours most faithfully, ALFRED C. HADDON.

This letter shows a real appreciation of Galton's services to Anthropology, but, as I have indicated, his lecture found no response in the writings of English anthropologists.

In announcing the award of the Darwin Medal to Francis Galton on Dec. 1, 1902, Sir William Huggins said it was conferred

"for his numerous contributions to the exact study of heredity and variation contained in *Hereditary Genius, Natural Inheritance* and other writings. The work of Mr Galton has long occupied a unique position in evolutionary studies. His treatise on *Hereditary Genius* (1869) was not only what it claimed to be the first attempt to investigate the special subject of the inheritance of human faculty in a statistical manner and to arrive at numerical results, but in it exact methods were for the first time applied to the general problem of heredity on a comprehensive scale. It may safely be declared that no one living had contributed more definitely to the progress of evolutionary study, whether by actual discovery or by the fruitful direction of thought, than Mr Galton "

And, now the letter which Francis Galton valued more than all! It runs:

TRINITY COLLEGE, CAMBRIDGE. *Nov.* 14, 1902.

MY DEAR FRANK, Many happy duties have come to me in my life, but few happier than that of now informing you, by the direction of our Council, that we have today elected you an Honorary Fellow of the College under the provisions of our Statute XIX, as a "person distinguished for literary and scientific merits."

We are electing at the same time Mr Balfour, Sir William Harcourt, Lord Macnaghten and Professor Maitland. Our other Honorary Fellows, since the deaths of Bishop Westcott and Lord Acton, are Lord Rayleigh and Sir George Trevelyan.

Need I say how it delights me to think that all your long and brilliant services in the cause of many a science should again link you in the later years of your life with the College to which, as I know, you have always been so loyal?

Believe me, very affectionately yours, H. MONTAGU BUTLER.

Can you kindly let me know by Telegraph whether you accept? I should like, if possible, to announce the five Fellowships *together*.

Since writing the above I have just seen the award of the Darwin Medal! Very delightful.

When a man is young, honours are a powerful incentive to further work, and as the years go by they test the judgment of those who conferred them. When a man is old—Galton was 80 years of age, and the wider world had long pronounced its judgment—honours mean far less to him, and need little exercise of judgment on the part of the givers*. There is a form of honour,

* Putting aside membership of learned societies at home and abroad and the holding of offices therein, I may note the following honours conferred on Galton: Gold Medal, Royal Geographical Society, 1853; Silver Medal, French Geographical Society, 1854; Royal Medal of Royal Society, 1886; Officier de l'Instruction publique de France, 1891; D.C.L. Oxford, 1894;

however, which gives most Englishmen intense pleasure. They feel bound, in a way that many foreigners find it difficult to understand, to their school, college, or university. These institutions have in many cases fascinating traditions, stately buildings and beautiful environment; they act on their students and inmates at a period when their minds are most impressionable, when they are learning to understand the value of friendship; when they first begin to realise all that life may mean for them. This is peculiarly true in the case of youths like Francis Galton who reach the free atmosphere of a University without the background of a great public school behind them. Too many public school boys miss half the joy of their undergraduate days by holding too tightly to their school traditions and friends, so that the College or University appears to them chiefly as a club where they can strengthen old associations. With Galton it was different, like Columbus he discovered the wonders of a new world, and what was largely due to his own mental growth he attributed to his College, to the intellectual and physical environments it provided; and, as so many have done, he felt a love for it, instinctive, like that we feel for the mother who reared us, or for our country. Such love is difficult to defend on rational grounds; the personnel of a college may be as "dull as the pictures which adorn their halls," our fellow-students may be mediocre—but blessed be the man unknown who put those two words together, *Alma Mater*, and applied them to the communal homes of our youth, those ever-verdant pastures, that we always look back to from the dusty highways of later life! Their honours are what we value most, even if their worth be little esteemed by the outer world;—an emotion no doubt of the heart, not a demand of the head, yet there are times when Rousseau gives greater delight than Voltaire. And the octogenarian was moved as he had scarcely been by other honours, much as his simple modest nature always rejoiced in any recognition, however long, as it seemed to us outsiders, postponed. Thus Galton wrote to his sister Emma about the Darwin Medal:

HÔTEL DES ANGLAIS, VALESCURE (VAR), FRANCE. *Nov.* 14, 1902.

You are so sympathetic that you will be glad to know that the Royal Society has awarded me the Darwin Medal for my "numerous contributions to the exact study of Heredity and Variation." It was established some few years ago, and is awarded biennially (or is it triennially) without regard to nationality. Grassi, the Italian, got it last time for his discovery of the life history of eels, whose early life had puzzled zoologists from before the days of Aristotle onwards. He found that some creatures that were fished up from the Straits of Messina (Sicily) were young eels and that eels alway go to *deep sea* waters to breed.—Well, I am very pleased except that I stand in the way of younger men. All well, except that my cough plagues me at night, a little before daybreak. No mosquitoes here. We are the only people in the hotel.

Ever affectionately, FRANCIS GALTON.

Wallace, Hooker, I think, and Karl Pearson are, besides Grassi, the previous medallists.

Hon. Sc.D. Cambridge, 1895; Linnean Society Medal (Darwin-Wallace Celebration), 1908; Knighthood, 1909; Royal Society, Copley Medal, 1910; and those recorded above. All, with the exception of the Geographical Medals, were conferred when Galton was well over 60 years, and in some cases over 80!

But about his *Alma Mater* he wrote:

HÔTEL DES ANGLAIS, VALESCURE, PRÈS ST RAPHAEL (VAR), FRANCE. *Nov.* 16, 1902.

DEAREST EMMA, Your letter has just come with the 2 extracts. Thank you much; I was sure that you and Bessy and Erasmus would all be glad to hear of the Darwin Medal. But there is even more to tell, of even yet more value to myself. They have elected me Honorary Fellow of Trinity College, Cambridge, which is a rare distinction for a man who has not been previously an ordinary fellow, or who is not a professor resident in Cambridge. The beautifully conceived and worded letter of Montagu Butler, the Master of Trinity, of which Eva has made a copy for you to keep, will explain much of this. Mr Balfour was, I think, a fellow, anyhow he was one of the most brilliant men of his year. Sir W. Harcourt and Lord Macnaghten were fellows, so I presume was Maitland who is a resident professor. Lord Acton was a professor. Sir G. Trevelyan was 2nd classic of his year, but did not wait long enough in England to gain his fellowship. It was given him after his successful administration as Irish Secretary. Bishop Westcott was of pre-eminent reputation as a theologian and as a classic, and had been an ordinary fellow. So had been Lord Rayleigh.

So I am in very good company indeed. Is it not pleasant? This is a sort of recognition I value *most* highly. All the more so, as I did so little *academically* at Cambridge, in large part owing to ill health. But I seem to owe almost everything to Cambridge. The high tone of thought, the thoroughness of its work, and the very high level of ability, gave me an ideal which I have never lost.

So much egotistically. I am getting much stronger here, and have made the discovery that much of my asthma has been due to warm and overcarpeted rooms. Mine here I have now had cleared of carpet and *underlying straw*. It feels so much purer and wholesomer. The first night after it was done I had no asthma at all. Looking to past experiences, I now see how commonly warm and carpeted rooms have been associated with my asthma, notably the drawing room of the Athenaeum Club, where I can rarely sit 10 minutes without beginning to cough. I am planning the taking up of carpets in my drawing, dining, bed and dressing rooms at home, and varnishing and staining the floors. I have two uncarpeted rooms there already where I have long noticed that I cough less than elsewhere (the bathroom and my workroom*).

The weather has been delicious here this morning. I took a good 4 miles walk without being tired, which is far in advance of what my powers were during the past summer. How I wish you† could get up and take walks too! We have a few friends already come back....

Bessy's journeyings for meals on account of kitchen repairs at her own house are amusing. So is V... B...'s consignment of *beetles*!

Loves to Bessy, Erasmus and all. What are Erasmus' walking powers now when at his best? How many miles does he think he could manage†?

Eva sends her love [here the handwriting changes]—and you will be glad to hear that Uncle Frank is looking remarkably well; this place has done a great deal for him mentally and physically; he can walk and eat and sleep like any ordinary person, but he does not present a very handsome appearance having a head still spotted with about 36 remaining bites from the mosquitoes of Hyères. We are so happy here, yr. affect. EVA. [Galton concludes] So much from Eva, who sketches and paints assiduously.

Ever affectionately, FRANCIS GALTON.

A characteristic letter showing two sides of Francis Galton's feelings, towards his *Alma Mater* and towards his "sibship." One further letter

---

* The "workroom" at Rutland Gate was a very depressing room, with a single window looking into a well or high-walled court. On deal shelves were placed boxes of pamphlets and papers; it gave one the impression of a store-room rather than a study. I think Galton chiefly worked, when on the ground floor at a writing table at the dining-room front window and when on the first floor at an oak bureau in the drawing-room.

† Francis was now 80, Erasmus 87, Emma 91 and Bessie 94!

concerning these matters may be printed here. It bears witness to the widespread admiration and affection felt for Francis Galton*.

TRINITY COLLEGE, CAMBRIDGE. 19 November, 1902.

MY DEAR MR GALTON, It was only today I heard, with very great pleasure, that your old College has done itself the honour of asking you to become one of its Honorary Fellows. We are proud of the distinction which you confer on the College, and we trust that you will not refuse to accept this mark of our sense of the great services you have rendered to science. To me the act of the College gives a personal pleasure, for I shall never forget your kindness to me at a critical time of my life, and I am happy and proud to think that I have enjoyed the privilege of your friendship ever since.

Let me take this opportunity of congratulating you on receiving the Darwin Medal. It is a high distinction, and I am sure you have richly deserved it.

Believe me, dear Mr Galton,

Yours most sincerely, J. G. FRAZER.

As I have said on p. 235 the current of Galton's thoughts in these years and his strong affection will be best made clear to the reader if I print here a small selection of the correspondence which passed between us in the years 1900–1902. The letters indicate Galton's essential generosity of mind, the close terms of intimacy he was on with Weldon and myself,—who were proud to feel ourselves in some measure his lieutenants,—and the keen interest he had in the early struggles of *Biometrika*. The feeling of the younger men among us, who got into close touch with Francis Galton, was something like that of Aristides to Socrates:

"I always made progress whenever I was in your neighbourhood, even if it were only in the same house, without being in the same room; but my advancement was greater if I were in the same room, and greater still if I could keep my eyes fixed upon you." It was not Galton's power of solving problems: suggestive as he was, his analysis often lacked power to cope with them. It was the atmosphere he cast round every scientific question; he carried his intimates into a rarefied air, where the one aim was to reach the goal of truth, not heeding who should get there first, or who should tell the tale of its discovery. I think the like conception expressed in different words is provided by Mrs Sidney Webb†:

"Owing to our [the 'Potter girls'] intimacy with Herbert Spencer we were friendly with the group of distinguished scientific men who met together at the monthly dinner of the famous 'X-Club.' And here I should like to recall that among these scientists, the one who stays in my mind as the ideal man of science is, not Huxley or Tyndall, Hooker or Lubbock, still less my guide, philosopher and friend Herbert Spencer, but Francis Galton whom I used to observe

---

* Sir James Frazer in kindly granting me permission to print his letter remarks "that the 'critical time of my life' referred to in my letter was in 1885, when my Trinity Fellowship would in the ordinary course have expired and the question of renewal came before the College Council. In the same year, shortly before, at Mr Galton's suggestion, I had read my first anthropological paper ('On some burial customs as illustrative of the primitive theory of the soul') before the Anthropological Institute, with Mr Galton as President in the chair, and when the question of the renewal of my Fellowship was raised shortly afterwards, I believe that Francis Galton and my ever-lamented friend Robertson Smith used their powerful influence to ensure the renewal and were successful. It was indeed a turning point in my life, and I shall never cease to be grateful to the two friends who stood by me at that critical time.......
...He [Galton] was indeed an admirable and lovable man from every point of view."

† Beatrice Webb, *My Apprenticeship*, pp. 134–5, 1926.

and listen to—I regret to say without the least reciprocity—with wrapt attention. Even to-day I can conjure up from memory's misty deep, that tall figure with its attitude of perfect physical and mental poise, the clean shaven face, the thin compressed mouth with its enigmatical smile, the long upper-lip and firm chin, and as if presiding over the whole personality of the man the prominent dark eyebrows from beneath which gleamed with penetrating humour, contemplative grey eyes. Fascinating to me was Francis Galton's all-embracing, but apparently impersonal beneficence. But to a recent and enthusiastic convert to the scientific method, the most relevant of Galton's many gifts was the unique contribution of three separate and distinct processes of the intellect: a continuous curiosity about and rapid apprehension of individual facts, whether common or uncommon; the faculty for ingenious trains of reasoning; and more admirable than either of these, because the talent was wholly beyond my reach, the capacity for correcting and verifying his own hypotheses by the statistical handling of masses of data, whether collected by himself or supplied by other students of the problem."

The following letters may serve to illustrate and deepen the above very admirable characterisation by a skilful artist in words!

(5) *Selected Correspondence between Galton and his biographer, illustrating the years* 1900–1902.

TEWFIK PALACE HOTEL, HELOUAN, CAIRO. *February,* 1900.

DEAR PROF. K. PEARSON, Thank you heartily for letting me see, as a New Year's gift, the important proof sheets. By much hammering, the bad part of the "law*" will be knocked out of it and the good, if any, will remain. You know probably that India ink (1) in water and common ink (2) may look alike, but if you pass the former through a filter of blotting paper the water alone comes through; not so with regard to ink. Now a mixture of (1) with water is not properly a blend, but a mixture with (2) is. When the particles in any case of "particulate" inheritance are small and *independent*, I do not see any sensible difference (within reasonable limits) between the behaviour of the two. But now comes in the consideration which I take to be the great problem, and that which as I conceive lies at the bottom of *stability of type*, viz.: regarding the imperfectly explored facts of group-correlation. Let, in a given "stirp," a, b, c, ... be *classes* of elements which develop in that order, the several classes consisting of a_1, a_2, ..., b_1, b_2, ..., &c. varieties. Now we find that a certain lineament, or trait, a_r, b_s, c_t, &c. tends to be inherited. If a, b, c, &c. were independent, the probability against the above particular combination would be enormous, whereas it is found to be frequent. What then is the cause? or, in default of knowing the cause, how can we represent to ourselves the character of the correlation? If a, b, c are developed in that order of succession, the particular and not improbable sequence of a_r, b_s must make the next step to c_t far more probable than if b_s had been preceded by say a_n or some other variety of a.

There must be an accumulating correlation of some kind. But how if a, b, c, &c. are simultaneously developed? Here I fail to make any picture to my mind of the way in which the needed group-correlation acts. I often watch the family traits in a party at church, trying to find out the beginnings and the ends in each inherited lineament of resemblance whether to the parents or to one another. They are usually indefinite, I think. My servant writes me word that your "Grammar of Science" has just arrived at Rutland Gate. Thank you sincerely. I must wait till my return, to read it.

We have had a very interesting and healthful journey to Wady Halfa and back, including a week's stay with Flinders Petrie at his diggings. The climate here near Cairo is far from being always benign. There are days of stormy wind with dust, and occasional down-pours of rain. I can't make up my mind as to the best places for an invalid—certainly neither Cairo nor Luxor. I have had two pleasant days in the desert with Prof. Schweinfurth the famous traveller.

I trust you have pulled through the wretched English winter fairly well.

Very sincerely yours, FRANCIS GALTON.

I hope to be back about Mid May.

* The Law of Ancestral Heredity, especially its application to alternative inheritance.

42, Rutland Gate, S.W. *June* 6, 1900.

Dear Prof. Karl Pearson,   On returning from a six months absence in Egypt and Greece, I found your valuable *Grammar of Science* on my table, and am reading it straight through at the rate of about an hour a day, with admiration at your thoroughness. It takes some time, as I find, to pick up dropped threads, so I have as yet little leisure.

I wonder if you have worked out the relationship between those who are cousins in a *double degree*, I mean the issue of the marriages in which 2 brothers have married 2 sisters. Their ancestry from Grandparents upwards, is identical. I should be very curious to learn what value you would assign to it in your "table of collateral heredity," p. 481 of the book.

I hope the past cruel winter in England has not hurt you. Weldon, whom I saw last week, spoke favourably of your health.

My tour has done me a world of good, besides being extremely interesting and pleasant.

Very sincerely yours, Francis Galton.

7, Well Road, Hampstead, N.W. *Dec.* 13, 1900.

My dear Mr Galton,   Your kind letter was very welcome tonight. 1 tried some year ago to sound people with regard to a journal of pure and applied statistics, but found a feeling pretty general that it might injure the R. S. S. Journal, although the sort of memoirs I had in view would I think not find a place in that journal. On the other hand I know a good many papers for which I hardly see a place and there is increasing material being gathered in Germany and America which is lost among masses of purely zoological papers or published in inaccessible proceedings. I think if a journal could survive its first two or three years there is a future for it of great service.

The thing came to an issue just now owing to doings at the Royal. My paper on Homotyposis was sent for some reason to Bateson as referee—he chose to tell me so himself, and also to tell me that he had written an unfavourable report. He came to the R.S. at the reading and said there was nothing in the paper—that it was a fundamental error to suppose that number had any real existence in living forms. That this criticism did not apply to this memoir only but to all my work, that all variability was differentiation, etc., etc.

Now all this may be quite fair criticism, but what is clear is that if the R.S. people send my papers to Bateson, one cannot hope to get them printed. It is a practical notice to quit. This notice applies not only to *my* work, but to most work on similar statistical lines. It seems needful that there should be some organ for publication of this sort of work and talking it over with Weldon, he drew up the prospectus, I gave a name,—the "K" was mine (K. P. not C. P.), —and we determined to see what amount of cork was forthcoming to float such a project. I don't think much can be done if we don't get 150 to 200 promises. But can we?—I fear not.

Yours always sincerely, Karl Pearson.

42, Rutland Gate, S.W. *Jan.* 2, 1901.

My dear Prof. K. Pearson,   Here is the MS. on Eye Colour, which I am delighted is of use to you still. I hope not to go abroad yet awhile, but it would be safer to write on the parcel when you send it back, "To await return." Tell me please, in time, whether the answers you have received relating to the new magazine or journal, are encouraging enough for a probable start.

Bateson's adverse views cannot be finally effective, being opposed to those of many other no less worthy authorities. But I presume from what you said, that they *are* effective as against the particular memoir on Homotyposis? Very sincerely yours, Francis Galton.

42, Rutland Gate, S.W. *Jan.* 7, 1901.

Dear Prof. Karl Pearson,   Thank you much for the "Lecture." It fits in with much that I habitually think about.—I wonder if this strikes you as reasonable:—

Probably zeal for military usefulness will cause many men to be physically examined as to fitness to serve. There are also medico-physical examinations for other services. Could any sort of *Degrees* be given to those (*a*) who simply pass the required standard for the particular purpose, (*b*) to those who pass as valid for purposes of hereditary transmission.

Two other examinations exist, that might be included in the (*b*) set:

1. That of a Life-Insurance Co. to certify a first-class life, which includes some facts about parents and brothers, together with local inquiries by their agents. I don't know what the cost of this may be in each case, but certainly the fact of being accepted as a first-class life by any notable Life Insurance Co. is an important fact, worthy of recognition.

2. Ordinary literary examination, to show that the man is not a real stupid.

Now fancy that Degrees are offered of a V. H. T. (valid for hereditary transmission of qualities suitable to a citizen of an Imperial Country) would they meet a want, and would they help in forwarding marriages of the fittest and discouraging others in any notable degree? If a well considered answer be "yes" I suppose the action would be to write an article upon it, with plenty of solid stuff in it and then if the idea should *take*, to follow mainly the direction in which "the cat may jump." If tried, it ought to be tried at first on a small scale, that is in a small community by a self-constituted board, laying down their own conditions and giving their certificate as a "Degree." One great question is that of self maintenance when once fully started and running. I should think the cost of the mere medical and physical examination would not be beyond the powers of, say, Cambridge Undergraduates and I fancy that (always supposing the idea to catch) it might be possible to get some help from the present examining authorities in respect to the (*b*) condition. I mean that arrangements might be made by which an Examination by one of these should be accepted by the Certificate or Degree-giving board.

I have thought over the subject a good deal and have more to say, but unless what *has* been said above seems reasonable to persons like yourself, the supplementary remarks would be useless. Will you kindly think this over at odd times during the next 2 or 3 days? I have written about it to no one else.

There is another important point of "what severity of selection should be aimed at." A very moderate one would, I think, meet the need. Say that $\frac{2}{3}$ pass and $\frac{1}{3}$ fail. The effect on the hypothesis that the successes alone intermarry and keep up the population would roughly put the output of children in the hands of the best half of all possible married couples—$\frac{4}{9}$ths of them. (Of course this is the rudest way of putting it; but it will do for present purposes.) If men, like cattle or Mormons, were polygamous a much severer selection would be wanted.

Very sincerely yours, FRANCIS GALTON.

7, WELL ROAD, HAMPSTEAD, N.W. *Jan.* 10, 1901.

MY DEAR MR GALTON, It would be a very great pleasure to me to know you were going to take the field with regard to what I am convinced is of the greatest national importance—the breeding from the fitter stocks. If one could only get some one to awaken the nation with regard to its future!—The statesmen, who really have the ear of the populace, never think of the future. They will not touch the question of coal supply nor that of fertility, and yet I am convinced these are far more important for the future existence of the nation than any question of local government, church discipline, or even technical education!—I think I told you we had nearly completed the reduction of our measurements on 1100 families, and one after another of the results confirm the higher series of values, about ·5 for parental correlation, that I found from the eye and horse colour data. I shall probably not publish these results for some time, as I have half made up my mind to accept an invitation to lecture at the Lowell Institute in Boston this year and these materials would be a good basis for lectures on Heredity. But they emphasise even more emphatically than your earlier value of $\frac{1}{3}$, the opinions you have expressed on the great part played by good stock in the community. Heredity is really more intense than we supposed it to be 10 years ago. Cannot this be brought forcibly home to our rulers and social reformers?

Now the difficulty in this case seems to me to be twofold. How can you (i) stop the fertility of the poor stock and (ii) multiply that of the good? The middle classes are I take it the result of a pretty long process of selection in this country, and I believe that they alone are the classes who largely insure. Your scheme therefore would at first apply only to them, and indeed to the best of them, for the others would not care a rap for a good bill of health, any more than they do for any moral suasion. You might influence by your health degree a small percentage of the whole community, say 4 per cent., but this percentage is probably identical with those you could equally well influence by moral suasion. I mean by preaching the gospel that the stability of the nation depends upon the good stocks breeding fully and the weak exhibiting restraint. But how are you going to get the better class workman to see that his checking the size of his family may make matters easier for him, but is at the expense of the nation's future? He is really unreachable by an assurance scheme, *unless you could attach your health degree to the proposals for old age pensions**. That appears to me a point worth thinking about. As I have said elsewhere it seems to me that only socialistic measures can touch this population question. Even if you can by moral suasion lead the better class artizans and the middle classes to see that limitation of the family may be anti-social (and I believe it might be possible) how are you going to check the unlimited production of the worse stocks? The "Neomalthusians"—as I know from sad experience—abuse any one who like myself ventures to criticise their doctrine of limitation, unless it be accompanied by the words "of the poor stocks *first*"; but this abuse is nothing to what one will arouse, if one ventures to assert that the huge charities providing for the children of the incapable are a national curse and not a blessing; that the "widow with seven children all dependent upon her, husband a clerk who died of consumption aged 35," and who seeks your aid to get her children into Reedham, is really a moral criminal and not an object for pity.

How can a health degree affect this source of rottenness? I fear hardly at all. Your only hope is to impress upon the few who really lead the nation, that the matter is one for legislation, that although we have got rid of Gilbert's Act, the workhouse and charity systems can still be sapping our national vigour, when coupled with a wide-spread neomalthusianism—due in the main to Bradlaugh—among the better working classes.

What then it seems to me we mostly need at the present time, is some word in season, something that will bring home to thinking men the urgency of the fertility question in this country. There is no man who would be listened to in this matter in the same way as yourself. You are known as one who set the whole scientific treatment of heredity going; no one has ever suspected you of being in the least a "crank," or having "views" to air. You will be listened to and it will be recognised that you write out of a spirit of pure patriotism. There is no one else, I believe, of whom this could be said, certainly no one who would be listened to in the same way. Let us have (*a*) known facts of heredity, (*b*) influence of relative fertility on national vigour, (*c*) actual statistics of birth rates of different stocks, and (*d*) proposed remedies (only, if they include the health degree, tack it on to old age pensions) brought home to those who think for the nation. Always sincerely yours, K. PEARSON.

If *Biometrika* be started Weldon and I want badly a paper however brief from you for No. 1.

7, WELL ROAD, HAMPSTEAD, N.W. *February* 1, 1901.

MY DEAR MR GALTON, I have several times planned to write and ask if I might come and see you, and now you are off before I have done so! I have been "crawling" through my work since December somehow, feeling mentally too tired to do more than get through my routine teaching and making no attempt beyond the day's necessary doings. My helpers go forward but I can only look on. I suppose one must pay eventually for all overwork, only one longs for a few more years to "finish up." Yes, I have settled on the American Lectures on Heredity and Variation for October. If any ideas on diagram-illustration occur to you, I should be very glad of suggestions. I have found a Genometer based on a suggestion of yours very useful at more than one popular lecture. It contains a gigantic lifeguardsman, a diminutive sailor and a "mean" man and illustrates the effect of any number of ancestors or collaterals of these types by means of a string working up and down. It always amuses people.

You will share my pleasure in the acceptance of the Homotyposis paper for the *Phil. Trans.* I hope we may float *Biometrika* so that one could to some extent relieve the pressure on the R.S. space, which I think is to some extent grudged. We had however only about 12 English acceptances, and we cannot venture even a first number without something like 100. We are now circularising everybody in America, Germany and Italy, but I am not very hopeful.

I suppose the Riviera is hardly a place where birds' eggs abound? I want to measure another 100 clutches of some species but hardly know which to select or where to go for it.

---

* [Galton wanted a medical examination such as the better insurance offices insist on extended to all classes of the nation. My suggestion was that a grading of lives was essential to a really sound national provision for sickness and old age pensions, proposals for which were then creating some stir. K. P.]

I fear that to ask for 100 thrushes or blackbirds nests in England would raise a scandal. I got much reproved for my 200 house sparrow nests last year. I trust that your journey may be a pleasant one and that you may escape the horrors of February and March, which my Wife tells me occasionally reach the South of France. You know Miss Shaen is at San Remo?—May I still keep the eye colour MS.? If you would prefer its return before you leave, just say so on a postcard. Always yours sincerely, KARL PEARSON.

7, WELL ROAD, HAMPSTEAD, N.W. *April* 18, 1901.

MY DEAR MR GALTON, I wonder if you are back from your winter journeyings. I want to tell you about the present state of *Biometrika*. We have about 60 promises of subscription, and we shall hardly get more now until the journal is definitely announced as coming out or until it has come. We have been talking over the matter with various publishers and printers, and so far the most reasonable terms seem to be those of the X— Press. Now it would be a great point to have the advertisement of this Press and the goodness of its get up, if we can. They are willing to take the journal on the same terms as they do the *Annals of Y*, which, with more expensive plates than we should think of, pays its way with some 270 subscribers. But they require a guarantee fund of £200. This they had in the case of the *Annals* and drew on pretty largely at first, but it is now refunded to the extent of £160. Whether we shall be equally successful is of course a very different matter, but I think there is no doubt that such a journal as *Biometrika* is wanted, and if we tide over the first few years, the journal will live. Weldon who was staying a few days with me this week wanted to take the whole risk on himself. This does not seem to me right. The natural thing would be for him and for me to share the risk, but with our very precarious condition at University College, this is out of the question. I can only guarantee a very modest sum. My view was that we should try and distribute the £200 about. Of course any one who subscribes may stand a very poor chance of seeing his money again, and to those to whom I have written I have said it must be looked upon as a loss until it reappears (if ever it does) as a stroke of fortune. I take it that the money would be banked and could be drawn only by joint order of Editors and Secretary of the X— Press.

Now I am writing to ask if you will aid to any extent in this proposal. I feel the less hesitation in frankly asking you because you are one of the men who I think can frankly say no, and the "no" would not affect our mutual relations.

Quite apart from this question, and I am sorry to refer to it in this letter, Weldon and I discussed two points: (1) The desirability, if you do not feel it involves you in worry and work, of getting you to join *in any way* the editorial committee. This consists at present of Weldon, myself and Davenport of Chicago, as American editor to collect material there. Of course we should be glad of any suggestion or aid you may care to give, but on the other hand we don't want to bother you with the hard work of the journal, and still less to make you in any way responsible for matter or method you might not sympathise with. (2) We want very badly to have a paper by you however long or short for our first number, a "send off" of some kind. Will you promise us this? You hardly know perhaps how much of weight your sympathy expressed in some form will carry with it, especially in America; it will be an uphill battle for some time with the biologists. Anyhow please let me know *first* your views as to my last two questions (1) and (2) and then rather more at your leisure whether you care to aid in the guarantee fund? I trust you have had a pleasant sojourn in the South. We are now having beautiful weather in Surrey. Yours always sincerely, KARL PEARSON.

HÔTEL BELLA VISTA, BORDIGHERA, N. ITALY. *April* 23, 1901.

MY DEAR PROF. KARL PEARSON, The straight-forwardness of your letter as to the probable total loss of the guarantee fund for *Biometrika*, is much more attractive to me than an enticing programme, for I like "forlorn hopes" in a good cause. I can just now spare the whole £200 and you shall have it, and I enclose the cheque, so you will be no longer bothered with that matter, and can give your spare energies wholly to starting the Journal.

As regards joining the Editorial Committee, if it could be done in a way that carried both in reality and in the eyes of the public no more responsibility and work than the position of "Consulting Physician" does in respect to a Hospital, I should be pleased to do so. Would

"Consulting Editor" *after* the names of yourself and Weldon as Editors do? Of other titles, "Referee" is almost the only one that occurs to me; probably you can suggest something. Of course a good-looking and well-printed title-page (not heavy-looking) is commercially helpful.

About writing a *short* "sending off" paper I think I could manage one on "Biometry,"—on its general aspect and principles. I have nothing serious enough in the way of original inquiry to give. Please send me a couple of copies (by return of post) of the programme, that I may better understand what may remain to be said. I trust you will see your way to make a considerable part of the contents of the Journal intelligible to those scientific men who are not mathematicians. It ought to be attractive to medical men and such like; also to statisticians of the better kind. Short notices of original work abroad always attract.

We stay here for a full week longer, I think,—and will leave address for letters that may arrive shortly after leaving. But 42, Rutland Gate will always find me in time.

Very sincerely yours, FRANCIS GALTON.

7, WELL ROAD, HAMPSTEAD, N.W. *April 27, 1901.*

MY DEAR MR GALTON, Your letter met me on my return home an hour ago. We have not any further programme printed at present than the circular I sent you some months ago of which I enclose two copies. Your letter made me very happy, partly because you so readily consented to my proposals as to editorship and giving us a "send off," partly because of the generally kind tone and sympathy it exhibited for our endeavours. As to your name as "Consulting Editor" and your proposed paper on the Aims of Biometry, these we may consider as settled, but I must consult Weldon before I reply fully as to your liberal offer. I think that he feels very much that you have done a great deal from the monetary side for biometry and that he would be unwilling to allow you to take so much of this burden on your shoulders. My view was to spread what I am unwilling until we have made trial to look upon as anything but a loss, over a number of guarantors, for I cannot carry my share of a moiety myself. But about all this I will write in a day or two when I have had an opportunity of considering the matter with Weldon. I don't propose to say what I personally feel about your readiness to aid, because it would be making into a personal kindness what I know is enthusiasm for the study of your life. I can only hope *Biometrika* will forward that, but it will have an uphill fight. Always yours sincerely, KARL PEARSON.

7, WELL ROAD, HAMPSTEAD, N.W. *April 30, 1901.*

MY DEAR MR GALTON, I have considered the matter of the *Biometrika* guarantee fund with Weldon and his view is that we should as frankly accept your offered aid as it is frankly given. It places us in a position to survive for at least four years and I think if we can survive the risk of infantile mortality we shall live on. At any rate we shall do our best to make the thing run and supply what we are sure is a real need. We want to make the science into a really great organ of discovery. It is almost pitiable to see how good material is wasted. I was reading a few days ago a paper by an American on colonies of statoblasts in which he had measured the variability in the general population and in the fraternity or colony. He introduced what he called a coefficient of heredity = (variability in population − variability in fraternity) ÷ (variability in population), and found this to be what he called small. Then he went into long reasons why heredity should be small in a colony of statoblasts. I found on working from his own data that the fraternal correlation came out ·44 or nearly *exactly what it is for stature of brothers in man*, or for their eye colour or anything else! In other words he had really demonstrated heredity in these lowly organisms to agree with its value in man and was yet searching about to show why it was so small! This is only one sample of dozens of like papers now being issued, and which must ultimately cast discredit on biometric processes, if we cannot indicate how these things ought to be worked out properly. Half the Editors' work will be to show authors *gently* how to use their own data! We will send you specimens of title-page as soon as we can. Also can you let us have your paper at a fairly early date—say before June 30 —so that we may not cover in any other part of the number the same sort of ground. Further any "Notes" that occur to you on possible biometric work, or notices of books or ideas you may come across, would be very welcome. Yours always sincerely, KARL PEARSON.

7, Well Road, Hampstead, N.W. *June* 29, 1901.

My dear Mr Galton, I am sending you the first proof of title-page and prospectus, etc. of *Biometrika.* You see it will be a capital size for tables and plates. The Syndics of the Press, Mr Wright told us, are keen on their own shield appearing, and he added, what I think is undoubtedly true, that it is effective as an advertisement. I felt in the face of this that it was not desirable to press for our own device. Will you let me have the proof back with any suggestions that occur to you? I hope you don't object to the Quaker-like simplicity of the names on the title-page.

I should have come to talk the whole matter over with you but this is my worst time— examinations etc. Yours always sincerely, Karl Pearson.

42, Rutland Gate, S.W. *Monday.*

My dear Prof. Pearson, You have arranged a capital title-page, severe in its simplicity, and the Cambridge Press symbol gives it additional weight. I quite approve. There is no note that I can see my way to contribute now. Very sincerely yours, Francis Galton.

I am just back from Cambridge, so excuse the few hours delay in reply.

7, Well Road, Hampstead, N.W. *July* 3, 1901.

My dear Mr Galton, I have been looking at one or two of Darwin's books to see if he anywhere emphasises *the value* of statistical inquiry. I can find nothing; and yet I feel quite certain he realised that value by undertaking, as he did, the long series of experiments in Cross- and Self-Fertilisation of Plants. In his book he states that he has appealed to you for an examination of his data from the statistical standpoint and for a report. It has struck me that although that letter is not in the *Life and Letters* it might possibly have survived. Do you think you have preserved it, and if so is there any apt remark as to the need of statistical method in solving such evolution problems?—I should be very glad, if you would let me know if there is. My address after tomorrow will be Manor House, Througham, Miserden, Cirencester, Glosters. Yours always sincerely, Karl Pearson.

42, Rutland Gate, S.W. *July* 4, 1901.

My dear Prof. K. Pearson, Darwin's letter has not I think survived but I recollect its terms well. They would not have helped in what you want. He began in his usual kindly and appealing way, apologising for the trouble, and implying that he had not confidence in his own power of making the best of the few "ipomaea" statistics, and then asked me to try what I could do with them. I doubt if he ever thought very much or depended much on statistical inquiry in his own work, in the sense that most members of the Statistical Society would have given to it;—though, as you know, he quotes statistical results that others had arrived at, not infrequently. Probably, or rather certainly, Frank Darwin would be the best authority on this. I am glad you have got away for a little into the country.

Very sincerely yours, Francis Galton.

42, Rutland Gate, S.W. *July* 8, 1901.

My dear Prof. Karl Pearson, I have just spoken to Frank and to Leonard Darwin, first separately and then together. Their views about their father's attitude towards statistics are the same as mine, except that Frank's was more strongly expressed. I fear you must take it as a fact, that Darwin had no liking for statistics. They even thought he had a "non-statistical" mind, rather than a statistical one. Very sincerely yours, Francis Galton.

I have temporarily mislaid your address, so send this via Hampstead.

42, Rutland Gate, S.W *Oct.* 25, 1901.

My dear Prof. K. Pearson, *Biometrika* has just come, and seems *most* appropriate in general get-up. The printing is beautiful and the size of page excellent. I heartily congratulate you. One small matter of great comfort to the possessors of a pamphlet, is to have its name printed along the back: Vol. 1. Part 1. *Biometrika* Oct. 1901. Do kindly have this done in future numbers. I have already had to write this along the back of mine as well as I could.

Herewith I send the Abstract of my Huxley Lecture—Oh! the trouble that the preparation of the lecture has given! It was so difficult to make a track free from bogholes, and to keep the stages in proportion. I hope it will further investigations by others.

I have one in view now, that I began upon some years ago, but found that enough years had not elapsed since the experiment begun to draw useful conclusions, but every year since has brought a fresh crop of data, and there ought to be enough now. It is the correlation in the Indian Civil Service between the examination place of the candidate and the value of the appointment held by him ? 20 (I forget the figure I used) years afterwards. It seems that the value of an Indian appointment is a very fair test of a man's estimated ability. Mr Tuppy, or some such odd name, wrote a capital analysis of the careers of Indian Civil Servants. I made great use of his book and could soon pick up the long-dropped threads. Nothing however could be successfully done without the cordial and confidential help of the authorities at the India Office. I dare say I may persuade them to help me again, as they did before.

I wish next Tuesday was well over. The paper will appear in full in *Nature* on Thursday.

Very sincerely yours, FRANCIS GALTON.

7, WELL ROAD, HAMPSTEAD, N.W. *Oct.* 25, 1901.

MY DEAR MR GALTON, Very many thanks for your kind letter. Certainly the back of· *Biometrika* ought to have been and shall in future be stamped. I hope No. II may be a little more varied. Macdonell's article on "Criminal Anthropometry" will be a contrast to Garson's in the *Anthropological Journal*!—Latter's on Cuckoos' Eggs will be interesting I think. He has measured and examined nearly 300. I hope to get also the Naqada Skull measurements in, and a good many more *Miscellanea*. Still I fear we shall not be popular enough for a wide range of subscribers.

I am quite sure your lecture has been a heavy piece of work. I know nothing which tries one so much as endeavouring to put scientific results in a form that the intelligent layman can grasp. I am just in the throes of producing two popular lectures for Newcastle—one on Natural Selection, and the other on Homotyposis—and I can appreciate from your abstract what yours has cost you.

Please remember *Biometrika* for the Indian Civil paper.

I have just been dealing with the Cambridge Graduates, correlating their degree with the shape and dimensions etc. of their head and physique generally. We have the full examination record of upwards of 1000 measured individuals. So far the relationship between size or shape of head and intellectual ability seems very slight, but the work is not yet completed. It appears to confirm the view I got from skull measurements, that size has very little to do with intellectual grade.

Next we have reduced the results for pairs of brothers measured in schools, and we find that vivacity, shyness, conscientiousness etc., are correlated precisely as stature, forearm, eye colour. I think this will be when finished as complete a quantitative demonstration of the inheritance of the mental qualities at the same rate as the physical as could be required. I fancy our method of using very simple classification (Memoir VII) would suit your Indian Civil data. Yours always sincerely, KARL PEARSON.

42, RUTLAND GATE, S.W. *Oct.* 31, 1901.

MY DEAR PROF. K. PEARSON, It would be very pleasant if we could meet and have a talk. On Sunday our routine is Lunch-dinner at 1; Tea at 4.30; Dinner-supper at 6.45. Could you come next Sunday for 2 or more of these meals and the intermediate time? If so, please say what you would prefer.

I should doubt whether the exchange of *Biometrika* on equal terms for the Anth. Inst. Journal would be a gain to *Biometrika*, as so very few of the members of the Institute would be likely to use it intelligently.

Quere defer the matter. But do as you think best. Very sincerely yours, FRANCIS GALTON.

5, BERTIE TERRACE, LEAMINGTON. *Nov.* 17, 1901.

MY DEAR PROF. K. PEARSON, Bravis-is-is-imo *re* like inheritance of physical and mental!! You have made a firm foot-hold here, well worthy of all the elaboration that you have and are giving to it. What a blessed feeling it is to come to solid rock, when floundering in yielding mud. I congratulate you most heartily. I write from the country but return by Friday, if not Thursday.

There is much to be talked over, amongst the rest the possibility of giving a summary of the contents of each No. of *Biometrika*, in language that a newspaper could copy, giving the net results obtained in the papers it contains, distinguishing between statement of facts that for the present go no further, and deductions from them. If you thought this feasible, the existence of such a résumé would greatly aid the reader.

You will have before long to give a glossary and definitions of technical words, and references to the places where they were first employed. Also, a very compact account of the chief processes used would be of great service to many (with references of course). Doubtless you have in view the eventual publication of a regular text-book on statistical operations.

I wish we could meet somehow. I could easily be at home next Saturday or Sunday if you cared to fix an hour and a meal, or meals. Dinner-supper on Sunday is always 6.45 to let the cook have time to put on her best bonnet for church. Such is the sex.

Ever sincerely yours, FRANCIS GALTON.

7, WELL ROAD, HAMPSTEAD, N.W. *Dec.* 26, 1901.

MY DEAR MR GALTON, I have been intending for some days to send you a line of sympathy on being laid up, but I wanted to enclose a New Year's Greeting from the workers in my statistical laboratory, and I could not get it finished until this morning. I have always felt we must go into the point more fully, since you laid stress on the view that ability was correlated with the size of the head in your criticism of Dr Lee's paper. There is still a chance that extreme genius may exhibit something *abnormal* in the size of head, but I think it is now pretty clear, if we are to look upon ability as normally distributed in the population, there is only a very small, practically negligible correlation between it and either the size or shape of the head.

We propose next to find out whether there is a higher relationship between ability and health, strength and general physique, and then to test its relation to temper and moral characters, from the school data schedules.

It is a shame to send a gift and then ask for it back!—But I have not had the chance of making a copy, and I might possibly find an abiding place for it in *Biometrika* or elsewhere. Please let me have also your criticisms and suggestions.

I am sending you besides a paper by Macdonell to appear in the next number of *Biometrika*. It is rather long and full of tables, but it involves nearly 18 months stiff work and the material is of value for a number of purposes. I think it shows that for many purposes the fourfold classifications we are now making can safely replace the old laborious tables of correlation.

With the best wishes for the New Year and with the hope that *Biometrika* may not during its first year of life disappoint you badly, I am, Yours always sincerely, KARL PEARSON.

42, RUTLAND GATE, S.W. *Dec.* 31, 1901.

MY DEAR PROF. K. PEARSON, The New-Year gift is indeed acceptable both in itself and as evidence of your continued zeal and power of influencing others to work with you. Hearty thanks and best New-Year wishes.

The non-correlation of ability and size of head continues to puzzle me the more I recall my own measurements and observations of the most eminent men of the day. It was a treat to watch the great dome of Sylvester's head. William Spottiswoode was another of the 5 or 6 largest; so was that encyclopaedic physiologist Prof. Sharpey. That most accomplished & many-sided official, Sir John Lefevre (formerly a senior wrangler), was the largest of all. Gladstone's head, which I myself measured, was very large. Again, comparatively the other day, I was one of a deputation of physicists to the Treasury about the National Physical Laboratory and sitting behind the front row I marvelled at their skulls. Lord Rayleigh, Stokes, Lord Lister, Lord Kelvin were all remarkable partly perhaps owing to the powerful moulding of their heads, irrespective of size. A Frenchman collected the recorded weight of brains of many eminent people and published them in one of the French anthropological periodicals many years ago. They contained remarkable weights. However I can say nothing against the validity of your results.

One thing ought to be remembered, that bigness of head and sturdiness of build go together. A judge (the late Sir Wm Grove), whose large head I often measured, told me that it came

PLATE XXXII

Francis Galton, about the age of 80.

before him on evidence that the hats of stablemen were markedly smaller than those of other people. He inferred that they were less intelligent; I, that stablemen are always light weights (in youth). A heavy boy would not do to exercise horses. Another of my 5 or 6 large heads was Admiral Sherard Osborn. He was very broadset. Also he was considered generally to be the ablest man of his day in the Navy and the accepted mouth-piece of reform. (He died of heart spasm while still young.)

I have been quite bad, this is by far the longest letter I have been up to for many days. I went to Brighton to shake off remains of bronchitis and brought it back increased 7-fold. What with phlegm and spasm I had a fight for it on Xmas Day, but am now mending fast. I dare not write more now or would have said something on Macdonell's paper. I wish he had seen his way to express the *magnitude* of the advantages of scattering the arrangement of the Register. One good reason for beginning with the head is that a criminal *must* have a head, but he need not have a finger or an arm—and these may be contracted.

E. R. Henry, who is now supreme over the identification department in Scotland Yard, is reclassifying the whole collection, primarily by finger-prints and secondarily only by measurements. He looks forward to abolishing measurements entirely in England, as he did in Bengal, stating that errors are more frequent than Garson thought and that they shield the culprit, whereas finger-prints cannot err. I think he overdoes the view, rather, but this is his attitude and he has the *power* to carry out his views. I was much pleased with the order and smartness he has imposed on the office. Garson's connection with it has entirely closed. He, unluckily for himself, took up a critical position towards Henry, who being his superior and a smart disciplinarian, would have none of it. If Dr Macdonell induces that vainest of men, Alphonse Bertillon, to remodel his cabinet it will indeed be a marvel.

I must rest now, with every good wish for you this coming year and for *Biometrika*.

Very sincerely yours, FRANCIS GALTON.

42, RUTLAND GATE, S.W. *Nov.* 2 (Sunday), 1902.

MY DEAR KARL PEARSON, I am just off to France, arriving on Wed. the 5th at Hôtel Continental—Hyères (Var) France and staying there a week certain, afterwards according to health and weather. I will thence write again. Don't post any thing to me there later than on Saturday next the 8th. I fear it would be too risky to send Beddoe's paper, of which you spoke. Your proof, that of your latest paper which you kindly sent me, goes with me. What fertility of mathematical invention you have!

I have recently attacked the finger-print problem (of natural relationship between the various patterns) in quite a new way (no mathematics in it, however), with most promising results thus far. It would be tedious to explain, but it will give me a couple of months happy occupation while abroad at the rate of 3 hrs. a day which is now my maximum of safe performance. Good-bye, Very sincerely yours, FRANCIS GALTON.

7, WELL ROAD, HAMPSTEAD, N.W. *November* 21, 1902.

MY DEAR FRANCIS GALTON, I have been hoping to hear your address so that I might send you a line of satisfaction with regard to the Darwin Medal. But as you must have left Hyères, and as I do not know how to reach you in Sicily, I send this via Rutland Gate.

It seems absurd for me to congratulate you! I can only just say what I said to Weldon when he announced the gift of the medal to me four years ago: "Francis Galton ought to have been given it, not I." To which he replied: "To you it means encouragement to go on, to him recognition of the achieved, which everybody already recognises."..."You get honour from the medal, he would give it honour"—or words to that effect. So it seems also to me that your receiving the medal will make it of greater value to younger recipients, but hardly give you that recognition which helps younger men with their work little known. I may write this now, for the fact that I received the medal four years ago has always had the feeling associated with it, that you ought to have received it long before I did. I trust, however, it may still give you pleasure, and for myself I can only say how it enhances the value of my own. I hope you have been having fair weather and maintained your health. You will have been lucky to escape the last ten days—the worst November I remember. Dr Beddoe has not yet sent me his article. I hope to have Vol. I. Part II wholly in type soon. Please remember me to Miss Biggs, and

Believe me, Yours always sincerely, KARL PEARSON.

(Postcard)   Hôtel des Anglais, Valescure près St Raphael (Var), France.
Nov. 27/1902.

Thanks, hearty thanks, for your very nice letter. My pleasure at the award is and was a little embittered by the thought of standing in the way of younger men. Of course I value the honour *very highly*. Also another most unexpected one of being just elected Honorary Fellow of Trinity, Cambridge, my old College. Thanks to the pure air here, I have wholly thrown off first the asthma and then the chronic cough! I never expected such good luck as this. We shall stay here a little longer and then to Italy. Very sincerely yours, Francis Galton.

Glad that the next No. of *Biometrika* is in type. You are making a success of it, to all appearance.

Address : Hotel Bristol, Piazza Barberini, Rome.

I shall be there for about 2 months beginning with Dec. 22d.

Dec. 8, 1902.

My dear Karl Pearson,   To my surprise the enclosed big cheque reached me this morning. I had quite forgotten it was part of the award. I cannot think of applying it to my personal use (as I have as much income as I want), but to some object in accordance with that for which the Darwin Fund was established, and can think of none more suitable than *Biometrika*. Please therefore take it as a sum to be paid in *relief*, so to speak, of the Guarantee Fund; not intended, even if it could be, ever to be repaid but to be swallowed up in the initial expenses. I am *very* glad to have the opportunity of thus contributing.

The pure air of Valescure has taken away the whole both of my asthma and of my cough, at least for the time, and I feel more *fit*, than for 2 or 3 years past. We leave Valescure to-morrow, reach Bordighera (Hôtel de Londres) next Monday, and Rome the Monday after.

I have no news that you would care about. The finger-prints give daily occupation. It is curious how many "blind alleys" one strays into, during any new course of inquiry. This one *seems* worth a good deal of trouble, but its merits may be more specious than real. Do please send me *Biometrika* news to Rome. Ever very sincerely, Francis Galton.

University College, London. *Dec.* 10, 1902.

My dear Francis Galton,   Your letter and its enclosure reached me this morning. I cannot tell you how I appreciate your kindness and thought in the matter. I am communicating with Weldon by this post. I know it will give him as much pleasure as it gives me. I think you know that finally we collected a fund of £400 to start *Biometrika* with, and that the total call on that fund as a result of initial expenses was under £70. Against this we have about 250 copies of Vol. I, which ought to be sold some day*, and which when sold ought really to recoup the Guarantee Fund as well as the smaller loss of the Press Syndics. What I would therefore propose to do, if it meets with your approval, would be to recoup the Guarantee Fund, so that we start the second year again with our £400 balance, and reserve the remaining £30 to help in the publication of any special memoir which is expensive on account of large tables or plates. I am not indeed at all sure that to devote the whole sum to one or two important memoirs as they come in, might not meet your wishes and the purport of the fund best. If so please let me know. The guarantors were besides yourself—Mr R. J. Parker†—the Attorney General's "Devil,"— Dr W. R. Macdonell, Weldon and myself, and I don't think any of us are very keen on seeing our money back again, if the Journal can be thoroughly established by its use. Hence, I think, we should look upon the recouping of the original Guarantee Fund rather as an omen that we had a longer definite life, than as a personal satisfaction. If we devoted £30, or any further sum to the publication of some extensive paper, please allow us to make a little note stating that help in the publication of that particular memoir has been obtained from your kindness with regard to the Darwin Fund.

I am so glad the change has suited you. I have not sent proofs because I thought your address so uncertain but I will write a "biometric" letter soon.

Yours very sincerely, K. Pearson.

* A prophecy fulfilled as several parts of these volumes have had to be reprinted.
† Afterwards he sat in the House of Lords, as Lord Parker of Waddington.

(Address)   HOTEL BRISTOL, ROME. *Dec.* 13, 1902.

MY DEAR KARL PEARSON,   Your letter awaited me here at Bordighera, on arriving this afternoon. The plan that most commends itself to me is that of paying off the £70, so as to leave the Guarantee Fund untouched up to the present time, and to use the £30, as you suggest, for getting good work done especially in plates, that would otherwise be left undone. But please use your *full* discretion.

I rather shrink from my name being used as you kindly propose. It is difficult to express what is wanted without any appearance of glorification, viz.: that I feel that the £100 could not be bestowed more appropriately than on *Biometrika*. It is especially difficult to express this without provoking the rejoinder that that is precisely the view that a Consultative Editor of the periodical might be expected to take! Don't put anything in type to the above effect without my seeing it first, please.

This blessed Riviera air! There ought to be a Goddess of that name and many temples to her, all along the coast.

I was amused to read long quotations from you, in the largest of type, impressed into doing duty as a puff for the Encyclopaedia Britannica, by the "Times." It was about the advantage of science to modern civilisation and consequently the advantage to everybody of buying that scientific encyclopaedia. Anyhow they found your weighty words very suitable to their own commercial object.

We stay 4 days here, 2 at Alassio, 2 at Pisa, and reach Rome on the 22nd. Wishing you all well through the horrid wintry weather. Ever very sincerely, FRANCIS GALTON.

I wrote the above in bad light, when I find both spelling and grammatic composition difficult on paper. Please on these grounds excuse the many corrections.

7, WELL ROAD, HAMPSTEAD, N.W. *Dec.* 27, 1902.

MY DEAR FRANCIS GALTON,   It is with a feeling of shame that I take up my pen, for I had fully intended to write you a letter to await your arrival in Rome. But a slight attack of influenza and a general feeling of inertia following on it have made me reluctant to do ought but the most necessary things. Now your letter comes to reproach me for not having bestirred myself to send you a Christmas greeting. I forward with this some *Biometrika* proofs for Part II of Vol. II. I expect you will have received Part I ere this. It is very late, but I sent the MS. to Press in August last! They are very dilatory. I have asked Yule to modify his article by giving a general popular account of association to start with. I think Lutz's paper is interesting as strengthening at least for one character the effect of a change of sex. The mouse paper in Part I is not quite definite enough, but I hope to get a second paper in Part II, on further results. The Shirley Poppy paper contains a great deal of work, and I wish it were more definite, but until we get a Biometric Farm where secular experiments of this kind can be carried out *under uniform conditions*, I don't think we can do much better. So far as it goes, it is quite in favour of plants obeying laws of inheritance very like those known to hold for man and horse. I hope to have a paper on the Law of Ancestral Heredity showing really what it assumes and how far we can at present assert it to hold.

It is pleasant to hear of breakfast out of doors in Alassio, and of the sun too hot to sit in at Baliano! I have just received 200 ants from Petrie's settlement and hear of 100 hornets in spirit coming. Please don't forget the celandines, if you get further south and find the collecting not too irksome. I shall hope to get the paper on the first series out in the next *Biometrika*.

Pray send me any point in the finger-print investigation which you think I might elucidate. I am much interested in its possibilities, and think it ought to be rendered available for heredity. Weldon is now in Sicily, most happy over snail finds. Yours very sincerely, KARL PEARSON.

(6) *Work and Correspondence of* 1903. In 1903, largely as a result, if indirectly, of Galton's influence, a Royal Commission was suggested for the purpose of inquiring into the asserted deterioration of the British race owing to bad environmental conditions. Galton grasped at once that a report of such a commission dealing only with possible degeneration would be of small service unless a larger object were kept in view in the course of the inquiry

itself, namely, the means by which any race can be improved, and these means were for him undoubtedly selective breeding. Accordingly he contributed an article to *The Daily Chronicle* of July 29, 1903, with the aim of propounding his views in a popular form. The article was headed (probably in the editorial office) "Our National Physique—Prospects of the British Race—Are We Degenerating?" As a matter of fact Galton in this article is more concerned with increasing our racial efficiency than with emphasising alarming reports of its deterioration, with regeneration rather than with degeneration. He states that he has no intention of confining his remarks to the wastrels and the slums :

"The questions I keep before me are whether or no the British race as a whole is, or is not, equal to its Imperial responsibilities, and again how far is it feasible to make it more capable of the high destinies that are within its reach, if it possesses the will and power to pursue them. I wish that each one of us should stand aloof from ourselves as a whole, and should watch the conditions and doings of our race, much as an authority of the Royal Agricultural Society might criticise the stock of his neighbour over the hedge. If we do so we may learn in what ways our own stock and its rearing are open to improvement and we may perhaps ensue it."

Galton has no doubt that the *pick* of the British race are as capable human animals as the world at present produces. He holds that their chief defects are to be found in their want of grace and of sympathy,

"but they are strong in mind and body, truthful and purposive, excellent leaders of the people of lower races. I speak more particularly of those who are selected to go abroad in various high capacities, whether by Government or by firms to carry out large undertakings under circumstances where they have to depend much on themselves."

The term "lower races" is very unfashionable at the present time, but it is a pleasing and emotional sentiment rather than real anthropological acumen which asserts that all men are of equal value at birth, or that all races are, physically, mentally and socially, of one standard of fitness. The distinctions between man and man, and race and race, are in the main inborn and not "innurtured"—I would say "inbred," but for the double meaning of that word*.

Of the "lower middle classes" Galton's judgment was very unfavourable. He finds the average holiday-maker and cheap-excursion tourist unprepossessing as compared with the like section of other European races. We may superficially, perhaps, but nevertheless with some justification, sum them up as mentally and physically litter-scatterers.

"As regards the physique of Britons, I think we brag or have bragged more than is right. Moreover we are not as well formed as might be. It is difficult to get opportunities of studying the nude figures of our countrymen in mass, but I have often watched crowds bathe, as in the Serpentine, with a critical eye, and have always come to the conclusion that they were less shapely than many of the dark-coloured people whom I have seen."

* Few teachers who have had to instruct young men of many races—and usually the best of the "lower races"—would deny that mentally at least they can be graded. Exceptional men may possibly arise in any race, but it is the averages we have to regard. It was greed that introduced the negro into North America; it was lack of insight which did not push him northwards in South Africa. In both cases the "lower race" now forms a grave and almost unsolvable problem for the future.

Galton gives an account of the Sandow competition in which the three best specimens were selected out of some eighty of Sandow's pupils. Galton was present when the trio was selected and thus states his impressions:

"I did not think these best specimens of the British race to be ideally well-made men. They did not bear comparison with Greek statues of Hercules and of other athletes, being somewhat ill-proportioned and too heavily built. I must say that I was disappointed with them from the aesthetic point of view, though in respect to muscular power they seemed prodigies. Sandow afterwards exhibited himself in a pose that brought out his chest and arms to full advantage, and in that statuesque position I placed him as far superior to all the competitors."

What Galton says about British physique and about the physical beauty of our trunk and limbs is probably very true. We have recently seen the foreigner our equal or even our superior at most of our national sports; he only needed the proper training to defeat us. Nor is the somewhat low standard of physical beauty confined to trunk and limbs—anyone who makes an extensive study of the English skull must be forced to the conclusion that aesthetically at least it is not of a high type. The stock-breeder "looking over the hedge" must conclude that these are not directions in which much can easily be achieved. Yet he would affirm emphatically and

"with justice that the whole of a race which was able to furnish the large supply which is produced in Great Britain of men who are sound in body, capable in mind, energetic and of high character, has the capacity (speaking as a rearer of stock) of being raised to at least the same high level."

This, Galton believes, could be attained by making use of both Nature and Nurture. Of the former Galton holds that if a strong and intelligent public opinion can ever be roused in favour of improving our racial breed, then there are a number of small influences which even now operate under existing sentiment and law and which are capable by co-operation and development of producing great results. He admits, however, that we have yet much to learn that lies well within the province of anthropology, before it would be justifiable to attempt a crusade; otherwise grave mistakes will be made and the movement will be discredited.

"My attitude, which has usually been misrepresented, is to urge serious inquiry into specific matters which still require investigation in the well-justified hope that a material improvement in our British breed is not so Utopian an object as it may seem, but is quite feasible under the conditions just named. But whatever agencies may be brought to bear on the improvement of the British stock, whether it be in its Nature or in its Nurture, they will be costly, and it cannot be too strongly hammered into popular recognition that a well-developed human being, capable in body and mind, is an expensive animal to rear."

It will be seen that here as elsewhere Galton places the acquirement of eugenic knowledge before eugenic action—Eugenics Research Laboratories must be developed before Eugenics can be safely preached as a popular creed. He illustrates this by propounding a problem concerning nurture: If a dole be available to help in the rearing of a child, at what period will assistance be most effective? Is it when it is growing most rapidly and most needs good feeding, or may irremediable mischief be done by withholding it until that age is reached? If the State has only a limited amount of money to spend on its children, let it investigate first when it is of most use in improving the breed—whether in infancy, at school age, or during the rapid development of youth.

The reader may think I have given too much space to an ephemeral newspaper article. It is not because of the suggestions it contains, but rather because it exhibits the cautious statements and the moderate proposals to which Galton gave expression even on a topic about which, as those who knew him well can testify, he felt with almost religious fervour.

During this year (1903) Galton had turned to finger-prints again, and was very busy trying to find a measurable character common to all patterns. He endeavoured to obtain this by what he termed the "interspace"—a diameter drawn across the core (of loops or whorls) so as to be perpendicular to both its upper and lower borders. The interspace was to be measured in a mean ridge interval of the core as unit, this mean ridge interval being obtained as the average of ten ridges taken along the interspace. The arches were a serious difficulty, for Galton concluded that they had no interspace, and they tended to lump up at one end of his frequency distributions. Galton's views are given in the accompanying letters; they were never published, although for the remainder of his life he occasionally returned to finger-print studies. As they may be suggestive to other workers, I reproduce them.

GRAND HOTEL, NAPLES. *March 2, 1903.*

DEAR KARL PEARSON, Your card of the 26th came all right yesterday, but the previous one which you mention, in reply to my letter enclosing Bicknell's, had and has miscarried. Hence my eagerness for tidings. You say that subscriptions are falling off—here however you will find one and probably two new subscriptions. I have written to Mr H. to say that I am forwarding his letter to you for reply and that I am ordering his book...to be forwarded to you also. Please answer to him his quere about the way of remitting his subscription. I know nothing of him.

It is to be regretted that biologists do not welcome *Biometrika*, but the welcome cannot yet be expected. Would it be possible to give a summary of work *done*, that *must* prove useful to biology and which without biometric methods could not have been done? We seem to need something of that kind more and more; something so free from technical language that newspapers could copy it, and their readers could understand and *like* it. Of course it could only contain cream and be in no way exhaustive, but it ought to be so far mentally digestible by the average biological intelligence as to leave some conviction upon it of the utility of biometry....

As regards the finger-prints I am in a little doubt, being not sure how far my collection of Bengal Criminals may be thought suitable, or even whether they are *strictly* non-selected. From the comparative absence of transitional patterns I fear that many of these may have been sorted out of the collection, which is one of a few hundred duplicates of some of the main collection of about 6000. They were used to enable Mr Henry (now Assistant-Commissioner at Scotland Yard) to show off the rapidity with which the original of any selected duplicate might be traced. It is possible that his clerks may have avoided troublesome transitional cases sometimes, but Mr Henry seems not to be cognizant of this. At all events I should prefer to work on my own collection, but that, alas, is *classified*, so I should have to go through the whole of it, 2600 odd in number, if I touched it at all. This would be a very tedious job, for I must not draw outlines on the patterns themselves,—which is easy but might spoil them,—but must *trace* them, which is very troublesome even with the best tracing paper and the best light. Would you however look at the enclosed table and tell me how it strikes you? Perhaps you might even get some one to work out the correlation index.

[Enclosure in Galton's Letter of March 2, 1903.]

Width of a Mean Ridge Interval in millimetres	Nº of cases
(omit Arches*)	7
·30 — ·34	1
·35 — ·39	10
·40 — ·44	42
·45 — ·49	67
·50 — ·54	60
·55 — ·59	13
	200

* There is no single interspace in an arch, as there is in all other patterns.

Arch	Loop	Single Spiral	Double Spiral

The Height of Interspace is measured between the dots

Height of Interspace in units of a Mean-Ridge-Interval

Left fore finger	Right fore finger ·0 / ·30	·40 / ·70	·80 / ·11	·12 / ·15	·16 / ·19	·20 / ·23	·24 / ·27	·28 / ·31	·32 / ·35	·36 / ·39	·40 / ·43	·44 / ·47	Totals
0 to ·30	10		2										12
·40 ,, ·70	1	1											2
·80 ,, ·11	1	1	3	3		2	1						11
·12 ,, ·15		2	3	1	3	1							11
·16 ,, ·19	2	3	2	4	6	6		1					24
·20 ,, ·23			3	2	2	8	10	3	1				29
·24 ,, ·27		1	2	3	4	12	2	4					28
·28 ,, ·31	1			1	1	4	9	10	3	1			30
·32 ,, ·35				1		7	8	6	2				24
·36 ,, ·39				1		2	7	5	2	1			19
·40 ,, ·43						1	2	2					5
·44 ,, ·47						1		2	1				4
·48 ,, ·51								1					1
Totals	15	7	14	13	17	25	4	26	25	8	6	3	200

Francis Galton March 1903

I am quite sure that my way of working on the 2 forefingers is the best for getting at the relations between the various patterns, and I have already learnt much that is new, but I shrink from more work on my present material. Finger patterns seem to me an ideally good subject, not only for heredity work, but for much else of evolutionary interest. If you think the enclosed table of 200 cases *full* enough, or nearly so, I should take pains to get that number, or double that number, printed off at some school or elsewhere, especially for this inquiry. They would have to be *rolled* impressions printed in *triplicate* at least. Such impressions are rapidly taken. I easily take 12 of my own fingers carefully in one minute, when all is ready, or in five minutes counting from the time of sitting down to the table with my apparatus in my pocket to that of rising with everything cleaned and packed in my pocket again....

Very sincerely yours, FRANCIS GALTON.

The weather is becoming cold, not good for travel further.

[Postcard]   UNIVERSITY COLLEGE, LONDON. *March* 6, 1903.

I have just had time to work out your correlation table. I find:

Mean, Left forefinger ·244   mean ridge interval.
S.D.   „   „   ·1047 „   „   „
Mean, Right   „   ·229   „   „   „
S.D.   „   „   ·1048 „   „   „
                    Correlation = ·8203.

The correlation of the distal phalanges of the R. and L. forefingers as given by Lewenz and Whiteley in *Biometrika* is ·79, i.e. within probable error of your result. KARL PEARSON.

I wrote further to Galton asking for information upon the "interspace" and upon the want of continuity due to the Arches being treated as of zero "interspace." One of the main difficulties in his restriction of the data to the two forefingers was that a rare type of print that appears on one forefinger may not appear on the other forefinger but on some other finger of that hand, and experience seems to show that the prints of all *ten* digits must be taken into consideration when judging the resemblance of relatives by means of finger-prints. I received the following illustrative letter from Capri:

March 16, 1903.

MY DEAR KARL PEARSON,   I have at last got your long letter of March 5 and enclosures at Capri, where we have been 9 days. After a very little more touring we turn homewards. The loss of 20 subscribers is bad, and so is the attitude of both biologists and mathematicians to *Biometrika*, but the second year of a new venture is always the most trying time*. The first flush of expectation is over, and the solid merits have not had time to assert themselves. It seems to me to want some cheery writing in good reviews to show in an intelligible form a few definite blunders into which biologists have fallen for want of biometric methods. I expect craniology would furnish topics. I recollect once that kindest of men, Sir Wm Flower, being on the verge of wrath because I pointed out the insufficiency of evidence drawn from the mean values of a few skulls of some savage race (I forget which) in determining the race to which a particular unknown skull belonged. Craniological literature would contain, I should think, many rash statements which could be assailed triumphantly by a facile writer and sharp critic. Dear old Beddoe is the most rambling of thinkers and writers as well as one of the most industrious of workers. I am not surprised to hear that his paper is far below the occasion, wrong in its criticism and wrong even in its arithmetic and generally slipshod. The photo-

---

* [Those were anxious days for the Editors of *Biometrika* as they watched the slipping away of their funds. Nowadays several parts of those early years have been reprinted, and complete series sell at very high prices!]

PLATE XXXIII

Illustration of Galton's method of measuring "interspaces" in terms of "mean ridge interval." From Letter of March 16, 1903. The "pricks," needle-pointed through the letter paper, are not visible in the reproduction; the original had to be held up to the light to see them.

graph of the skull that you send is exceedingly good, and is I presume (together with the rest) taken under standard conditions, and selected in some way free from bias, other than what may be clearly stated about them as intended to be conveyed by the word "English." English unless narrowly limited includes so great a diversity of type :—dark and fair, Cornish, Sussex, Midlands, Yorkshire, Welsh, Scotch, Irish, &c. &c.—ill-fed and well-fed, educated and uneducated, etc. etc.—that it is very difficult to deal with English as a whole, except by taking homogeneous subgroups*. I found this emphatically the case with my S. Kensington anthropometric data. Out of many thousand cases I failed to form a single homogeneous (quasi-homogeneous) group that satisfied me. If you think that your collection is fairly free from this difficulty, please tell me what you think the cost of printing them would be, and I will see if it be possible for me to afford it. It is most desirable that some standard and unquestionably useful work—obviously useful to biologists—should appear in *Biometrika*.

About the finger-prints, what I sent was a mere scrap and would require a great deal both of explanation and of collateral conclusions. The lump at the commencement of the series is to me of the greatest interest, for it emphasises the fact that the patterns do *not* form a continuous series, but a group or order composed of many sub-groups or species ; each of these has a curve of frequency of its own. They are in some sense convertible, and they form hybrids, but the arches are far more "pure blooded" (so to speak) than any of the others. They are antipathetic to whorls. An arch on one forefinger is associated with a whorl on the other only once or twice in a hundred cases, and then only imperfectly. Then there is the case of radial and ulnar slopes, and their connection with whorls. We have in fact a menagerie of different creatures, breeding promiscuously, and yet at all times divisible into a limited number of definite types, each with its own law of frequency, whose statistical proportions between themselves seem to be constant. Its study has therefore a very close bearing on the evolution of species (as indeed I pointed out in my first paper on Finger-Prints in *Phil. Trans.*). This study has the great advantages (1) that age has no effect on the patterns, when the ridge interval is taken as unit of measurement, and consequently (2) that it would be easy to get and to use family prints to 3 and even 4 generations, (3) that the data when once obtained are free from all error of measurement, for they are themselves the things to be measured †. I send prints of my own fingers, which are a *worse* example by far than the generality of those one might get, chiefly because the wrinkles of age leave numerous gaps in the form of white streaks, and also because I have smeared them by manipulations immediately after they were made, but they will serve to explain the dimension measured in the table I sent. The loops are troublesome only in the sense that the *very best* dimension is hard to define ; on the other hand many reasonably alternative dimensions give practically the same result. The measure desired

* [The skulls in question all came from a single 17th century pit in Whitechapel, and were reasonably homogeneous and close to similar series from Liverpool Street and Farringdon Street. The photographs were the first of the series of standardly orientated crania on a large scale which have since then continuously appeared in *Biometrika*. Galton's offer was spontaneous like several others, but not accepted. "Of course I could not think of your aiding us further at present. We made up the loss to the reserve fund with your Darwin Medal grant, and it left £30 to the good which might be reasonably expended on illustration if you approved. ...The photographs were all taken the same distance from the objective and in the same manner for each aspect, but different aspects had to be treated rather differently—a profile on a smaller scale than a frontal view, etc. The difficulty of getting a 'mean' focus on a solid body must cause some variation, however, even in the distance. On the whole, I think, photographs of skulls must be taken to represent qualitative characters, which are after all, if indescribable, realities. I have tried a good deal, but do not believe that cranial photographs will ever serve usefully purposes of measurement....I hope you will come back fit and well for climbing 'May hill,' which an old medical friend always describes as the great task of the year. I am going to Newbury to meet Weldon to-morrow to talk over Part III, while I hunt for Easter quarters. We want to be near Oxford, Weldon for the mice and I for Weldon." K. P. to F. G., March 20, 1903.]

† [I think Galton must mean here that the *stored* data are free from error of measurement. Whether we take head measurements or finger-print measurements (and Galton is speaking of quantitative not qualitative classification) the measurement must be taken *once*.]

is the magnitude of disturbance caused by the finger nail. When the disturbance is great compound patterns tend to appear as the "kernelled loop" (right ring-finger). [Galton gives sketch: see our Plate XI, Fig. 19.]

In the Arches the disturbance does not occur at any *one* place, but is distributed. [Sketch: see our Plate XXIII (1–6).] When I come back I must begin to collect data: viz. *triplicates*, rolled impressions of two forefingers, using a separate half-sheet of note paper for each person. I now understand quite what I want, and can use a clerk, working with comparatively slight supervision after he is well trained and started. The outlining is very distinct when done with the very black ink used by artists who draw for "process-work." I have contrived a wonderfully neat pocket-apparatus for printing, only the size of a *small* lucifer match box and value under 1^{d}. Very sincerely yours, FRANCIS GALTON.

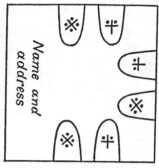

*† Prints of two forefingers in triplicate.

Galton, when he returned to England, circularised many folk, issuing small finger-printing apparatus, and asking for the prints of the two forefingers of as many relatives to be taken as possible. To aid him in the reduction of these and other data Galton desired to find an assistant. On the advice of Dr Alice Lee, he selected Miss Ethel M. Elderton—a most happy choice. She received her first training from Francis Galton, then became successively Secretary to the Eugenics Record Office, Galton Research Scholar in the Eugenics Laboratory, then Galton Fellow, and is now Assistant-Professor in that Laboratory. Perhaps this was the best result that flowed from the forefingers-print collection!

(7) *Work and Correspondence of* 1904. Two events of this year had importance in relation to Eugenics, the one dealing with scientific research and the other with popularisation. The first was Galton's gift of £1500 to the University of London for the furtherance during three years of the scientific study of Eugenics. I have already referred to the Galton Research Fellowship when discussing the definition of Eugenics. Our correspondence for the latter end of the year chiefly dealt with the various candidates for the Fellowship with some of whom I was acquainted as well as with their work. The selection committee ultimately recommended Mr Edgar Schuster, an Oxford student of Weldon's, who had already done good biometric work, and Miss E. M. Elderton was appointed as his assistant. University College provided rooms at 50, Gower Street, which at Galton's request were entitled the "Eugenics Record Office." In the same house were lodged for working purposes two or three post-graduates, an overflow from the Biometric Laboratory, but there was no other link between that Laboratory and the Office. Galton himself was in control, and the main scheme in hand was to form a register of "Able Families," of which only the portion dealing with Fellows of the Royal Society reached completion*. Schuster during his tenure of the Fellowship also wrote two memoirs, one on "The Inheritance of Ability" in conjunction with Miss Elderton and a second entitled "The Promise of Youth and the Performance of Manhood." These two memoirs

* See the present volume, pp. 113–121.

PLATE XXXIV

Francis Galton in 1904, aged 82.

were excellent pieces of work*, and I am the more willing to praise them as I had no connection whatever with the Eugenics Record Office. I was not on its Advisory Committee, and Galton, knowing how pressed I was at that time with work, did not as far as I can recollect ever consult me as to the research in his Office; once or twice only Schuster asked for aid in dealing with statistical matters. In the main it was Galton, with some aid from Weldon, who developed this first attempt at a Eugenics Laboratory. When two years later Galton asked me to take charge of the Office I was only too glad to publish Schuster's memoirs as the first and third of the new Eugenics Laboratory publications. These writings and a couple of papers on the inheritance of psychical characters and of deaf-mutism demonstrated that Galton's proposals for eugenetic research were feasible, and that his endowment was not being wasted. If in the future the question arises when and where did Eugenics as an academic branch of study take its origin, the answer can only be: In the autumn of 1904 in the two rooms at No. 50, Gower Street under the direction of Francis Galton, within a few yards of the house on the same side of the street where Charles Darwin started his married life when he returned from his voyage in the "Beagle." When Eugenics becomes a great factor of academic and political life—as important as State Medicine,—which I have no doubt it will be in the future, then that house will deserve to be commemorated!

The second important event for Galton and Eugenics in the year 1904 was really anterior to the foundation of the Eugenics Record Office. I have already noted that Galton had endeavoured, although not very successfully, to interest English anthropologists in Eugenics. He now turned with a somewhat greater degree of success to the Sociologists, and in particular to the newly founded Sociological Society. A lecture was given by him at a meeting of that Society held on May 16, 1904. It was exceedingly well-staged except in one unfortunate respect, the choice of a chairman. There was a reasonably well-directed discussion and there were written expressions of opinion upon Eugenics as science and art from a number of men with familiar names. Maudsley and Mercier were doubters and apparently ignorant of the knowledge already obtained; Francis Warner generalised on impressions; Weldon preached the sound doctrine "that there can be no doubt whatever that for the student of Eugenics or of organic evolution generally, the conclusions drawn from the larger mass of complex material are far more valuable than those drawn from the simpler, smaller laboratory experiment"; H. G. Wells† was of the opinion that more can be achieved in the way of improving the human race by the sterilisation of failures than by the selection of successes for breeding; Benjamin Kidd was dogmatic without being convincing; Palin Elderton

---

* Both now unfortunately out of print.

† This popular author set an absurd myth on foot by saying: "Eugenics which is really only a new word for the popular American term stirpiculture." "I wish," said the German Professor, "that Lord Rayleigh would more frequently acknowledge his indebtedness to Mr Strutt." Galton himself actually invented the word "stirpiculture" and changed it advisedly to eugenics!

considered that actuaries as a body hold that environment operates merely as a modifying factor after heredity has done its work; L. T. Hobhouse maintained that if the problem of stock is to be taken into consideration at all, then it ought to be by intelligently handling the question rather than submitting to the blind forces of nature, but until there is more knowledge and agreement as to criteria of conscious selection, " we cannot, as sociologists, expect to do much for society on these lines "; William Bateson held that " the ' actuarial method' will perhaps continue to possess a certain fascination in regions of inquiry where experimental methods are at present inapplicable," but urged that those who have such aims at heart (as Galton) would best further Eugenics by promoting " the attainment of that solid and irrefragable knowledge of the physiology of heredity which experimental breeding can alone supply"; he did not state the touchstone—faith in the research and the actuarial treatment—by which we can alone know that the knowledge is "solid and irrefragable "*; C. S. Lock obviously thought the proposals premature; W. Leslie Mackenzie thought that the effects of inheritance were so masked by nurture that in no individual case could we determine what was due to the former, and cited as an illustration that the modern movement for extirpation of tubercular phthisis could not become world-wide until the belief in the " heredity of tuberculosis" had been sapped; a view contradicted promptly by Archdall Reid who held that it was selection by consumption that made the Northern Races pre-eminently strong against consumption; J. M. Robertson evidently laid more stress on environment than heredity, and considered ill-feeding, ill-housing, ill-clothing and early profligacy on the one hand, and ignorance in child-bearing and begetting on the other, as the great forces of " Kakogenics"; Bernard Shaw agreed with the paper and went so far as to say " that there was now no reasonable excuse for refusing to face the fact that nothing but a eugenic religion can save our civilisation from the fate which has overtaken all previous civilisations." He held that " what we must fight for is freedom to build the race without being hampered by the mass of irrelevant conditions implied in marriage," and asserted that " a mere reduction in the severity of the struggle for existence is no substitute for positive steps for the improvement of such a deplorable piece of work as man." Shaw cleared away a good deal of the fog of previous contributors, but went further† than Galton certainly approved, and indicated methods of improving the race, for which, however biologically fitting, the time will not be ripe until the less drastic proposals of Galton have bred " under the existing conditions of law and sentiment‡ " a more highly socialised race. Galton's suggestions may seem very limited as compared with Bernard Shaw's attitude to race improvement, but he who would practically

* I can remember the day when certain so-called "Laws of Motion" were considered "solid and irrefragable"! Most of the progress in science consists in the passage from one "solid and irrefragable" law to a second.

† If a marriage is from the eugenic standpoint brilliantly successful "it seems a national loss to limit the husband's progenitive capacity to the breeding capacity of one woman," etc. etc.

‡ See the title to Galton's Huxley Lecture on our p. 226.

reform mankind must not begin by alarming it. We may remind the Editor of "Fabian Essays" that the doctrines of Eugenics will be best served, like those of socialism, by a slow process of impenetration.

The drift of the discussion as above indicated was to reveal clearly the past history, the narrow field of experience, the particular method of experiment or observation of the individual contributors. Impressions rapidly formed on a subject, which they had not thought over for years, like Galton, were produced without any foundation of facts or figures; my anticipations of what would flow from the various heterogeneous elements classed together as sociologists were realised. But Galton got an excellent advertisement for Eugenics, which he proceeded to follow up. The paper and the discussion on it were widely mentioned in the daily press. Sociology for the present biographer must be a study of man in the mass, the facts on which the science must be based depend upon averages, variations, associations and correlations—in short, sociology to become a science must be based upon the collection of data and the statistical treatment of those data. Such treatment I had found almost wholly missing in sociological memoirs. Sociology appeared to me to be like psychology before the introduction of the experimental method, like what physics would be without a mathematical handling, or insurance before there was an actuarial science; in the words of Galton, "Until the phenomena of any branch of knowledge have been submitted to measurement and number it cannot assume the status and dignity of a science." Until some sociologist should arise and grasp this fact and apply it to his studies, sociology in my opinion had not yet its founder*. Holding such a view I was somewhat astonished to receive a letter from Francis Galton dated April 12, 1904, running as follows:

My dear Karl Pearson, I hear they have been bothering you to take the chair at a Sociological meeting on Monday, May 16th, when I read a paper on Eugenics at 5 p.m.—However agreeable it might be to myself that you should do so, I beg that you will consult your own inclinations *entirely* in the matter, without the slightest regard to mine. I have just had a talk with Mr Branford who favourably impressed me with the idea that he had clear views of what the Society might do scientifically, and that he saw his way to give effect to them. The result is to ease my own mind in respect to offering the paper, or rather acceding to the request to send it.

What a slashing you administer to Professor Castle. He deserves it.

A book by Havelock Ellis "A study of British Genius" interests me. He has taken the "National Biography" as his store house, and shows forcibly the great contribution by English clergy to the ability of the next generation. That is a Eugenic fact for me, not unforeseen, however.

I trust you are all having a happy Easter at Rotherfield Greys. I fear addressing this so, therefore I send it to Hampstead. Kindest remembrances. Very sincerely, Francis Galton.

The actual meeting took place in the large hall of the London School of Economics, and the audience which the veteran of eighty-two years addressed was numerous and distinguished. The Chairman, in opening the proceedings, said:

"My position here this afternoon requires possibly some explanation. I am not a member of the Sociological Society, and I must confess myself sceptical as to its power to do effective

---

* The reader will appreciate my amusement when the Secretary of the Sociological Society, Mr V. V. Branford, spent much paper and energy in endeavouring to prove that Vico, Comte and Herbert Spencer were architects of a *science* of sociology!

work. Frankly, I do not believe in groups of men and women who have each and all their allotted daily task creating a new branch of science. I believe it must be done by some one man who by force of knowledge, of method and of enthusiasm hews out, in rough outline it may be, but decisively, a new block and creates a school to carve out its details. I think you will find on inquiry that this is the history of each great branch of science. The initiative has been given by some one great thinker, a Descartes, a Newton, a Virchow, a Darwin or a Pasteur. A Sociological Society until we have found a great sociologist is a herd without its leader—there is no authority to set bounds to your science or to prescribe its functions. This you must realise is the view of that poor creature the doubting man, *in media vita*; it is a view which cannot stand for a moment against the youthful energy of your Secretary, or the boyish hopefulness of Mr Galton, who mentally is about half my age. Hence for a time I am carried away by their enthusiasm, and appear where I never anticipated being seen—in the chair at a meeting of the Sociological Society. If this Society thrives, and lives to do yeoman work in science, which, sceptic as I am, I sincerely hope it may do, then I believe its members in the distant future will look back on this occasion as perhaps the one of greatest historical interest in its babyhood. To those of us who have worked in fields adjacent to Mr Galton's, he appears to us as something more than the discoverer of a new method of inquiry, we feel for him something more than we may do for the distinguished scientists in whose laboratories we have chanced to work. There is an indescribable atmosphere which spreads from him and which must influence all those who have come within reach of it. We realise it in his perpetual youth, in the instinct with which he reaches a great truth, where many of us plod on groping through endless analysis, in his absolute unselfishness and in his continual receptivity for new ideas. I have often wondered if Mr Galton ever quarrelled with anybody. And to the mind of one who is ever in controversy, it is one of the miracles associated with Mr Galton, that I know of no controversy, scientific or literary, in which he has been engaged. Those who look up to him, as we do, as to a master and scientific leader feel for him as did the scholars for the grammarian:

> ‘Our low life was the level’s and the night’s;
> He’s for the morning.’

It seems to me that it is precisely in this spirit that he attacks the gravest problem which lies before the Caucasian races—‘in the morning.’ Are we to make the whole doctrine of descent, of inheritance, and selection of the fitter, part of our everyday life, of our social customs and conduct? It is the question of the study now, but to-morrow it will be the question of the market-place, of morality and of politics.

If I wanted to know how to put a saddle on a camel's back without chafing him, I should go to Francis Galton; if I wanted to know how to manage the women of a treacherous African tribe, I should go to Francis Galton; if I wanted an instrument for measuring a snail,—or an arc of latitude,—I should appeal to Francis Galton. If I wanted advice on any mechanical, or any geographical, or any sociological problem, I should consult Francis Galton. In all these matters and many others I feel confident he would throw light on my difficulties, and I am firmly convinced that with his eternal youth, his elasticity of mind, and his keen insight, he can aid us in seeking an answer to one of the most vital of our national problems: How is the next generation of Englishmen to be mentally and physically equal to the past generation which provided us with the great Victorian statesmen, writers and men of science—most of whom are now no more—but which generation has not entirely ceased to be as long as we can see Francis Galton in the flesh.”

The Chairman then called upon Mr Francis Galton to read his paper on “Eugenics, its Definition, Scope and Aims*.” The theme of the lecturer was very similar to that of the address to the demographers of 1891, only there was no screening of the guns, and the word eugenics was freely used. Eugenics was defined as the science which deals with all influences which improve

---

* It will be found printed in *Sociological Papers*, published by Macmillan & Co. for the Sociological Society, 1905, pp. 45–50.

the inborn qualities of a race, also with those that develop them to the utmost advantage. Galton also limited himself to the inborn qualities of some one human population, i.e. to "national" eugenics. It will be seen that the definition is much looser than that of the University Committee of the following year, which limited the science to the "study of agencies under social control." The word "qualities" is used, but the study of "impairment" of racial qualities is only implicit, not expressed. The second paragraph of the address emphasises the fact that Galton would be utterly opposed to the word "moral" coming into the definition of his science; morality, goodness or badness of character, he tells us, is not absolute but relative to the current form of civilisation; the moment we begin to talk about a character as good or bad hopeless difficulty is raised. We must leave morals as far as possible out of the discussion. The essentials of eugenics may be easily determined; all would agree that it is better to be healthy than sick, vigorous than weak, well-fitted than ill-fitted for our part in life.

"There are a vast number of conflicting ideals, of alternative characters, of incompatible civilisations; but they are wanted to give fullness and interest to life. Society would be very dull if every man resembled the highly estimable Marcus Aurelius or Adam Bede. The aim of Eugenics is to represent each class or sect by its best specimens; that done, to leave them to work out their common civilisation in their own way. A considerable list of qualities can be easily compiled that nearly everyone except 'Cranks' would take into account when picking out the best* specimens of his class. It would include health, energy, ability, manliness, and courteous disposition. Recollect that the natural differences between dogs are highly marked in all these respects, and that men are quite as variable by nature as other animals in their respective species. Special aptitudes would be assessed highly by those who possessed them, as the artistic faculties by artists, fearlessness of inquiry and veracity by scientists, religious absorption by mystics and so on. There would be self-sacrificers, self-tormentors and other exceptional idealists, but the representatives of these would be better members of the community than the body of their electors. They would have more of those qualities that are needed in a State, more vigour, more ability, and more consistency of purpose. The community might be trusted to refuse representatives of criminals, and of others whom it rates as undesirable." (pp. 46–7.)

Galton then goes on to state what would happen if we could raise the average quality of our nation to that of its better moiety:

"The race as a whole would be less foolish, less frivolous, less excitable and politically more provident than now. Its demagogues who 'play to the gallery' would play to a more sensible gallery than at present. We should be better fitted to fulfil our vast imperial opportunities. Lastly, men of an order of ability which is now very rare, would become more frequent, because the level out of which they rose would itself have risen. The aim of Eugenics is to bring as many influences as can reasonably be employed to cause the useful classes in the community to contribute *more* than their proportion to the next generation†." (p. 47.)

* Some formal objection was taken to the use of the word "best," e.g. J. M. Robertson suggested that "the aim of Eugenics is to promote such calculation or choice in marriage as shall maximise the number of efficient individuals." There would, he said, always be *some* best and it was a contradiction in terms to say they represented their class. Possibly, but not certainly, "efficient" is easier to define than "best."

† Mr Robertson (see the previous footnote) seems to have overlooked this last sentence, it covers with greater generality his suggested aim of Eugenics. Under (2) above Galton actually speaks of civic *efficiency*.

Galton next sketches out what procedure an active society promoting Eugenics might adopt. It might, he considers:

(1) Disseminate knowledge of the laws of heredity as far as known and encourage their further investigation.

Incidently he emphasises the importance for Eugenics of the *actuarial* side of heredity, and remarks on its advance in recent years, and how the *average* degree of resemblance—the measure of kinship in each grade—is now obtainable, so that in the mass the effects of blood relationship can be dealt with even as actuaries deal with the birth- and death-rates. This actuarial side of heredity was ever present in Galton's mind, and was the topic of his Herbert Spencer lecture on Eugenics.

(2) Inquire into the present and the past rates of fertility of various social groups—classified according to their civic efficiency. Galton says that there is strong reason for believing from the history of ancient and modern nations that their rise and fall depends upon changes in this relative fertility. He considers that while there are causes at work which tend to check fertility in the classes of higher civic worth, nevertheless types of our race may be found which can be highly civilised without losing fertility, even as some animals become more fertile the more they are domesticated.

(3) Collect data as to large and thriving families. Galton considers that a "large" family may be taken as one in which there are at least three *male* children. His definition of a "thriving" family is important, and it seemed to me overlooked in the discussion; it is one in which the children have gained distinctly superior positions to those achieved by the average of their classmates in early life. It is clear that such a list of "thriving" families—a "Golden Book," of really noble stirps—must precede any attempt to encourage fertility in the classes of higher civic worth. But the formation of such a "Golden Book," even for a single social group such as the clerical, legal or academic professions, is a matter of extraordinary difficulty. Galton soon dropped the idea of making it depend on the children reaching "superior positions." He saw that it must depend upon the achievements of the stirp or stock as a *whole*. It was from the standpoint of this idea that Galton set Schuster to work on *Noteworthy Families* in modern science; that was to form the first section of the "Golden Book." Further portions of it were in part prepared and the "Register of Able Families*" was an offshoot from the same idea. Judged from the aim of the "Golden Book," *Noteworthy Families* (Modern Science) gains more meaning, if we cannot overlook its defects.

What Eugenics needs is a book of "Noble Families" in a modern sense; it could at first only apply to the upper classes, and there would certainly be numerous omissions and erroneous inclusions in the early issues. It would contain, just as a peerage does, a list of all families within which, inside a given range of ancestry and collaterals, a certain percentage of members had reached posts falling into a carefully selected list, or achieved results in politics, art, literature or science of a certain degree of worth. New families would

---

* See our present volume, p. 121.

always be coming in, old families dropping out, as the one reached, or the other fell short of the required percentage. Ultimately the book would be able to base itself upon its own inclusions. It could only be successful, if prepared by trained genealogists, eugenists and statisticians, working on pre-arranged rules. It would need an energetic and enterprising publisher, but it might in the end become as valuable a property as a peerage, the *Medical Directory*, or *Who's Who*. Such would be the final development of Galton's "Golden Book of Thriving Families," and to be recorded in it would be a higher patent of nobility than could be marked by any other directory or roll in the land.

"The Chinese, whose customs have often sound sense, make their honours retrospective. We might learn from them to show that respect to the parents of noteworthy children which the contributors of such valuable assets to the national wealth richly deserve." (p. 49.)

Achievements of their offspring would bring parents into the "Book of Noble Families."

(4) Study the influences which affect marriage. Galton discarded entirely the notion that the passion of love is so overpowering that it is folly to determine its course. Social influences and customs have immense power in the long run. If marriages which were unsuitable from the eugenic standpoint were socially banned, as marriages between near-kin have often been, such marriages would very seldom be made. From the origin of human marriage, and even before, restrictions and prohibitions have existed concerning the mating of human beings. Let us study how these customs have originated and what are their sanctions.

(5) Urge persistently the national importance of Eugenics. According to Galton there are three stages to be passed through: *First*, it must be made familiar as a branch of academic study. *Secondly*, it must be recognised that the subject demands serious consideration as an art. And *Thirdly*, it must be introduced into the national conscience, like a new religion.

Then follow what, in the biographer's judgment, are the most impressive sentences Galton ever wrote on the subject of Eugenics:

"It has indeed strong claims to become an orthodox religious tenet of the future, for Eugenics co-operates with the workings of nature by securing that humanity shall be represented by the fittest races. What Nature does blindly, slowly and ruthlessly, man may do providently, quickly and kindly. As it lies within his power, so it becomes his duty to work in that direction, just as it is his duty to succour neighbours who suffer misfortune. The improvement of our stock seems to me one of the highest objects that we can reasonably attempt. We are ignorant of the ultimate destinies of humanity but feel perfectly sure it is as noble a work to raise its level in the sense already explained as it would be disgraceful to abuse it. I see no impossibility in Eugenics becoming a religious dogma among mankind, but its details must first be worked out sedulously in the study. Over-zeal leading to hasty action would do harm, by holding out expectations of a near golden age, which would certainly be falsified and cause the science to be discredited. The first and main point is to secure the general intellectual acceptance of Eugenics as a hopeful and most important study. Then let its principles work into the heart of the nation, which will gradually give practical effect to them in ways that we may not wholly foresee." (p. 50.)

Galton stressed here as he always did the essential need to have a science of Eugenics before we make propaganda for its principles—the study is to come before the market-place.

A second paper by Galton is published in this volume of *Sociological Studies* (pp. 85–99). It is entitled: "A Eugenic Investigation, Index to Achievements of Near Kinsfolk of some of the Fellows of the Royal Society." It is a preliminary notice of the material later dealt with by Galton and Schuster in *Noteworthy Families**. As we have already very fully considered the latter work, discussion of this preliminary study is unnecessary. A few lines from the "Preface" indicating how confident Galton had become on certain points may, however, be cited here:

> "It is now practically certain from wide and exact observations, that the physical characters of all living beings, whether men, other animals or plants, are subject approximately to the same laws of heredity. Also that mental qualities such as ability and character, which are only partially measurable, follow the same laws as the physical and measurable ones. The obvious result of this is that the experience gained in establishing improved breeds of domestic animals and plants is a safe guide to speculations on the theoretical possibility of establishing improved breeds of the human race.
>
> It is not intended to enter here into such speculations, but to emphasise the undoubted fact that members of gifted families are, on the whole, appreciably more likely than the generality of their countrymen to produce gifted offspring." (pp. 85–6.)

Two more letters of this year—out of many others—may be printed here because they show not only the affection Galton bore to his lieutenants but also the encouragement he was continually giving them.

42, RUTLAND GATE, S.W. *May* 30, 1904.

MY DEAR KARL PEARSON, What an admirable paper you have just sent me. Such literature will help to unite many scattered forces of a higher order than journalists in the good cause. They *exist* and want to be found out and incorporated. I have been staying some days in a country house with Sir John Gorst, who is very keen and earnest about the degeneracy of the Board School Children. He thinks the Scotch Commissioners' Report, which I have not yet read, a very good one, but doubts the adequacy of the forthcoming (probably in July) report of the English Commission. When it is out he thinks that strong action of any or all kinds would be peculiarly effective. He does not seem to know much about heredity. I will send him your paper after re-reading it comfortably. He is or was a mathematician.

I never congratulated you on your wonderful show of skull photos at the R. Soc.

Very sincerely yours, FRANCIS GALTON.

42, RUTLAND GATE, S.W. *May* 31, 1904.

MY DEAR KARL PEARSON, Your remarks before the Eugenics lecture have just reached me in print. I had no idea at the time (owing to deafness) that you were saying such very kind—such over-kind things of me. I write at once fearing you may have thought my silence on the subject since, due to apathy; which it was not, but purely to ignorance.

Ever sincerely yours, FRANCIS GALTON.

(8) *Work and Correspondence of* 1905. The meeting at the Sociological Society in the previous year had undoubtedly been a success, it attracted a really widespread attention to Eugenics, and this among a circle less rigidly specialist and academically scientific than Galton's two earlier audiences. So pleased was he with the result that early in this year (February 14) he read a further paper on "Restrictions in Marriage†" before the Sociological Society with Dr E. Westermarck in the chair.

---

* See our pp. 113–121.

† *Sociological Papers*, Vol. II, 1905, pp. 1—53.

Galton considered that the public conscience as represented by tribal custom, law or current moral opinion had a powerful influence on conduct. This public conscience is usually reflected in sanctions enforced by the religion of the tribe or nation, often by appeal to the super-rational consequences of "sin," i.e. disobedience to the current social code. Occasionally social needs develop the public conscience more rapidly than the guardians of orthodox belief are able or willing to expand their religious creed, and there is friction, slight or grave, between what the forerunners call "progress" and the priests term "heresy." Somehow religion moulds itself to the developed public conscience, and all ends happily with the progressive "sinners" being canonised as saints. Noting the remarks of the speakers and correspondents which followed or resulted from Galton's paper, we may find the same type of statement unsupported by the only possible proof—that of statistics—again occurring. For example: "the defects of a quality seem sometimes scarcely less valuable than the quality itself," "it is highly probable that a very slight taint may benefit rather than injure a good stock," "marry Hercules with Juno, and Apollo with Venus and put them in slums, their children will be stunted in growth, rickety and consumptive," "in a low state of civilisation the masses obey traditional laws without questioning their authority. Highly differentiated cultured persons have a strong critical sense, they ask of everything the reason why, and they have an irrepressible tendency to be their own lawgivers. These persons would not submit to laws restricting marriage for the sake of vague Eugenics*," "at present the care for future man, the love and respect of the race, are quite beyond the pale of the morals of even the best," "the rise of intellectual qualities also involves under given conditions a danger of further decay of moral feeling, nay of sympathetic affections generally....Under existing social conditions it would mean a cruelty to raise the average intellectual capacity of a nation to that of its better moiety at the present day," with much more half-baked thought.

Some few speakers were more helpful; it may be that Galton, perhaps purposely, did not sufficiently emphasise the distinction between procreation and marriage, or indeed note that most primitive taboos concern mating rather than marriage; yet the distinction was in the minds of some of his supporters. Dr A. C. Haddon held that marriage customs among primitive peoples are not in any way hidebound, and that social evolution can take place. "When circumstances demand a change, then a change takes place, perhaps more or less automatically, being due to a sort of natural selection. There are thinking people among savages, and we have evidence that they do consider and discuss social customs, and even definitely modify them; but, on the whole, there appears to be a definite trend of social factors that cause this evolution. There is no reason why social evolution should continue to take place among ourselves in a blind sort of way, for we are intelligent creatures, and we ought to use rational means to direct our own evolution.

* Why should the precepts of Eugenics be "vague," if they start from scientific knowledge? Other critics asserted on the contrary that the more cultivated classes would reach eugenic conclusions, but the uneducated would pay no attention, and so the movement be idle!

Further, with the resources of modern civilisation, we are in a favourable position to accelerate this evolution. The world is gradually becoming self-conscious, and I think Mr Galton has made a very strong plea for a determined effort to attempt a conscious evolution of the race " (pp. 18–19). Dr F. W. Mott was strongly in favour of the segregation of defective children, and would encourage the State to set up registry offices, which could give a bill of health to persons contracting marriage, and these bills would have actuarial value not only for the possessors but for their children, and should enable them to obtain insurance at a lower rate*. Mr A. E. Crawley said that Galton's remarkable and suggestive paper indicated how anthropological studies can be made of service in practical politics. He considered that the science of Eugenics should be founded on anthropology, psychology and physiology—thus leaving out genetics, actuarial science and medicine, all equally if not more important! The part that Galton suggested religion might play in Eugenics seemed to the speaker excellent. "Religion can have no higher duty than to insist upon the sacredness of marriage, but just as the meaning and content of that sacredness were the result of primitive science, so modern science must advise as to what this sacredness involves for us in our vastly changed conditions, complicated needs and increased responsibilities " (p. 21).

Dr Westermarck thoroughly approved of Galton's programme, and said that Galton had appealed to historical facts to show how restrictions in marriage have occurred; he saw no reason why the restrictions should not be extended far beyond the existing laws of any civilised nation of to-day. He drew attention to tribes which made an exhibition of courage essential to the permission granted a man to marry, to German and Austrian laws prohibiting the marriage of paupers, and he saw no reason why similar laws should not be extended to persons who would "in all probability" become parents of feeble or diseased offspring. "We cannot wait till biology has said its last word on heredity. We do not allow lunatics to walk freely about even though there may be merely a suspicion that they may be dangerous. I think that the doctor ought to have a voice in every marriage which is contracted...men are not generally allowed to do mischief in order to gratify their own appetites."

Besides increased legal restriction Dr Westermarck thought that moral education would help to promote Eugenics†. Dr Westermarck concluded with

---

* This corresponds to the idea on p. 243 above of attaching medical certificates to the State sickness and old-age pensions scheme.

† This has, owing chiefly to the efforts of Galton, progressed largely during the past 25 years. Quite a number of persons have developed the eugenic conscience, and A seeks advice as to whether it is social to marry B; or C, having married D, as to whether it is anti-social to have further children who may turn out like E. The Galton Laboratory is not at present organised on a scale to answer such problems, although it does its best to do so: but the time is rapidly approaching when an institution above reproach from the medical standpoint, and equipped with a staff conversant with the various branches of human heredity and of genealogical study, might issue case-opinions and certificates. In the distant future it might hope to be self-supporting.

the following words, which could hardly have been better expressed by Galton himself:

"It seems that the prevalent opinion that almost anybody is good enough to marry is chiefly due to the fact that in this case the cause and the effect, marriage and the feebleness of the offspring, are so distant from each other that the near-sighted eye does not distinctly perceive the connection between them. Hence no censure is passed on him who marries from want of foresight or want of self-restraint, and by so doing produces offspring doomed to misery. But this can never be right. Indeed there is hardly any other point in which the moral consciousness of civilised man still stands in greater need of intellectual training than in its judgments on cases which display want of care or foresight. Much progress has in this respect been made in the course of evolution, and it would be absurd to believe that men would for ever leave to individual caprice the performance of the most important and, in its consequences, the most far-reaching function which has fallen to the lot of mankind." (pp. 24–5.)

It is worth while giving these expressions of opinion, because they indicate that Galton was beginning to make an impression, and on those whom it was worth while to impress. The purpose of Galton's lecture was to combat the objection often raised against Eugenics*, that human nature would never brook interference with the freedom of marriage. Galton wished to appeal from armchair criticism to actual facts. He stated that it is no unreasonable assumption to suppose that, when Eugenics is so well understood that its lofty objects become generally appreciated, they will meet with some recognition both from the religious sense of the people and from its laws. "The question to be considered is how far have marriage restrictions proved effective when sanctified by the religion of the time, by custom and by law." Galton next proceeds to show how monogamy and polygamy have each received religious sanction and religious condemnation in their place and turn; how celibacy has been a sin and a state of holiness †. If such customs do not arise from any natural instinct but from considerations of social well-being, may we not conclude that under pressure of worthy motives, limitations to freedom of marriage may hereafter be enacted by law or custom for eugenic purposes? Galton then turns to endogamy and exogamy, which in multitudes of communities have been enforced even under the severest penalties; he refers to the Levirate with its limitation on the widow's choice—he might have referred to the funeral pyre of the Hindoo widow—and to the strange custom adumbrated by the tale of Ruth and Boaz.

* To this word in the opening section is a footnote: "Eugenics may be defined as the science which deals with those social agencies that influence mentally or physically the racial qualities of future generations." This is not yet the definition of the University Committee, and a singular history attaches to the footnote. Mr Howard Collins read this paper in manuscript, and criticised the wording of the definition of the term "Eugenics"; and in a letter to Galton of Jan. 15, 1905, he proposed that it should read as follows: "Eugenics is defined as the science of those social agencies which influence mentally, morally and physically, the racial qualities of future generations." Galton adopted this wording, striking out, however, the word "morally." This indicates how far he was from accepting Sir Arthur Rücker's modification of the University Committee's definition.

† Galton enlarges a good deal on the celibacy of mediaeval Christianity and opines that pious efforts as great as those which founded monasteries and nunneries might under religious influence be directed so as to fulfil an exactly opposite purpose, thus homes or colleges might be endowed for young married couples from stock of high civic worth: see our p. 78 and the account of "Kantsaywhere" later in this chapter.

Marriage within the clan may be considered unmanly—a wife must be *captured*. Customs like these are not instincts, they have arisen from ideas of social profit. Yet they, like the complicated Australian marriage system, have religious sanction, nay, may be enforced by the penalty of death.

"Eugenics deals with what is more valuable than money or lands, namely the heritage of a high character, capable brains, fine physique, and vigour; in short, with all that is most desirable for a family to possess as a birthright. It aims at the evolution and preservation of high races of men, and it as well deserves to be as strictly enforced as a religious duty as the Levirate law ever was." (pp. 6–7.)

Next Galton refers to the influence of *taboo*.

"A vast complex of motives can be brought to bear upon the naturally susceptible minds of children, and of uneducated adults who are mentally little more than big children. The constituents of this complex are not sharply distinguishable, but they form a recognisable whole that has not yet received an appropriate name, in which religion, superstition, custom, tradition, law and authority all have part. This group of motives will for the present purpose be entitled 'immaterial' in contrast to material ones. My contention is that the experience of all ages and all nations shows that the immaterial motives are frequently far stronger than the material ones, the relative power of the two being well illustrated by the tyranny of taboo in many instances, called as it is by different names in different places." (pp. 8–9.)

The mere terror of having unwittingly broken a taboo may fill a man with the deepest remorse, or even kill him.

Under our own " civilised " law also and with religious sanctification, we meet the taboos of the prohibited degrees of marriage. They are in many cases not questions of instinct, but are primarily designed to preserve family life.

"The marriage of a brother and sister would excite a feeling of loathing among us that seems implanted by nature, but which further inquiry will show has mainly arisen from tradition and custom." (p. 9.)

Galton holds that a repugnance to inbreeding may have arisen from harm arising from too close inbreeding, but biologically the evil appears—when the stock is good—to have been much exaggerated. He thinks therefore that desire not to infringe the sanctity and freedom of the social relations of a family group has led to the taboo. " It is quite conceivable that a non-eugenic marriage should hereafter excite no less loathing than that of a brother and sister would do now." (p. 11.) Personally the biographer would consider the marriage of two individuals both members of unrelated stocks tainted with insanity as more heinous than the marriage of a brother and sister of sound stock—the risk of the latter to offspring depends on the existence of unrecognised and undesirable latent characters; there is almost certainty in the former case that a definite percentage of the children will either exhibit or transmit the taint. The thorough conviction by a nation that no worthier object can exist for man than the improvement of his own race is for Galton in itself the acceptance of Eugenics as a national religion. If we examine the reasons for such irresistible streams of popular emotion as are vaguely symbolised in respect for the national flag, in the King as personifying our country, indeed in all phases of patriotism, we shall discover that their springs lie in Galton's "immaterial motives," and it is in precisely such almost instinctive motives that he hoped to find ultimately a foundation for that highest form of patriotism, eugenic morality. Several of the contributors to the discussion emphasised the difference between "barbarous" and "civilised" peoples, suggesting that what anthropology tells of the former cannot be applied to the latter. To the careful student of mankind there are no rigid categories such as barbarism and civilisation ; to him the civilisation of to-day is the barbarism of to-morrow, and he can only smile when he is told that civilisation was born in and spread from Egypt. The man of to-day believes, of course, that his religions and his institutions are products of his "high" civilisation ; he does not see their growth through the ages and their roots in the fertile mud of what he terms "barbarism." He believes that the basal laws of his own psychic growth differ in some undefined way from those which controlled that of his far-distant ancestor. Galton thought otherwise :

> "The subservience of civilised races to their several religious superstitions, customs, authority and the rest, is frequently as abject as that of barbarians. The same classes of motives that direct other races direct ours, so a knowledge of their customs helps us to realise the wide range of what we may ourselves hereafter adopt, for reasons as satisfactory to us in those future times, as they are or were to them at the time when they prevailed." (p. 12.)

I have had several times to refer to Galton's views on religion in the course of this biography. The study of evolution had brought him freedom from the traditional faiths; like many of the leading men of science of his day he was an agnostic. But he was not an iconoclastic freethinker, he was willing that old faiths should remould themselves to new ideas, where some would have felt that it was futile to pour new wine into old skins. Even the ancient faiths in their old skins might help certain natures to-day. I well remember what he said to me when one of his closest relatives was received into the Catholic Church: "It may be a stable guide for emotional natures, it would be of no service to you or me." He was not only tolerant of others' views, but his sympathy induced him to satisfy where it lay in his power their religious cravings, even at the risk of his action being misinterpreted*.

---

* I venture to quote here a very characteristic and beautiful letter to his niece dated from 42, Rutland Gate, July 30, 1907:

"I should be *glad* to have family prayers as of old. The household needs a few minutes of daily companionship in reverent thought and ritual. The first morning when I had returned home after dear Louisa's death, we the remainder of the household reassembled as usual, but—oh the pitifulness of it—when half-way through the prayers, I lost all control of my voice, and fairly broke down, and dismissed the household. I never recommenced the custom ; partly shrinking from its memories ; largely because I felt that at least one of the heads should be able to join in the prayers *without any reservation*. This as I understand from your letter you would do now.
"I have again looked at the old and well-remembered prayer-book. It is sadly dilapidated and when last used required caution in handling. I will bring it with me. It might be replaced with advantage. Both Louisa and I felt that the psalms became monotonous, and that it would be well to read alternatively or otherwise parts of the rest of the Bible. I will get a Bible for the purpose of marking out suitable passages, also a prayer book. (It interested me much to find that the published list of Mr Gladstone's favourite psalms was almost identical with my own selection.) It would also be well to increase the variety of the prayers. Mine were 14 collects, two for each week day. We will consider all this at Hindhead. You know and will respect *my* limitations in selecting passages. I *must* be true to my own convictions as you will be to yours."

Galton's conviction was that prayer is subjective in its influence and should be an inspiration, not a petition. I may quote extracts from three letters to his nieces, which support this statement. April 9, 1907: "I think in earnest prayer of you and poor Fred, for I *can* pray

Galton had in reality a deeply religious nature, and in this sense we must read the concluding sentences of this memoir in which he again emphasises the conception that Eugenics will hereafter receive the sanction of religion, even of the present Christian doctrine.

"It may be asked 'how it can be shown that Eugenics falls within the purview of our own faith.' It cannot, any more than the duty of making provision for the future needs of oneself and family, which is a cardinal feature of modern civilisation, can be deduced from the Sermon on the Mount. Religious precepts founded on the ethics and practice of olden days require to be reinterpreted to make them conform to the needs of progressive nations. Ours are already so far behind modern requirements that much of our practice and our profession cannot be reconciled without illegitimate casuistry. It seems to me that few things are more needed by us in England than a revision of our religion, to adapt it to the intelligence and needs of the present time. A form of it is wanted that shall be founded on reasonable bases, and enforced by reasonable hopes and fears, and that preaches honest morals in unambiguous language, which good men who take their part in the work of the world and who know the dangers of sentimentalism may pursue without reservation." (pp. 12–13.)

Such was Galton's view on the need for the reform of religion. There are several addenda to this paper which I will briefly note here.

In his Reply to speakers (pp. 49–51) Galton remarked that Eugenics is a wide study with an excessive number of side paths into which those that discuss it are apt to stray. Such was essentially the case in the present instance where Galton in his paper had limited himself to the question of whether communities will submit to restriction in marriage. The subjects dealt with in the reply were:

(i) *Certificates.* These were to be of *mens sana in corpore sano* and were to regard ability, physique and hereditary factors. Of these Galton says that such Eugenic certificates could only be issued at some future time dependent on circumstances. He admits that mistakes may be made at first in devising a satisfactory system but is hopeful for the future. As we shall see later, Galton in the following year actually drafted a scheme for Eugenic certificates. In the surviving fragmentary chapter of his utopia " Kantsaywhere," dealing with the College of that place, there is a very full account of the examinations for Eugenic certificates.

(ii) *Breeding for Points.* Critics had suggested that breeding of domesticated animals is successful because they are bred for individual points.

and *do* pray conscientiously and fervently, though probably in a different form from that you yourself employ." May 12, 1907: "Did I ever tell you that I have always made it a habit to *pray* before writing anything for publication, that there be no self-seeking in it, and perfect candour together with respect for the feelings of others." And again, Jan. 20, 1910: "I have read a most interesting article in the *English Review* by Prof^r Murray, the Professor of Greek at Oxford, on the working religion of the Pagan Greeks at about A.D. 400 (Marcus Aurelius' time). He gives extracts from two writers of that date beautifully expressed. One of them is a man named Eusebius (not *the* Eusebius) which is in the form of a prayer such as I would employ. It is not 'give me this or that,' but 'may I not fall into this or that faulty conduct.' It is an aspiration *not* a solicitation. The prayer in question would be a valuable addition to *any* prayer-book to say the least. I should like it and others like it to replace almost all that are there."

This is the opinion of a man whose paper on prayer of 1872 had led to his treatment as a very flippant freethinker! See our Vol. II, pp. 115, 175, 258, etc.

Galton says that some contributors to the discussion had been unnecessarily alarmed. No question had been raised by him of breeding men like animals for particular points, to the disregard of all-round efficiency in physical and intellectual (including moral*) qualities and in the hereditary worth of their stock. (Personally also I very much doubt whether most breeders select animals for individual points without close regard to other characteristics.) Galton remarks that

"Moreover, as statistics have shown, the best qualities are largely correlated. The youths who became judges, bishops, statesmen, and leaders of progress in England could have furnished formidable athletic teams in their times. There is a tale, I know not how far founded on fact, that Queen Elizabeth had an eye to the calves of the legs of those she selected for bishops. There is something to be said in favour of selecting men by their physical characteristics for other than physical purposes. It would decidedly be safer to do so than to trust to pure chance." (p. 50.)

(iii) *The Residue.* Galton does not make here a very strong reply to those who objected that, after the selection of the fitter, the residue would interbreed and grow increasingly inferior†. He appears to overlook his own point, that it is essential to create a differential fertility, so that the better stocks increase at a greater rate.

(iv) *Passion of Love.* To the argument that "Love is lord of all," and will not be restrained, Galton replies that a slight inclination and falling thoroughly in love are two different things, and it is against the former that taboo applies, whether it is due to rank, creed, connections or other causes. "The proverbial 'Mrs Grundy' has enormous influence in checking the marriages she considers indiscreet." (p. 51.)

(v) *Eugenics as a Factor in Religion.* Here Galton adds to his memoir two additional pages (pp. 52–3) as a short essay on this topic. He considers that Eugenics strengthens the sense of social duty in so many important ways—for it promotes wise philanthropy, the notion of parentage as a serious responsibility and a higher conception of patriotism—that its conclusions ought "to find a welcome home in every tolerant religion." There follows a vivid description of "mechanical" evolution—one of the finest word-paintings that perhaps anyone has made of the world's history—and then the statement that man has already largely influenced the quality and distribution of organic life on the earth and that if he will only recognise it, it largely lies in his power to influence the evolution of humanity itself. The brief essay concludes with the lines that occupy a place of prominence in the Galton Laboratory of National Eugenics as among the most stimulating words of its Founder:

"Eugenic belief extends the function of philanthropy to future generations, it renders its action more pervading than hitherto, by dealing with families and societies in their entirety,

---

* This confirms my view (see p. 224) that Galton would have included the moral with the mental characters.

† He supposes that in the future there would be a freer action for selective agencies, e.g. there would be a reduction of indiscriminate charity, but this seems a return, with emphasis, to the crude processes of natural selection.

and it enforces the importance of the marriage covenant by directing serious attention to the probable quality of the future offspring. It sternly forbids all forms of sentimental charity that are harmful to the race, while it eagerly seeks opportunity for acts of personal kindness, as some equivalent for the loss of what it forbids. It brings the tie of kinship into prominence and strongly encourages love and interest in family and race. In brief, Eugenics is a virile creed, full of hopefulness, and appealing to many of the noblest feelings of our nature." (p. 53.)

Besides the two memoranda to which we have just referred, Francis Galton presented a short paper entitled "Studies in National Eugenics," which was appended to his "Marriage Restrictions." He refers to the appointment of Mr Schuster to the Research Fellowship, and sketches out the various inquiries which the new Eugenics Record Office might undertake. They form indeed an excellent scheme for any laboratory proposing to undertake eugenic research. Most of Galton's problems still remain unsolved owing to the difficulty of procuring accurate and adequate data, and they will remain so until the public at large is willing first to fill in at all, and secondly to fill in veraciously investigators' schedules, and until the State recognises how important it is that school, asylum and prison should be treated as laboratories, where under suitable regulations men of science may work*.

As confirming Galton's view that probability is the basis of Eugenics we may note that the bulk of his suggested problems demand the collection of data and their statistical treatment.

I. The first problem is the estimation of the average quality of offspring from that of their parents and ancestry, and this covers questions of relative fertility. Under this heading Galton includes genealogical work on (*a*) Gifted Families; (*b*) Capable Families; (*c*) Degenerate Families; (*d*) Extent of social class interchanges, to what extent do "castes" rule in modern civilised communities; and (*e*) Possibility of obtaining valuable eugenic data from Insurance Office records.

II. The Effects of Action by the State and by Public Institutions. Under this heading we may deal with (*a*) Habitual Criminals, and the problems of their origin and segregation; (*b*) Feeble-minded and Insane, their origin and the restriction of their reproduction; (*c*) Grants for higher education, how far these are advantageously used, and to what extent they might be employed to encourage fertility in eugenic marriages; (*d*) Indiscriminate charity, including out-door relief and perhaps we may now add "the dole." Have they, as there is reason to believe, tendencies other than eugenic?

III. What factors in religion, custom or law, and what social influences tend to restrict eugenic marriages or reduce their fertility?

IV. Heredity. "The facts after being collected are to be discussed for improving our knowledge of the laws both of actuarial and physiological heredity, the recent methods of advanced statistics being of course used" (p. 16). Galton suggests two special problems of great interest: (i) Effect on offspring

---

* Schoolmaster, medical officer and governor of prison have no time for statistical inquiries. Too often on retirement they publish statements based merely on *impressions*, and none knows better than the statistician how fatally inaccurate these may often be.

of *differences* in parental qualities* and (ii) a thorough study of characters in Eurasians in order to test the applicability of the Mendelian hypothesis to man.

V. A Bibliography of papers bearing on Eugenic topics is desirable. Many papers already exist, published in scientific transactions and journals, which bear on the Eugenists' problems; such a bibliography should include papers of breeders and horticulturists. Considering the enormous development nowadays of Genetics it would probably be well to treat separately Genetics and Eugenics.

VI. Co-operation between students of Eugenics. Probably Galton had in mind here special journals, societies, and congresses.

VII. Certificates of Eugenic fitness. To these we shall return later.

It will be seen that Galton's programme did not lack comprehensiveness.

Another event of this year was the invitation to Galton to accept the Presidency of the British Association at the York Meeting in 1906. It is desirable to indicate that it was not from want of asking—and even of gentle pressure—that the Association missed the honour of numbering Francis Galton among its past presidents. In this he stands with his cousin Charles Darwin; the names of two of the most original scientists of the Victorian epoch fail to appear on the presidential roll.

The following letters received by Galton on May 8 and answered on May 9 explain the situation.

BRITISH ASSOCIATION FOR THE ADVANCEMENT OF SCIENCE. *May 5*, 1905.

DEAR MR GALTON, At the meeting of the Council of the British Association held at Burlington House this afternoon, it was unanimously resolved that you be nominated as President of the British Association for the meeting to be held at York in 1906. The proposal was received by the Council most cordially, and the officers were instructed to communicate with you and ascertain whether you will agree to the nomination.

---

* I do not know on what Galton's suspicion rested of a marked influence on the characteristic (*c*) of a child, if there was a great difference (δ) between the paternal (*f*) and maternal (*m*) characteristics. Theoretically, if ϵ be the coefficient of assortative mating, r of parental heredity supposed the same for both parents, σ a standard deviation, and $r_{\delta c}$ the correlation of δ and *c*, then:

$$r_{\delta c} = r \frac{\sigma_f - \sigma_m}{\sqrt{(\sigma_f - \sigma_m)^2 + 2\sigma_f\sigma_m(1-\epsilon)}}$$

Since the coefficients of variation are nearly the same in man and woman, we have, if M_1 and M_2 are mean values in father and mother,

$$r_{\delta c} = r \left/ \sqrt{1 + \frac{2M_1 M_2 (1-\epsilon)}{(M_1 - M_2)^2}} \right.$$

In the case of absolute measurements in man and woman, $M_1 = (1 + \frac{1}{12}) M_2$ and $\epsilon = {\cdot}2$ roughly.

Accordingly $\qquad\qquad r_{\delta c} = r \times {\cdot}069 = {\cdot}03$, approximately.

Hence, statistically, there is no significant influence of the *difference* of parental characters on the character of the child. Physiologically, of course, there may be some influence of extreme differences, but such being rare it may not be detectable in the statistical treatment.

35—2

May I add that I am sure it will be a matter of rejoicing and gratification to Biologists generally if you see your way to accept this position and become our President at the next meeting in this country. I am asking Professor George Darwin, the President-Elect for this year's meeting, to write to you also—so I hope you will receive a letter from him in the course of a day or two. I am, dear Mr Galton, Yours very sincerely, W A. HERDMAN, Gen. Secretary.

This letter was backed up by one from George Darwin.

NEWNHAM GRANGE, CAMBRIDGE. *May 6, 1905.*

MY DEAR GALTON, You will perhaps already have received an official intimation that you were yesterday unanimously nominated Pres^t of the B.A. for the York meeting. I had the pleasure of proposing your name, and I pointed out that you ought to have been nominated years ago, and that the fact that men of science were formerly somewhat blind to the great work that you have done gave no excuse for omitting even this belated recognition. That you may not think that this is merely my personal opinion, I should add that speaker after speaker endorsed what I have said. We all hope that you may feel yourself able to accept the nomination. It was pointed out as an objection that your deafness would be a difficulty in as much as presiding at the Council meetings could hardly be carried out efficiently by you. To this most, perhaps all, considered that there was a complete answer—you have only to absent yourself from Council meetings. During the present year Balfour never comes—as we knew he would not—and we get through our business with the aid of the V.P.'s.

I hope that you will not allow this consideration to deter you from acceptance, and, if you will take it, my advice to you would be that you should not attend any Council meetings during your year of office, when you would have to take the chair, or at least should ask a V.P. to preside. I cannot of course judge whether you will feel yourself disposed to undertake the duties, but I can only very heartily express the hope that you will feel you have the strength to do so. Yours very sincerely, G. H. DARWIN.

To this letter I add Galton's reply:

42, RUTLAND GATE, S.W. *May 9, 1905.*

MY DEAR GEORGE DARWIN, It was only last night that I returned and found your very kind letter and that of Prof. Herdman to whom I have just written. I am deeply sensible of the proposed honour and fully recognise the unique opportunity afforded to the President of the Brit. Assoc^n of drawing the attention of the whole scientific world to such views as he may put forward. Also I am cordially grateful to the thoughtful way in which you propose to make the work less laborious and independent of my deafness. But the fatal fact remains that I am not strong enough even under all these alleviations. The preceding excitement would be enough to upset me. I cannot stand even a moderate amount of flurry. It is of no use for me to fight against impossibilities. Long since I have learnt to renounce many tempting pleasures, and must do so now. The only chance I have of doing useful work during the remainder of my life, lies in doing it quietly and living very simply much like an invalid, and in never undertaking to tie myself to a day when I might prove quite unfit. Once before when Sir William Flower was President and the names of possible persons were to be considered at a Council meeting at which I was present, he with the previous assent of the other General Officers, emphatically proposed me at the first. I immediately begged to be left out of account, being too painfully conscious even then of the limitations of my strength. Notwithstanding kindly pressure, I persisted in the refusal. It would be foolishly rash if I made the venture now.

Ever sincerely yours, FRANCIS GALTON.

P.S. I have had a pleasant and healthful 2½ months in the Riviera (Bordighera), but missed your sister. I saw Miss Shaen during her brief visit there. What an eventful August you will have at the Cape. I heartily wish you every possible success and pleasure. But what a racket it will be !

During this year Galton was very busy with the superintendence of his Eugenics Record Office and many of his letters relate to proposed work, to developments at the office, or to suggestions and criticisms touching the biographer's researches. I give three illustrations to show how keenly alive he still remained to all going on in our joint field of work.

42, Rutland Gate, S.W. May 31, 1905.

My dear Karl Pearson,   If your timely and most useful article on Dr Diem's material in the *Brit. Med. Journ.* is intended to start an organised inquiry, towards which I can *in any way* help, pray command me. It is just one of the things I want to see done. Quere, a reasonable plan would be to reprint your article in a pamphlet form, with tables to show exactly what is wanted, and after preparing the way a little to circulate it judiciously. Is there not an error—at all events the sentence requires explanation—in "Dr Diem's tables show that nervous disorders are *more numerous* in the parentages of the sane than in those of the insane"? What are "nervous disorders"? Or are sane and insane transposed *? If a pamphlet were circulated the meaning of the phrases 1 to 5 in Dr Diem's and your list should be defined in it. As, for example "want of mental balance"! We are all of us so mad! How mad must we be to justify the epithet of "unbalanced mind"? Parental and fraternal histories ought to be easily accessible among the insane and feeble-minded, and among the sane still more so. But in the latter case there are often skeletons hid in closets. One seems to want corroboration of what is said by others who have known the family intimately. Biographers fib so much. I have just been reading one that includes two letters praising a man as a gentle angel, whom I recollect as a red pimple-faced obstreperous and most eccentric schoolmaster in my very early days. Where is truth to be found?   Ever yours sincerely, Francis Galton.

This research was not at the time pushed further. What is essential to the effective study of the heredity of insanity is a register of the persons in the kingdom who have at *any* time in their life been in an asylum (and of course it must state from what type of mental disease); at present we can only *guess* what percentage of the population has been certified at any time as insane.

42, Rutland Gate, S.W. July 27, 1905.

My dear Karl Pearson,   I kept your letter the last to open, as I dreaded it would contain a grave and well merited rebuke, but it did not, and the motive for the rebuke is happily dissipated. It was the announcement by Murray that he was about to publish eugenic matter for the University of London before he had received authority to do so. It was a stupid blunder of his, for which he wrote a most penitent letter that was laid before the members of the Senate yesterday, who have condoned it—for their resolution in the University Intelligence, p. 7, of to-day's *Times* puts all on a solid footing. The material in question consists of 65 Noteworthy Families in Modern Science, and "is to appear as Vol. ɪ of the publications of the Eugenics Record Office." This is a big recognition in my opinion. Murray is pleased to publish on the ½ profit system....I envy the old biometric teas†, but everything "dehisces." I go north on Saturday towards and then to Westmoreland; Eva Biggs goes south to Devon, and in the 3ʳᵈ week of August we reunite at Ockham.

Last Monday and Tuesday evenings we spent at that wonderful air-cure Hindhead, where I had the great pleasure of seeing again Mrs Tyndall, who lives in the house her husband (Prof. Tyndall) built.

---

* If "nervous disorders" be used in the sense of slight nervous troubles, hysteria, excitability or depression, far short of insanity, the explanation may be that in the case of stocks tainted with insanity, these cases are intensified and the sufferers become actually insane.

† For some years Francis Galton and his niece had come within reach of the biometric holiday workers for a few weeks in the summer. We were often some distance from each other as at Bibury, Witney and Oxford. The morning was given to work, then the victoria carried our leader and bicycles the remainder of the party to some inn, in a village if possible with a beautiful church, and there was a biometric tea, at which discussion turned not wholly on work.

I am rather pleased at the way that has occurred to me of explaining why the men of highest genius have so few able descendants and these often cranks, viz. that there is negative correlation between their faculties,—sensitiveness and dogged work, imagination and good sense, etc.—so that the inheritance of such an unstable combination is improbable. There is much to say, this is only a notice, so to speak.

Ever very sincerely, with kind remembrances to Mrs Pearson, FRANCIS GALTON.

Even letters which touched chiefly on personal matters were sure to contain at least a few sentences as to work.

THE RECTORY, OCKHAM, SURREY. *Aug.* 25, 1905.

MY DEAR KARL PEARSON, It was with self-restraint that I did not write to say how grieved I was at your domestic sorrow, and how deeply I sympathise with you. I feared to extract a reply and knew you were overworked. This note is merely to enclose my brand-new circular, which I begin to distribute among friends, and hereafter I hope much more widely. If you think any of your lady co-operators especially are likely to help and take interest, I would gladly send circulars to them. Miss Elderton is established now at the "Eugenics Record Office" and at work there*.

This is a pretty and healthy place, and friends are near. Sir H. Roscoe has a beautiful garden, 600 and more feet above the sea, where everything flourishes. Kindest remembrances to you both. Eva Biggs is at this moment sketching or choosing a sketching place by an artistic but foul pond. Ever sincerely, FRANCIS GALTON.

(9) *Events and Correspondence of* 1906. During this year I do not think that Galton published any papers, except the Memoir on Resemblance and the humorous little note in *Nature* on the cutting of a cake (see Vol. II, p. 329, and above, p. 124). But it was full both for Galton and his biographer of new and sad experiences which, as they were to some extent common to them both, brought them closer together and ripened their friendship. To the one the loss of a sister†, to the other of a mother; to both of an irreplaceable friend and colleague, a death rendered the more bitter by its unexpectedness, and by attendant circumstances, which touched both with nearly equal sorrow. I had started with a keen appreciation of Galton as a scientist, I had learnt to value him as friend and counsellor; I now understood and deeply admired the strength of his humanity and his generosity of mind. The following letters may give some idea of the warmth of feeling that existed between Galton and his two lieutenants, even as the tripartite relationship was dissolved.

7, WELL ROAD, HAMPSTEAD, N.W. *Jan.* 24, 1906.

MY DEAR FRANCIS GALTON, May I send just a line of very heartfelt sympathy with you in the loss of which I have just heard? I know it will be the greater in that you were not in England at the time. I am at the age when these losses begin to be more frequent, and deprive life of much of its old "go"; and just at present one lives a day at a time, with two or three of one's own generation and some of the generation above almost more than threatened. Hence one feels very strongly the closeness and the mystery of death; and sympathy—which one is helpless to express—goes out to a friend in like case. I have often thought the only real expression of a feeling like this is given by the hand and eye, and not by the tongue, which is so helpless that we had better go on with the old routine of life, speechless on such points.

* As Secretary. Francis Galton hesitated about a woman taking part in academic matters, although he had begun to realise the good work of the women in the Biometric Laboratory. He was comforted by the Principal's opinion, "Sir Arthur Rücker speaks highly of lady secretaries, and generally agrees with what we talked about." Letter to K. P., June 20, 1905.

† "Bessie," Mrs Wheler.

I have sent Schuster's paper to press. Hartog has paid the account. I was seeing Dr Pearl yesterday and put my head into Miss Elderton's door; she seemed bright and fresh, and said she had plenty to do; so I hope the work of your Eugenics Office is going forward.

The enclosed letter may amuse you. I think that X. is a very dangerous person, if his notion of eugenics is the intermarriage of consumptive stocks! Very many thanks for your long letter. I wish there were some simple colour register. I don't expect it is easy to get colours like terra-cottas and salmons out of Abney's apparatus. I shall send you a copy of the poppy plate when it comes. I hope Miss Biggs will not be too scornful about it!

Weldon will have told you about Y. and Z.'s attack at the R.S. Weldon had a good paper last Thursday and Z. drew as usual red herrings across the track.

Affectionately yours, KARL PEARSON.

Feb. 1, 1906.

MY DEAR KARL PEARSON, Thank you very much for your letter of sympathy. I have now lost the last tie that brought the family's interests together as to a common focus, and kept each member informed by letter, weekly or otherwise, of the welfare of the rest. To what an enormous amount of grief do the tombstones of any churchyard bear witness!

The "slasher" against X. is right well deserved. I had always a faint misgiving of his Oriental ways and fluency, which steadily deepened until I have come to look upon his aid as unreliable and dangerous. He strikes me as an interesting evidence of the danger of entrusting political power to Oriental subjects—Indian, Egyptian and others.

I will venture shortly to ask you to do me a very great favour, namely to look over a short type-written paper on "the Measurement of Resemblance," and tell me what you think of it. The thing has, as you may remember at Peppard*, been often taken up by me, puzzled over and temporarily laid down. It is at length worked out, I think, fully and practically, but before venturing on publication, I should *greatly value* criticism. At this moment it is only in an uncorrected draft, and I do not wish to hurry before putting it into a corrected form and sending it to London to be typed. The typist will then be instructed (say in a week or a fortnight) to send you a copy†.

We go to-day to "Hôtel de la Rhune, Ascain, Basses Pyrénées," for a week. It is a picturesque Basque Village, four miles from here. Then we *probably* return to where I am writing from, "Hôtel Terminus, St Jean de Luz, Basses Pyrénées," for a day or two, and afterwards according to conditions not yet determined on which we are dependent, to somewhere else. These may lead either to a dip of a fortnight into Spain or to another Basque village, I cannot foresee which. I will send address later on.

The Basque orderliness, thorough but quiet ways, and their substantial clean-looking houses, tug at every Quaker fibre in my heart, and I love them so far. As to their wonderful language unlike in syntax to any other, the virtue of these parts is accounted for by the legend that Satan came here for a visit, but finding after six years that he could neither learn the Basque language, nor make the Basques understand him, he left the country in despair. With kindest remembrances to Mrs Pearson. Ever sincerely yours, FRANCIS GALTON.

It would be an interesting problem to determine what *is* the degree of likeness of a man to himself, by correlating the habits and modes of thought of individuals at selected ages. We might thus obtain a measure of the permanence of individuality. How far is one the same man at 20 and 60 years of age? Galton at least in his love of travel at 18 and 84 exhibited a marvellous sameness. His love of ingenious mechanical apparatus also remained fully as strong, and his humility was not a whit less.

"How curious it is to see," remarks Lord Minto, "how exactly people follow their own characters all through life."

* The long vacation of 1903 was spent at Peppard, the Galtons on the Green, the Pearsons at Blount's Court Farm, the Weldons at the far end of the village, and various biometric workers round about. It was a delightful and fertile summer.

† See the section in Vol. II on the "Measurement of Resemblance," pp. 329–333.

If Galton's character seemed to me at first to change between 1890 and 1910, it was only because with ever increasing intimacy I learnt to understand him better and better.

7, WELL ROAD, HAMPSTEAD, N.W. *Feb.* 16, 1906.

MY DEAR FRANCIS GALTON, Very hearty thanks to Miss Biggs and yourself for your consoling words as to the plate of poppy petals. I feared you would be as disgusted with them as I felt, but you have not the originals to place beside them. I think, however, we shall succeed in getting something better in the final proof. Your paper reached me safely the day before yesterday, and I have read it through thrice. It seems to me most suggestive and I want very much to be making "isoscopes" and practically trying how it works. It would be most satisfactory to find it giving a higher average degree of resemblance between relatives than between strangers. You use I suppose one eye only to see both objects simultaneously? Would it not be well to get a simple instrument made by Beck or Baker from your drawings with an ocular micrometer, and test on photographs? or are you thinking of finger-prints? Would you like the paper in *Biometrika* or do you want a wider audience? I need not say we shall be most pleased to have it. Affectionately yours, KARL PEARSON.

Thus matters seemed to be slipping back into their old channels, with work in the foremost place. Easter was to be spent by us at Longcot with the Weldons near at hand in little Woolstone inn at the foot of the hill marked by the White Horse (or rather "White Dragon"). There were the usual plans for further work, visits to Oxford to see the mice and cycle-rides to make lay studies of church architecture. Weldon was not in good health, he was depressed and thought a visit to a picture gallery in London would be a relief. He went, and from the gallery passed to a nursing home, and died within twenty-four hours of double pneumonia.

42, RUTLAND GATE, S.W. *April* 16, 1906.

MY DEAR PEARSON, Weldon's death is a terrible and disastrous blow, so utterly unexpected. Few if any men will feel it more deeply than you who were so intimately associated with him, not many more than I do. We have lost a loved friend, and Biometry has lost one of its protagonists. I feel intensely miserable about it and shall feel the void he has left for probably the rest of my life. I should greatly have liked to pay the last tribute of friendship to his remains by attending the funeral, but I dare not risk it. Among other things an incipient mild phlebitis in a leg prevents my standing during many minutes and my doctor is strict on this.

I do indeed pity Mrs Weldon from my heart. How deeply your Wife will feel it all, and how helpful she is sure to be, as you are. Give my kindest remembrances to her. We go to the country on Wednesday but letters will be forwarded from here. It will be a sad day.

Affectionately yours, FRANCIS GALTON.

The first part of the funeral service was in the chapel of Merton College, and to my surprise I saw Galton there.

THE AVENUE HOUSE, BISHOPTON, STRATFORD-ON-AVON. *April* 19, 1906.

MY DEAR KARL PEARSON, The card of invitation showed it was possible for me to attend the first part of the funeral without harm, so as you saw I went, and came on here by a later train. It is inexpressibly sad. I do not myself yet fully know all the circumstances, but the more I know the more pity full* it seems. I should be very grateful for tidings about Mrs Weldon, into whose sorrow I could not venture yesterday to intrude. If you or Mrs Pearson have the

---

* So Galton wrote, and the words express more than "pitiful."

opportunity of doing so pray express my sincerest sympathy with her.......I do pity you both in losing so dear and intimate a colleague and in so tragical a manner....I am staying with my niece, Mrs Studdy, here at her house, until Saturday morning, then I go to my nephew Edward Wheler, Claverdon Leys, Warwick. These two, another dear niece Mrs Lethbridge and my ageing brother, now exhaust the list of my near relatives. Kindest remembrances to Mrs Pearson. I heard of her movements from Professor Clifton and knew she was not in Oxford yesterday.

Ever affectionately, FRANCIS GALTON.

The blow struck us both severely; there was much to think over, and some things had to be done immediately, *Biometrika* reconstituted, an eloge written, a memorial to Weldon instituted and many papers sorted. Without Francis Galton's continuous sympathy, aid and counsel, it would have been impossible in that year to continue my work.

First, as to the Weldon memorial; largely by the aid of two or three generous donors, of whom it is needless to say Galton was one, enough money was eventually obtained for a marble bust by Hope Pinker, to be placed in the Museums at Oxford, and a biennial Weldon medal with premium for the best biometric memoir published in the immediately previous years—the medal to be awarded by Oxford electors, but not confined to that University nor to British subjects. The scheme, as finally drafted and accepted by the Hebdomadal Council, was largely Galton's creation. I had felt very strongly that biometry was destined eventually to take an important place in biology, especially in researches into evolution and that, for an international prize of this kind, at least in the more distant future, the Council of the Royal Society would be the fittest judges.

Secondly, sheet after sheet the eloge on Weldon went to Francis Galton and was returned with criticisms and suggestions. He was especially dissatisfied with my brief references to Weldon's part in the attempt to remodel the University of London, and to his work in relation to the Evolution Committee of the Royal Society. As some history, little recognised, is conveyed in this interchange of letters, I have ventured to insert several of them here. They will illustrate the help Galton gave to his younger friends and the sympathy he felt for all their difficulties.

The wrapper and title-page of *Biometrika* had to be hastily rearranged, and I wrote to Galton for advice. His suggestions, very closely followed, ran thus:

CLAVERDON LEYS, WARWICK. *April* 25, 1906.

MY DEAR KARL PEARSON,   Friday, May 4, after your College meeting, will quite suit me to all appearances...but I can foresee only a short way, and have to mould my plans upon others. I go to London to-morrow, and am away in Essex, Saturday to Monday, but have no further engagements. About the future of *Biometrika*, would not the simplest plan be for you to edit it *solely* in your name? Weldon often said that he wished you would do so, for all the work had been and will be yours. You suggested that "founded in 1901 by Weldon, yourself and myself" should be inserted. You must not give so much prominence to me. Why not keep to the existing formula and say: "Founded in 1901 by Professors K. Pearson and W. F. R. Weldon in *consultation* with Francis Galton." Then simply "Edited by Karl Pearson"? A list of coadjutors would scarcely add weight to your name...........Affectionately yours, FRANCIS GALTON.

7, Well Road, Hampstead, N.W. *April* 29, 1906.

My dear Francis Galton,   The scrapbook with the photograph* reached me just before leaving Longcot, and the other book was awaiting my arrival here. I shall endeavour to get an enlargement, for as you say the attitude is very characteristic, but I fear it will not stand much enlarging. Please tell Miss Biggs I will take all care of the book. The other book shall go back to its place on the shelves at Oxford, when I next go down. I found the finger-print books and the letters in going through the papers at Oxford. I shall keep myself free on Friday and you will tell me whether you are able to see me. At times there seems so much to talk to you about and then again it all passes from me. It was possible to go on as long as I was attempting to put the papers at Oxford in order, but I seem now quite dazed, and for the first time in all my teaching experience the idea of facing my students and lecturing seems positively repellent,—at times impossible. I feel wholly without energy to start the term, and if I could only see the man able to do my work, I would ask for 6 or 9 months leave of absence. I have only sounded this personal note because I want you to pardon me, if I say or do anything stupid at present. Yours always sincerely, Karl Pearson.

42, Rutland Gate, S.W. *April* 30, 1906.

My dear Karl Pearson,   The account of your overwrought spirits and energy quite distresses me. I look forward greatly to seeing you here on Friday. If there are hopes of your coming *earlier* than 4 p.m. on that day please send a postcard that I may not be out. My time is quite at your disposal.......Anyhow I look forward to some quiet conversation with yourself alone. Ever affectionately, Francis Galton.

42, Rutland Gate, S.W. *May* 7, 1906.

My dear Karl Pearson,   My attempts have been fruitless to put anything down that you are not already familiar with, about Weldon's characteristics. The extraordinary fulness and accuracy of his letters astonished me. He would write almost a treatise, and insert long tables with apparent ease and as a work of supererogation, which would be a large labour to most men. I suppose too that a certain *pertinacity*, in the favourable sense of the word, was one of his most marked peculiarities. The extraordinarily wide range of his accurate, not superficial, knowledge, was another feature. He was too kindly a critic of things that I asked him to criticise to be of value to me on those occasions, I am sorry to say. Rightly or wrongly my impression always was that he needed some one very strong scientific end in view to compel him to concentrate his remarkable powers more steadily. But I may be judging incorrectly here. I wish I could think of more, this much is I fear useless to you.

Affectionately yours, Francis Galton.

7, Well Road, Hampstead, N.W. *May* 13, 1906.

Dear Francis Galton,   I want to ask your opinion about resigning my fellowship of the Royal Society. You will remember that the last paper I contributed to the Society met with a great deal of difficulty in getting accepted—probably was accepted only on account of your nice little speech. But the Secretaries communicated a resolution to me that I should be requested in future contributions not to mix statistics and biology in the same paper. This of course was equivalent to the intimation that they would not accept future biometric papers from me. I was at the time—I think it is more than three years ago—sorely tempted to resign, but did not do so under the impression that it might be looked upon as personal dudgeon. I have not communicated any paper of *my own* to the R.S. since. The one case where I presented a paper was an application by Miss Cave of our statistical methods to a problem in meteorology. In that case the Secretary wrote and suggested that I should withdraw the paper as the meteorologists did not approve the methods used†. This I declined to do and after some controversy the paper

---

* Of Weldon; at his death, but few, and those unsatisfactory, portraits could be found.

† A commentary on this judgment is that the Meteorological Office recently sent round a circular to various persons, including myself, asking if we could provide further correlations of barometric pressures! Still the pioneers of correlation work in meteorology were hardly treated.

was printed. I have always meant, however, to test the biometric question again, and when Dr Pearl gave me what appeared to me a really noteworthy paper—showing for the first time that even the Protozoa do not mate at random, but assort themselves—the very important result wanted to show how species can be differentiated, even if all members are fertile *inter se*— I presented it to the R.S. as a test case. The Secretary wrote to Weldon who was then Chairman of the Zoological Committee and stated that it would be much better to print it in *Biometrika* than in *Phil. Trans.* The paper is a long one with much illustration and as neither Weldon nor I saw why the Royal Society should be closed to biometricians, the suggestion was therefore refused. Unfortunately Weldon's death left the matter to be finally decided by the committee under a new chairman and they have now settled not to publish the paper "mainly on difficulties felt as to the biological significance of the quantity measured." The quantity measured is the correlation coefficient in three characters for conjugating protozoa, and it appears to demonstrate that physiologically in the lowest types of life, like is compelled to mate with like by structural conditions. It appears to be the first clear demonstration of Romanes' physiological selection, and supplies the need Huxley felt for evidence that differentiation can arise inside a species fertile-*inter se*. It is the old tale that men are set to express an opinion on a biometric paper when they do not know what is the significance of a coefficient of correlation! But I think I have really good ground now for doing what Weldon and I more than once talked about, retiring from the R.S.

My chief work and interests now lie in biometry. If the R.S. will have nothing to do with it, and publishes papers and reports of which the writers lack the most elementary knowledge of statistics, then the Society ceases to appeal to me in any way. I cannot see that I shall do any harm in raising my protest, however feeble it may be. It is not, I trust, a personal point, for I have sent nothing for three years to the Society, but I do not care to sit still and see a really fine piece of work consigned to the Archives, when the Society ought to have felt it an honour to publish it. However, tell me frankly your views and I shall abide by your advice.

Always yours sincerely, KARL PEARSON.

42, RUTLAND GATE, S.W. *May* 14, 1906.

DEAR KARL PEARSON, I fully understand and sympathise with your feelings. It is a disgrace to the biologists of the day, that their representatives in the Royal Society are incapable of understanding biometric papers, and of distinguishing between bad and good statistical work. To that extent I am entirely at one with you, but I do not on the above grounds see that your resignation would mend matters. It is a very general rule of conduct *not* to withdraw when in a minority, because a vantage ground is surrendered by doing so. It is far easier to reform a society while a member of it than when an outsider. The object is, by direct and indirect means, as occasion may offer, to *educate* biologists in the elements of higher statistics. This is being slowly but surely done by *Biometrika*, and would be done more quickly if you could find somebody with a light touch to show as much of the way as biologists should go, and of the false ways in which they have strayed, as the conditions of the case permit, by writing in popular magazines. I do not see why the *meaning* of correlation should not become familiarised, though the methods of work are technical. And similarly for much more, the objects aimed at may be explained, though not the processes. There is an Arab proverb about the ease with which the greater part of the honey in one pot may be transferred to another, and the extreme difficulty of transferring the whole.—So much for mere elementary education.

As regards higher work, you may be driven to make *Biometrika* a seat of judgment, *not* to argue or to enter into controversy (which I know you hate), but simply to pass sentence with reason given just like newspapers do. A fair lash, on the proper quarters, at the right

---

I can recollect two bad cases; at one university a memoir on barometric cross-Atlantic correlations was refused a prize because the methods were not "original"; at a second university an elaborate piece of work, showing that there was no correlation between the position of the moon and any meteorological phenomena, failed as a thesis for the doctorate on the ground that its results were *negative*—as if negative results were not in this case as important as positive— and besides, as the meteorological expert put it, the candidate had omitted to inquire how far thunderstorms are subject to lunar influence!

moment, bestowed whenever it is deserved, would soon be dreaded, and become a check on charlatans; it would afford a motive to others towards acquiring biometric knowledge in order to appreciate the punishment. You can do all, or any part of this, with more effect as a fellow of the R.Soc., than otherwise, so I should say *don't resign*, but abide your time, and give a good and well-deserved slash now and then to serve as a reminder that your views are strong, though not querulously and wearisomely repeated. Ever yours, FRANCIS GALTON.

The above two letters, relating to matters now of the fairly distant past, are not printed with a view to renewing old differences, or justifying past phases of feeling, but to indicate how close was Galton's relation to his younger scientific friends, and how he aided and counselled them in all their scientific relations. In another matter also he was both materially and advisorily most helpful. The Weldon memorial fund was certain to be sufficient to provide a bust of Weldon for Oxford, but I was ambitious that it should do more, and this in the special manner that I thought Weldon himself would have most approved. I wanted something that should form a permanent encouragement to biometricians the whole world round, and I had specially in mind the younger men. I wanted to see besides the bust, the institution of an annual or biennial medal and premium. In order to obtain a greater range of subscribers, I proposed that the three universities with which Weldon had been associated should in turn adjudicate the proposed medal and premium, and I drafted the first appeal for the Weldon memorial fund to this effect and sent it to Francis Galton for his criticism.

42, RUTLAND GATE, S.W. *May* 27, 1906.

MY DEAR KARL PEARSON, I am heartily at one with you in your object, but see difficulties in the proposed method of attaining it. They are:

1. The experience of like attempts shows how difficult it is to raise as much money as you want. I could tell you my own, but being personal do not like to write it.

2. The Royal Society fails to find competent referees in biometry, much more would the three several universities fail to do so. The dignity of the body which awards medals is of less consequence than the assurance that the award is just.

3. An annual or biennial medal and premium, to be awarded to each of the three universities in turn, does not seem a very attractive bait.

I write with much diffidence as to what I think would be preferable:

(*a*) Mention a medallion as a possible alternative to a bust*. It would be cheaper, and would serve as an appropriate design for the medal.

(*b*) Supposing that the assurance of an annual sum of £— would justify the issue of a medal I should be prepared to give as much as would purchase £— consols for that purpose....But it must be *anonymous*....

(*c*) If this plan seems acceptable I would at once send the sum with an accompanying letter to this effect: "I enclose the sum in question for instituting a periodical medal or premium in memory of Prof. Weldon to be awarded to the author of the most valuable biometric publication of recent date, on the understanding that you will consult biometric friends on the conditions that are to regulate the award, and more especially to determine whether it should be limited to one class of English biometricians, to all classes†, or be independent of nationality...."

* Probably a bust would need to have been produced to get the medallion, as no portrait in profile existed. K. P.

† Galton was rightly desirous that the award should not be confined to biological biometricians but should embrace sociology, anthropology, etc. K. P.

Galton next expressed his desire that the medal should be associated with *Biometrika* and suggested how this might be done. To this idea, I was strongly opposed ; it would not have attracted outside support and sympathy, the journal might cease to exist, its vitality was not then fully established, and there would be no trustees for the fund. With Galton's gift the medal was assured ; I had no doubt that the remaining sum needful would be forthcoming from Weldon's personal friends, and there was no occasion to make a broad appeal to the three universities. It was possible to stress the international character of the medal, which I had much at heart.

Please at this present stage, consider nothing of the above as *final*. I only put it forward in the form that *now* occurs to me which would doubtless be much improved by your and other criticism. Pray give it freely. To-morrow I go out of Town for three nights so excuse me if I miss a post or two; my letters will be forwarded, of course. Enclosed I return your draft. With all good wishes for the final success of the plan. Ever very sincerely, FRANCIS GALTON.

UNIVERSITY COLLEGE, LONDON. *May* 28, 1906.

MY DEAR FRANCIS GALTON, Yesterday I was looking at a letter from the Weldon series, dealing with the foundation of *Biometrika*. I had just written to tell him that the complete guarantee fund was forthcoming, and the journal could really start. He begins "Dear good old Galton, dear good old everybody," and that is somehow just how I feel, when I write now to you! This is not an answer to your letter because I want to think it over and reword my original proposals, but I feel quite certain that the annual medal you suggest would not only be invaluable as an inducement to men to strive their best for biometric research, but would indirectly produce some good papers for our journal. I quite agree that it should be open to sociological as well as purely zoological inquiry. I will write again in a few days. My heart is very full just now. We have had Mrs Weldon with us for two days going through papers and letters. I am beginning to see the lines of my memoir a little better.

Affectionately yours, KARL PEARSON.

P.S. I am not at all sure that it would not be of great value in the future to publish some at least of Weldon's letters. They are full of suggestion for research, and represent his scientific spirit far more effectively than his published papers.

[Undated, but early *June*, 1906.]

MY DEAR FRANCIS GALTON, Many thanks to Miss Biggs for all the trouble she has taken in hunting up those letters from Weldon, and you for letting me read them. You need not fear any criticism of my work by *him* will influence me. Our friendship had gone through the fire and nothing could modify my judgment or affection now. But this is a hard week, I have been at Oxford sorting papers for three days, and I have brought the memoir down to the early biometric papers. I will send the result to you soon. It is hard now to distinguish exactly what was yours and what was his, but I don't think you will feel hurt if I have not always put the praise where it should be. It is easier to praise the dead than the living. Please just stick to your life, till mine is gone; I can't do all this again. It is the fourth time I have had to throw all my energy into a dead man's papers and work, and three times the man has been so to speak a part of my own life. How can one tell the tale? Affectionately yours, KARL PEARSON.

7, WELL ROAD, HAMPSTEAD, N.W. *June* 28, 1906.

MY DEAR FRANCIS GALTON, A good and a bad piece of news. In the first place another anonymous donor wishes to add a second £— Consols to yours. This is good because we might hope the fund would go up eventually to £1000 and this would be very good indeed.

In the next place I wrote to Lord Rayleigh asking him whether he thought it at all likely that the R.S. would consent to act as trustees of a Weldon medal and premium for biometric work. He replied that he "would sound the officers." I have his reply to-night, which I am sending to you. You will see that it is distinctly unfavourable. In the first place, I did not do more than ask him his opinion as to what the Council would be likely to do, if the proposal were made to them. You will see that he speaks of referring it to the Zoological Committee. Now that is hopeless—that body has just refused Pearl's really good bit of biometric work "principally on the ground that they do not see the biological significance of the quantity measured," i.e. they do not see what is meant by a correlation coefficient. Further, the idea of the Evolution Committee having anything to do with the matter is too absurd*. That Committee is now merely a body for running Mendelism and the last thing to commemorate Weldon would be to assist that movement.

Now I want you to tell me what to do. Whether: (1) to let Lord Rayleigh put the matter before the Zoological Committee: in which case the offer will probably be rejected. (2) To write to Lord Rayleigh and point out that the Zoological Committee—as it does not contain a single biometrician—can hardly express a useful opinion on the point. I believe it is simply a method of shifting the decision on to another body than the Council. (3) To ask him to consider the proposal as withdrawn. (4) To ask him to bring the matter directly before the Council, so that we may know that they and not the Zoological Committee are responsible for the decision arrived at.

Kindly let me know what your views may be. Of course other trustees can be found, e.g. the University of London. But I feel that for the distant future the R.S. would have been the right trustee for an international thing of this kind. Affectionately yours, KARL PEARSON.

Please return Lord Rayleigh's letter. If you could by any means let me have a reply by to-morrow, Saturday, night, it might save the matter going further, if that is your advice.

42, RUTLAND GATE, S.W. *June* 30, 1906.

MY DEAR KARL PEARSON, I think that the R.S. ought to be left severely alone. Their official representatives repudiate biometry and their Council is already overtasked in awarding medals. I can quite imagine their doing what the R. Geograph. Soc. have already done, viz. refuse any offer to found a new prize. Oxford University seems to me far more suitable in many important respects, and its list of Professors (as given in Whitaker) affords at least 10 suitable electors,...and there could be no valid objection, I should think, to specifying certain names in addition. I have not however an Oxford Calendar by me to refer to, for precedent, but will go to the Club and if there be time, will write again, thereon, to-day. The 10 [? 11] Professors are

Anthropology	Medicine
Astronomy (Law of Error)	(?) Natural Philosophy—(I don't know in the
Botany	least what this is)
Comparative Anatomy	Physiology
Geometry	Pure Mathematics
Human Anatomy	Zoology

I should suggest a short printed circular, enclosed with a few lines of written letter, to each of these, saying that so much money is already in hand, that it is proposed to found a Weldon biometric Medal, or other award,—that it is suggested that the University of Oxford would be the most suitable body to bestow it,—that there are at least 10 professors with whose subjects biometry has some connection, from among whom a suitable board might be selected by the University to adjudge the award....Finally to ask for their suggestions and whether they are willing to co-operate in furthering the proposed plan.—That, after answers shall have been received, the question of approaching the University will be considered.

Would it be convenient if I called on you to-morrow (Sunday) afternoon? I would suggest at 2 o'clock, but any other reasonable hour would suit me equally well. Will you telegraph? and I will abide by what you tell me. Affectionately yours, FRANCIS GALTON.

I have an engagement *here* at 4.15.
I am delighted to hear of the additional £—. I return Lord Rayleigh's letter.

* The President in his letter had suggested that if the medal were accepted, the Evolution Committee might be a suitable body to award it. K. P.

42, RUTLAND GATE, S.W. *July* 6, 1906.

MY DEAR KARL PEARSON, The first thing that I heard of the Evolution Cttee was from Michael Foster who said that the C. of the R. Soc. had been asked to form one, and that they would on the condition that I would act as Chairman, to which I assented.

The offer of a big sum to help in founding a Darwinian establishment for plants and animals was made by me tentatively on many occasions, on the condition that the large balance needed for such an institution could be raised elsewhere. I repeated it more or less formally during the existence of the Cttee, but the response was quite inadequate. The offer of Charles Darwin's house in Down at a moderate (? nominal) rent was made by the Darwin family to the Cttee, but the double event of cost of maintenance and the practical impossibility of visiting it from London on Sundays owing to the awkward hours of the trains, made it impossible to accept the offer. No one benefited by my offer ; " no jackals came down for the spoils*."

The work of the Cttee was a great disappointment to me. For one thing, I had hoped that it would be sufficiently authoritative, or rather that its weight would suffice to weld numerous bodies that have gardens or menageries into common action, to allow some plots or cages, &c. for research. The Clifton Zool. were prepared to do this, but Thiselton-Dyer said that even he could not depend on the gardeners at Kew to carry out any experiment accurately, so that plan fell through. I knew that the Zool. were untrustworthy helpers—I mean the keepers.

The Cttee talked more than worked, and Z. was very boring, writing very long letters to me and always averse to compromise. V., whom he brought in as an Associate, was to my mind, distinctly objectionable in using the name of the Cttee when he had received no sanction to do so. On the whole, the Cttee seemed to be doing so little and working with so much friction that I did not care to be longer connected with it, so I resigned. Weldon did so too, guided by much the same motives.

This is all I have to say. It necessarily relates chiefly to myself but indirectly perhaps to Weldon, whom I then found very helpful, as he always was.

Miss Biggs and I have spent a long day in Henley—Peppard—Stoke Row, etc. We saw Mrs Grey at the Manor House† and the boat races were going on at Henley. It was a glorious day for us—We passed Blount's Court Farm.—I trust you are now well placed at Winsley Hill.

Ever yours, FRANCIS GALTON.

WINSLEY HILL, DANBY, YORKSHIRE. *July* 11, 1906.

MY DEAR FRANCIS GALTON, I enclose two things. First, a sympathetic card (which please return) from the Vice-Chancellor, Oxford, as to the Weldon Prize. Secondly, the proofs of the part of the memoir which I think you have already seen, and also the MS. of the London period. I hope to get the Oxford period done this week. I want you to let me have the MS. back, if you can by return, it must go to Press as soon as possible. I have found it very difficult indeed to write the London part, because the Evolution Committee formed such a very large part of Weldon's life in those years, and I cannot think it was good for him. You were most kind and sympathetic, but he felt that he had to do something of moment and to do it quickly. Further it had to be done under constant fire of unfair criticism. I have found piles of papers about this, that I knew nothing about before, and it is heartrending to think that I was worrying him about his mathematics at the same time. Reading the papers through now it seems to me that a definite plan was formed about 1896 to eject the biometricians and take possession of the Evolution Committee. But all that Weldon wrote, and he wrote and spoke strongly about the R.S. publishing the Mendelian Reports in a semi-official way, may be applied equally to his own work in the early stages. Z.'s attacks did not start until Weldon had reviewed Z.'s book in 1894 or 5, and then they became incessant and ceased only with the death of Weldon. The book was, I think, faulty, but I looked up Weldon's review (in *Nature*) the other day, and it in no way

* See, however, our p. 134 above.

† The house occupied by Galton in 1903 during our Peppard stay. We were at Blount's Court Farm. K. P.

justified those years of unceasing nagging which led to the capture of the Committee. I suppose I shall have my years of it now*!

Please write quite frankly and I will endeavour to modify anything which you think must be altered. You will I know understand that I am placing Weldon alone in the centre of my stage. Affectionately yours, KARL PEARSON.

* This forecast was confirmed in the same year:

" Of the so-called investigations of heredity pursued by extensions of Galton's non-analytical method and promoted by Prof. Pearson and the English Biometrical School it is now scarcely necessary to speak. That such work may ultimately contribute to the development of statistical theory cannot be denied but as applied to the problems of heredity the effort has resulted in the concealment of that order which it was ostensibly undertaken to reveal. A preliminary acquaintance with the natural history of heredity and variation was sufficient to throw doubt on the foundation of these elaborate researches. To those who hereafter may study this episode in the history of biological science it will appear inexplicable that work so unsound should have been respectfully received by the scientific world. With the discovery of segregation it becomes obvious that methods dispensing with individual analysis of the material are useless. The only alternatives open to the inventors of those methods were either to abandon their delusions or to deny the truth of Mendelian facts. In choosing the latter course they have certainly succeeded in delaying recognition of the value of Mendelism, but with the lapse of time the number of persons who have themselves witnessed the phenomena has increased so much that these denials have lost their dangerous character and may be regarded as merely formal." Mendel's *Principles of Heredity*, Edition 1906.

The attacks made on the early papers of the Eugenics Laboratory were largely encouraged by writings of the above character (see our pp. 399, 406 and 408). It is, perhaps, needless to say that it was Galton in his *Natural Inheritance* and neither Weldon nor myself who were "inventors of those methods." The author of Mendel's *Principles* failed to realise that (i) Evolution by Natural Selection depends upon mass-changes, i.e. on selective death-rates which demand actuarial methods, and (ii) that all scientific knowledge is *relative*, there is no absolute truth in science; we seek the best *description* of the phenomena we observe, and as there may be more than one effective description of the group of events we are investigating, there is no necessary opposition between an analysis of individuals and an analysis of mass-changes. The one may have as great scientific value as the other.

It is difficult to see how admiration for Francis Galton, and even for parts at least of his *Natural Inheritance*, was compatible with a complete contempt for biometric methods, but William Bateson's view, as expressed in the following letter (which Mrs Bateson most kindly permits me to publish), seems to indicate the source of our divergence. For me there is no absolute truth in scientific knowledge or in religious creed, the one provides conceptual models of more or less descriptive exactness of our sensations of phenomena, the other fits itself to the emotional needs of differing races, periods and individuals.

Mendelism is only a truth as long as it is an effective description; a continuous or a discontinuous conceptual model of a group of natural phenomena may be equally valid as " scientific knowledge."

MERTON HOUSE, GRANTCHESTER, CAMBRIDGE. 7. vii. 09.

DEAR MISS BIGGS, I have been in Paris a week and only found your letter on my return. Of course I will now send the book to Mr Galton, and I am delighted to do so. It did not occur to me that you might have mentioned our conversation to him. I had greatly wished to send it but came to the conclusion that the simpler course was to refrain.

I don't think many people admire, or can admire, Mr Galton more than I do. The novelty of his thoughts and the freshness of his outlook on nature are not to be found in any other living writer, so far as I know. I often remember the thrill of pleasure with which I first read *Hereditary Genius* and the earlier chapters of *Natural Inheritance*, and every year when I read aloud pieces of those books to my class, as I always do, I can see what excitement they have the power to cause.

You ask whether "Just as all creeds are right," may not "all the paths of Science and Art" be right? Hardly, I think, if for the words we substitute the things themselves. In Art, yes: all are surely right which are sincere; for to the individual artist that which is sincere is, by that very prerogative, to him, Art. The multitudinous forms of art are the product of our manifold natures, and no one may decide for another. But in the natural world of Science, or in the supernatural of the creeds for that matter, I cannot see how there can be more than one right, nor how the path which ends in the wilderness can, outside the language of compliment, be called right. How often have I regretted that Mr Galton has not been with us in the past ten years! It has been indeed a strange perversity of chance. Please see to it that he does not trouble in any way to acknowledge the book for, as I said to you, I shall quite understand. Yours truly, W. BATESON.

42, RUTLAND GATE, S.W. *July* 13, 1906.

MY DEAR KARL PEARSON,   I return the papers. They greatly interest me. I have put trifling marks on pages 5, 6 of the proofs and on 61 of the MS. The only remarks I would make on the MS. are that (1) perhaps the University of London part might be clearer, briefer and more emphatic, and (2) that I think more might be made of the possibilities of an Evolution Cttee than is alluded to on p. 64. For my own part, I thought at first, and this was my main motive in joining it, that the numerous bodies engaged in horticulture and zoology might in one aspect of their work, be co-ordinated by the Cttee and that research of a scientific kind might be introduced into the proceedings of each of them. A Cttee would help to keep them up to the mark, and prevent overlappings. But the desire for this seemed too faint to produce any such result.

I cannot recall the meeting mentioned at the Savile Club, and doubt in consequence whether I was really present at it. I am almost sure that Michael Foster's asking me to take the chairmanship was the first thing that I ever heard about the proposed Cttee. Dear! dear! what a list of efforts are included in the life of an actively minded man like Weldon—successes and failures.

I return the Vice-Chancellor's letter, which is excellent so far as it goes.

Heron's admirable paper reached me *after* I last wrote. Is he the excellent man you spoke to me about, who was not then quite ripe for the Eugenic Research Fellowship. He seems just the man to hold such an appointment.

We have just returned from a brief country visit. It is delightful to hear that you are so pleasantly placed among old Quaker associations. They—the Quakers—were grandly (and simply) stubborn. I think we shall go again to Ockham for August but to another house—negotiations are pending. Affectionately yours, FRANCIS GALTON.

WINSLEY HILL, DANBY, GROSMONT R.S.O., YORKS. *July* 14, 1906.

MY DEAR FRANCIS GALTON,   Your letter and suggestions are very helpful. Your corrections to the proof shall be made. The other points I will refer to one by one.

University of London.   It is awfully difficult for *me* to give the full account of this. I had got many men to join the Association, George Meredith, Hardy, Besant, etc., by a more or less personal appeal stating that we wanted to found a university absolutely homogeneous with a professor at the head of each department on the lines of a Scotch or German university. Huxley was elected president *after* this scheme had been adopted and brought his enormous force to work on a small executive committee of which I was secretary to carry out a plan of *his own* in which we were to compromise with colleges, night schools and the existing university to get a *federal* body. He arranged meetings with each of these institutions. The first with the University of London was to come off in a few days. I protested that this was not the policy on which the Association had been built up and that the executive committee could not go beyond its instructions. Huxley with all the force of an old hand completely confused me—all I know is that I resigned the secretaryship and that the members of the committee asserted that I had promised not to take action against Huxley's scheme. Personally I don't think I made any definite promise, but I know that Huxley saw danger to his project and engineered me into a state of confusion. When I had time to think it over I saw that he had left me in an absolutely false position. I must either be entirely untrue to the men of weight and name who had joined the association on the basis of a genuine professorial university or break through Huxley's entanglements*. This I did by an open letter to him, sent to the *Times* and to him at the same time. I put myself right with the members of the Association but entirely in the wrong with regard to Huxley. Ultimately the Association reversed the whole of Huxley's policy, but these doings (1) had killed its effectiveness, (2) hurt Weldon fearfully and (3) made people believe me impossible on committees. Huxley *must* be right and such a small person as myself must be wrong.

* In my opinion to-day Huxley by his action destroyed all the chance there then was of a real university for London, and left us with the miserable pretence of a university that still exists. The "Association for promoting a Professorial University in London" had practically united all the teachers of weight in London and many other men of mark as well. It was wholly impossible to carry through any pettifogging federal scheme without its sanction. Huxley had no real academic ideals, and a suspicion of all universities controlled by the professoriate. His error was to accept the presidency of an association whose programme was entirely opposed to his own views.

Looking back on it now, I think Huxley was morally wrong; he used all the force of his name and position to get a younger man, who was really responsible for the movement, out of the way in order that he might carry out a different scheme. I was formally wrong, but morally right, and nobody saw, not even Weldon, that I, having taken a false step, was doing what was painful to me to put myself right with men whom I had induced—often by much talk and persuasion—to join a movement for a great ideal of academic reform.

Now you will see that I cannot put all this directly into Weldon's Life. But it was a remarkable instance in which his admiration for his hero, and personal affection for a friend came into opposition, and he succeeded in preserving both, and this although I never gave him as I have given you the grounds for what I did. It is this element in the whole matter which makes the account of Weldon's relation to the University movement, as you find it, obscure.

I have put in six more lines about the Evolution Committee emphasising what your aims were and how they were rendered unavailing by the members pulling in different directions and the struggle of different schools. To my mind the absence of such an experimental farm as you suggested has been the great drawback of the past years. We want a land "Marine Biological Association." But it would never have been possible to combine the thoroughness of Weldon with the slipshod character of the rival school. Friction would have destroyed everything. The only hope is that a Dohrn may arise some day, a man with the energy and force of character to carry it out which marked him. The worst of it is that the Americans have already got such a station under the Carnegie Institution, but so far they have done nothing very profitable with it; it needs as chief a very clear strong thinker. The success of these things always lies in the strength of the individual who dominates the whole. Dohrn must have been splendid....

I enclose a letter to you, which seems to me to confirm my version of the first origin of the 1893 Committee. In 1896 Nov. or Dec. you were so weary of Z.'s incessant letters to the Committee—the originals or copies occupy an entire box in Weldon's papers—that you suggested Z. should be added to the Committee. Now was the old Committee *dissolved* and a new one formed, or as I suggest were Bateson, Dyer and myself* added to the old Committee and shortly after many others? There is no definite statement in Weldon's letters, but between Nov. 1896 and February 1897 the Committee appears to have taken a new lease of life, the old statistical object is dropped, many new members appear and the whole scheme of breeding and inquiry by circulars to breeders comes into being. Can you throw any light on these points? I enclose the circular that Weldon in his letter says he has sent to Darwin, Poulton and Macalister, and received their assent to. Weldon in a letter of Dec. 4, 1893, says:

"I am writing to ask people to meet on Saturday at 3.0 (Dec. 9th) as you (F. G.) suggested, but at the Savile Club, 107 Piccadilly." He states that as the Royal Society is not available on Saturdays he has chosen the Savile. Perhaps the locus was changed later?

Might I have the enclosed back, so that all the papers may be in order and together, if there is need for any further reference? Also will you return the enclosed poem in W. F. R. W.'s handwriting? Is it a translation and if so of what? It reads rather as if it were. If not, what made him choose this metre, and what is it the prologue to? It is the only poem I have found. What is the reference to Macrobius? Affectionately yours, KARL PEARSON.

42, RUTLAND GATE, S.W. *July* 16, 1906.

MY DEAR KARL PEARSON, I have found my (scanty) diaries of 1891–1897, and have been to the R. Soc. to read the minute book of the Evol. Cttee and refresh my memory. The sequence of affairs was I think this, so far as I was cognisant.—*First* Michael Foster's call on me— I have no record of this,—about the then *talked of* Cttee. *Second* the Savile Club meeting, of which I have no recollection, but believe it must have been just an informal ratification of views previously well discussed. My diary notes the engagement. *Third* the appointment of a R. Soc. Cttee, in the Minutes of whose first meeting Jan. 25, 1896, a letter was read from me to the R. Soc. "suggesting the desirability of appointing a Cttee for conducting statistical inquiries into the measurable characteristics of plants and animals." Also, a letter from the R. Soc. appointing us, myself (as chairman), F. Darwin, Profs. A. Macalister, Meldola, Poulton and Weldon, giving us £50 to start with, and recommending us to apply to the Govt Grant Cttee for any further sums we might think necessary.

* The R.S. records show that I was added in 1896: see p. 126 above.

Jan. 1897, Bateson, Godman, Heape, Lankester, Maxwell, Masters, Salvin, were elected members, and Bateson attended.

Feb. 26 (clearly of the same year, from the above facts) *when Bateson, etc. were present*, it was resolved to ask that the objects of the Cttee should include " accurate investigation of Variation, Heredity, Selection and other phenomena relating to Evolution." In this year it was briefly called (? for the first time) the " Evolution Cttee."

June 15, 1899, the question was raised " whether the Cttee ought not to cease to exist."

Nov. 29, 1899, Discussed and read a letter (about to be sent?) from me to the Sec. R. Soc. expressing my view "that the Cttee would not serve any useful purpose by continuing to exist," but asking reappointment for one year.

Jan. 25, 1900, Dyer, Meldola, Pearson, Weldon and I all resigned. (The Cttee still lingers on and meets about once a year.)

There is no indication of any previous Cttee for this or any allied purpose, but Weldon had many grants, personally, for his shrimp experiments, etc. Neither was there any break in the continuity of the Evolution Cttee.

I quite see your difficulty about the history of proceedings connected with Huxley and the University of London,—how to satisfy the reader and yet not be too explicit on painful subjects.

The allusions to the poem· (which I return with Weldon's letters) are not understood by me. I do not even yet recall who "Macrobius" was—(not a *Macrobe*, the inverse of a "Microbe"!).

I still think that I must have a lot of Evolution Cttee correspondence *somewhere* in my cupboards, etc. If I can find anything worth sending you shall have it, of course.

Affectionately yours, FRANCIS GALTON.

7, WELL ROAD, HAMPSTEAD, N.W. *Oct.* 22, 1906.

MY DEAR FRANCIS GALTON, I am rather distressed that I have heard no more of you. I trust no news is good news and that with mild weather you have had a quick recovery. Please let me have a line as to your locus and "status."

I still want to talk to you about many things of which it is almost impossible to write. *Item.* The Weldon Medal and Premium. This now amounts to £870, but I have endeavoured to adopt your original wishes as to the conditions. As I understood them, they involved: (i) The institution of a biennial prize and medal— not for an essay *ad hoc*, but for some piece of published work during the previous four years, which forms an advance in biometry. (ii) That the prize should not be confined to any one nation or men of any one university. (iii) That the paper must consist (*a*) of the application of statistical theory to the study of special problems in Zoology, Botany, Anthropology, Sociology or Medicine, or (*b*) of such extensions of statistical methods as may be of value in such investigations. These are in general terms what I have put before the Hebdomadal Council and that body will discuss the point to-day. I think the perfect openness of the medal and premium is what Weldon himself would have wished. It would hardly be possible to find a fitting man every two years in Oxford itself. If Oxford finds it impossible to give prizes outside its own body, it will be best, will it not, to try London?

I have been somewhat surprised to hear from Schuster that he was resigning the Eugenics Fellowship *immediately*. I wrote at once to him, saying I was sorry to hear it, and feel that he would have done better to talk it over with you first. At the same time, I think I see his position, he feels that his work, which he very frankly says is limited to certain directions, is not on the lines you want. I think what he has recently done on inheritance of mental characters is very good, but it will not attract much attention, and much popular attention is going in the next few years to be attracted to Eugenics. The difficulty will be to get a man who is really sound and yet can catch the popular ear. I don't know where to find the right man for you, although men who will do good bits of work, if one suggests them, are always forthcoming. I hope you won't be worried about it all....

Affectionately yours, KARL PEARSON.

The death of Weldon was a terrible disaster to Francis Galton and his biographer, but while equally felt by both, the effects of the shock were more intense and lasting in the case of the much older man*. He seemed to me from this date less able to take independent action, and to find reliance on others more needful.

(10) *Eugenic Certificates.* Among Galton's papers I have found a manuscript entitled *Eugenic Certificates* which belongs to this year; I also found typewritten copies of this manuscript, some of which had clearly been circulated for criticism and advice. I expect, although dated June, it had been written in the early part of the year, as it is the natural sequel to the memoir on Eugenic Studies read before the Sociological Society†. I had not seen the manuscript before I found it among Galton's papers after his death. Our correspondence in May‡ will I think explain why he did not show it to me, although for some time past he had shown me most of his writings. He may very probably have thought that I should hold the time for issuing Eugenic Certificates not yet ripe. But I do think it important for the future progress of Eugenics that the manner in which Galton visualised Eugenic Certificates should be recorded.

Galton's unpublished MS. on Eugenic Certificates.

Private for consideration.

FRANCIS GALTON, 42, RUTLAND GATE. *June* 1906.

EUGENIC CERTIFICATES.

The time seems to have arrived when the question should be seriously discussed, whether it be practicable and advisable to issue Eugenic Certificates that would and ought to be accepted as trustworthy and that would be inexpensive and yet self-supporting.

The subject is full of difficulties, but I think they can all be met if certain restrictions be permitted, of which the following are the chief:

1. The purport of the certificate to be that in the opinion of the Judges, the achievements of the holder and those of his near kinsmen prove him to be distinctly superior in Eugenic Gifts to the majority of those in a similar position.

2. That certificates be granted at first only to men, and these between the ages of 23 to 30 inclusive and belonging to the educated classes. At an earlier age they would have hardly had sufficient opportunity of proving their powers, at a later age the memories of the youthful achievements of their kinsfolk in the previous generation are difficult to verify.

The practicability of giving certificates to women would require a special discussion. It will not be alluded to again in the following remarks.

3. That the qualifications for a certificate be limited to facts that are permanently recorded in some accessible form, so as to be verifiable. They must be described on a ruled schedule that will be supplied on application.

4. The achievements are to be drawn from the results of some of the numerous competitive trials, whether in sport or in earnest, in athletics, in literature or otherwise, to which nearly every young man of the educated classes is now subjected; also to such prizes, awards or appointments, etc. as may have been gained.

---

* It is noteworthy that Galton's general correspondence, which for most years was voluminous, is much reduced in 1906; apart from my letters to him, very few other letters appear to have survived.

† See p. 272 above.

‡ See pp. 282–284 above.

5. Evidence will be required of a somewhat higher* order of achievement than that to which the certificate testifies, lest undue weight should be assigned to success due to especial aptitude rather than to all-round ability, or to a success won under exceptionally favourable circumstances. The hardship of a certificate being sometimes withheld from a deserving man through want of convincing evidence is a lesser evil than the occasional grant of them to the undeserving. In the first case, an individual fails of his due; in the latter case, the credit of the certificates is shaken.

6. The ignorance of particulars concerning a man's ancestry is usually so great that inquiries concerning hereditary gifts must perforce be limited to his nearest kindred, namely to (1) his (whole) brothers and sisters, (2) his father and the father's (whole) brothers and sisters, (3) his mother and the mother's (whole) brothers and sisters, all of whom are usually within the scope of information easily procurable by persons aged 23 to 30. Half brothers and sisters are not taken into account. (The questions concerning the kinsfolk of the applicant, while they are framed to extract really useful replies, have to be much less detailed than those which concern the applicant himself.)

QUESTIONS TO BE ANSWERED BY EACH APPLICANT FOR THE EUGENIC CERTIFICATE.

1. Your name in full, place and date of your birth, your address, your occupation. Maiden surname of your mother; full name and address of your father, his occupation.

2. Have you undergone a physical examination as to fitness instituted by any branch of the public service? If so, state particulars and date.

3. Can you refer to any physical competition of which records are accessible, in which you were ranked distinctly higher than the average of your competitors? If so, give not more than three of the most notable instances.

4. Have you performed any physical feats that were distinctly beyond the powers of men of the same age and of equal training to yourself? Give particulars of not more than three of the most notable instances.

5. Have you been classified in any important literary examination, whether at school or college, for a public service or otherwise? Give particulars of the three instances in which you were especially successful and mention the number of competitors in each case; if they be not known with exactness, give limits within which their numbers certainly lie.

6. Have you been awarded prizes or other distinctions at any large college, school, or university, or by any literary or scientific body? If so, state the particulars of not more than three of the most noteworthy ones.

7†. Have you been elected to any coveted post of trust, paid or unpaid, in any school or college or in any association, whether it be athletic, scholastic or other? State particulars of not more than three of the most important of these as evidence of the trust placed in you by your comrades.

8. Have you received any and what appointments? If so, do not mention more than two or three of the principal ones.

9. State anything else that you may think favourable to the conclusion to which a certificate testifies.

Note. Weight will be given by the Judges to the general character of your replies, which should be appropriate and satisfactory, but brief.

It will be observed that an accumulation of small instances is reckoned superfluous, when a few prominent facts suffice to carry the desired conviction.

Nothing whatever is to be written that cannot be quoted and cannot be verified.

---

* The greatest successes are due to more than the average powers as the greatest failures are due to less. F. G.

† Can any further good test of *character* be suggested? F. G.

For Private Information only.

QUESTIONS *A*, CONCERNING THE KINDRED OF THE APPLICANT.

On the Fraternity consisting of the Applicant and of his *whole* Brothers and Sisters. Write A by the side of the figure in the first column that refers to the Applicant.

Register	Names or Initials in order of Birth	Sex M. or F.	If deceased	
			Age at Death	Cause of Death
1 2 3 ⋮				

Register	Initials	Notable achievements of any brother or sister of the Applicant that fall within the purview of this Certificate

We certify that to the best of our knowledge the above account is correct, also that with the exceptions mentioned below, no member of this Fraternity has ever suffered from Insanity, Epilepsy or other severe form of nervous disease.

Exceptions giving full particulars. If no exceptions write the word "None"	

Signatures of $\begin{cases} \textit{Writer of the above notice.} \\ \textit{The Applicant.} \end{cases}$

The last paragraph and the corresponding paragraphs of Questions *B* and *C* must be on a separate sheet marked "confidential." F. G.

QUESTIONS *B*, CONCERNING THE KINDRED OF THE APPLICANT.

On the Fraternity consisting of the Father of the Applicant and of the Father's *whole* brothers and sisters. Write F by the side of the figure in the first column that refers to the Father, and describe his achievements more fully than those of his brothers.

Register	Names or Initials in order of Birth	Sex M. or F.	If deceased	
			Age at Death	Cause of Death
1 2 3 ⋮				

Register	Initials	Notable achievements of any brother or sister of the Father of the Applicant that fall within the purview of this Certificate

We certify that to the best of our knowledge the above account is correct, also that with the exceptions mentioned below, no member of this Fraternity has ever suffered from Insanity, Epilepsy or other severe form of nervous disease.

Exceptions giving full particulars. If no exceptions write the word "None"	

Signatures of $\begin{cases} \textit{Writer of the above notice.} \\ \textit{The Applicant.} \end{cases}$

QUESTIONS *C*, CONCERNING THE KINDRED OF THE APPLICANT.

On the Fraternity consisting of the Mother of the Applicant and of the Mother's *whole* brothers and sisters. Write M by the side of the figure in the first column that refers to the Mother, and describe her achievements more fully than those of her sisters.

Register	Names or Initials in order of Birth	Sex M. or F.	If deceased	
			Age at Death	Cause of Death
1 2 3 ⋮				

Register	Initials	Notable achievements of any brother or sister of the Mother of the Applicant that fall within the purview of this Certificate

We certify that to the best of our knowledge the above account is correct, also that with the exceptions mentioned below, no member of this Fraternity has ever suffered from Insanity, Epilepsy or other severe form of nervous disease.

Exceptions giving full particulars. If no exceptions write the word "None"	

Signatures of $\begin{cases} \textit{Writer of the above notice.} \\ \textit{The Applicant.} \end{cases}$

As far as I am able to trace from Galton's correspondence only two men gave expressions of opinion upon Galton's proposed Eugenic Certificate. Mr Havelock Ellis, having seen Galton's proposal in the paper at the Sociological Society, wrote to ask Galton whether he had taken any further steps in the matter; his letters seem to indicate his sense of the difficulty of the project rather than the strong enthusiasm which surmounts difficulties. Mr J. Tracey, Tutor of Keble College and an authority of the Oxford University Appointments Board, was, according to Galton's notes of an interview, distinctly favourable. He said that many examinations covered practically all the personal questions Galton wished to be answered. Therefore having passed any one of these examinations would *so far* be a sufficient justification for a Eugenic Certificate. Some appear to enter a short way into family history. Indian Civil Service, Woolwich, Egyptian (Soudan) Service are especially notable. Most certificates take no cognizance of hereditary ailments, if there be any in the family. Could such ailments be properly ignored? Mr Tracey thought there need be no fear (under reasonable precautions) of false returns. Also he did not think Galton's estimate of 10s. per certificate unreasonable, if rooms were allowed and an unpaid board of referees could be had. *Upshot* (as drawn up by Galton) "I must collect material about the chief existing examinations from G. G. Butler, David Mair and others, and write an article based on it to show what could at present be easily done *re* Eugenic Certificates." I do not think this article was ever written, but a fuller account of Galton's views was later provided in "Kantsaywhere."

(11) *Reconstitution of the "Eugenics Record Office."* We have seen that Mr Edgar Schuster resigned his post of Research Fellow in Eugenics, and although he was willing to continue holding the post for a short period, coming up two or three days a week from Oxford, he wished to be relieved as soon as Galton could make new arrangements. Our leader was ailing, the death of Weldon and certain home troubles had depressed him sadly, also the wintering in England—contrary to his custom hitherto—and at such a place as Plymouth*, undoubtedly checked that vigour of action which had hitherto characterised him. He felt it impossible to cope with the search for a new Eugenics Fellow and the direction afterwards of his work. Our experience in the Biometric Laboratory had taught us the serious length of time it takes to collect statistical data and afterwards to reduce them fully by modern statistical methods, whereas Galton was undoubtedly eager for quick returns; he approved brilliant essays in the monthlies, and wanted to see marked progress in the acceptance of Eugenics in his own day; he had not yet fully differentiated Eugenics as a science from Eugenics as a creed of social action. He was not urging hasty action†, but he did not, I think, fully realise that all eugenic research was of a very laborious and lengthy kind.

---

* From October to March Galton passed this winter in Plymouth on the Hoe (at various addresses).

† See our pp. 220–21.

He wanted his Eugenic Office to show immediate results; and just for this reason I had stood as far as possible aloof from it, except when he or his assistants directly consulted me on statistical points. Further, who was I to advise him? You cut off all the suggestiveness, all the power of original productivity of a man of genius, if you recommend him to follow your own dull, laborious and commonplace methods of attaining truth. But matters were now reaching a crisis; there was certainly no obvious successor to Schuster, Galton felt incapable of further personal supervision, and there was a possibility that the seedling he had planted, which might otherwise develop great academic branches of study, might perish as a sapling for lack of careful tending. I felt that my only chance of giving effective aid was to put clearly before him the difference in our modes of approach to the same goal.

7, Windsor Terrace, The Hoe, Plymouth. *Oct.* 24, 1906.

My dear Karl Pearson, This afternoon I have (1) moved into the above lodgings, (2) received your letter, (3) received Schuster's reasons for resigning. I am far from fit, but the bronchitis is quite gone. I expect to be here for at least Nov^r & Decr. I have lent 42, Rutland Gate to some relatives during these months. The £— for the Weldon medal and premium (? is the bust to come out of this) is a substantial sum, and I congratulate you on your persuasive powers. Don't *now* let any conditions that I made at the beginning hamper your action. I feel quite sure that you will do the right thing. If Oxford refuses, and then London University accepts, I am not at all sure that it would not be a gain to the cause.

Schuster's brief letter of resignation surprised me, so I wrote nicely to ask for reasons, which he has given fully in the sense of what you wrote to me. I am not fit now for effort, and am inclined to ask the Senate not to fill the vacant appointment yet. I wish that somehow it [the Eugenics Record Office] could be worked into your Biometric Laboratory, but I am far too ignorant of the conditions to make a proposal. If any feasible plan occurs to you, pray tell me; it is almost sure to have my hearty acquiescence. I have of course followed with all possible interest Lister and you, and look forward to his answer in to-morrow's *Nature**. He will probably try to raise a different issue, but I am sure you are far too cautious to follow any red herring dragged across the path. Also I have just read your letter in the *Times* on Sidney Webb's topic. Excuse more as I am rather tired.

Affectionately yours, Francis Galton.

I am so glad you approve of Schuster's recent work, which he will send me in due time.

University College, London. *Oct.* 25, 1906.

My dear Francis Galton, I am sorry indeed to hear you are still ailing, and trust you are taking all care of yourself. I want to add one or two points to my letter of yesterday.

Weldon Memorial Fund. The bust fund is now about £240; the medal and premium fund about £870. I want to raise the former to £300 and the latter to £1000. Personally I should like to maintain the condition which I think you originally suggested that the prize should be *international*. I believe that not only shall we thus get good men, but that the subject will attract new workers everywhere. If the prize be confined to the members of one university, we shall get very little but small academic essays.

Next as to Schuster: you will remember that you wrote to me when I was in Yorkshire, asking if I could suggest any work for him, as he was coming to the end of his material

---

* Mr Lister, as President of the Zoological Section of the British Association for this year, had made a strong attack on Biometry. This was clearly within his competence, but as illustration of the futility of biometricians, he cited matters from Dr Pearl's paper on Paramecia, which he had only seen confidentially as a referee, and which the Royal Society had settled not to print, nor had it at that date been published elsewhere. It was a repetition of the Homotyposis memoir indiscretion.

I wrote and suggested the insanity data to him, as I felt the problem was one of some importance, and I knew I could probably get some good material. But I told him *very distinctly* that I made the suggestion with hesitation, and he must consult you.

My letter then pointed out that any problem which is of first class importance—such as that of the relative influence of heredity and environment in the case of insanity—requires a long time for the collection of data and as long a time for the reduction of them, and next I ventured to break my own views to Francis Galton.

Now there arises the difference between the biometric work here, and what it seems to me, if I interpret your views rightly, you want done in the Eugenics Record Office. We have many irons in the fire, there are about a dozen workers always engaged, and one inquiry often goes on for five or six years through two or three generations of students; but it gets done and published at last. It seems to me that this "secular" progress is almost impossible without continuity. If your Fellow during his term of office is to collect and reduce data, and publish pretty frequently work of a striking kind—and this appears to be needful to make the subject popular and keep it in view—then he cannot take up a *big* statistical inquiry. It is not always easy to find a fairly rounded easy bit of work such as I set A. There is on the other hand always plenty of the heavy continuous work. B. is not a man of striking originality, but he is a very safe man; find him a problem, give him help and advice and he will do sound work. His tendency has been, however, more and more to the biometric side. I feel that this is not, perhaps, what you want for Eugenics at present, and that you hold that there is room for more than the biometric treatment of sociological problems. I have had great hesitation in taking any initiative at all in the Eugenics Record Office work, because I did not want you to think that I was carrying all things into the biometric vortex! When Schuster informed me that he was resigning the Fellowship, I at once asked him to reconsider his position, and talk it over with you first. He then said he had fully determined to undertake more purely biological work. I suggested to him that if he felt he must give up Eugenics, he might take up the problem for which Dr Mott has got material, namely the convolutions of the brain in the sane and the insane. But while an inquiry as to environmental and hereditary influence on insanity does seem to me eugenetic, I am not clear that the relation of brain complexity to mental grade is; and accordingly my suggestion was only to be definite if Schuster found himself on resignation wanting a problem. Personally I should like to see him going on with the fellowship, until you are able to consider what had best be done. If he wants eugenics work, I think I could provide him with the data for 300 tuberculosis cases, and show him how to get more. The brain convolutions form the more fascinating problem and well done might produce a good deal of stir; but this is all I can say about Schuster's resignation. You can appoint a man like A. to succeed him, but will he find his problems *for himself*, and then make something of them? I am uncertain, and a good popular problem might not be discoverable every year. He would probably come to me and all I could give him would be some of the "secular" work which was nearing completion; that might be a rather dull and commonplace process for him.

Now my *personal* idea of the Eugenics Record Office is that it should continue steadily to collect data bearing on the effect of environment, of heredity and of intercaste marriage upon man; that the Fellow should go on with annual or biennial appointment, and should live in London and work daily at the office; that the results accumulated should be published, like A.'s paper, at irregular intervals, when a bit of work was completed, and be issued from the Eugenics Office. I think great results could be obtained ultimately in this way, but it would have to depend on my idea of "secular" accumulation. You will understand what I mean when I say that our investigations on school-children took five years to collect and two to reduce; that our measurements of families took four years to collect and two to reduce; that our present inquiry on the inheritance of disease has been more than two years in progress and it may be more than another two before reduction can be begun*; our inquiry as to

* This was written in 1906, the full reduction was only begun in 1927 and is still in progress! K. P.

albinism in man has been going on for three years and is far from completed; it is precisely so with the tuberculosis and insanity researches. No Fellow in his one or two years of work could attempt to complete a six years' research of this kind, but he could help in carrying on such work, and publish during his period of office such researches as happened to be nearing completion. I think in this way he might keep the eugenics idea before the public; but the scheme is essentially based upon the "secular" accumulation of data and continuity in the direction of the office such as we have had here in our biometric work. If you think I could aid in such a plan I will do so willingly, and am ready to place at the disposal of your office such inquiries as we have in hand relating to man, but I should need to control the manner in which the data were reduced, and see that the material which has taken considerable time and much energy to collect was properly dealt with.

On the other hand I am conscious that much may be done on eugenic lines apart from biometric methods, if you can only get the right man, but it is doubtful whether there would be continuity in the work or any permanent collecting of records. Still an able man would advertise the subject much better than I can do with many other claims on my energies, and I do not wish to minimise this aspect of the matter. I will always under *any* conditions of your Office, give it what aid lies in my power and you may wish for. My hope would be that, if you let matters slide for a little now, you will be ready and able to take up the directive work again in a few months' time.

I have no idea what Lister will say, but I expect it will be a protracted fight! There is another man asking all sorts of questions in the *Times* to-day!

Affectionately yours, KARL PEARSON.

It will not be needful to print more than one other letter from me regarding the association of the Eugenics Record Office with the Biometric Laboratory, namely, that of placing before Galton the draft plan of the "Galton Laboratory for National Eugenics" (see below, pp. 304–307). I do this to indicate that the lines on which that Laboratory has been run since Galton's death were settled by the letters which passed between us in 1906, and to remind the reader that the Galton Laboratory was actually started early in 1907 under its present Director, and except in the matter of greater power and activity was not modified in any essential manner by Galton's death in 1911. It is, perhaps, unnecessary to state that the text of Galton's will of 1908 and the codicil to the same will dated 1909 were not those which the following letters indicate that he put before me for criticism in 1906; the latter belonged to a will of the same year. The substance of the final paragraph of the codicil of May 25, 1909, was first made known to me by Galton's executors.

While in 1911 I was glad and proud to be elected Galton Professor, especially as it was in accordance with Galton's wish, it was with much hesitation that I took over in 1906 the voluntary task of supervising his Eugenics Office. Above all things I dreaded that any difference of view between us as to work should in the slightest impair what to me was a most perfect friendship. It is a sign of Galton's generosity and large-mindedness that although he remained thoroughly interested in what we did, he never once attempted in the least to control us or to express anything but keen satisfaction in our proceedings. When I recall that Galton was always a man of very definite opinions, that the science of Eugenics was his creation not mine, and lastly that in those four years of my supervision Galton passed from 84 to 88 years of age,—a time when the majority of those who survive grow querulous and

38—2

are apt to criticise younger men,—it seems to me that none can need stronger proof of how his sagacity and power of self-sacrificing friendship lasted to the end.

(Confidential.)   7, WINDSOR TERRACE, THE HOE, PLYMOUTH. *Oct.* 25, 1906.

MY DEAR KARL PEARSON,   We are substantially in such close agreement *re* Eugenics that I can write very briefly. I quite agree to the "secular" work, but with occasional "Chips from the Workshop," to use Max Müller's and Bunsen's phrase. The Eugenics Fellowship was avowedly an experimental venture, so this seems a proper opportunity to reconsider its constitution.

As regards ways and means (this is confidential) I am prepared to ensure £500 a year to its maintenance during my lifetime, and fully £30,000 clear on my death for a professorship. What is best to be done during my lifetime, considering my age and precarious health and powers? The "Fellow" should work under continuous direction and in London as you say, and not in too solitary a fashion. Could he be made to lecture or to demonstrate, in connection with the Biometric (or even the Economic) School? A Professor would have a class, which would keep him to the collar.—Anyhow, it would be convenient if Schuster continued as a stop-gap, working as you suggest at tuberculosis, for that would retain Miss Elderton and the rooms. I would ask him if you thought well.

Of the few younger persons whom I know, none seems to have a larger portion of what is desirable than C.'s son, the statistician, who has now a Government post. He is full of ideas. I do not know whether what could be offered to him, *including a post-obit Professorship*, would tempt him to give up his not well-paid Government appointment. If you thought well, and could suggest a scheme that the Senate would be likely to approve, I could ask him or any other good man that might be suggested. This is of course quite confidential. So it is on these points I want advice.

I ought to explain about B.'s letter. It was so brief and dry that I was unable to appreciate the merits of the case, which I subsequently did when *you* wrote.......His second letter in reply to mine wholly removed that impression. I personally like him much. He wants juiceyness (jucyness? I can't spell it!).

So Lister is silent this week in *Nature*. Excuse bad writing in an armchair.

Ever affectionately yours, FRANCIS GALTON.

7, WINDSOR TERRACE, THE HOE, PLYMOUTH. *Nov.* 19, 1906.

MY DEAR KARL PEARSON,   I am most sensible of your helpfulness and kindness, and find myself so much at one with you that I can now write briefly. Understanding that whatever is done now should be with reference to the "post-obit," I will begin with a revised codicil, *see enclosed.*—After you have corrected it, and it is otherwise put into order, I propose to send it to Hartog for his suggestions ; and finally to my lawyer.

The work of the Office should now I think be directed towards this end by thoroughly working the new Fellow or Student in statistics of a kind that *you* approve, but having a eugenic tendency like so many of your own biometric papers.

Next for the choice of Schuster's successors. Your very kind proposal of undertaking the supervision of the Office for a year or 18 months removes from my mind a great weight of responsibilities that I have not health to fulfil. If you undertake it, clearly the choice of the men ought to lie wholly with you. If fairly good luck attends the venture we may find a man by that time (18 months) sufficiently trained and prepared to grow into a good Professor. A. seems to have excellent stuff in him and to be in every way of a suitable disposition, but as I said in my last letter, he should be encouraged to interest himself in the sociological problems and the collected data of the day, and leisurely to prepare a *provisional* or rather a *suggestive* programme of future office work. Too much of pure mathematics will be harmful to him from the present point of view.

It is most desirable that the future Professor should be on easy social terms with the executives of various societies and departments, and A. seems quite capable of that position before long. As I said, we both liked him much. He inspires confidence, too.

Affectionately yours, FRANCIS GALTON.

7, WINDSOR TERRACE, THE HOE, PLYMOUTH.   *Nov.* 29, 1906.

MY DEAR KARL PEARSON,   I have now heard all I wanted to hear about the bequest. As mentioned, I enclosed the draft in a "private and confidential" letter to Hartog. He answered with full approval and with that of Rücker and wished to show it also to the V.Ch., but Sir Edw. Busk was out of town. He has since seen it, made some minor but important suggestions, and I will now send it to my lawyer to draw it fully out in duplicate for a final review before incorporating it as a Codicil. It is so important that it deserves this trouble.

I am most desirous to learn the conclusions of your Oxford meeting. The *Times* had no reference to it.

Now about the immediate future of the Fellowship. I quite agree with all you say about A. When I wrote I did not understand the importance to him of a D.Sc. All that I wished to convey was that if he were to go in for Eugenics he ought not to give too much time as a *diversion*, so to speak, to pure mathematics to which, I dare say, his heart turns. I did not know of your Library at University College and had for the moment forgotten the Foxwell Library.

What steps must be taken to secure a Fellow, or two Students, etc., to succeed Schuster? I must wholly turn to you to represent me in this matter as I said before, and will write to-day to Hartog to tell him so. It is very good indeed of you to undertake the work for a year or 18 months; better the latter, as it will carry on to Mid-summer when I am much more likely to be serviceable, if wanted, than in winter. See P.S. to this letter and please read the enclosed to Hartog and forward it with any remarks of your own.

I feel the want of congenial talk now, but as Plymouth is not London must make up my mind accordingly. I look forward to the Report of Rayleigh's Presidential Address on the 30th. Do you know of an apparently very striking Report of the Inspector of Inebriates to the effect that they are naturally degenerates and near insanity, and that the women have huge families? There was an Abstract of the Report in a Devonshire newspaper. I have written for the Report itself. Also, has the Japanese, K. Toyama, sent you his elaborate Mendelian experiments on Silkworms?   Ever affectionately yours, FRANCIS GALTON.

P.S. I may as well quote from Hartog's letter of Nov. 14:

"I think that the Senate could have no objection whatever to the Association of the work with the Biometric Department under Prof. Pearson whose guidance will, as you say, be of very great value."

7, WINDSOR TERRACE, THE HOE, PLYMOUTH. *Dec.* 4, 1906.

MY DEAR KARL PEARSON, *Biometrika* reached me yesterday and I have read with the fullest appreciation your excellent and affectionate memoir of Weldon. It appears to me a model memoir, so well proportioned and so graphic. One sees in every page the great care you have taken. It is a worthy monument to a life prematurely closed. If he could only have written more of his book on Heredity! He was so familiar with such a mass of biological facts.

I return Prof. Turner's letter, and have written to him in aid of what he wants in respect to B., giving such help as I can. I wish him every success. I grieve to gather from Turner's letter that besides all your other worries and the 'flu, there has been the illness of one of your children.

Enclosed is my lawyer's draft of the Codicil. I should be most obliged if you would look it over and pencil any suggestions you think proper upon it, and return it to me. I will then re-consider it and send it (a clean copy) to Hartog. It grieves me to add this straw to your over heavy load.

I quite agree with you about offering the bust to Hope Pinker to make for £250.

Ever affectionately yours, FRANCIS GALTON.

7, WINDSOR TERRACE, THE HOE, PLYMOUTH. *Dec.* 6, 1906.

MY DEAR KARL PEARSON, Let me begin by saying how interested I am in the new *Biometrika*, and having plunged into your Intelligence paper got myself into a desire to work out the classes on the Centile principle, and hope to send the results to-morrow. Alas, I work so slowly now! You will have some day to discuss the *slowing down* of all functions with *age*.

Thank you very much for your very judicious suggestions about the Codicil, I will go carefully again through it and hope to send it off to-night to Hartog. University College charges *no* rent for the rooms occupied by the Office, but pray, as you kindly propose, talk over the matter for the future with the Provost.

Will you then, please, provide work after February and see to carrying on the Office? I simply feel myself powerless, as I said before, and have no wish to meddle in and to mar whatever you may do for me. I leave it quite to you to arrange with Hartog and the University, about selecting Schuster's successor or successors and giving them work.

Your news about the inheritance of the tuberculous diathesis is good and very important.

I am grieved to hear of the pain and anxiety you have gone through about Helga. Turner writes to me saying that my letter to him about B. was just what he wanted—I am glad.

Ever affectionately yours, FRANCIS GALTON.

7, WINDSOR TERRACE, THE HOE, PLYMOUTH. *Dec.* 12, 1906.

MY DEAR KARL PEARSON, Excuse delay in reply, my bronchitis has been troublesome but the attack is now passed.

What wonderful papers yours are, and how conspicuously they show the need of high mathematics in order to deal rigorously with apparently simple questions. I have now read your "Relationship to Intelligence of...&c." not once only, but more or less minutely more than three times (I am so slow, now!), but as to the "Random Migration" I have only read the conclusions and am awe-struck at the mathematics.

It is delightful to hear that you are already well enough to take part in the quartet dinner of successive occupants of the same chambers*. You ought to be proud of one another. The day-dreams of boyhood and youth are never fulfilled, or overpassed. Napoleon was no exception.

You had better I think tear up that centile paper I sent, which cannot be amended sufficiently for publication in any form. The diagram ought to be changed considerably. I have been improving on it and think I may make a little paper, suitable to some minor publication, that would be useful as a *first step*, and that would give the results of the kind in question with much ease, though only roughly. But I won't bother you with this now.

How well you have arranged the Title, etc. on the cover of *Biometrika*. I am very glad that you retain Weldon's name as you do. It is good news about Hope Pinker.

The Codicil after final revision by Hartog and Sir E. Busk has now been executed. I posted it yesterday to my lawyers.

The weather to-day is about as vile, with squalls and driving rain, as weather can be!

Ever affectionately yours, FRANCIS GALTON.

7, WINDSOR TERRACE, THE HOE, PLYMOUTH. *Dec.* 14, 1906.

MY DEAR KARL PEARSON, You really misread my "hearts of hearts" *re* mathematics. I worship and reverence them†, though in their application I have a tendency towards economy in their

---

* The men who in succession shared my chambers in Harcourt Buildings in the Temple were W. M. Conway, afterwards Sir W. M. Conway, art critic and M.P., Robert J. Parker, afterwards Lord Parker of Waddington, and E. C. Perry, afterwards Sir E. Cooper Perry, Principal Officer of the University of London.

† It was difficult to convince Galton that any higher mathematics were needful for statistical work than the percentile method of treating the normal curve and the linear regression graph. Perhaps the following sentences extracted from a letter to his sister Bessie (Mrs Wheler) concerning the education of her son Edward may be fitly quoted here. They are dated Feb. 6, 1866 and show the value Galton set on some mathematics:

"The value of a solid substratum of elementary mathematics is I can assure you of an importance almost equal to that of a new power in every profession in life. I see it at every step. Ingenious men without the thoroughness and precision, which mathematics alone are sure to give, sink below their natural level when competing in life with those that have it."

Tests for the significance or non-significance of the differences between samples of populations, which essentially require higher mathematics, he had not been forced to consider in his pioneer work.

use, under the ever-haunting fear lest the exactitude of their results may not outrun the trust-worthiness of the data. That is all. My fundamental misgiving is concerned with a too free use of the statistical axiom: "that unspecified influences tend to neutralise one another in a homogeneous series." My doubt always dwells on the questionable assumption of homogeneity, believing that extreme values are liable to be often caused by an heterogeneous admixture, present and active though undiscerned. So I love the ruder but theoretically correct statistics overmuch, feeling always *safer* within their moderate limits of one or two decimal places. All this is quite harmless, is it not? It is a purely general statement quite without reference to *Biometrika*.

I will with pleasure send a *revised* note about the centile matter, for it wanted re-writing.

I should be very glad to pay for the calculation of centiles and one tenths of centiles (in other words of mill-iles if such an awkward word existed) in terms of the P.E. and to *3* decimal places for printing in *Biometrika*, together with the revised matter. The present centile table is not quite minute enough to save trouble in interpolation. It is printed in *Natural Inheritance*. The form of it might be improved and if you approve I will draft what I mean and send it. Would you care to insert it in *Biometrika* if calculated? And can you find a calculator? I look forward to seeing your table of amounts of pigmentation.

About the Eugenics Office—what can I now do? I am still in a position to make terms with the University, before paying into their hands next month the promised £1000 to carry on the Fellowship for two additional years. Why should I not make it a condition that it should be treated as a department of the Biometric Laboratory and be wholly under your control? This would suit my views perfectly. Hereafter when the Professorship is established and a tradition of accuracy has been formed the Office could take a wider scope, as already arranged. Do tell me what you think I had best do.

I have not seen Lock's book yet. Murray wrote me saying he was about to publish a book by a Caius man (I forget the name but presume Lock) on Heredity, who wished to insert a portrait of me together with others and asking if I could spare a photo for the purpose. I sent three to choose from; the one he selected was the non-copyright one by Mrs Brian Hodgson, the same photo as that in *Biometrika*, which has also been published elsewhere more than once.

I hope you will pitch into any errors this writer may have made in his book.

Montague Crackanthorpe has been writing to me about his very readable paper in the current *Fortnightly*. He has many irons in the fire, I must urge him to keep the Eugenic iron red-hot.

Ever affectionately yours, FRANCIS GALTON.

Copy of a rough Draft of a Letter to Sir Arthur Rücker.

Dec. 16, 1906.

I feel in some difficulty about the immediate future of the Eugenics Office. Prof. Karl Pearson expresses himself as most desirous to give help to myself *personally* but fears a greater loss of time and energy than he can spare if the management of the Office by him requires his frequent attendance at S. Kensington for Cttee meetings and discussions on points of detail. There must of course be two opposing views which it is hard to harmonise: (1) The reasonable desire of the University to strictly control a department that avails itself of its prestige and occupies rooms that it provides; (2) The desire of the man who works it to do so with the minimum of "red-tape" entanglement. The question is the more difficult because so far as I know there is as yet no one available for the immediate appointment as Fellow, who can fill the Office properly without considerable statistical oversight, such as Pearson almost alone could give and would I think be willing to give if he had a free hand. Personally I should be quite satisfied if the management of the Eugenics Office could be put for a while under the complete control of K. P. as a branch of his Biometric Laboratory, but whether you would approve of this I do not know, neither do I know whether K. Pearson would accept the charge though I have hopes that after persuasion he might do so. Invalidism sadly hampers me. My intention though not my promise is to continue the £500 annually during my life and, as you know, I have made provision for a considerably larger support after my death. But before paying the £1000 to the University thus prolonging the annual grant up to 3 years from now, one year being already provided for, I should like to be assured that the immediate future of the Office is arranged for in a way likely to give good results without that assistance which I feel no longer able to give.

I have no evidence of the final form of Galton's letter to the Principal of the University, but I judge it was at the very least written in the above sense, and I know that during the four remaining years of his life the University certainly troubled me with a minimum of what Galton describes as " red-tape entanglement." It is a sign of his insight into men's characters, that he emphasised what I largely dreaded, but had not insisted upon as a great difficulty to him.

7, WINDSOR TERRACE, THE HOE, PLYMOUTH. *Dec.* 20, 1906.

MY DEAR KARL PEARSON, The enclosed from Sir A. Rücker has just reached me. I am most grateful that you consent to supervise. I can assure you that you need not have the slightest fear that the direction you may give to Eugenic work will disappoint me, for I know that what is done will be *thorough* and such as could not have been done by other means as effectively as in connection with your laboratory. Please return me Rücker's letter.

I send back your newspaper cuttings. It is astonishing to watch the difficulties that intelligent people—of "Child-Study" stamp—have in taking in new ideas. You recollect how it was in Darwin's time.—Poor humanity!

I have just been reading a daring book by Dr Rentoul, *Race Culture or Race Suicide* (Walter Scott Publishing Co.). It is full of shortcomings but very suggestive, and shows how much has been written and done here and in America. There is a massive movement going on out of the public sight. He mentions a fact that may be new to you, as it was to me, that a certain "Malthusian League" sends pamphlets "Why have any more children?" to the parents mentioned in the Birth Columns of Newspapers. Have you seen any of these? I recollect something of the sort long ago by (?Annie Besant).

You shall have the proposed form of the *Table* (Centiles etc.) in a day or two*. Once again I am most grateful to you. Affectionately yours, FRANCIS GALTON.

(12) *Scheme for the Francis Galton Eugenics Laboratory.*

As a result of our interchange of letters in the last three months of 1906, I was able to put on to paper a scheme for the small beginnings of a Eugenics Laboratory—for so the Eugenics Record Office was re-named—which more or less satisfied both of us, and which on February 1st, 1907, was accepted by the University practically without modification. Galton presented an additional £1000 which with the balance of £600 or £700 from his first gift enabled the little Laboratory to run for three years. In 1909 and 1910 Galton made further gifts of £500 for the years 1910 and 1911.

Final Form of Scheme for a Eugenics Laboratory for the University of London.

HAMPSTEAD. *December* 22, 1906.

MY DEAR FRANCIS GALTON, I had my last day of College work yesterday, and I have been trying to put into form my thoughts on the Eugenics Laboratory work, as they have been settling down in my mind during the last few weeks, for your suggestion and approval. Let me just explain a number of points first. I want to make the Eugenics Laboratory a centre for information and inquiry. I want to extend the tendency which is growing up for outside social and medical workers to send their observations to the Biometric Laboratory. But to do this I think we ought to try and associate some half dozen men (in the first place, say) with the Laboratory as an advisory committee or as associates. (I mean by this men from whom we can seek advice on points in their own fields of research.) I have not yet consulted the individuals I have in mind.

Now as to personnel. I don't think we shall do better than Heron. He is very keen....He is doing some mathematical teaching at present of which he has to give a month's notice of termination. So that if you approve, I would suggest his name at once to South Kensington.

* This paper in the form Galton desired with Centiles calculated by Dr W. F. Sheppard was published in *Biometrika*. See our Vol. II, pp. 401–2.

I would suggest that Miss Elderton be no longer spoken of as a clerk, but be made a Francis Galton *Scholar*. She is quite capable of doing original work. I should give her a little additional instruction in statistical methods, and set her on to research work either alone or in conjunction with Heron, so that her name would appear on the publication of it. I would further suggest that her stipend be raised. My reasons for this are as follows....She is very competent* and is now fairly well trained, and it is very desirable that we should retain her services. She is keen on the work. Further, if we are to get really good workers, we must give them a method of insuring to some extent their future. Now to have published something and been a Francis Galton Scholar, not merely a clerk, will give Miss Elderton a better chance if she passes later into social work of any kind. It is most desirable that people trained in the Eugenics Laboratory should pass into work in public or municipal service of some type, as in dealing with mental defectives or invalid children, etc. We shall thus develop into a training school for practical eugenic work.

My next point is that the office should if possible have a paid computer. We cannot afford more than, I think, £— for this. We shall not get for this the services of a man, or the whole time of a first class woman. ...Miss Barrington is the only person I can think of who is thoroughly trained and who would possibly be willing to give three or four days a week to computing for the Office. You will, perhaps, know her from her conjoint papers on inheritance in Greyhounds and Shorthorns. If we got her services, we should have a staff of three who would push through a lot of work.

I should suggest a continuous series of Eugenics Laboratory Publications. Even if we cannot publish an independent series, they should be published with continuous numbering and volumes of offprints made up and distributed to the Press to show the activity of the Laboratory. If the funds admit and there seems a possibility with the unexpended balance, an independent series of memoirs might be issued. ...Anyhow the important point is that, wherever and however published, there should be a single title "Eugenics Laboratory Publication No. —" and continuous numbering.

I think the Eugenics Laboratory ought through its Fellow, and with our aid in the Biometric Laboratory, to give instruction and aid to students and research workers in Eugenics and on this account some more detailed entry should be made in the University Calendar and occasional advertisements appear in one or two journals.

Next as to the purchase of reports, journals and books. I am very keen on the formation of a good library, and anything you get, pamphlet or book, that you would weed out of your own library pray turn over to the Eugenics Laboratory. Also it would be most valuable if you would send us the titles of any books or reports that we ought to look up. I shall certainly read Rentoul.

As to the additional room in Gower Street: I have two rooms there adjacent to the present Eugenics Office and can give up one, at any rate until October, to the Laboratory, because Heron will want a room for himself. Please pardon the enormous length of this letter.

Affectionately, KARL PEARSON.

If you approve, we might send something like the enclosed scheme and explanation of it to Sir Arthur Rücker.

Proposed Draft Scheme for the Francis Galton Laboratory
for the Study of National Eugenics.

UNIVERSITY OF LONDON.

The term National Eugenics is here defined as the study of the agencies under social control that may improve or impair the racial qualities of future generations either physically or mentally. The Laboratory is at present established at No. 88, Gower Street, and the Staff consists of:

(i) The Francis Galton Research Fellow. (*a*) The Fellow is appointed by the Senate on the recommendation of a special Committee reporting through the Academic Council. (*b*) The value of the Fellowship is £— per annum; that it be tenable for one year in the first instance, and

---

* The opinion of the first Galton Fellow may be cited here: "Miss Elderton has certainly been a remarkable success at the Eugenics Office; but I think her marvellous energy and quickness to learn anything new would have enabled her to succeed at anything she undertook."

for two subsequent years on favourable report from the Committee at the end of the first and second year's tenure respectively. (*c*) That the duties of the Fellow be to devote the whole of his time to the study and teaching of Eugenics. He shall report annually to the Committee on the nature of his researches during the year, and send to each of its members from time to time copies of such publications as he may solely or conjointly with other members of the staff have issued. (*d*) The chief object of the Fellowship is Research, but the Fellow will be expected to acquaint himself with statistical methods of inquiry and to give instruction to students or inquirers in Eugenics or allied problems, such instruction not to occupy more than a couple of afternoons a week. As chief executive officer he will be responsible for the general conduct of the Laboratory.

(ii) The Francis Galton Scholar. The method of appointment and the duties of the Francis Galton Research Scholar shall be similar to those of the Francis Galton Fellow, the stipend being such as shall from time to time be fixed by the Committee.

(iii) The remainder of the annual income of the endowment shall be devoted to assisting the general work of the Laboratory, to accumulating statistical material bearing on National Eugenics, and to the publication of researches made by the Laboratory or its associate members.

(iv) Members of the Staff are expected to work during the academic session daily (Saturdays excepted) in the Laboratory. The vacations will be about 3 weeks at Christmas, 3 weeks at Easter and 6 weeks in the summer, but it may be necessary to arrange the latter so that the Laboratory is not entirely closed for so long a period as 6 weeks.

Proposed Budget and Personnel......................

There is an unexpended balance from the original grant for each of the first two years. I would suggest (i) the purchase of a much needed Brunsviga Calculator. (ii) Various books of tables, etc., in current use for statistical work; a slide rule or two, etc. (iii) Providing with suitable furniture, cupboards, bookcases and lockers, an additional room in Gower St.* (iv) Incidental and unforeseen expenses. It is most desirable that any further balance should be allotted to (i) Purchase of Reports, Journals and Books, (ii) Issue of inquiry schedules and pedigree forms, (iii) Publication Fund, (iv) The gradual formation of a collection of instruments useful for observing the mental and physical conditions of children and adults, so that information and practical object lessons can be given to inquirers on these points.

Proposed Statement for Insertion in the University Calendar, or for Advertisement. (It is most important that persons engaged in social and medical inquiries should know of the existence and work of the Laboratory.)

The Francis Galton Eugenic Laboratory.

University of London (temporary address: 88, Gower Street, W.C.). The Laboratory is under the supervision of Professor Karl Pearson, F.R.S., in consultation with Francis Galton, F.R.S.

Francis Galton Fellow in National Eugenics: ....

Francis Galton Scholar . . . : ....

Computer . . . . . . : ....

Advisory Committee: The following have kindly consented to aid the Staff of the Laboratory in special forms of inquiry†: ....

It is the intention of the Founder that the Laboratory shall act (i) as a store-house for statistical material bearing on the mental and physical conditions in man and the relation of these conditions to inheritance and environment, (ii) as a centre for the publication or other form of distribution of information concerning National Eugenics. Provision is made in association with the Biometric Laboratory at University College for training in statistical methods and for assisting research workers in special Eugenic problems. Short courses of instruction will be provided for those engaged in social, anthropometric, and medical work, or desirous of applying modern methods of analysis to the reduction of their observations.

* This will be necessary if the staff be increased to *three* members. But the whole of this should be of a nature which will be useful when the Laboratory is transferred to permanent quarters.

† The Advisory Committee or list of consultants originally suggested consisted of: (i) a Commissioner in Lunacy, (ii) a R.A.M. Professor, (iii) an Actuary, (iv) an Anthropologist, (v) a Zoologist, (vi) a Pathologist and (vii) an Ophthalmologist.

7, WINDSOR TERRACE, THE HOE, PLYMOUTH. *Dec.* 22, 1906.

MY DEAR KARL PEARSON, I can hardly tell you what comfort and relief your letter has given me, feeling that your views and mine are in close accord, and that you have such a masterly grip of the situation.

I send back your letter for convenience to you of future reference, also Sir A. Rücker's to me, which had better be kept with it. I have made a few notes by the side of your letter to me, which will save additional writing. There is little to add to these. I am very glad that you retain Schuster on the Advisory Committee, not only on account of his own merits but as evidence of the continuity of the work, and am particularly glad that you feel that the appointment of Miss Elderton as "scholar" is feasible from the University point of view, and that you propose to raise her salary. She has always seemed to me an invaluable member of the staff. The computer will prove a real help and a relief to the future work of the Office.

The funds cover the estimated expenditure very narrowly. I wish I could undertake to give more. Possibly I may in a few months be able to give help towards the library and other non-recurring expenses, but I can't promise.

The "Eugenics Laboratory Publications" may greatly help by drawing attention to calculations stored in the Office but not printed as yet on account of cost; being in that respect a sort of glorified statement of similar matter to that which you insert on the red slip in each *Biometrika*. It could also contain lists of books received, and perhaps of memoirs wanted. The Form etc. of it would have of course to be carefully considered.

When I sat down I thought that a longer letter than this would be required, but on again looking over the notes I have made in the margins of your letter it seems that I have exhausted my say. I will write another letter about other things.

Affectionately yours, FRANCIS GALTON.

We may conclude this topic by reproducing a personal letter from Sir Arthur Rücker of thanks to Francis Galton.

UNIVERSITY OF LONDON, SOUTH KENSINGTON, S.W. *Jan.* 28, 1907.

DEAR MR GALTON, I must write a private line to you thanking you most sincerely for your generous gift of £1000 and the still more generous schemes you are developing. Your endowment is certainly the most original as I hope it will be one of the most useful of those that have been made since the University was reorganised, and it is a great pleasure to me to find that so old and kind a friend has selected the University of London for his gift. It is these external signs of approval which lighten a task, which like all work worth doing, involves detail which sometimes amounts to drudgery. In proof of the fact that you have chosen a progressive body may I tell you one fact. The income of the University 5 years ago was £29,000. This year it is £95,000 and including University College (now incorporated) it will next year be about £120,000. But we want not only public bodies like the L.C.C., the Goldsmiths' and Drapers' Companies to help us; we want distinguished individuals to recognise the existence of the University amidst the welter of London life. You are one of the few who have done this, and none have done it with more originality and generosity. Please accept my warmest personal thanks and Believe me,

Ever yours sincerely, ARTHUR W. RÜCKER.

One out of several further letters of Galton to his biographer dealing with other matters may be reproduced here in order to show how the octogenarian still retained interest in photographic problems. Cf. our Vol. II, p. 313.

7, WINDSOR TERRACE, THE HOE, PLYMOUTH. *Dec.* 25, 1906.

MY DEAR KARL PEARSON, More than one justifiable cause has prevented my returning the beautiful mouse-skin pictures till now. Do you need to use a coloured glass in making comparison between the coloured mouse skin and the black and white picture? I return the plates. I should like to have sent you a fully worked out picture of a method I have often thought of by which the mean tint of a variegated black and white *rectangular* portrait ought to be got. But I have

39—2

wasted too much time already on what may be of no use. So I simply send the enclosed. Abney used a rotating cylinder with a black and white drawing wrapped round it, in order to get the photographic equivalent to each combination of black and white*.

Cylinder
with paper round it

What good news about Pearl. I return his letter. I will shortly send the centile paper and diagram. Some delay has necessarily occurred about it. I am not idle but get through things now so slowly.

Best Christmas wishes to you all, and may you enjoy cake with the F. G. cut. My *Nature* of last week has miscarried so I have not yet seen my own paragraph, though others, like yourself suspecting me, have written to me about it. The post is just going out so I conclude now,

Ever affectionately, FRANCIS GALTON.

To obtain the mean tint of a rectangular picture.

Mount the picture on a [rotating] cylinder with axis vertical [? horizontal], in front of a camera. The dark slide of the camera to have a narrow vertical slit. Take an exposure—then cap. Move screen the width of the slit and take a second exposure; again cap. Repeat the process until the sum of the widths of the slits is equal to the length of the picture. Print off. The print will be streaky and of same width as the original was long. Mount the print crossways on the same cylinder as before and proceed as before. The result will be a plate of a uniform tint, the mean tint of the original. F. G. *Dec.* 25, 1906.

Galton's plan to get a mean tint is suggestive although it is not quite clear how he proposed to carry it out in practice, especially in dealing with the *relative* mean tints of engravings, say, of different sizes, or of piebald skins. Would a whole series of cylinders be needful to fit subjects of different heights, or must the subjects first be reduced to a standard size by photography ? How in practice would such reduction affect the *relative* tints of the two engravings? Again I do not follow the necessity for the slit, or how it is to be moved. A photograph of the engraving on the rotating horizontal cylinder would give vertical streaks on the plate. A print from this, which must be taken under stringently standardised conditions, could then be put crosswise on a cylinder of proper size and again photographed to obtain a uniformly tinted negative and thence a print. The "greyness" of this print would have—with absolutely standardised conditions—some relation to the average tint of the engraving, but I cannot see that they would be the same. Supposing the lighting always (artificially) the same and the exposures identical, it would be possible to compare the "greyness" of the prints thus defined with those obtained from known amounts of black and white on the cylinder, and thus form a scale. As I have said, Galton is here suggestive, and the problem is of some practical importance, but it needs much experimental work before it can be considered solved.

(13) *Work and Correspondence of* 1907. The year 1906, owing to reasons in part indicated, had been a year of stress and change for both Galton and

---

* A similar arrangement was adopted in the Biometric Laboratory for tint comparison judgments in 1894 (see *Phil. Trans.* Vol. 186 A (1895), p. 392). It is still used in the Anthropometric Laboratory attached to the Biometric Laboratory.

his biographer. In 1907 Galton, though he still had cause for anxiety, re-covered something of his usual mental activity and hopefulness. He had been asked and with some hesitation had consented to give the Herbert Spencer Lecture at Oxford on June 5th ; meanwhile his biographer had been invited to give the Boyle Lecture on May 17th, and in accepting had taken for his topic : "The Scope and Importance to the State of the Science of National Eugenics." On the other hand, Galton chose as his subject-matter : "Probability, the Foundation of Eugenics*," although the most interesting part of his lecture strayed somewhat from that topic. By the title Galton chose for his lecture he definitely gave forth as his opinion that his new science of Eugenics ought to be based on the actuarial treatment of man. For him the selective mating-rate, the selective birth-rate, the selective death-rate, and heredity in man were fundamentally mass-problems, to be solved statistically, by actuarial methods; shortly, Eugenics was a branch, the most important branch, of Biometry. For evolution the important matter is the changes that are taking place in the type or average of a species, and the variations that render these changes possible. We may never be able to predict what the individual child C of a given A and B will be like, but we can state the *probability* that he will be so and so ; in other words, we know the average distribution of character in the children of all parents like A and B. If A and B both come of stock tainted with insanity, we can predict with con-siderable accuracy the percentage of their offspring which will be insane or transmit insanity. It is no argument against the eugenic principle—that A and B ought not to have had children—to tell us that their particular child, C, is sane—he may be indeed a genius. The aim of Eugenics is to improve the race as a *whole*—to raise our nation above its present low level—not to breed one sane man at the cost of producing one or more bred insane. No farmer would be content with his flock, if with every white lamb, however fine its wool, he added at the same time a black sheep to his flock ! I think this is the meaning of Galton's statement that probability is the foundation of Eugenics, and of his opinion expressed in his letter to Bateson that "an exact knowledge of the laws of heredity" would scarcely help us in the problems of Eugenics†.

A few letters here may throw light on the trend of events.

March 2, 1907.

MY DEAR FRANCIS GALTON, Just a line to say that I have been asked to give the Boyle Lecture at Oxford this year (May 17th) and have settled to take "The Scope and Importance to the State of the Science of National Eugenics" for my topic. I expect my views will not wholly satisfy you, but they may help to push forward the whole movement and lead some of the younger Oxford men to think over and possibly take up the subject. If any ideas occur to you before Easter I should be glad of jottings or of suggestions for lines of thought. My idea is to indicate what we know already, what again we need to find out, and how much all these

* Oxford, at the Clarendon Press, 1907.

† See p. 221 above. It is, perhaps, needless to remark that with all the thousands of pounds and of pages devoted to genetic research during the last 25 years we seem to-day scarcely nearer the *exact* knowledge of the laws of heredity ; the further we advance the more complex does the problem show itself.

matters bear on national welfare, indeed that right views on them are essential to healthy patriotism.

I trust no further news is good news. I had a line from Mrs Weldon in which she speaks with much warmth of feeling of having received a very kind letter from you.

Affectionately yours, KARL PEARSON.

I see you are an optimist and believe in the existence of eleven men of judgment to one crank; I am inclined to think we too often get eleven foolish to one wise, and this average gives the latter a chance!*

3, HOE PARK TERRACE, PLYMOUTH. *March 7*, 1907.

MY DEAR KARL PEARSON, It is now fixed that both Eva Biggs and I return on the 13th [?14th], this day week, to 42, Rutland Gate. I have been much better of late. Illness seems gone out of me, and I walked a total of between 2½ and 3 miles yesterday, which l have not been able to do for many months.

If you have time, it would be useful, I think, to your Boyle Lecture, to read Sir John Gorst's *Children of the Nation*. At first I only read a chapter here and there, which seemed superficial and inadequately proven, but knowing that he has far from a superficial mind, I re-read it from beginning to end to my great profit. He shows the awful waste of good human material by bad administration, such as is not only preventable, but is prevented in some other countries. He wants to rouse public opinion. Eugenics as I pointed out when first adopting the word has the two-fold meaning of good stock and good nurture; in short "well-bred" in its fullest sense. *If* you include "nurture" in your lecture, Sir J. Gorst's book might give useful hints. Also it shows the machinery (with its drawbacks) that exists for carrying through new plans.

When we meet I should like to ask a question which you and Mrs Pearson could greatly help in solving, namely *if* I were to give up my house in Rutland Gate (being unable to live in London except for a few months in the year) would Hampstead be a good place to go to? I don't want more than a smallish house, but with a sunny (small) garden, and everything healthy. Such things seem to abound in Hampstead. Is it so?

Ever affectionately yours, FRANCIS GALTON.

BIOMETRIC LABORATORY, UNIVERSITY COLLEGE. *March 9*, 1907.

MY DEAR FRANCIS GALTON, I am so glad to hear that you are feeling the revivifying influences of spring. It is glorious to feel the sun again and to realise that summer is approaching. I am glad beyond that to hear there is even the slightest chance of your coming to Hampstead! I don't want my joy, however, to run away with me; I must not be too biased by it in telling you all the "ills" and "wells" of Hampstead as I know it.

[Here follows a rather too detailed account for reproduction of the advantages and disadvantages of Hampstead, especially the latter in the winter.]

Of course to me personally it would be a very great gain, I can hardly say how great, because I don't want that to weigh with you when health and environment generally must be considered in the first place;—but Sunday afternoons would be more of a possibility, and *you* know what nearness means in an overcrowded life!

The Eugenics Laboratory goes steadily along. Miss Elderton has got the ten types of cousins worked out for two characters and will soon have them done for four. The general conclusion is that cousins have almost exactly the same degree of resemblance as grandparents and grandchildren, i.e. correlation equal about 0·3. This seems to me of very considerable importance, because (i) cousins are contemporaries and can be more easily and accurately investigated than grandparents, (ii) there are far more of them than in the case of grandparents, and hence a closer estimate can probably be made from them †. Further, as far as I can judge, the marriage of first

* From a postcard of nearly the same date. This had reference to Galton's "Vox populi" letter to *Nature* of March 7, 1907. See pp. 403–4 of Vol. II of this *Life*.

† I may add that two cousins can of course be found who have no correlation *inter se*, only with the subject, which is an advantage not possessed by brothers. Thus the multiple correlation of four properly chosen cousins (·53) does not fall much short of that of two brothers (·58).

cousins must stand on exactly the same footing for good or ill as the marriage of a descendant with an ascendant in the second degree. About 75 °/₀ of the asylums in the country have already sent their reports with statistics and Heron has had a good deal of work in arranging and sorting them. The net result is that there appear to be about half-a-dozen men who might be both capable and willing to take elaborate family histories of the insane in their charge; the collection of such histories should be started at once. Always affectionately, KARL PEARSON.

Francis Galton did not come to live in Hampstead. I do not think, however delightful it would have been for me, that his health would have stood the winters there as well as at Hindhead or Haslemere, where he passed the winter months after he had ceased to feel able to spend them abroad. Thus 42, Rutland Gate remained his house to the end, and now—largely at the suggestion of his faithful servant Gifi—carries on the balustrade above the porch a tablet with the following words:

SIR FRANCIS GALTON

1822 – 1911

EXPLORER   STATISTICIAN

FOUNDER OF EUGENICS

LIVED HERE FOR FIFTY YEARS

42, RUTLAND GATE, S.W. *April* 8, 1907.

MY DEAR KARL PEARSON, I am anxious to learn that you are well. You wrote of digestive troubles—which sap strength—are they past? Yesterday I came across a long-missing packet, which I had labelled "Old Papers of the Evolution Committee, R. Soc.—of probably no present value. Might be useful if a Darwinian Institute were ever founded*." It would have been valuable to you when writing the Memoir on Weldon, containing as it does many letters from him. Whether you would care to see the packet now, I cannot guess. You shall have it at once if you like. I have only looked cursorily through its contents, but find them decidedly interesting as giving the various opinions of a variety of experts. What is of more present interest to you now is the enclosed letter from Leonard Darwin. I send it without comment. It may at least suggest a phrase in your forthcoming lecture at Oxford. Please return it when done with.

The Vice-Chancellor has asked me through Poulton to give the "Herbert Spencer" Lecture, but I have declined on the ground of infirm and uncertain health....I was delighted to hear yesterday that the Petries expect to add a unit to the forthcoming generation.

Ever affectionately yours, FRANCIS GALTON.

Major Leonard Darwin's letter referred to a brief talk with a distinguished politician about the Poor Law Commission; the conversation was very unsatisfactory. Darwin had spoken of the eugenic aspect of the matter, but the minister showed no interest and said doctors did not believe in heredity, or words to that effect. Major Darwin was very anxious that a combined effort should be made on the part of scientists to place their aspect of the

* See pp. 128–135 above.

case before the Commission. He considered that the biographer might give good evidence as he had thought on this side of the question. Anyhow Galton was not to answer his letter, but "to set something going to stir up such muddy opinion."

7, WELL ROAD, HAMPSTEAD, N.W. *April* 8, 1907.

MY DEAR FRANCIS GALTON, May I come and see you? I expect I have been rather foolish. In order to work I have been staying at home for the Easter vacation; it is the first time for more than 20 years, and I expect it won't pay. But the arrears were so overwhelming that I ought to stay at home, I think, for a year. There is material which has been waiting for years to be put into paper-form, and it is not fair to some of my co-workers to go on leaving it untouched. College teaching—a good deal of an elementary kind, but of the bread and butter sort—has gone on increasing year by year, so that I get little time for research work during term. I have, however, taken a bold step and written to the authorities suggesting that I ought to have some of the work taken off my hands. If one were in Germany, or had accepted one of the posts that have been offered in America, one would by fifty be able to do the work one is best fitted for. But this in England is only possible at Oxford and Cambridge; at all the newer Universities one has to undertake endless teaching work, which has no relation to the field of one's greatest efficiency. Now I have had my grumble out, and you can put it down to solitude and dyspepsia!—My wife and bairns are away and I have had several days of solitary meals, and 10 to 12 hours a day of solid work.

At meals I have read: (i) Rentoul: his exaggerations and fallacious statistics spoil an otherwise strong case. He is wretchedly careless also in expression—rather a medical Bernard Shaw. In fact he displeases me. (ii) I skimmed Plato's *Republic* and *Laws* again for eugenic passages, but they don't amount to much beyond the "purification" of the City by sending off the degenerates to form what is termed a "colony"! (iii) I read, much to my own pleasure, George Gissing's *Private Papers of Henry Ryecroft*. It is quite different from anything I have previously read of Gissing; over and over again false and annoying, but it is really literature, and there are some fine passages—pessimistic though they be. If you do not know it, it is worth considering.

Sheppard has computed the first half of your table and is progressing with the second, but for high values it means a good deal of work. He promises it in a few days now. The Press has been taking holiday and I have no proofs of the Eugenics paper yet. We have reduced the wasp material and I have written up the paper; we find that *workers* are the most variable, then the *drones* and lastly the *queens*. This is noteworthy because the drones are said to be more variable than the workers in the case of bees. I have also written a reply to Y.'s attack on my Huxley Lecture. It is the hardest task possible to have to reply to an old pupil and friend, who smites you without talking it over with you first!

Now when may I come and see you? When will you be alone? I could come Thursday about 3.30, if that would suit you. Affectionately, K. P.

P.S. I return Leonard Darwin's letter and enclose the others which you need not return. There is much to be done yet before statesmen of the Lord —— type will realise what statecraft has to learn from the Science of Man!

I wish you had felt up to the "Herbert Spencer" Lecture.

42, RUTLAND GATE, S.W. *April* 9, 1907.

MY DEAR KARL PEARSON, By all means come on Thursday about 3.30 as you propose. If you care for a change, and to dine quietly and to sleep here, you would be *most* welcome; but send a line in time to prepare bedroom. Don't bring dressing things. I fear you are much overworked. You shall see the packet of Evolution Committee papers when you come, and you can then settle whether I shall post them to you after you leave. I have not seen Y.'s attack. If you come to sleep here, why not stay on? You shall have all the comfort and privacy I can give you. I am quite alone now, and should be delighted if you would do so. There is an extraordinary battalion of family griefs just now, the two appendicitis cases of my niece's sons are going on very badly, and she herself is utterly overwrought and ill. ...Nay, there is still more to the bad—but I won't croak and bore. Ever affectionately yours, FRANCIS GALTON.

7, WELL ROAD, N.W. Sunday, *April* 14, 1907.

MY DEAR FRANCIS GALTON, It was quite a holiday seeing you on Thursday, and I came back with fresh vigour to my task. I have got about a quarter of the lecture now done. I am sending you by parcel-post (*a*) Pollock's *Spinoza*, a fine book, which some day you will let me have back. (*b*) A series of my own Essays, which please do not return. In mitigation of anything which may offend you in them, I may say that most of them were written 25 years ago and all of them more than 20. The only ones that I suggest you should look at are Nos. 6 and 7, possibly No. 10 might interest you in a spare moment.

I enclose the proof of the wrapper for the Eugenics Memoirs. I hope you will approve it. Will you return it to me with suggestions of any changes you would like? I shall have to send it to the University for approval....By the bye, I was amusing myself by trying to draw up a pedigree of Darwins and Wedgwoods on the basis of *Noteworthy Families*, pp. 18–19. On p. 18 Josiah Wedgwood is said to be George Darwin's *me me fa*, and on p. 19 his *me fa fa*. Hence his mother's father and mother's mother must have been brother and sister! On p. 19, l. 6, I read: "*me fa fa* (she was her husband's *fa bro dau*)." Now the "she" is I suppose the *me*, hence the great Charles' wife was a *Darwin*, his father's brother's daughter, but her father's father was a Wedgwood. Hence she was a Wedgwood. Something seems to have gone wrong on pp. 18 and 19.

Will you put the W + D pedigree for me on a bit of paper? I have got very confused over it. Can you send me —'s address? It has occurred to me that it might possibly do good, if I sent a few lines. I think, perhaps, I am the only person, who knowing so much, could effectively say something more. It might not help, but I don't think it could harm. If you advise me not to, of course I shall not attempt it. But sometimes a call to the immediately obvious duty is really helpful. Affectionately, KARL PEARSON.

42, RUTLAND GATE, S.W. *April* 16, 1907.

MY DEAR KARL PEARSON, I am so glad that the pleasant visit you gave me was no hindrance to your work. Excuse delay in replying to your questions....I must postpone for another day the Darwin pedigree. The original papers are, I think, at 88, Gower Street, but I may succeed otherwise in working it out. The books are safely come! Many thanks. I will read both of them leisurely.

As regards the entries on the wrapper, they seem to me to be quite clear and appropriate, except that the address given to applicants to exchange publications should be to some *person*. I have put "to the Editor" as a suggestion. As regards the colour of the wrapper, it may have distinctive merits, but not in the sense that the printing on it is distinct. At this moment I cannot read it in a darkish corner of the room, and I have often noticed in the heaps of periodicals on the tables at the clubs that the printing on the blue cover of the *Edinburgh Review* is by far the most indistinct of any. As regards size you naturally want to be constant to that of your other publications, so I say nothing against it, though my own unbiased feeling would be strongly in favour of Royal 8ᵛᵒ.

The Vice-Chancellor of Oxford has attacked me about the Herbert Spencer lecture with such a kind and thoughtful letter,—assuring me that if when the time comes I should feel unequal to delivering it personally, or even of being present, he would arrange for its being read in my absence,—that I felt obliged to cancel my previous refusal. So I shall have to hold forth towards the end of May. I see that the first of these lectures was given in 1905 by Frederick Harrison. What may have occurred in 1906, I do not yet know....I will be able to tell more when I write about the Darwin Pedigree. Ever affectionately, FRANCIS GALTON.

[HAMPSTEAD.] *April* 19, 1907.

MY DEAR FRANCIS GALTON, I should have written yesterday only I was hoping to hear possibly from you again. I want to say how glad I am to hear you are going to undertake the Herbert Spencer Lecture after all. The only point I feel some compunction about is whether I have not, unwittingly, taken your subject from you. I had no idea at the time I sent them my title that you would be lecturing yourself in Oxford, and I would change it even now, if they had not posted it about the place. At least I judge they must have advertised it in some way, because I have received one or two letters already on the title. Now can you look upon me as your John the Baptist, making the way straight? I am getting my lecture typed so that I may send you a copy. Will you let me know, if there is anything that trenches too much on what you have in view, and I will cut it out? Of course it is all *you* in a certain sense as it deals with Eugenics from beginning to end; still you must see it and give me your views....The Eugenics folk are back, at least Miss Barrington was up with some problems yesterday.

Bateson has edited a vast work—the Report of the Hybrid Conference—wholly Mendelian. I come in for my fair share of abuse! There is just one paper of 1½ pages which would have pleased Weldon. It is by a Canadian on the inheritance of bearded and beardless wheat—one of the "striking Mendelian illustrations." He very quietly demonstrates by aid of illustrations that the Mendelian theory does not work. Affectionately yours, K. P.

P.S. Your letter just come and I have re-opened this. Your Darwin pedigree is, I think, clear but there is still, I believe, a slip. You say: Mrs Darwin was her husband's *fa bro da*. Her husband's *fa* was a Darwin, and therefore his *bro* was a Darwin, and his *bro da* would be a Darwin and not a Wedgwood in maiden name. I think it should be she was her husband's *me bro da*.

42, Rutland Gate, S.W. April 21, 1907.

My dear Karl Pearson, It is amusing that at Oxford we should both be proclaiming Eugenics as one of the large progeny of the University of London! Really the study is gaining an academic *status*! I do not think we shall clash as, though the title of my lecture is "Probability, the Foundation of Eugenics," there are new points in it, and for the rest, when you send me your typed copy I shall have time to revise my own lecture by cutting out anything that appears as duplication. I should be most grateful for your free criticism of mine, which, owing to my slow work, won't be written out even, much less typed, by the end of this week. It shall be sent to you as soon as ready. What is your date? Mine is towards the end of May, but I do not yet know more precisely. You are quite right, the passage ought to have been she was her husband's *me bro da*, the "she" being of course transformed into a more intelligible expression....After much discussion with relatives, I have determined to safe-guard my interests by engaging (as soon as I can find one) a "Nurse-housekeeper," that is, an upper servant (not a lady), age about 40, who could manage *well* the household, mend my things and be able to write letters in an emergency, which were fairly well-spelt, etc., and also nurse me well when I am next ill. Such persons exist in abundance but are hard to find. If Mrs Pearson knows of any such I should be grateful to her to tell me. I should give the "Nurse, etc." good wages, fully up to her "market worth." Ever affectionately yours, Francis Galton.

The reader of the letters of Galton, 1906–1907, will realise that while he was mentally as active as ever, clear and concise in his judgments, his physical strength had begun to fail him, and he became more and more conscious of the need to be cautious about himself. This need was emphasised by two accidents which he met with in the course of this year.

Extract from a letter of April 22, 1907, of K. P. to F. G.:

My lectures are both at Oxford. I lectured at Cambridge last term on statistical methods. I give the Boyle Lecture on May 19th to the "Undergraduate and Junior Graduate Science Club," but I believe others attend....On May 21st I lecture to the Philosophical Club (a club of Oxford lecturers and dons) on "The Possibility of a wider Category than Causation." This lecture starts from the idea that no two physical entities are exactly alike, e.g. not even two atoms are precisely identical. They form a class with variation about a mean character. Hence even in physics the ultimate basis of knowledge is statistical—the category is of course correlation not causation. The main difference is that in physics the correlation coefficients are nearly unity but in biology they diverge considerably from unity. Except that in this second lecture I shall assert that Probability is the basis of *all* knowledge (not only of Eugenics!), it will not touch on your topic at all. But I am rather sorry if I trespass on your field in my first lecture. All I can say is that you must read it before delivery and allow me to be, if possible, your way-straightener.

42, Rutland Gate, S.W. *April* 22, 1907.

My dear Karl Pearson,   I telegraphed in order to save a post....It was purely a blunder of mine about Cambridge instead of Oxford for your second lecture. I wish you all success on May 19th and again on May 21st. Have you any *proof* that the ultimate atoms are unlike, other than by inference? But I shall see what you say in good time. Of course it is most probable that they differ. I think my lecture will not trespass at all on yours except as far as the title may suggest. You are very good about the Albinism and the Eugenics publications. I like to feel that the Eugenics Laboratory is a sort of annexe to your Biometric Laboratory, using the same methods and working with similar precision under your guidance. I do not a bit understand the *Royal Soc. Proc.* memoir just out on the constitutional peculiarities of albinos. Anyhow it seems that their blood behaves differently in the presence of "proteids"—a mere name to me—from that of pigmented people. (Can people of *piggish minds* be properly styled *pig-mented*? I crave pardon!!)   Ever affectionately yours, Francis Galton.

[Hampstead.] *May* 3, 1907.

My dear Francis Galton,   Here at last is my lecture typed by Miss Dickens's Office! It was hastily written and the tables have yet to be added. I should esteem it a great favour if you would write on the blank facing sheets any suggestions that occur to you, and let me have back the copy for emendation. I fear the whole thing is very laboured, but I am writing under much pressure and feel a good deal the want of a holiday. I hope all goes well with you.

Affectionately, Karl Pearson.

Our letters for the next fortnight chiefly cover the last stages of the final drafting of the Weldon Prize regulations. Then they touch again the Oxford lectures. I will cite first the letter which reports my own lecture.

7, Well Road, Hampstead, N.W. *May* 29, 1907.

My dear Francis Galton,   I think you may care to hear how my Oxford campaign has passed off. My lecture on Friday was fairly well attended. It was in Balliol Hall, and I soon found that I must throw up my manuscript and take to talking. Of course this made me slip many points, but that won't so much matter as the lecture is to be printed. On Saturday I went through the mice with Mrs Weldon, had a talk with Schuster about his brain-work, and wrote about half my lecture for Sunday. That was given in Magdalen Summer Common Room to the Philosophical Club. The members seemed to me mostly groping in the field of obscure definitions. The metaphysicians did not understand me, and the few science folk present were hostile. They could not grasp how much wider the correlation category is than the causal. However, I think I did some good, although these Oxford dons did not impress me as a group of very clear and powerful minds*.. It was quite different when I faced in January the Cambridge mathematical lecturers—then one felt in the presence of men of superior intellectual power, and was rather ashamed of oneself. I hope at any rate I have done some Baptist work, and you will find the way straightened. They know now, or ought to, what Eugenics signifies and what the word correlation denotes. I had an interview with the Vice-Chancellor and hope the Weldon memorial will shortly now be settled. I trust this bitterly cold weather will not get a hold on you; it makes me at times feel very incapable and inert. I hope your lecture has got written without too much effort. I hear it is to be given in the Sheldonian Theatre, which, I fear, will want more volume of sound than Balliol Hall.   Always affectionately, Karl Pearson.

The following letters deal with Galton's lecture.

42, Rutland Gate, S.W. *May* 25, 1907.

My dear Karl Pearson,   Here is my lecture, but without the 9 diagrams on one page, and without the references to them in the text. They have been redrawn and are being "processed." I send them thus as there is not too much time. Any suggestions in the text would be most welcome.   Ever affectionately yours, Francis Galton.

* Looking back on the discussion now, I think we were really speaking different tongues, wherein the same words carry different atmospheres.

40—2

7, WELL ROAD, HAMPSTEAD, N.W. *May 26, 1907.*

MY DEAR FRANCIS GALTON, I wish the Hampstead dream had been realised and that I could first have run in and spoken to you, instead of having to trust to the written word!... Now to the lecture. I like your opening and your finishing *extremely*, and your centre I should like also, if I heard you deliver it with the manuscript thrust aside, while you talked to the audience in Froebel fashion. I quite realise your point, that it is possible to make these biometric conceptions part of the average man of culture's ideas. Every word you have written would be telling, if you were teaching the teacher to teach. But I am not certain how far your very condensed five object lessons will be acceptable when you bring them in the Oxford June week to your *child*, not to your teacher. What you must do later is to expand them into a small primer of biometry. Now what I feel is this, that if you do not attempt to read these elements of a primer from your manuscript, but just talk a bit about them in the middle of your lecture, you will lead your audience to read these parts afterwards in print, while you fascinate it meanwhile personally as you have the power to do. That is really my sole criticism—an Oxford June audience is the child and not the teacher.

These other points involve merely suggestions of slight changes: (i) Surely you have inverted the order of our Huxley Lectures. My lecture was in 1903, but I think yours was two years earlier and not the year after. In fact you put the right date on the top of p. 7. So here you will see, you, not I, led the way! (ii) Will you think me ungrateful, if I ask you not to praise me quite so much? It is natural that I should feel and speak strongly about your work, because I owe so much to it for method and suggestion, but if you praise me 'tis as you branded your own herring as of peculiar virtue. Please re-read in this sense pp. 2—3 and 9. I know you will grasp how much I appreciate all your praise, but others possibly will not see it from the same standpoint. (iii) Would it not be well to free yourself on p. 21 from your unit by measuring your A and B in terms of their standard deviations? You thus avoid the difficulty which occurs to the mind coming fresh to the subject of the index of correlation * depending on the units used—lbs. weight, inches of stature, etc.—and thus providing no comparable ratio, but one varying with the units. If you agree to measuring in terms of your standard deviations as units, all values of the index of correlation are comparable and lie between -1 and $+1$. All this is, of course, very familiar to you [see, indeed, our pp. 5, 51, and Vol. II, p. 393, but it passed from Galton's mind when preparing his manuscript].

You would bring it home to your hearer and save him some difficulty, if you gave a hint that the coefficient of correlation lies arithmetically between 0 and 1, and has only a numerical value, being independent of scales, such as those of weight, length or units of pigment intensity.

I wish I could come to Oxford to hear and possibly help you. I would if it were July, but I am under rather high pressure, and one of my ears is giving me much trouble and exciting the neck in some way. Affectionately, KARL PEARSON.

I shall hear how the lecture goes, I have no doubt; but I should like to hear when you have a chance how the lecturer gets through the exertion, which is another matter.

Galton was not fit to speak at Oxford, one of the reasons being the accident referred to in the following letter. The lecture was read by Mr Arthur Galton.

42, RUTLAND GATE, S.W. *May 27, 1907.*

MY DEAR KARL PEARSON, I have now a bout of ill fortune. Feeling particularly well I went on Friday to Bushey Park and returned a bit tired but nothing more. However a horrid bout of bronchitis came on and on Saturday night 12.30 on getting out of bed I rested in the dark on an insecure table with crockery and tumbled on the floor with such a clatter and bound with the bed-clothing dragged after me. I had not the strength to free myself so there I lay till 6.30 when the household stirred and the united strength of three maids got me into bed with a very sharp sciatica. It is possible that I may be fit to go to Oxford on June 5 but I feel practically sure that my lecture must be read for me.

---

* I use here the term employed by Galton in his lecture; by 1907 the name "*coefficient* of correlation" was in general use.

Thank you much for your suggestions, but I can't conscientiously adopt those that relate to yourself. The errors of date of the two Huxley lectures were serious (I can now trace how it occurred; I had bothered over it). I dare not think of Hampstead now, feeling that I mayn't be fit for more than a bath chair, hereafter. An oak floor makes a hard bed for an invalid, as my ribs, etc. loudly proclaim in their language of feeling. It is a good biblical phrase "the iron entered into my soul"; that is just what the oak has done—also Hudibras'

> "Now am I out of Fortune's power.
> He that is down can fall no lower*."

I wonder whether, when the lecture is over, I could persuade Miss Elderton to write a primer of the proposed lessons. If the idea takes, it would be worth her while. Ladies often do these things better than men. Ever yours affectionately, FRANCIS GALTON.

Thanks for the appreciative account of your doings at Oxford†. I return it.

Galton was beginning as the result of his experience of the women workers in the Biometric and Eugenics Laboratories to have a higher opinion of the contributions of academically trained women to science. (See Vol. II, pp. 132–4.)

7, WELL ROAD, HAMPSTEAD, N.W. May 27, 1907.

MY DEAR FRANCIS GALTON, I am so very sorry indeed to hear of your accident, although I am glad you can be humorous as to its incidents. But you really ought not to be, so to speak, out of range of the household and unable to summon them for six hours! You must have some-one in your dressing-room within call. You ought at least to have bells and sticks within reach.

I shall still hope that it may be possible for you to deliver the lecture yourself, for although I would not have you make any effort that would have risk to health in it, I still know what a great pleasure it would be to many at Oxford to hear you speak yourself. As soon as you have got this over, you must see Miss Elderton and talk your project over with her.

Always affectionately, KARL PEARSON.

You must not let anything I have said induce you to attempt more than you feel quite capable of, but it would be were it possible so fine to speak to Oxford in one's 86th year!

It is high time that we turn to the Oxford Lecture itself; the letters above printed will suggest to the reader how much time and thought its preparation cost Galton. Strange are the vagaries of chance, the outward plumage of Galton's lecture on Eugenics approached the wrapper-colours of the *Edinburgh Review* and *Eugenics Laboratory Publications*! (See p. 313 above.)

After a vivid and brief characterisation of Herbert Spencer:

"Spencer's strong personality, his complete devotion to a self-imposed and life-long task, together with rare gleams of tenderness visible amidst a wilderness of abstract thought, have left a unique impression on my mind that years fail to weaken" (p. 5),

Galton passes to the aid which Spencer gave him personally by discussing with quick sympathy and keen criticism in the old smoking room of the Athenaeum Club, while waiting for a game of billiards, the ideas with which Galton at the time was teeming. We may imagine that the process was scarcely mutual; it is hard to think of Herbert Spencer *seeking* criticism of his ideas, although they naturally met with it, when he gave expression to them (see *Memories of My Life*, pp. 178, 257–8). For Galton, Spencer was

---

* See Vol. I, p. 64.
† *The Oxford Magazine*, May 23, 1907, p. 345.

a whetstone whereon he could give his conceptions greater sharpness and clarity, and he confesses in the present lecture that he misses this much in his old age. And yet looking back on all that correspondence of some twenty years, re-reading our letters, it seems to me that both Weldon and I were ever seeking to guide our master into what we thought the straight and narrow path*. But the following passage shows how badly we had failed:

"Among the many things of which age deprives us, I regret few more than the loss of contemporaries. When I was young I felt diffident in the presence of my seniors, partly owing to a sense that the ideas of the young cannot be in complete sympathy with those of the old. Now that I myself am old it seems to me that my much younger friends keenly perceive the same difference, and I lose much of that outspoken criticism which is an invaluable help to all who investigate." (p. 6.)

After this preliminary reference to Herbert Spencer, Galton began with a section on the *History of Eugenics*. He referred to the accident that the word "Eugenics" should have occurred in the titles of both Boyle and Herbert Spencer lectures and passes that praise on the Boyle lecturer to which I raised objection in my letter reproduced above (see p. 316). He then mentioned the coining of the word "Eugenics," in his *Human Faculty* of 1883, and recapitulates his creed wherein man is to control organic evolution, as he controls physical nature, and eugenic conceptions are to attain a religious validity—are indeed to become phases of a "categorical imperative." In this creed he emphasises

"the essential brotherhood of mankind, heredity being to my mind a very real thing; also the belief that we are born to act, and not to wait for help like able-bodied idlers whining for doles. Individuals appear to me as finite detachments from an infinite ocean of being, temporarily endowed with executive powers. This is the only answer I can give to myself in reply to the perpetually recurring questions of Why? Whence? and Whither? The immediate 'Whither?' does not seem wholly dark, as some little information may be gleaned concerning the direction in which Nature, as far as we know it, is now moving. Namely towards the evolution of mind, body, and character in increasing energy and co-adaptation." (p. 8.)

Galton re-states the view that we men may very likely be the chief, perhaps the only executives on earth, and that as such we are responsible for our success or failure to further certain obscure purposes, which we must strive to ascertain†. Our instructions, if obscure, are yet "sufficiently clear to justify our interference with the pitiless course of Nature, whenever it seems possible to attain the goal towards which it moves by gentler and kindlier ways" (p. 9). Galton admits that in 1883 the idea of directed evolution did not appeal to investigators, "it was too much in advance of the march of popular imagina-

---

* I have before me at this moment a long paper by Galton in manuscript dated April 1890; it is on the topic of "Sexual Generation and Cross Fertilisation." It appears to have received the *coup de grâce* from a letter of Weldon's which is attached to it, suggesting that Galton should make a study of modern cytological ideas before proceeding further. It seems to me that the criticism of youth, bursting with the newer knowledge, may not always be of advantage to the inspirations of enthusiastic age with a riper practical experience and a much longer period of close observation. Youth makes its mistakes regardless of the counsel of age, and sometimes those very mistakes bring to it "la gloire." Let old age blunder without restraint from the young, and possibly after-generations may see in those very blunders not the least luminous rays in the aureole of genius.

† Jonathan Hutchinson asked what was his religion replied: "I am a good planetarian." So might Galton have asserted.

tion." It had to wait till the publication of *Natural Inheritance* in 1889 ; then Galton found the lieutenants he stood in need of:

"The publication of that book proved to be more timely than that of the former. The methods were greatly elaborated by Professor Karl Pearson, and applied by him to Biometry. Professor Weldon of this University, whose untimely death is widely deplored, aided powerfully. A new science was thus created primarily on behalf of Biometry, but equally applicable to Eugenics because their provinces overlap [i.e. in Man]. The publication of *Biometrika*...began in 1901." (p. 10.)

Galton then refers to the Huxley Lectures of 1901 and 1903, and to his own papers of 1904 and 1905, to the establishment in the latter year of the Eugenics Record Office with its Research Fellow, and to the foundation in the year of the lecture of the Laboratory for National Eugenics. It is a brief, but adequate history of the small beginnings of the new science, concluding with its definition, that of the University of London Committee.

I have so far passed over the earlier portion of this section which does not really belong to the History of Eugenics, but rather to that of Evolution. Galton refers to that wondrous creation the *Hyperion* of Keats, to the succession of deities ; Chaos ; Heaven and Earth ; the Titan brood ; the Olympian Gods. Each ousting their parents, and forming a notable advance, physically and mentally, on their predecessors. Thus Galton would have each generation of men advancing by their self-constituted control of evolution through heredity to higher qualities :

> "So on our heels a fresh perfection treads,
> A power more strong in beauty, born of us,
> And fated to excel us, as we pass
> In glory that old Darkness." (ll. 212–15.)

Thus in his 86th year Galton showed how little he had lost of that poetic imagination, which always marked his fertile mind. He could read into the barbaric theogony of primitive Greece a lesson for the men of to-day.

The second section of the lecture is entitled: *Application of Theories of Probability to Eugenics*. It commences with the statement that Eugenics demands *quantitative* results. It is not content with such vague words as "much" or "little," but seeks to know "how much" or "how little" in precise and trustworthy figures. Given, Galton says, that we know that a certain class of persons, *A*, is afflicted with some specified degree of degeneracy we wish to find out how many of their offspring, *B*, will also be afflicted and to what extent. Further we want to find out: "What will be the trustworthiness of the forecast derived from averages when it is applied to individuals?" Galton then turns for a measure of untrustworthiness to the average deviation, *D*, from the forecast.

"The smaller *D* is, the more precise the forecast and the stronger the justification for taking such drastic measures against the propagation of class *B* as would be consonant to the feelings, if the forecast were known to be infallible. On the other hand a large *D* signifies a corresponding degree of uncertainty and a risk which might be faced without reproach through a sentiment akin to that expressed in the maxim 'It is better that many guilty should escape than one innocent person should suffer*.' But that is not the sentiment by which natural selection is guided, and it is dangerous to yield far to it." (p. 14.)

Galton admits that a thorough investigation of the kind referred to, even if it were confined to a single grade of a specific degeneracy, is in itself a very serious undertaking:

"Masses of trustworthy material must be collected, usually with great difficulty, and be afterwards treated with skill and labour by methods that few at present are competent to employ. An extended investigation into the good or evil done to the state by the offspring of many different classes of persons, some of civic value, others the reverse, implies a huge volume of work sufficient to occupy Eugenics laboratories for an indefinite time." (p. 14.)

It will be seen how thoroughly Galton's mind was imbued with the conception that the science of Eugenics has to deal with mass-phenomena, that it is essentially based on statistics and must adopt the actuarial method, i.e. that it is based on probability reckoned on past experience. This conception leads him directly to his next section: *Object Lessons in the Methods of Biometry.* He proposes to speak of those fundamental principles of probability, which are chiefly concerned with the newer methods of Biometry, and consequently of Eugenics. " Most persons of ordinary education seem to know nothing about them, not even understanding their technical terms, much less appreciating the cogency of their results " (p. 15). Galton accordingly sets out to sketch in outline a series of lessons of a Kindergarten type, which a teacher may fill in, and thus lead the ordinarily intelligent person, though he be ignorant of mathematics, to a knowledge of the fundamental ideas on which probability is based. He fears that this will scandalise biometricians†, but he has previously softened their wrath by saying that no man can hope to achieve much in Biometry without a large amount of study, the possession of appropriate faculties and a strong brain !

I do not propose to enter into the nine pages of the Lecture (pp. 15–23) which draft this scheme of " Object Lessons." They have, as I shall indicate later, been developed by W. Palin and Ethel M. Elderton into a primer of statistics. Most of the ideas have already been considered in this biography; the scheme proceeds in the main from " median " and "quartiles," and covers the simpler forms of variation and correlation.

The final section of the lecture is entitled : *Influence of Collective Truths upon Individual Conduct.* Galton commences by noting that probability will provide a solid foundation for action in the matter of Eugenics. But the

" stage on which human action takes place is a superstructure into which emotion enters, we are guided on it less by Certainty and by Probability than by Assurance to a greater or lesser

* This is the terrible dilemma in which the tender-hearted Condorcet found himself when he came to analyse the probability of criminal trials leading to correct judgments. There, however, life had come into being ; here it need not be called into existence.

† I think this was a little poke at his friend, who had really criticised the *occasion* not the matter of Galton's "object lessons." The friend had indeed already in the " 'eighties" given several Kindergarten courses on experimental probability at Gresham College to City clerks and Government employees, who afterwards became statisticians, and besides to a considerable number of bookmakers and professional gamblers who entered keenly into the spirit of the demonstrations, and whose gratitude took the form of free gifts of "tips" for the Derby and schemes to break the bank at Monte Carlo ! ˙

degree. The word Assurance is derived from *sure*, which itself is an abbreviation of *secure*, that is of *secura*, or without misgiving. It is a contented attitude of mind largely dependent on custom, prejudice, or other unreasonable influences, which reformers have to overcome, and some of which they are apt to utilise on their own behalf. Human nature is such that we rarely find our way by the pure light of reason, but while peering through spectacles furnished with coloured and distorting glasses." (p. 24.)

The general drift of this final section, if not so clearly put as Galton has elsewhere expressed it, is that the principles of Eugenics must be made part of the social code, a collective truth of society at large, whose power over the individual can scarcely be overrated.

"The enlightenment of individuals is a necessary preamble to practical Eugenics, but social opinion is the tyrant by whose praise or blame the principles of Eugenics may be expected hereafter to influence individual conduct." (p. 26.)

Galton considers that the opinion which holds particular social codes of conduct to be unchangeable is like the conviction of lovers that their present sentiments will endure for ever. Love is notoriously fickle and so also is public opinion. Galton illustrates this by the fashion of hair on the male face. In the days of his youth the "assumption of a moustache was in popular opinion worse than wicked, it was atrociously bad style." During the Crimean War the infantry were relieved from shaving, and on their return to England beards spread to the laity, but stopped short of the clergy. Then a distinguished clergyman "bearded" his Bishop on a critical occasion, the Bishop was so overcome that he yielded without protest, and "forthwith hair began to sprout in a thousand pulpits where it had never appeared before in the memory of man" (p. 27). Once mould public opinion to consider a non-eugenic marriage atrociously bad form, and the victory is won—the law, as Galton indicates, follows limpingly the growth of the public conscience.

"Considering that public opinion is guided by the sense of what best serves the interests of society as a whole, it is reasonable to expect that it will be strongly exerted in favour of Eugenics when a sufficiency of evidence shall have been collected to make the truths on which it rests plain to all. That moment has not yet arrived. Enough is already known to those who have studied the question to leave no doubt in their minds about the general results, but not enough is quantitatively known to justify legislation or other action except in extreme cases. Continued studies will be required for some time to come, and the pace must not be hurried. When the desired fulness of information shall have been acquired, then will be the fit moment to proclaim a 'Jehad' or Holy War against customs and prejudices that impair the physical and moral* qualities of our race." (pp. 29–30.)

That the Herbert Spencer Lecture, notwithstanding fine passages, is not fully up to Galton's best work may strike the reader†, but he cannot see it in the same aspect as those of us who knew the extreme stress, not then fully ended, to which the old man in his 86th year had been for twelve months subjected. It was a surprise to some of us that he ventured to lecture at all, and we rejoiced that the lecture could be as good as it was.

* It is startling to see this word reappear here after the use of "physical" and "mental" in the definition of Eugenics on p. 12 of the lecture!

† See my remarks at the top of p. 316.

Six days after the Herbert Spencer Lecture, Galton wrote to me:

42, RUTLAND GATE, S.W. *June* 11, 1907.

MY DEAR KARL PEARSON, An invalid's days creep by so uneventfully, that the passing of time is little felt and one gets too easily in arrears. I owe you many thanks for your kind interest and inquiries. For my part, I have practically said "goodbye" to bronchitis, for the present,— and feel plucky enough to venture on a visit to a niece in Leamington on Thursday, and, if that proves successful, to go a little further afield to a nephew for the next week. Letters here will always be forwarded.

I am very curious about your new method of determining correlation. When you publish, don't forget me.

Crackanthorpe's "Population and Progress" interests me much. His last chapter (VI) opens out quite a new horizon to me, and suggests a subject for discussion at some future Hague Conference—viz. limitation of *populations*! The pullulating nations have ever been the *primum mobile* of invasions. If a country breeds more than it can provide for, there is bound to be an outburst. It is Germany's difficulty and temptation at the present moment. My head is full just now of such ideas, and of encouragement to entertain them, derived from that excellent article of Mrs McFadyean's in the *XIXth* [*Century*], showing how the *women* all over the world are now becoming enlisted in furthering the limitation-of-families question. They have so far less temptation to be imprudent than their husbands, and suffer so far more acutely from imprudence than the latter do, that their awakening to the question seems of the higher importance. I had never looked at the matter before from the woman's point of view, as Mrs McFadyean does. It does not now seem to me nearly so hopeless as it did to limit the families of male degenerates, if the purely selfish feelings of their mates can be worked on and aroused.

Ever affectionately, FRANCIS GALTON.

This summer the biographer and his family were at River Common near Petworth, and Francis Galton took a house at Haslemere; but although I cycled over to see him, it was not possible to resume the old "biometric teas," the shadow of Weldon's death still hung over both households, and I had in addition much anxiety about the illness of my Father. Galton gave me sound advice and in my letter to him of July 7 I find the words: "I always feel about you, as I felt about Henry Bradshaw, that if I put a personal difficulty, I shall get the help of a contemplative man of riper experience. I think it is a trace of the old Quaker blood in both of you."

I give a few of the letters which passed between Galton and myself in the remainder of 1907.

THE GALTON EUGENICS LABORATORY, UNIVERSITY COLLEGE, GOWER ST. *June* 20, 1907.

MY DEAR FRANCIS GALTON, I am writing a number of letters about the Eugenics Laboratory and feel I must send one to you while I sit invigilating here with my examinees. I have a good bit of news for you. The Education Committee of the L.C.C. has consented to place its material—observations on the mental and physical condition of London school-children—at our disposal. I believe there are 8000 cases to be dealt with. This has relieved my mind a good deal, for I was growing very anxious as to whether I could provide the Laboratory with enough material to work at. Miss Elderton is simply a cormorant! We are slowly collecting several series of data, but the time to get them up to a number big enough for safe conclusions must be long, and some of the data that have been sent to us have not proved very good. Heron's memoir on "Inheritance of the insane Tendency" will go to press as soon as I have the time to throw into shape his rough draft, I hope early next month. Miss Elderton's work on cousins is practically done; Miss Barrington's on inheritance of defective eyesight and the influence of home environment on eyesight (overcrowding, etc.) is nearly complete. I think this will form a good six months' work to start with. We shall then get on to the data for children from Manchester, Birmingham and now London.

You will be amused to know how general now is the use of your word *Eugenics*! I hear most respectable middle-class matrons saying, if children are weakly, "Ah, that was not a eugenic marriage!"

We are going two or three miles from Petworth for the Long Vacation. Have you made your plans? I hope you have been feeling quite well again and well over that unfortunate slip. Your lecture is doing much good. I expect mine will be out by the end of the month and we shall be able to get up quite a talk concerning Eugenics in the journals.

I want to ask you about the new rooms for the Laboratory. The College is ready to give two rooms next the new Biometric Laboratory, which will be open in October in the main buildings. Of course this would make matters much easier for me, and easier for the Eugenics folk, who have to come and see me, but you must let me know your views. I propose a social gathering of some kind, when the new laboratories are opened in October, to bring biometric and eugenic folk together, and to advertise the whole thing. Always yours affectionately, KARL PEARSON.

Has the new nurse appointment proved a success?

42, RUTLAND GATE, S.W. *July* 12, 1907.

MY DEAR KARL PEARSON, Thanks for Palin Elderton's letter which I return. My domestic servants' insurance is through the X. Society, and absolves me from troubles like yours, for I have to pay *retrospectively* at the close of each year for extra servants. Leonard Darwin, who is in touch with politicians, has again urged me to ask you to offer "hereditary" evidence before the Poor Law Royal Commission. He fears that the subject will otherwise be wholly ignored in what is likely to become the basis of legislation for many years to come. I suppose the point is to afford evidence: (1) that the undesirables contribute largely to the *naturally* undesirable portion of the population, (2) that natural undesirability is a *fact*, (3) that various forms of charity unnecessarily *promote* the propagation of the less fit, and (4) that the methods of *restraining* it are important to consider. It seems to me that (2) ought to be "rubbed in," also (3). Can you not do something in this way by writing to the Secretary of the Poor Law Royal Commission enclosing a programme of what you are prepared to testify? According to Leonard Darwin the present occasion is a most important one to interfere in. Excuse *my* interference! Ever affectionately yours, FRANCIS GALTON.

Last night I wrote to accept one of the houses that have been inquired about. It is in Hindhead (Haslemere), and rejoices in the name of "Yaffles." "Yaffle" in the patois of the district means, I am told, a green woodpecker. The garden of the house adjoins that of Mrs Tyndall who has lived there ever since her husband died. She will be a nice neighbour. I go there on Aug. 1 for 6 weeks. There is a railway connection, I see, between Haslemere and Petworth, and the distance *direct* between the two places seems on Bradshaw's map to be only about 10 miles. So I trust we shall meet as heretofore not infrequently.

ROCK HOUSE, RIVER COMMON, PETWORTH, SUSSEX. *July*, 1907 [after the 12th].

MY DEAR FRANCIS GALTON, Your letter to hand. The "Yaffle" is a fairly common name for the green woodpecker in the South. I have heard him a good deal here and we always call him the "Yaffle." You will be open and high up, but I hope Mrs Tyndall has removed her Husband's big screens!

To-day I feel incapable of writing to any Secretary of a Royal Commission, for I am hit by a slight attack of 'flu which is I find just dying out here. As I have no fever, I think I may write to you safely. If Major Darwin would send me the Secretary's address I would write to him and forward a copy of the Eugenics lecture which will reach me shortly. I hate, however, suggesting myself to anybody; I suppose if they wanted my evidence they would ask for it.

. . . .

Affectionately, KARL PEARSON.

42, RUTLAND GATE, S.W. *July* 29, 1907.

MY DEAR KARL PEARSON, It *seems* long since I wrote, for I have had a most interesting stay in a large moated house in Suffolk. Clear water all round it, no smell, stables away from it, draw-bridges raised each night as they have been for some hundreds of years, etc. etc. It is Helmingham Hall, where the widow of Lord Tollemache lives. The son and heir is at a more important house in Cheshire. On Thursday I meet E. B. at Waterloo Station and we go down together to *Yaffles, Hindhead, Haslemere*, for 6 weeks.

Before leaving town I called at the Home Office to learn the address of the Poor Law Royal Commission. The porter wrote out the enclosed, R. G. Duff being the Secretary. I hope you will send them the copy of your lecture with the *passages marked* on which you think evidence ought to be taken, and which you are prepared to give. It is really an important crisis, as I am assured.

It will be very pleasant if we could occasionally meet, much as of old. *Yaffles*, if you could be persuaded to bicycle so far, is very prettily situated with a terraced garden and two out of door sheds,—in one or other of which I hope to spend much of the day. How are you all thriving? A word as to the outcome of your own trouble about your Father's health and the doctor's opinion would be very welcome. With kindest remembrances to your Wife,

Ever affectionately yours, FRANCIS GALTON.

YAFFLES, HINDHEAD, HASLEMERE, S.O. *August* 13, 1907.

MY DEAR KARL PEARSON, All goes well, and I expect that all will be re-established. I can't write details, but the more hopeful prospect seems destined to be fulfilled completely.

I began here with ill-luck. A second tumble at night, with some slight concussion and a considerable attack of sick headache. All this has passed happily away and I am quite well. This and what I had to learn as regards the first paragraph above, was the cause of my delaying so long to write.

This house is perfectly charming. The grounds cover 4 acres of hill side, and are partly wild, partly terraced, with seats everywhere and distant views. The house itself is a sort of bungalow, just large enough to hold us two and Eva's half-sister, Mrs Macintyre, who lives at Penang and is over for a short holiday, with her baby. She is an acquisition in more than one way. The house is beautifully clean and fresh, very artistic, and many shelves-full of excellent readable books. I have done but little owing to the above-mentioned reasons—in fact I dared not even read a line for two days.

Do tell me about yourself and yours. Petworth is within the reach of a long drive from here, and I see there is railway connection of a sort, but I fear roundabout and by different railways. I should be delighted to meet you anywhere not further than Petworth. Jonathan Hutchinson, the surgeon, has established quite a large museum in Haslemere, which forms a scientific centre. On Saturday he gave lunch to a School-Hygiene-Congress party and invited us. Preparations for 50 hungry people, and only 25 (including ourselves and his own party) came to the lunch. He has a medical museum in Gower St and is going to live there and catalogue it.

Ever affectionately yours, FRANCIS GALTON.

ROCK HOUSE, RIVER COMMON, PETWORTH, SUSSEX. *August* 26, 1907.

MY DEAR FRANCIS GALTON, It was a great pleasure to see you on Saturday, the more so as you looked so fit and bright. I cannot but think that the return to your old home habits has been good for you, and I feel sure that it will gladden Miss Biggs to be conscious of this also. I should have sent you a line on Sunday but I buckled to and got Heron's draft memoir off hand, and sent it for his consideration. It ought not to be long now before it is out, and I think it will produce some effect. I suggest that you see it in slip, as it is then quite easy to adopt criticisms and modifications and is far easier for you to read.

I hear from one of my folk, that Captain Hurst at the B.A. meeting asserted that in human eye colour, blue is a Mendelian unit, *all* the other shades forming the opposite allelomorph. That those cases we have shown in which two blue eyed parents have other coloured offspring are solely due to our not properly examining the colour of the eyes, and if we had done so a small amount of pigment, orange or brown, would always be found. It is the old Mendelian trick, if you may pick your individuals you can prove anything. It is of course perfectly true that if you take two blue eyed parents, both of whom come of blue eyed stock, you will get blue eyed children. The test of Mendelism lies in two blue eyed parents of other stock, always giving blue eyed offspring. But if they don't they will be dismissed as having a small but unrecognised amount of orange pigment! No doubt, if we settled beforehand the blueness of the eyes of both parents, and some of the children were and some were not blue eyed, we should be told that the wife's virtue was not beyond question and that she had had a fancy for a heterozygous paramour! That point has indeed already been suggested in this inquiry!

I have been rather pleased. In my Homotyposis paper I dealt with sweet-peas and felt pretty certain that they must be *cross*-fertilised, because of the numerical constants. Of course it looks commonsense from the blended forms one sees everywhere. But Darwin in *Cross and Self Fertilisation of Plants* strongly believes that in England they are not so. Now I have watched the whole process here. The bee works in a sort of frantic manner, pushes both flaps down and the pistil rises from its case, and usually he sweeps both sides of it with his hind legs. The bees I have seen have their belly and the whole of their hind legs covered with the pollen of the sweet-pea, and there is not the least doubt that there must be a great deal of cross fertilisation. Darwin speaks of the difficulty of access of the bee, but it is singular that with his great accuracy of observation he should have missed the simplicity of the whole thing. It is really rather striking to watch the bee at work. If you have any sweet-peas in that beautiful Yaffles garden, do try and confirm my observation. Affectionately yours, KARL PEARSON.

I am not sure that this bee is the ordinary hive bee; it looks a somewhat stouter insect, but of much the same type. I have not seen more than two working at the same time on a long row of sweet-peas, although there might be 5 or 6 at the same instant on a small lavender bush, but these bees would, I found, very quickly visit 20 or 30 flowers*.

YAFFLES, HINDHEAD, HASLEMERE, S.O. *August* 30, 1907.

MY DEAR KARL PEARSON, The caricature of you is uncommonly good, though of course not flattering. Even the upper part of the back is distinctive, but the remainder of the dwarfed body is not good. I will keep it, if you don't want it back.

Schuster's paper in the *Eugenics Laboratory Publications* reached me yesterday and very interesting it is. I will write to him. I shall be very glad to see Heron's paper "in slip."

About the sweet-peas, when I reared them all those years ago, I selected them on the advice of both Hooker and Darwin, and was assured also that in nursery gardens rows of peas of different colours were often planted side by side, and that no cross fertilisation was ever observed. But I have with my own eyes seen, as you have, bees (of some kind) visiting flowers in succession without, or with little, regard to their colours and supposed their visits to be innocuous, though why, I have never been able to understand. There are only a few sweet-peas here, at the bottom of the garden, and no *hive* bees anywhere about, but bees of alien kinds, so I cannot easily repeat your observation in respect to hive bees.

It was a very great pleasure to see you last Saturday, and to have a long talk. To-day, we drove to Linchmere and saw in the church a brass tablet to Salvin (the S. American botanist, who had a property near here). You may recollect him at the meetings of the R. S. Evolution Committee. He was usually reticent but very helpful on occasions and always a thorough gentleman. Ever affectionately yours, FRANCIS GALTON.

On and after Thursday Sept. 12—QUEDLEY, SHOTTERMILL, HASLEMERE.
YAFFLES. *Sept.* 8, 1907.

MY DEAR KARL PEARSON, I have rented the above house for 2 months certain, with option of continuing through the winter. It is pretty and has 1¾ acres lawn and garden with a well-warmed greenhouse into which the drawing-room opens. So I have a fair chance of pulling through the winter in it. What "Quedley" means, I don't yet know. I gather from a letter from Gifi that the new part of *Biometrika* is out and has been received in Rutland Gate. If so, it will soon reach me. I see that Schuster's article has attracted favourable newspaper notice. The enclosed (don't return it) is a good example.

All goes on quietly here. I have at last got into good working order a method of "lexiconising" silhouettes. I can't conceive why artists and anthropologists have never succeeded in sharply determining points of reference in the human features, when it is so easy to obtain them by the intersection of tangents. The enclosed (don't return it) shows my primary triangulation. The *C*, *N* and *F* (obtained by intersections) are closely approximate expressions for the tip of chin, of nose, and of "nasion" (to adopt the word you used). With a *small* repertory of descriptive symbols, I find it feasible to give a formula for any profile, whence a very respectable duplicate of it can easily be drawn. Types of races ought to be readily defined and compared

* See the present volume, pp. 6–7.

on this principle, but I have not yet attempted to do so. However, I have a book of racial portraits at home, which I will get here to experiment with. How do you all get on? When do you return to dear smoky London? Ever affectionately, FRANCIS GALTON.

Primary triangle
of a profile.

$CF = 100$ "cents."

Measurements are
all in cents.

CX, CY are the axes
for rectangular coords.

F.G. Sept. 8, 1907.

ROCK HOUSE, RIVER COMMON, PETWORTH, SUSSEX. *Sept.* 9, 1907.

MY DEAR FRANCIS GALTON, I am extremely glad to hear that you are going to try Haslemere for the winter and I hope most sincerely that it will prove a success. I fear "Quedley" is not Sussex dialect but personal to the owner. I hope it will be as sunny and bright as the "Yaffles." I have written to Sir Robert Parker asking him if he will let me lunch with him one day this week. He is, I know, in London on Wednesday, so that it will be towards the end of the week. May I come and see you afterwards? I would let you know the day. I should have to start back at 5, as I do not care to cycle after dark, but I should like to see you again before I get back to work. My Wife goes to Oxford on the 17th and we all go back on the 21st. I do not begin lecturing until the 1st, but I want if possible to get everybody arranged in their new quarters, and we shall hope to give our inaugural tea-party when you are again in Town. I am probably in for two controversies; one in the *British Medical* on the inheritance of the tuberculous diathesis, and one (possibly, in *Nature*) on the correlation of stellar characters. This, in reply to attacks at the B.A.

I think your profile scheme is quite good, only you must measure and find names for the *angles* of your fundamental triangle. Would it be also worth while taking the projection on the median sagittal plane of the centre of the auricular orifice, or of some point on ear? This reminds me that I have had some idea of measuring such of the University College students as will consent thereto. If you thought well we could set up a profile-taker in the dark room with magnesium wire and sensitive paper and soon get a large number. What do you think?

I am glad to see the favourable notice of Schuster's paper. I think on the whole we must be well content with our "First Fellow."

I find there are two kinds of bees. I have captured specimens of both to-day. One sort certainly gathers honey, but never touches the pistil or pollen of the sweet-pea, the other is never content and does not leave the flower until he has swept the whole of the pollen from the stamens onto his belly. I am sending the two kinds of bees to be identified.

Affectionately, K. P.

QUEDLEY, HASLEMERE. *October* 2, 1907.

MY DEAR KARL PEARSON,   Enclosed I return Heron's paper, with suggested verbal corrections in pencil. The proposal (even *if* it be not wholly his) of a General Register of the Insane deserves all emphasis and would be a good subject for the Eugenics Office, as such, to "agitate" about. One first step would be in writing an article (by Heron himself, or by someone else) to appear shortly after the publication of his memoir. I wrote to him and have received this morning a clear ground-plan of the new rooms in University College. I am so glad that all is now so compactly under your wing. Shall you have an opening *tea-party*, and when? I should like to come up on purpose, but doubt its wisdom as I feel that fiend Bronchitis is hiding just round the corner, ready for a spring. I have indeed had premonitory symptoms already. Still, if the weather continues fine, I would come up on purpose.

As yet I have not met either of your two friends, Mr Justice Parker or Nettleship. But cards have been interchanged. This place grows upon me and seems more suitable for the winter than any other that I know of. All goes well here. *At least* one doctor is said to be so good that it seems a waste of opportunity not to be ill while here, and to send for him! You will be head and ears over in work, so I will not write more now.

Affectionately yours, FRANCIS GALTON.

QUEDLEY, HASLEMERE. *October* 10, 1907.

MY DEAR KARL PEARSON,   I hope the delay of two days in my reply has not inconvenienced you. I found it very difficult to put some of ——'s scarcely intelligible sentences into readable shape. But it is a most interesting paper and the diagram is very striking to the eye.

I have, thanks to you, seen much of Mr Justice Parker and of Nettleship, both at their several houses and at mine. You have too much now in hand to think of new things, but a suggestion thrown out in conversation with the former deserves bearing in mind, namely a discussion of the *parentage* of the *unemployed*, which *may* prove to be of degenerate quality. It does not seem very difficult to carry it out. Another topic which I discussed with Sir Alfred Lyall and an Indian friend of his, is the feasibility of testing *promise and performance* in the Indian Civil Service, where appointments go very much by merit. I worked at this some (? 20 or more) years ago and have lots of MS., but I then published nothing, because the data were too few. Now, they are fairly abundant. I should like to talk this over with you some time. About the opening *tea* at the new rooms, I am quite at your service and would come up for it (and for other things, for two or three nights) whenever you may appoint. There is always uncertainty as to my impending bronchitis, but I will come if I can and you must excuse me if I fail. Will it be an evening conversazione or a late afternoon tea? It is growing autumnal here, but very little of the foliage has yet changed in tint. Where it has changed the effect is beautiful. Haslemere continues to commend itself as a winter residence. I am very grieved at your domestic anxiety, which *must* increase rather than diminish. You have all my sympathy.

Ever affectionately yours, FRANCIS GALTON.

[HAMPSTEAD.] *October* 16, 1907.

MY DEAR FRANCIS GALTON,   I feel I must write you a line to tell you that my Father died at two o'clock yesterday. The operation had been delayed too long and his whole system was so weakened that the slight shock of the operation was more than he could stand. I feel my energy will be for some time very fully taken up with executorship and trustee work, although I shall have valuable aid in my co-executor, Sir Robert Parker. But it is a difficult and lengthy business to close up a home. There is no one now naturally to continue it, as my Sister and I already have made our own homes and environments. It is difficult to disperse all the hundred and one things one has known from one's childhood; almost sacrilege to sell them, and yet nobody wants the bulk of them.

My Father was a man of immense will and endless power of work, with a wonderful physique. A cripple from a fall from his pony when a boy, he was yet a splendid shot and a good fly fisher, striding over the fields gun in one hand and stick in the other in a way which out-tired me as a boy. Then he would be up at 4.30 to prepare his briefs, take a standing breakfast at 9, and rush into his brougham; back at 7 o'clock, dinner over at 8, he was in bed at 9, and so for month on month, we only saw him at these hurried meals, when speaking was scarce allowed.

Even in the vacation time he would take his sports in the same way ; 6 a.m. was the right time to be out on the river and the day went on till dusk, because the fish bit best after sunset! Long days for a fidgetty boy, who was only allowed to use his rod when there were no fish to be frightened! Even these last 15 years when he has been working on Domesday Book, accumulating immense piles of MS., my Father on my entry would sometimes point to a chair and forget me if I stayed. An iron man with boundless working power, who never asked a favour in his life, and never really got on because he forgot to respect any man's prejudices, and never knew when he was beaten. I learnt many things from him, and know that I owe much to him physically and mentally. But we were too alike to be wholly sympathetic. He thought my science folly and I thought his law narrowing,—the view of both of us being due to an inherited want of perspective in the stock! Still he was a man of character and strength. I never saw him give in charity, yet I know now from his papers that more than one of his relatives owe to him their success in life—"Loan barred by the Statute of Limitations" is the quaint way in which he docketed the documents relating to the expenses of a college education for a nephew, or the starting in life of a brother! I am rambling on when I ought to be thinking of other things, but just now all other matters seem small, when one is taking stock of a completed life, which no other has seen or can now see so closely, nay, who seeing would judge to be at all significant.

Affectionately, KARL PEARSON.

[HAMPSTEAD.] *November 23, 1907.*

MY DEAR FRANCIS GALTON, I have been wishing much to write a line to you, but I have been very pressed, and troubled also with a severe cold on my chest. However I must send you one little line now. First, Schuster was with me on Wednesday. He is arranging for an Anthropometric Laboratory for the Oxford students and came up to ask about instruments and other points. I had a sort of half idea that your old instruments went to Oxford from South Kensington. If this were so, can you tell me who has charge of them? It might save purchasing certain things. Schuster seemed to think that there were possibilities in Oxford, which wanted pressing now that we had sown the seed of Eugenics there.

Miss Elderton has been away with a bad cold. The radiators in the rooms have proved incapable of doing their work and we have had great difficulties. So bad indeed that Dr Alice Lee has resigned, which will be a great loss to me, although she had recently been a little difficult to work with. I know only one person her equal in rapid and correct calculation and that is Miss Elderton; we *must* keep the latter at the Eugenics Laboratory, if we can. I passed her memoir for press finally to-day. She has worked out about 60 correlation coefficients for Uncles and Aunts and this mass of material shows that the intensity of resemblance is much the same as for Cousins. I have advised her to write a second paper on Uncles and Aunts, and discuss the whole point as to this paradox. She has put in a reference to this in the Cousin paper.

I hope Haslemere is proving a good winter resort, and that you are not so low down as to get the valley frosts. I think I told you, did I not, that I paid £1000 into the Oxford University Chest for the Weldon Memorial recently? I have asked for copies of the final scheme to send to the donors. Affectionately, K. P.

QUEDLEY, HASLEMERE. *November 26, 1907.*

MY DEAR KARL PEARSON, I was becoming anxious through not hearing from you, knowing that you were not well and are overworked. This is bad weather for your cold and for that of Miss Elderton. I grieve that you are losing Dr Alice Lee. It is most desirable that the paradox of almost identical intensity of kinship to an uncle and to an uncle's son should be faced, as you propose, by Miss Elderton, and I am very glad that the intention is referred to in her Cousin paper.

As regards the S. Kensington instruments I gave them all to Professor Thomson for use at Oxford, in the Cavendish [? Anatomical] Laboratory. Schuster would do good work if he could show the exact *importance* of each measurement proposed and could arrange a system that is of real and proved value and at the same time simple. Correlation would play a large part in devising this, for if A is closely correlated with B and C, it may be sufficient (under limitations of time, trouble and expense) to observe A and to neglect B and C. I look forward to receiving a copy of the final scheme for the Weldon Memorial and am very glad that so substantial a sum as £1000 is available. I wish I had "radiators," even poor ones, in this house, which is becoming cold notwithstanding many fires. A sharp winter would be felt severely in it.

Methuen's "literary adviser" has written to me a sweet and fetching letter for an auto-biography to be published by them. I am disposed to write it, for it will give daily occupation for some time and will revive many memories. So I am discussing with him a single volume nicely got up, on the half profits basis. Oddly enough a common friend to myself and Methuen (whom I do not yet know personally) was spending last Saturday to Monday here under an engagement to lunch on Sunday with Methuen; so I gave him the letter to show and talk about. Methuen proposes to call, but is now much invalided as the result of an operation last summer.

Ever affectionately, FRANCIS GALTON.

(*Antescript*. This is a "business" and not the personal letter which I want soon to write.)

THE GALTON EUGENICS LABORATORY, UNIVERSITY COLLEGE. *December* 1, 1907.

MY DEAR FRANCIS GALTON, I fully appreciate your point as to the facing what people will say about cousins being at least as alike as uncle and nephew. When Miss Elderton did the cousins' work, we had only my eye-colour work (based on your material) to compare with it. For 8 series of eye-colour correlations each embracing about 1200 cases we found a mean cor-relation value ·265. Miss Elderton has worked out 32 series from my General Family Records—for uncles and nephews, etc. For Health and Intelligence we get mean of 16 series of about 1000 each, ·272, practically the same as for my eye-colour work. Temper and Success which involve more doubtful judgments give about ·20. You will see that these are comparable with Miss Elderton's cousin resemblance of ·267. You ask how does it come about? Frankly I can't say. But I want to draw your attention to another point. When you first started this correlation work, you expected parental correlation to be $\frac{1}{3}$ and brothers' to be $\frac{2}{3}$. My view in the "Law of Ancestral Heredity" paper, *pure theory*, was $\frac{1}{3}$ and ·4. These values would also arise from simple Mendelism. Now you see *I* still thought the brothers would be *more alike* than parent and off-spring, because the other parent would disturb the relation of one parent to the child; just as we might suppose the uncle's wife would do. But when we have worked out long series of parental correlations and fraternal correlations, what is the result? Why that it is very difficult to show that they differ from equality. I think my Family Measurements were very reliable and yet for long series the parental correlation came ·46 and the fraternal ·50, and probably this difference was due to comparing different *generations* of adults, i.e. father and son do not live in the same environment as two brothers. My position at present is that we have to find out the correlations from observation and when they are definitely known, turn back to theory. *Alternate inheritance* would, perhaps, give fraternal = parental correlation and would, I think, make cousins and uncle and nephew equal. It is, I think, in some such "determinant" direction that we must look for light in this matter. I will add a note to Miss E.'s paper.

Affectionately, K. PEARSON.

QUEDLEY, HASLEMERE. *December* 20, 1907.

MY DEAR KARL PEARSON, How nice Miss Elderton's paper looks. The Laboratory publications make a most respectable show. I am very glad you inserted the paragraph you did, at the end, showing that the paradoxical result of cousinly likeness being the same as avuncular, has not been unnoticed. The more I think about it the more amazed I am that an uncle's wife or an aunt's husband should exercise no appreciable effect. Facts are of course the supreme authority, but it is hard to bow before them here.... Ever affectionately, FRANCIS GALTON.

QUEDLEY, HASLEMERE. *December* 28, 1907.

MY DEAR KARL PEARSON, The *Tribune* article is clever and only too true. I have been long desiring to start some movement to raise the deplorably low standard of scientific literature and have corresponded about it privately. Sir Archibald Geikie, whose family are now settled here, is still more emphatic than myself, and we had a good talk yesterday. We both belong to the R. Soc. of Literature, and I hope to induce it to take the "improvement in style of current scientific literature" as a serious duty. A man ought to feel as ashamed of publishing a slovenly memoir as he would of appearing at a public ceremony dirty and ill-dressed. But it is not only of an aesthetic but of a matter-of-fact trouble one has to complain—viz. of the length of time that is wasted by the reader in trying to understand what ought to be expressed by more vivid language, simpler expressions and more logical arrangement. I am now writing on this very subject to Sir Edward Brabrook who is the chief working authority of the R. Soc. of Literature, to enlist his interest and to get advice. Geikie and I did form a provisional scheme of action.

This house, Quedley, really is not cold. Nettleship, who was here yesterday, and whom I asked, found no fault with its situation. The valley fogs do not as yet reach it, while I hear great complaints of cold and fog at Hindhead. In fact I really think I have fallen upon the most suitable house in the whole place, for my particular needs.

I am now busy, as long as I can work, day by day, over my "Reminiscences." It is curious how the sense of "past" disappears. All my life from 5 years to 85 is beginning to seem to me "present," like a picture on the wall. Ever affectionately yours, FRANCIS GALTON.

7, WELL ROAD, HAMPSTEAD, N.W. *December* 30, 1907.

MY DEAR FRANCIS GALTON, I am very glad my alarm about the cold at Quedley is false. I certainly did not mean to disturb you needlessly. It was only my short experience of the valley some way above Shottermill where we had a house for four weeks.

I should rather like to talk over the point of scientific literature with you, because I think there is danger of two distinct factors being confused. In the first place every paper ought to be written in lucid English. With this I am the more in sympathy, because I realise to the full my own difficulties in this matter. We want far more essay writing from the science student, although this must not be driven to the Oxford extent of making the discovery of fitting words the main occupation of the student. On the other hand every science must have its special terminology, and its symbolism and short-hand. These can be interpreted into long-hand and simple English in popular lectures and reviews, but in the scientific memoir written for a scientificly educated public the terminology and short-hand of the special branch of science concerned must be preserved for the brevity and lucidity they provide. You might, I think, as well demand of a mathematician a definition and explanation of dy/dx in a *Phil. Trans.* paper as ask in a scientific memoir on heredity for an explanation of the fundamental equations

$$(DD) \times (RR) = 2(DR), \quad (DR) \times (DR) = (DD) + 2(DR) + (RR)$$

of Mendelism. This symbolism is now known and accepted by all students of heredity whether they believe in Mendelian theory or not. Similarly such terms as "somatic" and "gametic" are to be found in every biological textbook. When therefore the *Tribune* cites such things as these and calls them "jargon," it is merely stating that its writer was incompetent to review the memoir because he was ignorant of the terminology of the branch of science he was discussing.

This is quite apart from the possible want of lucidity of the English, or from any demand for a popular exposition of the results reached by more elaborate memoirs. These may be desiderata, but they are not to be confused with a mere absence of scientific terminology: and I think we have now reached an epoch when the popular exposition of heredity should be taken more fully into consideration.

In February it will be a year since our régime began, and the appointments of Mr Heron and Miss Elderton will come up for consideration, as well as my own relations to the Laboratory. I feel my own limitations very keenly, and it might well be that other supervision would give the scheme more go and a more popular character. I need hardly say that I am ready to fall in entirely with your views, either to make way for a man of more leisure and activity, perhaps more in touch with the outside world, or to go on as we have been doing for one year more. As for the Galton Fellow and Scholar, I think we ought to give them some notion as to the future. The Fellow has done good work, but has not at present quite as much initiative as I shall look for later; the Scholar has much impressed me, and is even more able than I anticipated. Taking the difficulty of finding new and efficient workers, I think we shall not readily find better instruments even if we agree that they need a more active guide. If you agree, there ought to be some report to the University and perhaps a meeting of your Committee. I will very readily draft something, if you will quite frankly send your views on the immediate future. Whatever is done now ought to be done so as to terminate definitely in February, 1909. I think the present people are too good for one year only of work, but they ought to understand that you may want to remodel the Laboratory scheme in 1909.

Have you considered the possibility of resuming the reins yourself this year? I only came in default of any obviously better person to supply your place, and I am only a *locum tenens* ready to move on when you say the word. Affectionately, KARL PEARSON.

I think Galton's opinion of the falling off in style of scientific, especially R.S., memoirs was very well founded. He possibly did not realise some of the factors that had contributed to it. In the early history of the Royal Society the responsibility for the issue of papers seems to have rested with the Secretary (or Secretaries), and, I think, some of the feeling of this responsibility for editing lasted up to the days of Sir George G. Stokes, who must have spent endless time and energy over the verbal and critical emendation of authors' papers. Failing this editorial work much must depend on the printers' readers. My own—now fairly considerable—experience suggests that it is only at the University Presses of Cambridge and Oxford that one can be certain of the highest efficiency not only in proof-reading and suggestion, but in ensuring that corrections are properly made. The glory of a press depends as much on the general culture of its readers as on the beauty of its type. A second factor which I believe has largely escaped notice lies in the change of the class from which the writers of papers are now drawn. With the system of education as now developed the majority of men of science are springing from humbler and less cultured homes than formerly. Many of them have never passed through the literary training of public school and university, but have been " educated" in secondary schools and science laboratories, and have only exceptionally an appreciation of style, or any power of lucid expression. Add to this, and anyone who examines statistically the recent list of the fellows of the Royal Society will confirm the statement, the men of leisure and culture who occupy themselves with science, while formerly numerous, are now a vanishing minority; thus we see how it is that the hurriedly written papers of the modern professional scientists lack the lucidity of expression, sometimes the grammatical English, of the more leisurely savants of the middle of last century. Galton was keenly alive to the result, if possibly he had not studied fully the causes of the change. I think that Sir Archibald Geikie in the discussion which followed Galton's paper at the Royal Society of Literature came nearer to pointing out the inevitable evolution which has taken place in the scientific world. He said:

"It seems to me that no candid reader can compare the scientific memoirs published at the present day with those which appeared a hundred years ago without coming to the conclusion that in average literary quality the modern writings stand decidedly on a lower level than their predecessors, and that the deterioration in this respect is on the increase. The earlier papers were for the most part conceived in a broader spirit, arranged more logically, and expressed in a better style than those of to-day. They show their authors to have been generally men of culture, who would have shrunk with horror from the slipshod language now so prevalent.

"If it be asked what reason can be assigned for this change, various causes may be suggested. In former days, the number of men of science was comparatively small, and they belonged in no small measure to the leisured classes of the community. They were not constantly haunted by the fear of losing their claims to priority of discovery, if they did not at once publish what they had discovered. They were content to wait, sometimes for years, before committing their papers to the press. And no doubt the printing of their papers was likewise a leisurely process, during which opportunity was afforded for correction and improvement. But this quiet, old-fashioned procedure has been hustled out of existence by the more impatient habits and requirements of the present day. The struggle for priority is almost as keen as the struggle for existence." (*Trans. R. Soc. of Lit.* Vol. XXVIII, Part II, p. 10.)

Sir Archibald might have added that in many cases it *is* a struggle for existence, since the chance of appointments too often is made to depend in the case of young men rather on the quantity than the quality of their published papers.

(14) *Events and Correspondence of* 1908. The events of this year have been to some extent foreshadowed in the letters of 1907. We have seen that Galton was busy with two projects, namely (i) with an endeavour to improve the literary style of scientific memoirs, and (ii) with the writing of his volume of memories. There are three other matters to which we shall also refer; they are (iii) the proposal to found an association for promoting Eugenics—the Eugenics Education Society, (iv) his papers before this Society, when founded, and (v) the Darwin-Wallace celebration at the Linnean Society. We will take these in a somewhat different order, interpolating correspondence which may throw light on their origins.

(a) *On the Literary Style of Scientific Memoirs.*

QUEDLEY, HASLEMERE. *Jan.* 1, 1908.

MY DEAR KARL PEARSON, The first thing in this my first letter written in 1908, is to wish you and yours the happiest and most fruitful of New Years that it is reasonable to desire.

I think you have read more into my letter than it was intended to hold. We are fortunate in having Heron and Miss Elderton, and it would be natural to continue their appointments, unless *you*—and I understand that you do not—wish otherwise. I am sure *I* should be sorry to lose them. Then as to yourself, the idea of your ceasing to superintend that which you have built-up so powerfully on solid foundations, simply makes me shiver. Pray not a word or thought further about this !!

Now as to what I want the Royal Soc. Lit. to do. You and I are at one in respect to the necessity of strong action to put a stop to obscurity of expression, to bad grammar, and to faulty logical arrangement. The remaining question regards *technical language*. My own feeling is to restrict it so far that capable scientific men, who are familiar with *cognate* branches, shall be able to understand memoirs without difficulty. At present they are not able to do so without great labour. Here I have in view the publications of the Royal Society, which, and Geikie feels at least as strongly as myself, are faulty in this respect, besides being uncouthly and barbarously written. Heaven knows that I am only too willing to have my own faults of writing roughly corrected, and how much I feel indebted to the slashing of a friend, who kindly read the MS. of some of my early writings. He treated them ignominiously, I saw they deserved it and was grateful. You, I think, are more inclined to consider those memoirs which are addressed somewhat exclusively to specialists. But even here more caution seems required. A technical word does not quickly acquire the exact technical meaning it is intended to convey. Take your own useful expression "sib," which you apply to the children of the same parents. I see Skeat in his Dictionary defines it as nearly related, and he shows Gossip = God-sib, to be equivalent to God-parent. So when "sib" is used in your limited sense, the addition once for all in the same paper of the words "children of the same parents" would be helpful and prevent puzzling*. It is certainly well to *minimise* the use of technical words. The English language is a powerful weapon in skilful hands, and much more can be expressed briefly in it without technical language than is generally attempted. I heard last night from Brabrook and find

---

* I am puzzled by this paragraph. I had introduced the word "siblings" not "sibs" to cover a group of brothers and sisters regardless of sex and equivalent to the sense lost to modern English of "Geschwister," or "Søskende." "Sib" stands to me as an equivalent for kin, and I was somewhat vexed when Nettleship cut my "sibling" short into "sib." It would appear as if I must have sinned by somewhere using "sib." Speaking from memory, I should say that on the early occasions on which I used "sibling," I had defined the meaning I attached to the term.

that my letter to him has fallen aptly. They are very shortly going to consider seriously what to do. Complaints are so wide and loud, not by any means from non-scientific people only, and the Royal Soc. Lit. feels that criticism falls within its province. All goes well here. The Hope Pinkers lunched with us yesterday. He told me about the progress of the bust of Weldon and that you had seen it. Ever affectionately, FRANCIS GALTON.

7, WELL ROAD, HAMPSTEAD, N.W. *Jan.* 2, 1908.

MY DEAR FRANCIS GALTON, I am very busy to-day, but I must send you some few lines in reply to your very kind note. I reciprocate heartily your wishes for the New Year, and these include my desire for the success of your proposed attack on the citadel!

I am quite ready to continue superintending the Eugenics folk, and you must not suppose I am not interested in the work. All the same I am quite prepared to surrender the reins whenever you feel another man would achieve more in the particular directions you have most at heart. We are both ultimately of Quaker stock, and I want you to talk quite frankly when the time comes, remembering that I shall not be hurt by any decision you may take. I have so much in hand, that to close one phase of my work only means more progress in other phases. I should only feel sad if something were to happen which closed *all* phases of my work. Why, if Eugenics and even Biometry were closed down, I should turn to Astronomy with all my energy and time; I know how badly statistical knowledge is needed for problems therein! I will send you a little formal note shortly as to the re-appointment of the "Eugenicians" (that word shall not go further!), which you can forward with any further comment to the University authorities. As a mathematician I must emphasise my view that symbolism is an enormous gain to any branch of science. Just think where we stood in statistics without the theory of total and partial correlation coefficients! But how in the world can we express in any brief and decent English the formula

$$\rho_{12.3} = (r_{12} - r_{13}r_{23})/\sqrt{(1 - r_{13}^2)(1 - r_{23}^2)},$$

for the influence of the mother (2) on the son (1) for a character constant in the father (3)?

I think you are wholly right to demand good grammar and *clear* expression, but I believe your movement will fail in these demands, if you attempt to drag terminology and symbolism into the fight. My ideal scientist in this respect was Clifford; every educated man can follow his popular addresses, yet how few but mathematicians his scientific memoirs. Discovery and popularisation are distinct aspects of scientific work. They were excellently combined in Clifford and Huxley, and largely in Darwin; but you must not expect to find this combination frequent nowadays. Your battle will be the easier, if you avoid arousing the wrath of the specialist in this respect. You have him in a cleft in the matter of English, but I fear you court failure, if you assert that the average man of science ought to be able to follow all the specialist papers in the *Phil. Trans.* If the terms accepted by every student of a specialised branch of science and the whole of its symbolism—its "short-hand"—are to be classed as jargon, and given short shrift, I sadly fear the Royal Society of Literature will find itself prostrate, Don Quixote-like, before the windmill! Affectionately, K. P.

How I shall rejoice to see the "Reminiscences"!

Extract from a letter of Francis Galton.

QUEDLEY, HASLEMERE. *Jan.* 25, 1908.

...The same morning that brought your reply to my letter, brought also the typed copy of your Report from the University of London, which I signed as approving. What a very good report you have made! I wish I could see any glimmer of light in the cousinal = avuncular correlation. It seems almost equivalent to fraternal = nepotal correlation, and quite incredible *a priori.*

7, WELL ROAD, HAMPSTEAD, N.W. *Jan.* 27, 1908.

MY DEAR FRANCIS GALTON, Being kept to the house and sofa to-day—not 'flu, or I would not write to you—I have some chance of getting letters off hand. Many thanks for your kind and helpful letter of this morning. There are now some 40 to 50 avuncular correlations worked out and they fully confirm the view that the relationship of cousins is as high as that of uncle and nephew. There are several points that need to be thought out carefully. The cousins are generally of the same generation and approximate age ; the uncle and nephew belong to different generations and may be of considerably different ages. But if anything the avuncular correlation is less than the cousinal, and accordingly I am not sure that the age and environmental differences would do more than equalise their values. Again we should expect brothers to be more alike than parent and offspring, but the fraternal correlation is only very slightly greater than the parental, and this again is due possibly to the age and nurture influences being more effective in the latter case *. As an illustration of what might happen, let us adopt as hypothesis an alternative inheritance in which $\frac{1}{2}$ the offspring follow one parent and $\frac{1}{2}$ the other. In this case 50 $^{\circ}/_{\circ}$ of the offspring are like a given parent, but only $33\frac{1}{3}$ $^{\circ}/_{\circ}$ of the brothers are like a given brother. Thus the parent has *greater* resemblance to his offspring than the brother to his brethren. Now let us look at the grandchildren of a pair, A and B, on the assumption of this alternate inheritance :

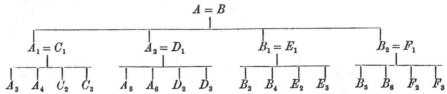

With regard to the original grandparents, the 16 grandchildren are either like one or other of them, A or B, or unlike them, taking after their daughters or sons-in-law, C_1, D_1, E_1, or F_1. Thus 25 $^{\circ}/_{\circ}$ of the grandchildren are like a given grandparent. Now consider an A_1 uncle, he has 12 nephews or nieces and 2 of these are like him, i.e. 16·6 $^{\circ}/_{\circ}$. Each individual cousin like A_3 has two out of 12 cousins like himself, again 16·6 $^{\circ}/_{\circ}$. It would thus appear that on such a theory we should have as great a resemblance between cousins as between uncle and nephew. Now I don't suggest that this scheme is actually at all representative of what takes place, but it seems to me to indicate that we can invent schemes in which it does not follow that uncle and nephew have a greater measure of resemblance than cousin and cousin, nor brother and brother a greater measure than parent and offspring. We must first observe and obtain our correlations and then endeavour to interpret them. Affectionately, K. P.

The divergence of view between Francis Galton and myself with regard to the use of technical terms is well illustrated in the following letter. I had sent him a paper in proof which was shortly to appear in *Biometrika*. Of this he wrote :

QUEDLEY, HASLEMERE. *March* 8, 1908.

MY DEAR KARL PEARSON, I would strongly urge a footnote to the first page † headed— "Technical words used," including *Chromomeres*, *Chromosomes*, *Determinants*, *Mytosis*, also even *Cytology*, *Somatic* and *Zygote*, with definitions of each. *Allogene* might be dismissed with the remark "explained in text." Thinking of the men who ought to read the memoir with interest, —Yule, MacMahon, G. Darwin, Burberry, etc.,—there is hardly one who would know the meaning of these words, or would care to read the memoir unless they were first defined. This or some analogous plan would often be a *great help* to readers of *Biometrika* articles. It is a most interesting investigation of yours. I had long had a vague idea that something of the sort was needed, but could not phrase it satisfactorily to myself. You must indeed feel the void left by Weldon.

* Galton's argument was that in the case of cousins (sons of two brothers) there were *two* wives, the cousins' mothers producing variability, whereas in the case of uncle and nephew there was only *one* mother, the sister-in-law of the uncle, to be considered. So in the case of two brothers, we might argue there is no source of difference in descent, but in father and son the mother comes in as a cause of additional variability.

† This refers to the proofs of a memoir "On a Mathematical Theory of Determinantal Inheritance from Suggestions and Notes of the late W. F. R. Weldon," ultimately published in *Biometrika*, Vol. VI, pp. 80–93.

It is good news that you have taken "Moorcroft" for Easter. It will tempt me to stay on longer here than I had in my mind. I am glad the Eugenics Education Society's meeting was hopeful. Crichton-Browne may make a useful president, but he has many irons in the fire. However it is all in his way, and if he is hopeful about it, he will throw energy in. I wish I could see your show at the University College soirée. My book is nearly finished in *draft*, and is typed, but much has yet to be done to it, in verificating [*sic*] and the like, which will be troublesome. May you have a healthy relief from your excessive work here in Hindhead!

Affectionately yours, FRANCIS GALTON.

QUEDLEY, HASLEMERE. *March 16, 1908.*

MY DEAR KARL PEARSON, In reply to your card asking me for something to exhibit at the U.C. soirée, I have thought of an effective, yet somewhat absurd thing. But I have failed to get it. It is a *Punch* cartoon, published I fancy in the early '70's, of a weedy nobleman addressing his prize bull:

Nobleman—By Jove, you are a fine fellow!
Bull—So you would have been, my Lord, if they had taken as much pains about your ancestors, as you did about mine.

I wrote to *Punch* to make inquiries, but they have not succeeded in identifying the picture. It would have been a capital thing to frame and to let lie among other exhibits. I should have been much disposed towards utilising it in some way farther on my own account. I cannot think of anything else suitable. Your Tables of the Coefficients of Hereditary Resemblance ought to be shown somewhere. A model of the old kind but differently arranged, like this perhaps, would be effective. [Here is inserted a rough drawing of a geniometer without figures (see our p. 30 and Plate I) working by aid of a lever to indicate the average regression of an individual on various ancestors.] Heron might devise one, say 2 ft. high, to stand on the table, and to be *worked there and explained*. If so, it ought to be *rough*. People would understand it quicker.

I am reading J. Arthur Thomson's new book on Heredity. The first part seems forcible and good. I had no idea that there was so much to be said about Acquired Faculties. I am curious to get on with it, but am obliged to be slow, and am now just at Mendel. By the way I find that I had the honour of being born in the same year, 1822, as he was. All goes on well here. I trust that "Moorcroft" will be a great success and no "April 1st" venture.

Ever affectionately, FRANCIS GALTON.

7, WELL ROAD, HAMPSTEAD, N.W. *March 26, 1908.*

MY DEAR FRANCIS GALTON, I am hoping to see you so soon that I should not write, were it not to tell you that I may be rather later in coming to Haslemere than I intended, and somewhat more inclined to be discontented. My youngest child has got whooping cough, so our party must be broken up. My Wife stays to look after her, and my Sisters-in-law come down with two of my children on Saturday to Moorcroft. I hope to get down early next week. I shall stay here as long as my Wife will allow, as I am not very easy about my bairn. She does not take these sort of things lightly, and I dislike whooping cough more than most diseases for its sequelae.

We have had a busy day, or rather three days. Lord Rosebery—the Chancellor—came to open the new wing to-day and walked round our new rooms. I showed him our skulls and the Eugenics Laboratory. He said: "Now how do you pronounce that word? I shall call it Eughennics," i.e with a hard *g* and a short *e*. And so he did in his speech afterwards! Then to lead him back to his past I showed him hair from mane, tail and flanks of nearly 100 chestnut horses. But he looked solemn and said: "Ah, Mr Gladstone had a great interest in chestnut horses, owing to the coloration of the Homeric steeds." In his speech later he paid you and your Laboratory quite a pretty compliment. We had many guests, but whether there were any worth showing things to is another question.

You would see that —— has been convicted of an indecent assault. The whole thing is so improbable, and sounds so impossible that we must wait for the appeal. But it must at present be a bad blow for the Eugenics Education Society. He was giving six lectures on Eugenics! Luckily that word has not been mentioned and I hope may not be, and I can't think this charge can be true*. I don't like several of these Committee-men, but this appears wildly unlikely. If it were proved, I should think the Society would go to pieces, but it would also be bad for us, if the word Eugenics were to get smirched in the beginning in this way.

Affectionately, KARL PEARSON.

QUEDLEY, HASLEMERE. *March* 27, 1908.

MY DEAR KARL PEARSON, I long to see you again, and hope that your mind will soon be easy about the whooping cough. Poor child! I have taken this house up to April 14th, so shall well overlap your stay here.

——! what a name that man has. It is enough in itself to make ridicule out of Eugenics. I know nothing more about the accusation yet than you have told me.

The *Times* gives a cheering account of University College and Lord Rosebery's speech. Of course the *g* in Eugenics is properly hard, but we say it soft in Genesis, Genus, Generation, etc., even in Prince Eugène. Ever affectionately, FRANCIS GALTON.

We tried to tempt both Heron and Miss Elderton to come here for this week end, but both happen to be engaged. Au revoir!

BIOMETRIC LABORATORY, UNIVERSITY COLLEGE, GOWER STREET, W.C.
April 28, 1908.

MY DEAR FRANCIS GALTON, I am back and at work! I hope the paper will be a success and that there will be some interesting discussion. I am *wholly* with you as to the great need of condensation, of improvement of English and of style. I am not wholly with you as to use of scientific terms, or as to the possibility of making all scientific papers intelligible to the educated, but not specialised reader. I hope that the former can be pressed without unnecessarily attacking the extensive use of terminology....

Has the map scheme progressed at all? Let me know if I can do anything further by aid of photography or otherwise. Yours, always affectionately, KARL PEARSON.

Think of a good name "Thesaurus rerum ad hereditatem pertinentium," "Thesaurus facultatum humanarum," or what†?

Galton's paper was read by Mr Pember—he was not able himself to deliver it—on April 29, 1908, before the Royal Society of Literature‡; it is entitled: "Suggestions for improving the Literary Style of Scientific Memoirs." In my opinion it is of more value from the standpoint of the biographer, than from any influence it had, or alas! is likely to have, on "the simplicity of language, clearness of expression or the logical arrangement" of scientific memoirs. Galton's remedies were: (i) That the Councils of Scientific Societies should not be left in the dark as to the goodness or badness from the literary standpoint of the memoirs they are asked to publish, and accordingly should directly ask the referees of papers whether they consider the memoir referred to them (1) clearly expressed, (2) free from superfluous technical terms, (3) orderly in arrangement, (4) of appropriate length, (5) if it introduces any new terms (to be cited) has used necessary and appropriate words, and (6) generally has an adequate literary style. This is to suppose that the referees will be men of sound literary taste, whereas in nine cases out of ten they would be selected for their specialist knowledge, and the barbarous would sit in judgment on the barbarian. (ii) That in order that scientific societies might be

---

* The conviction at the Police Court was quashed on appeal to Sessions.

† This has reference to *The Treasury of Human Inheritance*, the prospectus and materials of which were then being prepared.

‡ *Transactions*, Vol. XXVIII, Part II, pp. 1–8.

made to realise the occurrence of literary faults in the memoirs they publish, occasional articles might be issued "containing a selection of passages that are conspicuous for short-comings."

I must confess that (ii) seems to me a method more likely to produce effect than (i), and it might still be worth the combined efforts of a stylist and a natural philosopher, could they meet—after a satisfactory dinner—on the common ground, like Galton and Spencer, of the "old smoking-room of the Athenaeum*."

Galton was far too modest to pose as a literary critic. Of himself he writes :

"I am far too sensible of my own grave deficiencies to assume that position. But a man need not be a cobbler in order to know when his shoe pinches. My standpoint is merely that I find many scientific memoirs difficult to understand owing to the bad style in which they are written, and that I am conscious of a rare relief when one of an opposite quality comes to my hand." (p. 2.)

Galton does not give any actual illustrations of bad grammar and faulty syntax; probably he considered that to do so was to pillory individuals, where the whole herd was to blame. When he passes from such errors to other literary defects he does cite a couple of cases, i.e. the contrasted terminations of the two Mendelian terms *dominant* and *recessive* (which should be *recedent*), implying a distinction which does not exist, and the use of such words as "Dimethylbutanetricarboxylate" by modern chemists.

"It is of course understood that these are what have been termed 'portmanteau' words, in which a great deal of meaning is packed, but they are overlarge even for portmanteaux; they might more justly be likened to Saratoga trunks, or to furniture vans." (p. 4.)

The chemists certainly do seem to be rather lacking in imagination, but it would be impossible to make any suggestion to them without a very full understanding of their needs. As to the Mendelian term "recessive," the fault, as far as English is concerned, lies with those biologists who first translated Mendel's papers. It was the discovery of a fit English equivalent, not the invention of a new scientific term†.

Galton then quotes his favourite English poet Tennyson to show how much power there is in the English tongue to express clear ideas in words of few syllables.

"Long English words and circuitous expressions are a nuisance to readers and convey the idea that the writer had not that firm grasp of his subject which everyone ought to have before he takes up his pen." (p. 4.)

But is not the real problem a harder one than Galton admits ? The whole force of the poet's lines lies, not in clear cut definition of the words used, but in their linked atmospheres; it is just the width of meaning, the long train

---

* Perhaps a still more effective method, which did not occur to Galton, would have been to have drawn up a petition to the Council of the Royal Society, signed by as many Fellows as possible, drawing attention to the literary quality of scientific memoirs. Probably every Fellow would have signed, not wishing to be thought a *vir obscurus*.

† Mendel actually uses "dominirend" and "recessiv." I can find no previous history of the latter word in German, nor has that language a form like "recedent."

of memories and associations, which enables us to see the picture before us. Take one of Galton's quotations :

> "One show'd an iron coast and angry waves.
> You seem'd to hear them climb and fall
> And roar rock-thwarted under bellowing caves,
> Beneath the windy wall."

All the adjectives are used in figurative senses, and the beauty of the passage lies not in the use of clear and narrowly defined terms, but in the atmospheres which experience and usage have attached to the words in the memory of the reader. It is precisely these atmospheres which form the staple of the poet's craft. They are a grave danger to the scientist, and he strives to meet them by coining new words with stringently limited meaning, or, less advantageously, using old ones in a new, narrowly defined sense. Every time a really great poet uses a word he enlarges its atmosphere, while the object of the scientist is—at any rate for the time being—to circumscribe a word's atmosphere ; he can often achieve his end by adopting little used words*. I would not weaken by a jot Galton's criticism of bad grammar, careless writing, or sheer pedantry in terminology, only I do not believe it feasible to write scientific memoirs with simple English words like Tennyson used in his *Palace of Art*. As an editor and teacher I agree with Galton that

> "The preliminary culture of students of science seems usually to have been very imperfect";

and again :

> "The comparative rarity among the English of a keen sense of the difference between good and bad literary style is a great obstacle to the reform I desire. It is especially noticeable among the younger scientific men, whose education has been over-specialised and little concerned with the 'Humanities.' The literary sense is far more developed in France, where a slovenly paper ranks with a disorderly dress as a sign of low breeding." (pp. 5-6.)

I would have every postgraduate training in a laboratory for research write at least a monthly essay on a topic bearing on his branch of science. Yet grant all this, and still I feel that it was not only the "slovenly papers" which agitated Galton. Unconsciously behind it was the importance he felt of keeping abreast with the half-dozen branches of knowledge, in the early nurture of which he had taken part. His paper is the swan's song of the last of the great Victorian leaders in science. In his youth he had followed and contributed to the early growth of Anthropology, Meteorology, Evolutionary Biology, Genetics, the Theory of Statistics, and Psychology; but these sciences had outgrown their infancy, had become highly specialised, and teemed with new terms with which he could not keep in touch. It would have been a very great task for a younger man ; for the octogenarian, however outstanding his intellect, the task was impossible. Galton was, perhaps, over-inclined to attribute this incapacity to follow, as he longed to do, all new developments in half-a-dozen sciences to the obscure use of language or to the introduction

---

* Thus "conjugation" is a better word than "mating"; "dominance" than "mastery"; "probability" than "chance"; "evolution" than "unrolling"—the simple English words before scientific adoption would have too wide customary atmospheres.

of what he held to be unnecessary mathematics into discussions where he felt certain elementary theory could have provided a solution.

Galton's physical strength was indeed waning, he was seriously unwell during the Easter of 1908. His mind still remained as fertile as ever in ideas, he was continually planning new projects, but the mental energy needed to carry through serious investigation was failing him.

42, RUTLAND GATE, S.W. *May* 18, 1908.

MY DEAR KARL PEARSON,   At length, I am to be allowed an hour's drive—after quite a long bout of bronchitis and asthma. It began here in Easter week and has kept me for 10 days or so mostly in bed, and quite invalided. I have contrived twice to get people here to dine, half on business, but though leaving them early it rather overtaxed me. The doctor declares that I am fast getting well at last. You may judge how incompetent I have been by the fact that even yet I have not tackled the last part of *Biometrika*. But I have nearly got my "Memories" off my hands.

A letter of yours, April 28, has only come into my hands this morning. The housemaid had dropped it, and so it lay unopened behind a box in the hall.

*   *   *   *   *   *   *   *

That Eugenics Education Society promises better than I could have hoped. Crackanthorpe is serious about it, and Professor Inge has joined it! I can't find that Crichton-Browne has as yet done much. A. acts as a *restrainer*, but is very eager, and they have got a particularly bright lady Secretary who acts and works hard for the love of the thing. I have not *yet* ventured to join it, but as soon as I am assured it is in *safe* management, shall do so.

I hope you are all the better for Hindhead. I am eager to get (in half an hour) my first out-of-doors view of this May time. Ever affectionately, FRANCIS GALTON.

The appearance of the Eugenics Education Society—another child of Galton's fertile mind—in these letters may be best explained by printing here the rough draft of a letter of Galton to Montague Crackanthorpe, dated so far back as December 16, 1906. Having established his Research Institute, Galton now turned, as he had done in the case of Finger Prints, to the popularisation of the principles of Eugenics.

May I consult with you on the following?

Is not the time ripe for some association of capable men who are really interested in Eugenics, and might not the existing Eugenics Office of the London University serve as a centre? If you think so and cared to suggest the outline of a working plan and a few good names, I should be grateful. I am too much of an invalid to push forward any undertaking except by letter. Still I think something useful might be done even in that manner. I do not yet see the way clearly and am desirous of fresh ideas.

Edgar Schuster has resigned his Research Fellowship, the future of the Office is just now uncertain. One idea is to have a "Fellow" at £250 a year, a Student at £100 in addition to the very capable Secretary, of good actuarial blood, who is already there and is familiar with the ways of the Office. Do you know of any capable man who would be a likely candidate for the vacant Fellowship? Hitherto it has been an annually renewable post. The Office is in Gower St, in rooms rented by University College and near to the Biometric Laboratory of Prof. Karl Pearson, who is a pillar of strength.

(b) The Darwin-Wallace Celebration of the Linnean Society of London, 1st July, 1908.

Two things remain impressed on the biographer's mind as memories of that day. I first felt the strong need Francis Galton had for a supporting arm. By the time the medals had been distributed, and the recipients* had spoken, the fatigue had so tried Galton that he had to leave the meeting. I saw that he rose with difficulty, and leaving my seat also, saw him home. He had spoken well, but the exertion and the closeness of the day had severely taxed him.

The other memory is also a sad one; we had met to do honour to a great English leader of scientific thought, one whom I take it we all respected, and to whom many of us felt we owed a deep debt of gratitude; he had given us, as Galton said, a keen sense of intellectual freedom. It was, as it were, a memorial service of thanksgiving, which all men of science could join in together, irrespective of divergence of scientific creeds. Some wag on the Linnean Executive had placed William Bateson in the chair adjacent to mine. I awaited his coming with expectation, determined that our greeting should disappoint the wag. But Bateson refused it, sat sideways on his chair, with his back to me, the whole of the medal distribution, and no doubt the wag was amused by what was simply pain to me—pain, that a distinguished biologist should refuse to join harmoniously with a biometrician, however despised, in a common service of reverence to one so immeasurably greater than either of us.

Dr Dukinfield H. Scott, the President, addressing Galton, spoke as follows†:

"Evolution, as understood by Darwin and Wallace, depends upon three factors, Heredity, Variation and Natural Selection. In the study of the first of these factors, Heredity, the work of the present day is characterised by the application of exact methods, whether on biometrical or Mendelian lines. It was you, Dr Galton, who first showed the way by which exact measurement could be applied to the problems of evolution and heredity, and indicated that their laws must be susceptible of proof. You have pointed out a new method, and the possibility of a more logical treatment of evolutionary questions. By establishing such principles as that of 'Regression to Mediocrity' you have added new laws to evolution, and under the name of 'Cessation of Selection' you have suggested an explanation of degeneration following disuse, anticipating that afterwards independently proposed and elaborated by Weismann‡, and called by him Panmixia.

"The ingenuity of your methods, your energy and enthusiasm in applying them, and your constant interest in the work of others, and readiness to help them, have made you a great

* Alfred Russel Wallace, Sir Joseph D. Hooker, Ernst Haeckel, August Weismann, E. Strasburger, Francis Galton and E. Ray Lankester, all of whom but Haeckel and Weismann were then present; the last remaining leader, Lankester, died just ten days before I wrote these lines.

† *The Darwin-Wallace Celebration, held on Thursday, 1st July,* 1908, *by the Linnean Society of London,* London, published by the Society, 1908. The work contains admirable portraits of Darwin and of the medallists, pp. 24–26.

‡ I think it desirable to publish the following letter from August Weismann. It admits the priority of Francis Galton in the main idea involved in the continuity of the germ-plasm.

Francis Galton, Esq., London.   Freiburg i. Br. 23 *Febr.* 1889.

Sir, You had the kindness to send to me your new book "Natural Inheritance" and a whole series of smaller papers you published before on the same subject. I thank you very much for your kindness and I am indeed very glad to have now all your memoirs at once at hand for consulting them. Till now I did not know all of them, but some ones, for instance "A Theory of Heredity" from 1875. It was Mr Herdman of Liverpool

PLATE XXXV

Two portraits of Charles Darwin : on the right at age 31, from a water-colour painting by Richmond, formerly in the possession of his daughter, Mrs Litchfield ; on the left at age 33, with his eldest son William, from a daguerreotype, in the possession of Lady George Darwin.

power in the advancement of evolutionary studies: a power which has only been strengthened by your characteristic open-mindedness and willingness to accept new views.

"You have shown, throughout the wide range of your work, that exactness of method is consistent with the charm of style; and we may recall the words of your cousin, Charles Darwin, in speaking of your famous book on *Hereditary Genius*, 'I do not think I ever, in all my life, read anything more interesting or original.'

"The new departure which you inaugurated in the study of Evolution, has been previously recognised by the award of the Darwin Medal of the Royal Society. We desire to add our recognition of the originality and importance of your work by asking you to receive the Medal which commemorates the united discoveries of Darwin and Wallace."

This speech, while admirably characterising some of Galton's work, misses entirely its fundamental aim, namely, by an accurate knowledge of the laws of evolution, as expressed in the three factors referred to above, to make man master of his future development, to give him control, biologically as he now largely has physically, of Nature herself*.

Galton with his usual modesty made no reference to his own work; the occasion for him was one for reverence towards those who had emancipated our minds. He said:

"I thank you for your remarks, Sir. You have listened to-day to many speakers, and I have little new to say, little indeed that would not be a repetition, but I may say that this occasion has called forth vividly my recollection of the feelings of gratitude that I had towards the originators of the then new doctrine that burst the enthraldom of the intellect which the advocates of the argument from design had woven around us. It gave a sense of freedom to all the people who were thinking of these matters, and that sense of freedom was very real and very vivid at the time. If a future Auguste Comte arises who makes a calendar in which the days are devoted to the memory of those who have been the beneficent intellects of mankind, I feel sure that this day, the 1st of July, will not be the least brilliant."

who—some years ago—directed my attention to this paper of yours, after having read my own papers on heredity, on continuity of germ-plasm and others. I regret not to have known it before, as you have exposed in your paper an idea which is in one essential point nearly allied to the main idea contained in my theory of the continuity of germ-plasm. You will find in the English translation of my essays just now appearing, a note by Mr Poulton, which draws the attention of the reader to your ideas. I shall profit by the next occasion which offers itself to me to give a more extensive account of your views and to point out the differences between our views. Heredity is a very complicated and difficult matter and I am afraid that none of us, who have thought about it, will have solved the whole mystery of it. Nevertheless it may be, that we have touched upon the right way which leads to the solution of the riddle.

Please excuse my bad and perhaps not always intelligible English! But as you tell me, that you do not read German with fluency, I thought it more convenient for you, if I wrote in bad English than in good German.

As soon as my book shall appear you will receive a copy of it, which I beg you to accept as a sign of my high esteem. Believe me, faithfully yours, August Weismann.

The "continuity of the germ-plasm" almost universally attributed to Weismann is really, as I have indicated in Vol. II, pp. 81–2, 169–171, 186–7, a product of Galton's inventive mind.

* Among the speakers, I think only Lankester recognised the part which our knowledge of evolution must in the future play with regard to human society:

"'Darwinism' must in the future guide statesmen and politicians as well as men of science. It is in its application to the problems of human society that there still remains an enormous field of work and discovery for the Darwin-Wallace doctrine. The science of heredity, of fecundity and sterility, of variation and adaptation, has yet to be far more completely studied and developed in its application to man and to human aggregates than it yet has been; at the same time a true psychology has to be arrived at and made, together with a knowledge of heredity, the basis of education, of the government and of the prosperity of the modern state. How far we are from any satisfactory progress in this direction, the words and the actions of political leaders of all parties at this moment fully demonstrate" (*loc. cit.* pp. 29–30).

But Galton, besides holding this view, had stepped into the public arena, and proclaimed a science which by providing a creed should control the biological evolution of man—Eugenics.

It was indeed an impressive meeting, the last occasion on which the "Old Guard" of Darwinism answered to the roll-call. Galton and Hooker died in 1911, Strasburger in 1912, Wallace in 1913, Weismann in 1914, Haeckel in 1919 and Lankester in 1929.

WINSLEY HILL, DANBY, GROSMONT R.S.O., YORKSHIRE. *July* 6, 1908.

MY DEAR FRANCIS GALTON, I have got back to my Yorkshire moors and their fresh air,— if it be cold,—and I hope to do a good three months' work! Let me have a line to say you were none the worse for the Darwin-Wallace Celebration, and I hope none the worse for this cold bout that has followed it.

I smelt the good smell of the turves and the bracken and the young heather and saw the first young grouse yesterday. The only grief is to come back after two years and find those one left hale now in the churchyard. When you know nearly the whole country-side, there are sure to be big gaps in the ranks. In London where one does not know even the names of one's neighbours within fifty yards of one's house, one does not get into touch with other folk's sorrows. I shall be here, if you write at any time, the whole holiday, except perhaps a couple of days to the South of the moors to see a tablet we have put up to my Father in his birthplace. Here we belong to those who have "gone over the moor," and have thus passed out of memory. As one of my ancestors of 1680 says in his will "Let my son Henry take my black mare and ride across the moor." That meant he was to go and seek his fortune south. My Father remembered as a boy the Quaker relatives from this Dale riding pillion with their wives across the moor, and stopping at his grandfather's house on their way to York Quarterly Meeting. That was his last touch with Danby. Four years ago I saw a farmer riding pillion with his wife over the moor on what is still called the "Quakers' Path." Four miles up on the moor is the solitary hut which used to be a meeting-house, whence Gregory Pearson was taken to York Gaol in 1684, where he died. My other forebear, George Unthank, came back alone a year afterwards across the moor. You will understand why I like the smell of the moor.

Affectionately, K. P.

42, RUTLAND GATE, S.W. *July* 10, 1908.

MY DEAR KARL PEARSON, I delayed writing to get news, if any, from A. who came here the day before yesterday, and Heron yesterday (both to dinner). That Eugenics Education Society seems really promising. The prospectus has been re-worded and members are coming in. Mrs Gotto is marvellous in her energy. I have been doing rather too much, with the usual penalty in consequence, of $\frac{3}{4}$ of this day in bed, but no real harm. Next week I go into Oxfordshire and Worcester to a great nephew and to a niece respectively, and then back until August 1 when we go to a house in the neighbourhood of Petersfield for a month, whence I will write to you (with address of it).

You must feel like *Antaeus*, who was revived by touching his mother-earth. The Quaker associations must be at times almost overpowering, where you now are.

I expect the first batch of the proofs of my "Memories" every day. They have done all the little illustrations, and two portraits of me—that which you know well, and one from Furse's picture. I shall be glad when the book, index and all, is finally off my hands. I called at University College and found them at full work in the Eugenics Laboratory. I wish I could think of a good way of measuring the power of "Mrs Grundy," in some one important social usage. "The force of popular opinion" would be a good subject for an essay, if numerically assessed.

That Linnean Medal has been nicely mounted in an unpretentious little round wooden frame, with glass on both sides. What kind care you took of me that day,—Hooker had a large luncheon party on the morrow, none the worse for the ceremony!

Affectionately yours, FRANCIS GALTON.

WINSLEY HILL, DANBY, GROSMONT R.S.O., YORKSHIRE. *July* 16, 1908.

MY DEAR FRANCIS GALTON, Thank you for your helpful suggestions and corrections about the *Treasury* circular. I think something might be done to gradually give a value to words in current use. I have endeavoured to do so in the case of correlation, defining "high" 1 00 to ·75, "considerable" ·75 to ·50, "moderate" ·50 to ·25 and "low" ·25 to ·00. I found writers were always speaking of "high," "moderate" or "low" relationship and I thought it worth while to make a start with more exactitude.

It is extremely good of you to undertake a "Butler" pedigree*. Can you do one of the "Pollocks" also? The "Darwins" are already done and we shall want ten at least for the first number. Can you think of any families that we might look up in the Eugenics Office Records? We could put together two or three science ones from the *Noteworthy Families*. We want one or two "governmental" or "executive" families. If you can think of any names, we will see if they are already done in the Laboratory Records.

Will you do us another favour, i.e. write a page, or, if it must be, only half a page of preface to the first part? If the thing starts well, it will go on through the years, until it will be the great mine for searchers after nuggets of heredity, and it would be pleasant to think of a few lines from you starting what will be a great monument, I hope, of your inspiration. I am sending you back a prospectus (rough form) to be a slight guide as to the nature of the work. I am asking Professor Osler of Oxford to write me a few lines of appeal to the younger medical men to aid from that side. He is the one man before whom the profession bows down, and if he aids it will be a great gain. I must not write more as I am rather invalided with a four days' attack of neuralgia. I got the doctor in to-day, but we have not yet succeeded in getting to the root of the trouble. Affectionately, KARL PEARSON.

[A leaf from a calendar was attached containing the words:]

July 13, Monday.   "Gather, then, each flower that grows,
                                      When the young heart overflows,
                                      To embalm that tent of snows†."
                                                  Maidenhood.

† Did any man of science ever write as wildly and carelessly as this famous poet?

SHIRRELL HOUSE, SHEDFIELD, BOTLEY, HANTS. *August* 11, 1908.

MY DEAR KARL PEARSON, I long to know how you are faring, and that the neuralgia has ceased. All is abundantly right with myself; moreover my publisher's men have been active and during the past week have sent me the whole of my "Memories" in proof, which has occupied all my working hours to revise. But this is done now. I had hoped to hear from you something more about the "Thesaurus," and to see your circular which probably contains a specimen of *how* you wanted biographies drawn up—whether in respect to a single character or how far generally? In your letter and in the lithographed page I see no provision for symbolising school-boy or university success, obtained when the person has not *completed* his opportunities. Thus, how to symbolise a youth of much school promise, and who has gained an open scholarship, but is not old enough yet to compete for higher things. I have at least three such cases in the Butler family. The same kind of difficulty of classification may occur in other subjects. Thus :—"not affected but still within the danger zone."

You ask about whom to apply to for the Inge, Buxton and other families. Ask Professor Inge himself. I can't recollect his address (I think in Brookside) but "Cambridge" would surely find him. The pedigree at the Eugenics Office was sent by him and he is very willing. Sir X. Y. would either do, or get the thing done, for his family. He has some near lady relative who is versed in pedigrees, but it would be awkward to address him about tuberculosis for instance. I fear the maladies of that family would be like skeletons in their cupboards. Both Sir Vernon Lushington and Sir Ed. Fry, heads of their respective families, would be likely to contribute.

You will probably have seen Crackanthorpe's letters to the *Times* about the Feeble Minded Report, one on Friday and one to-day. I have not yet procured the Report itself but am writing for it to-day. (I read the *Times* extract of it.)

You will gather from the above that I have done *nothing* last week in respect to the proposed "send off" or to the Butler family. I have been working up to my full strength

---

* Pedigrees of families distinguished for scholarly, literary or executive power were being compiled in the Eugenics Laboratory.

elsewhere, but at length am fairly free. It would be easy to get what you want as regards scholastic success in the U. and V. families, but in respect to health and character it might be otherwise there as elsewhere. I know that U. *shrinks* from anything like a medical pedigree, not because his own is other than good but on more general grounds of not alarming the young with the terror of impending, hereditary disease. Under any pseudonym, his family history would be recognised by some one, and so become generally known. I fancy that you will get the medical information you mostly want from un-related bystanders rather than from members of the family. Send me a line to remove my present difficulties that I may set to work for you. I am happily housed and gardened here. I gave a day to see my brother in the Isle of Wight, which by road, rail and steamer is about two hours off. All the rest of the time I have stuck to my books. Affectionately yours, FRANCIS GALTON.

I return the page of the *B.M.J.*

WINSLEY HILL, DANBY, GROSMONT R.S.O., YORKSHIRE. *August* 17, 1908.

MY DEAR FRANCIS GALTON, Your letter I think will have crossed one of mine. It would have been answered sooner, but it went astray, through no fault of your addressing but owing to postal blunders, which seem characteristic of this district! I am sending you a piece of the albino memoir. Will you please let me have it back, as I have not yet corrected it, and I want to return it for Press. I have not a spare Plate of Fig. 61, but send one of Plate XXXVI. You can use the symbol ↘ to mark non-adult brilliancy. It is well in the pedigree to stick to a single character, but in the account of the pedigree, put in all points bearing on this character. Thus look at Fig. 61 in proof sent. You will see hair and eye colour given as far as possible; mentally deficients and deaf mutes are cited, also any other cases of weakness or degeneracy. You will see also that age is frequently stated.

I have not yet printed a revise of the "Thesaurus" prospectus because I wanted first to see what helpers we could get. I think we shall be all right on the medical side.

I am so glad to hear your quarters are comfortable. You are certain to find nice neighbours. Are you within driving distance of Cowdray Park? It is perhaps the most beautiful park in England, if you get up to the north from the motor road through it. There is an aged oak with a seat to it on a path which strikes north after passing the dower house (?) near the west gate, which is to my mind typically English in its environment.

Here the hills are glorious purple and the "Grouseler" has not yet begun to disturb the peace. The cold, I suppose, has kept the birds back. I am glad the "Memories" are done; how exciting it will be reading them! Yours always affectionately, KARL PEARSON.

SHIRRELL HOUSE, SHEDFIELD, BOTLEY, HANTS. *August* 27, 1908.

MY DEAR KARL PEARSON, Our stay is so near its end that there is barely time for a to and fro letter, so I have written and send the enclosed at once. I have sad misgivings about being able to make out here a good U. pedigree. It is an eminently sane and healthy stock, and very athletic, but the first wife of U. (I think you will gather whom I mean) died of uterine cancer. He was most unwilling this should be known to her children and contrived that the Register of Death should ascribe it to a true but secondary cause.

Professor V comes in a few minutes to dine here. His is a noted family, I believe. I will see what can be done about it, and if favourable will write. In great haste.

Ever affectionately, FRANCIS GALTON.

On *Monday* we leave here, and tour about for a week or fortnight. 42, Rutland Gate will then be my address but letters may be delayed in reaching me.

42, RUTLAND GATE, S.W. *Sept.* 7, 1908.

MY DEAR KARL PEARSON, Having failed to satisfy myself about the U. pedigree, I wrote to C. U., who handed my letter to his very capable son, who sends me the enclosed cards and "tree." Will they, subject to a few pencilled and other corrections, do? I replied to him that I had sent them on to you, that the *names* would be struck out, but that his name would be wanted for authenticity. When I hear from you, I will redraw the tree on a larger scale and see to its revision before returning it to you. Will that do? By when do you want it?

I hope the weather has so far mended with you as not to bring your holiday change earlier to an end, than was originally intended. It is pleasant enough here. I sat out yesterday in my bath chair in the park, for an hour or more.

I have secured a pretty little house in Brockham, just south of Box Hill, with the Mole River for its meadow boundary. It is called "The Meadows." We go there at the end of October. My own matters get on. The whole of the text of my book is in the printers' hands "for Press" and the index is in their hands too, but not yet in type. I shall be glad to have wholly done with it.

Eugenics gets on. I have drafted an Address for the October meeting of the new Society of which I enclose the prospectus (*No*, I don't. I can't find one!). The address takes up fresh ground and I must ask Crackanthorpe to smash it into shape as soon as it is type-written. I see that in the President of the Anthropological Section, Ridgeway's, address, there is a good deal of *platitudinous* appreciation of Eugenics towards its close.

What do you think of Frank Darwin's Address? I must read it carefully yet again, but at present it seems to me that he asks for too much *tenacity* of memory from each of innumerable units. The forgetfulness of one of them would create a havoc in the orderly development. But I write crudely. Ever affectionately, FRANCIS GALTON.

Your tale about Churton and the mad college porter is very amusing*.

42, RUTLAND GATE, S.W. *Sept.* 24, 1908.

MY DEAR KARL PEARSON, I returned yesterday to London and the new No. of *Biometrika* arrived shortly after. I am glad that you have that off your hands. Your last letter, which describes your health as run low and the quantity of work ahead, made me feel sad, and fearful that the residue of your scanty holiday may have been far short of what your health needs. How I wish I could be of service to you in any way. It is a shame that your powers and zeal should be used up by comparatively small details of not the most advanced tuition†. I did not write before, being unwilling to add to your work. Now when you have time, a line would be very acceptable just to say how you are.

The U. pedigree is not even yet such as I could wish. The V. U.'s, on whom I relied, were out of town and when they returned just before I last left it, could not find the required notes. I will now try a different way.

I have let this house for the winter, beginning with Nov. 1, and have taken "The Meadows," Brockham, Dorking, for that same time. It is small but very well appointed, and is pretty. Moreover it stands high, notwithstanding its name and the fact that the river Mole bounds its adjacent meadow. Box Hill is just to its north and is said to shelter it.

I address this to Hampstead, thinking that you may have returned by now.

Ever affectionately, FRANCIS GALTON.

I am a little busy with the new Eugenics Education Society. Also I have just read the proof-sheets of Saleeby's forthcoming book on race improvement. It has some new things, but too much denunciation. However he rubs certain elementary truths strongly into the reader.

7, WELL ROAD, HAMPSTEAD, N.W. *September* 25, 1908.

MY DEAR FRANCIS GALTON, Many thanks for your sympathetic note. We came back last Saturday and I am trying to get back into harness again. I enclose the final form of the prospectus of the *Treasury*. I do not propose to issue it just yet, until we are a little farther forward with Part I, but we began drawing the plates for it to-day. I think we shall have a good first number. I have got a good Pollock Pedigree; Sir-Edward Fry answered very nicely and I hope to get fully the data from him. Mr Vernon Lushington has not yet answered; I have

---

* Alas! Galton's letter to me concerning Churton, the abnormally shy College dean of my undergraduate days at King's, Cambridge, and my reply citing the incident of the under-porter mistaking him for the devil have alike perished.

† At this time the biographer was giving 24 hours a week to teaching and demonstrating, apart from aiding research workers, supervising Galton's Eugenics Laboratory and much heavy editorial work.

one or two other heavy pedigrees in hand. I see the "Memories" announced. By the by, I picked up a privately printed "Pedigree of the Family of Darwin" issued in 60 copies only; it gives your pedigree pretty fully. It will be helpful in doing the Darwins. I must come and see you soon. Affectionately, KARL PEARSON.

7, WELL ROAD, HAMPSTEAD, N.W. *October 7*, 1908.

MY DEAR FRANCIS GALTON, I was indeed sorry to hear yesterday that you had called the day before, and I had not been down at College. It was, indeed, a disappointment, because I want to see you for your own sake and to talk about several things. I have been working very hard in my last few days of freedom to get my Appendices to the memoir on Albinism done. Not the text, that has yet to be written, but the descriptions of the 550 pedigrees and the bibliography through the Press. We shall do the statistical part from this printed Appendix. I have also been gradually getting the plates of photographs printed off. It will be my biggest piece of work should I live to complete it.

Meanwhile to-day all the rush of the term has begun. I have four new postgraduate biometricians of good type, one a doctor working at plague bacilli and opsonins; another a biologist from Harvard, and a third who is taking up the influence of earlier judgments on later judgments.

In Eugenics we are all hard at work. The memoir on the inheritance of eye characters and the influence of environment on sight has been delayed, because Nettleship thought we ought to give more account of earlier work. Some weeks have been spent in studying such work, but it really is of very little service for our purposes.

Heron is nearing the end of his London children and Miss Elderton of her Glasgow children. She finds the employment or non-employment of mothers influences sensibly but not very markedly the physique of the child, but the employment of the father as measured by the mortality of that employment is also influential, though not so sensibly. Perhaps the greatest difficulty is that the employment of the mother is correlated with the mortality rate of the father's trade. If he follows a bad trade with a high mortality rate, then the mother generally has employment out, or home work. So the wheels of the whole machine are interlocked and it is very difficult to get the simple *independent* causes either of degeneracy or of physical fitness in children.

Your subject looks very good. Can you send me a ticket or two more for people I know would like to hear you? I shall certainly hope to be present.

Yours always affectionately, KARL PEARSON.

I cannot get to Oxford for the Weldon ceremony to-morrow. I should have liked to be there, but it meant risking a breakdown.

The first plates of the "Thesaurus" are nearly ready for the engraver, i.e. the drawings are ready and I hope to get it out in November.

42, RUTLAND GATE, S.W. *October 8*, 1908.

MY DEAR KARL PEARSON, It was just a chance visit on the spur of the moment that I paid on an exceptionally fine day to University College. I knew well how busy you would be and shrank from offering myself, but I am very free and could come almost any day and hour you might suggest*.

It will be a *grief* as well as a great pleasure to me if you come on the 14th to the lecture. I have asked the Secretaries to send you cards. But don't think of coming if you are tired. You have indeed both hands full and overfull of work. Thanks for all you tell me about the Eugenics work and the Biometric.

To-day one thinks much of the Weldon ceremony. I could not venture to attend it, however gratifying it might in itself be to do so.

You will probably have received, or will receive almost immediately, my *Memories of my Life*. The reading of it will keep; don't think you are expected to do that now, in the midst of all your other work. Methuen has got it up, I think, very well and legibly. What an immense deal must be omitted in any autobiography and that not the least important!

Ever affectionately yours, FRANCIS GALTON.

* Galton had called without warning and found me out.

HAMPSTEAD. Friday, *October* 9, 1908.

MY DEAR FRANCIS GALTON,   It was a great pleasure to receive the *Memories* this morning. There is nothing which delights one more than to realise a little better that part of the life of a great friend in which one has had no share. I am only at Chapter III yet, but the reading so far suggests many points. First and foremost that you must make me a pedigree for the inheritance of longevity in the Galtons. I think it would be very suggestive. I shall hope to *see* you on Wednesday, if circumstances don't, as they probably won't, allow me to speak to the chief performer.

Would Saturday, October 17, next be a possible day to come and have a chat with you? I could come in immediately after lunch, or any time up to 5 o'clock that would suit you.

Affectionately, KARL PEARSON.

My Wife joins with me in demanding that in the Second Edition of the *Memories* there shall be a portrait of F. G. as a young man.

(c) *Papers read before the Eugenics Education Society.*

We must now return for a time to Galton's scheme for a Eugenics Society: see p. 339 above. This had in the earlier part of 1908 been vigorously pushed by Montague Crackanthorpe and the "Eugenics Education Society" had come into existence. At Mr Crackanthorpe's suggestion, Galton read a paper on Eugenics at the former's house, 65, Rutland Gate, on June 25, 1908. The paper was printed in *The Westminster Gazette* of the following evening. The paper is of much interest and is in part autobiographical.

Galton starts with the statements that the word Eugenics is pronounced with a soft *g* and that the Science of Eugenics is based on Heredity. He points out that the latter word does not appear in Johnson's Dictionary, and he says that forty years previously he had been chaffed by a cultured friend for adopting a French word*. Notions about human inheritance were very vague and confused, and the subject had never been squarely faced. The prevalent notion was that inheritance existed in animals and plants, but men were in another category. It was admitted that physical characters were sometimes inherited, but the heredity of mental characters was stoutly denied by many—as it still is denied by some. This sprang partly from theological grounds. "There was much talk about men being equal and masters of their own fate." Galton tells us that his first opinion was formed in 1840, when he was at College:

"Where competitions of all kinds showed most clearly to an unprejudiced eye that men were *not* equal in their natural powers, but most diverse in mind as well as body. It was also noticeable that high gifts of both of these tended to run in families."

* Galton used the adjective "hereditary" as early as 1863 and 1864 in his papers on "Domestication" and "Hereditary Talent" written in those years (see our Vol. II, p. 70). He used "heredity" in his *Hereditary Genius*, 1869 (see p. 334). According to the *New English Dictionary* the word had been used in the sense of estate, property or succession, as early as 1540, but apparently it was given for the first time a biological sense by Herbert Spencer in his *Principles of Biology*, 1863, see §§ 80 and 82. I do not know whether Galton adopted it from Herbert Spencer or from the French writers. On the other hand, "hereditary" was used of disease in both the 16th and 17th centuries, although without considering whether the disease was truly hereditary or conveyed by infection *in utero*. Thus "hereditarie lepresie" in 1597 and "hereditary gout" in 1699.

The first evidence that strongly impressed Galton, even in those early Cambridge days, was that of the Senior Classics. To be Senior Classic was scarcely less a feat than to be Senior Wrangler in the good old days when "Seniors" existed. Yet out of forty-one Senior Classics Galton found six who had a father, son or brother who was Senior Classic, or in one case a Senior Wrangler. He remarks that no mere tuition could account for this, they must have been born with exceptional capacity. He found that in every form of bodily and mental activity the same rule applied—those who achieved most had more achieving kinsmen than chance or good teaching could account for.

We thus recognise the birth of the ideas which came to fruition in *Hereditary Genius* as occurring when Galton was at Cambridge, surveying unnoticed the academic phenomena around him. At that time, he remarks, there were "no means such as we now have—thanks to the development of statistical science—of measuring with numerical exactness the closeness of the various kinships."

From these observations the lecturer said he had concluded that man was not an exceptional creature in respect to heredity, and that what applied to other animals and to plants applied also to him:

"I perceived that the importance ascribed by all intelligent farmers and gardeners to good stock might take a wider range. It is a first step with farmers and gardeners to endeavour to obtain good breeds of domestic animals and sedulously to cultivate plants, for it pays them well to do so. All serious inquirers into heredity now know that qualities gained by good nourishment and by good education never descend by inheritance, but perish with the individual, whilst inborn qualities are transmitted. It is therefore a waste of labour to try so to improve a poor stock by careful feeding or careful gardening as to place it on a level with a good stock.

"The question was then forced upon me—Could not the race of men be similarly improved? Could not the undesirables be got rid of and the desirables multiplied? Evidently the methods used in animal breeding were quite inappropriate to human society, but were there no gentler ways of obtaining the same end, it might be more slowly, but almost as surely? The answer to these questions was a decided 'Yes,' and in this way I lighted on what is now known as 'Eugenics.'

"Eugenics has been defined as 'The study of those agencies which under social control may improve or impair the racial qualities of future generations, either physically or mentally.' It aims at showing clearly how much harm is being done by some one course of action, and how much good by some other, and how closely connected social practices are with the future vigour of the nation. Its procedure is the reverse of fanatical; it puts social problems in a clear white light, neither exaggerating nor underrating the effects of the influences concerned. It is probable that even democratic governments will hereafter appreciate the value of Eugenic studies, and deduce from their results recognised guides to conduct. Such governments would be compelled to do so in their own self-defence, if not on higher grounds; otherwise they would come to an end, for a democracy cannot endure unless it be composed of capable citizens.

"The influence of public opinion, together with such reasonable public and private help as public opinion may approve of and support, is quite powerful enough to produce a large, though gentle, Eugenic effect. It is already becoming possible through Eugenic study to foresee with much assurance that such-and-such proposed action will influence a definite percentage of the population, though we cannot at present, and probably never shall be able to, foretell whether the individuals so affected will be A, B, C, or X, Y, Z.

"To the statesman this individualisation is unimportant, since individuals are only pawns in the great game which he plays. The true philanthropist, however, concerns himself both with society as a whole and with as many of the individuals that compose it as the range of his affections is wide enough to include. If a man devotes himself solely to the good of the nation as a whole, his tastes must be impersonal and his conclusions appear to a great degree heartless, deserving the ill title of 'dismal' with which Carlyle labelled Political Economy. If, on the other hand, he attends only to certain individuals in whom he happens to take an interest, he becomes guided by favouritism, oblivious alike of the rights of others and of the well-being of future generations. Statesmanship is concerned with the nation; Charity with the individual; Eugenics is concerned with and cares for both.

"A considerable part of the huge stream of British charity furthers, by indirect and unsuspected ways, the production and support of the Unfit. No one can doubt the desirability of money and moral support, now often bestowed on harmful forms of charity, being directed to the opposite result, namely, to the production and well-being of the Fit. For the purpose of illustration we may divide newly married couples into three classes according to the probable civic worth of their offspring. Amongst such offspring there would be a small class of 'desirables,' a large class of 'passables,' and a small class of 'undesirables.' It would surely be advantageous to the country if social and moral support, as well as timely material help, were extended to the desirables, and not monopolised, as it is now apt to be, by the undesirables.

"Families which are likely to produce valuable citizens deserve at the very least the care that a gardener takes of plants of promise. They should be helped when help is needed to procure a larger measure of sanitation, of food, and of all else that falls under the comprehensive title of 'Nurture' than would otherwise have been within their power. I do not, of course, propose to neglect the sick, the feeble, or the unfortunate. I would do all that available means permit for their comfort and happiness, but I would exact an equivalent for the charitable assistance they receive, namely, that by means of isolation, or some other less drastic yet adequate measure, a stop should be put to the production of families of children likely to include degenerates."

Galton then referred to the newly founded Eugenics Education Society and the previously founded Eugenics Laboratory, and concluded as follows:

"I will only add to this brief address that my purpose will have been fulfilled if I have succeeded in impressing on you the idea that Eugenics has a far more than Utopian interest; that it is a living and growing science, with high and practical aims. I would ask you to make the Society known to your friends, and to persuade them as best you can to help on its good work."

It was a thoroughly good paper for a man in his 87th year, and expresses in a marvellously brief space the creed of Eugenics. It is perfectly true that a democracy cannot endure unless it be composed of capable citizens, but did Galton fully appreciate what follows, when, as is the usual case, a democracy *starts* with a majority of incapable citizens? A government which drew a line between capable and incapable would rapidly perish; for the incapables care nothing for the future of the race or nation, but seek from their necessarily subservient governments *panem et circenses*—more time to pillion-ride, more leisure for cigarettes, chocolates and cinemas—at the cost of the capable. Eugenics—however sturdily we preach its creed, and we have no preacher to-day like Galton—must be unsuccessful if we *start* with such a democracy. We might as successfully ask the weeds in a garden to make way of their own accord for the flowering plants whose development they choke. Let my readers think what a gardener could achieve, if his tenure of office depended on the consent of the weeds!

I will now reproduce some of the letters of the autumn of 1908.

42, RUTLAND GATE, S.W. *Oct.* 13, 1908.

MY DEAR KARL PEARSON, I see no reason against the Eugenics Laboratory publications including similarly *solid* work to its own, especially of a statistical kind which cannot easily find a home elsewhere. On the contrary it seems to me advisable. For a more popular kind the Eugenics Education Society might afford a home. As to F.'s work I gather that it is hardly up to the mark of a Eug. Lab. publication. If you think it to be on the border line and would send it to me, I would do my best to give a casting vote. I should be quite prepared to *exact* a revision of the paper in accordance with your suggestions to him, *before taking it into consideration at all*. He might be told this definitely.

I fear that Mrs Gotto may have bothered you about speaking to-morrow. Please absolve me from the charge of having incited her.—Quite the contrary, I have insisted that you must not be troubled, but for all that I believe she has been irrepressible in her zeal.

Affectionately yours, FRANCIS GALTON.

42, RUTLAND GATE, S.W. *Oct.* 15, 1908.

MY DEAR KARL PEARSON, I have read F.'s memoir and return it with a few remarks, which you can if you like send to him. What you said last night was *excellent*, and very helpful to the Society, as showing what valuable work they might do as collectors of facts, and organisers of local inquiry into family histories. I wholly go with you there.

Affectionately yours, FRANCIS GALTON.

Galton's paper, to which reference is made in the preceding letters, was read before the Eugenics Education Society at the Grafton Galleries on October 14, by the author himself. It was, I think, the last time I heard him address an audience, but he spoke clearly and well, and seemed less fatigued than at the Darwin-Wallace celebration. The paper is entitled "Local Associations for Promoting Eugenics," and was printed in the issue of *Nature*, Oct 22, 1908 *.

Galton begins by stating that he only proposes to consider what steps can be taken by local associations in the large field of *positive* Eugenics, namely in favouring those especially fit for citizenship; for the time being he put on one side the topic of restricting the production of undesirables, which has been sometimes termed negative Eugenics†. The problem before Galton was the nature of the furtherance of Eugenics that local associations more or less affiliated to the Education Society could provide. He writes:

"It is difficult, while explaining what I have in view, to steer a course that shall keep clear of the mud flats of platitude on the one hand, and not come to grief against the rocks of over-precision on the other. There is no clear issue out of mere platitudes, while there is great danger in entering into details. A good scheme may be entirely compromised merely on account of public opinion not being ripe to receive it in the proposed form, or through a flaw discovered in some non-essential part of it. Experience shows that the safest course in a new undertaking is to proceed warily and tentatively towards the desired end, rather than freely and rashly along a predetermined route, however carefully it may have been elaborated on paper.

"Again, whatever scheme of action is proposed for adoption must be neither Utopian nor extravagant, but accordant throughout with British sentiment and practice.

---

* Vol. LXXVIII, pp. 645–647.

† The term is not very satisfactory. "A-eugenics" is worse, "Cacogenics" is cacophonous, dys-genics should I fear be dys-eugenics, for it would signify without the "eu," I take it, absence of any generation, whereas it is to represent that branch of our subject which studies what may control misbreeding in man. Further the word used must be such that the study of cures for misbreeding is not confused with the practice. For example, what is the opposite of a eugenic marriage, i.e. one approved by the principles of Eugenics? If it be an "a-eugenic" marriage, then "a-eugenics" sounds rather like the practice of misbreeding, than the body of principles which we propound to minimise it. Galton in this paper uses the term "anti-eugenic" for an undesirable mating—the word is correctly formed, but "anti-eugenics" might signify propagandism against the principles of eugenics rather than the study of the causes making for anti-eugenic matings, or the factors which might minimise them.

"The successful establishment of any general system of constructive eugenics will, in my view (which I put forward with diffidence), depend largely upon the efforts of local associations acting in close harmony with a central society, like our own. A prominent part of its business will then consist in affording opportunities for the interchange of ideas and for the registration and comparison of results. Such a central society would tend to bring about a general uniformity of administration, the value of which is so obvious that I do not stop to insist on it.

"Assuming, as I do, that the powers at the command of the local associations will be almost purely social, let us consider how those associations might be formed and conducted so as to become exceedingly influential."

Galton supposes that in any district a few individuals, some of local importance, desire keenly to start a local association. After forming themselves into an executive committee, and nominating a president, officers and council, they would form the association although it has no legal corporate existence. This committee should next with the aid of the central society provide for a "few sane and sensible lectures" on Eugenics and on the A, B, C of heredity. They would seek the co-operation of local medical men, of the public health officers, of the clergy, of lawyers, and of all officials whose duty brings them into touch with various classes of society. The new association would embrace everybody likely to have sympathy with the eugenic cause; it would be thus much like any political or philanthropic agency. Then we reach something more original. The committee is to seek out "worth" in their district; by civic worthiness Galton understands the value to the State of a person as it would be assessed by experts or fellow-workers. Each class is to choose its own men of worth, students to be chosen by students, artists by artists, business men by business men and so forth*. These men of worth are to be invited to social gatherings. "The State is a vastly complex organism, and the hope of obtaining a proportional representation of its best parts should be an avowed object of these gatherings." Clearly Galton was considering that the local association would be a mixture of social classes, and he cites the meetings of the Primrose League at one end and those in Toynbee Hall at the other end as illustrations, given considerable tact, of what such reunions might achieve for the eugenic cause. He thinks the committee by its inquiries into " worthiness " would obtain a large fund of information as to the notable individuals in the district, and their family histories. These could be used for eugenic studies; the histories should be tabulated in an orderly manner, and the more significant of them communicated to the Central Society.

Speaking for himself only Galton states that in classifying persons as to " worth," he should consider them under three heads: in the first place physique, in the second ability and in the third character; subject, however, to the provision that inferiority in any of the three should outweigh superiority in the other two. Galton admits character as the most important but it is not so easy to rate as the other two. " The tenure of a position of trust is only a partial test of character, though a good one so far as it goes." From this Galton passes to a conception that he had broached many years earlier†, associations of the well-born—the " Eugenes "—for mutual aid;

---

* See p. 231 above.   † See our Vol. ii, pp. 78–9.

the "worthies" are to become a caste, with a just pride in their common worthiness, and with a feeling such as the soldier has for his regiment, or the boy for his school.

"By the continued action of local associations as described thus far, a very large amount of good work in eugenics would be incidentally done. Family histories would become familiar topics, the existence of good stocks would be discovered, and many persons of 'worth' would be appreciated and made acquainted with each other who were formerly known only to a very restricted circle. It is probable that these persons, in their struggle to obtain appointments, would often receive valuable help from local sympathisers with eugenic principles. If local societies did no more than this for many years to come, they would have fully justified their existence by their valuable services.

"A danger to which these societies will be liable arises from the inadequate knowledge joined to great zeal of some of the most active among their probable members. It may be said, without mincing words, with regard to much that has already been published, that the subject of eugenics is particularly attractive to 'cranks.' The councils of local societies will therefore be obliged to exercise great caution before accepting the memoirs offered to them, and much discretion in keeping discussions within the bounds of sobriety and common sense. The basis of eugenics is already firmly established, namely, that the offspring of 'worthy' parents are, *on the whole*, more highly gifted by nature with faculties that conduce to 'worthiness' than the offspring of less 'worthy' parents. On the other hand, forecasts in respect to particular cases may be quite wrong. They have to be based on imperfect data. It cannot be too emphatically repeated that a great deal of careful statistical work has yet to be accomplished before the science of eugenics can make large advances.

"I hesitate to speculate further. A tree will have been planted; let it grow. Perhaps those who may hereafter feel themselves or are considered by others to be the possessors of notable eugenic qualities—let us for brevity call them 'Eugenes'—will form their own clubs and look after their own interests. It is impossible to foresee what the state of public opinion will then be. Many elements of strength are needed, many dangers have to be evaded or overcome, before associations of Eugenes could be formed that would be stable in themselves, useful as institutions, and approved of by the outside world."

These associations would be standing examples of the benefits which flow from following eugenic rules and the evils which arise when they are disregarded. Ultimately a public opinion would be created in the district selected as a eugenic field.

"The power of social opinion is apt to be underrated rather than overrated. Like the atmosphere which we breathe and in which we move, social opinion operates powerfully without our being conscious of its weight. Everyone knows that governments, manners, and beliefs which were thought to be right, decorous, and true at one period have been judged wrong, indecorous, and false at another; and that views which we have heard expressed by those in authority over us in our childhood and early manhood tend to become axiomatic and unchangeable in mature life.

"In circumscribed communities especially, social approval and disapproval exert a potent force. Its presence is only too easily read by those who are the object of either, in the countenances, bearing, and manner of persons whom they daily meet and converse with. Is it, then, I ask, too much to expect that when a public opinion in favour of eugenics has once taken sure hold of such communities and has been accepted by them as a quasi-religion, the result will be manifested in sundry and very effective modes of action which are as yet untried, and many of them even unforeseen?

"Speaking for myself only, I look forward to local eugenic action in numerous directions, of which I will now specify one. It is the accumulation of considerable funds to start young couples of 'worthy' qualities in their married life, and to assist them and their families at critical times. The gifts to those who are the reverse of 'worthy' are enormous in amount; it is stated that the charitable donations or bequests in the year 1907 amounted to 4,868,050*l*. I am not prepared to say how much of this was judiciously spent, or in what ways, but merely

PLATE XXXVI

Francis Galton, aged 87, on the stoep at Fox Holm, Cobham, with his biographer.

quote the figures to justify the inference that many of the thousands of persons who are willing to give freely at the prompting of a sentiment based upon compassion might be persuaded to give largely also in response to the more virile desire of promoting the natural gifts and the national efficiency of future generations."

Was it only the idle dream of an old man? Scarcely! Galton had grasped the truth in his early youth that man would respond to careful breeding even as other animals; he had propounded his gospel in full manhood, as early as 1864, when nobody had listened to him; he had repeated his doctrine in 1883, when he was sixty years old, with scarcely more effect. And now in his last years he called on his fellow-countrymen once more to have faith and act on that faith. There is a hereditary nobility, an aristocracy of worth, and it is not confined to any social class; it is a caste which is scattered throughout all classes; let us awaken it, that it may be self-conscious, and realise how the national future lies incontrovertibly in the feasibility of making it dominant in numbers and submitting the rest to its control. Those who imagine that Eugenics as a national faith was the dream of an octogenarian, have failed to understand the whole trend of Galton's intellectual development; he preached and waited, he waited and taught. The dream of his youth, he endeavoured to the extent of his ability to make practice in his old age. As in the case of Finger-prints, he took the precaution of first establishing a science, and then followed it with his appeal for public recognition of the principles of his science through all the channels at his command. We shall see that he did not think them exhausted by newspaper articles, eugenics education societies and associations, or by public lectures.

What he might have achieved had he been ten years younger, or the English public ripe for his teaching a decade earlier, it is not possible to say. For two more years he fought for his creed, but his physical strength was failing. In his earlier days his chief recreation had been walking alone and thinking; his best thoughts came to him on these occasions. We can follow the change in the truthful record he gives under *Recreations* in successive editions of *Who's Who*. We find:

In 1898,   " Chiefly solitary rambles,"
in 1904,   " Solitary rambles,"

but in 1908, the year we have now reached:

" Sunshine, quiet, and good wholesome food."

He gave a literal interpretation to the word "recreate," and we find him from 1908 onwards seeking, well wrapt in rugs, sunshine and quiet in a sheltered garden corner, or on the "stoep" of a fitly chosen winter home.

Sitting thus, Galton's thoughts rambled through the past eighty odd years and they became again actual to him. As he says in a letter to his biographer: "How much an autobiography must omit," and this, although in a lesser degree, is true of a biography, if it be compiled within fifty years of its subject's decease and its writer would not pain survivors!

Memories of my Life*. Galton's letters indicate how busy he was during the latter half of 1908 with this book. It would not be fitting—were it indeed feasible—to give an analysis of his work here. Our biography has, indeed, endeavoured to give a picture of Galton's personality, his deep affection for his relatives and for his friends; it has been able to say what he could not say of himself. An autobiography can only indirectly characterise its subject, unless its writer be as unabashed as Benvenuto Cellini, or as self-soddened as Jean-Jacques Rousseau. But beyond this characterisation, we have endeavoured to lay stress on Galton's contributions to science and to reproduce his thoughts in his own words. The reader will find little of this in the *Memories*; they deal not wholly, but chiefly, with the men—many of them noteworthy in their day—whom Galton had known in the course of a long lifetime. They are delightful reading, full of anecdotes and reminiscences, but the Galton of our volumes—the scientific originator, the modest inquirer, the intensely affectionate and reliable friend—is not easily recognised in the pages of his autobiography.

There are, however, two or three passages I should like to quote here for the benefit of those who are unable to read the *Memories*—now, alas, out of print. The first illustrates the depth of Galton's feelings for his friends. He is speaking of his college friend, Henry Fitzmaurice Hallam, born in 1824, only to die when he was 26 years old. He was the younger son of the historian, and brother to Arthur Hallam, who died at 22 and was the subject of Tennyson's *In Memoriam*.

"Henry Hallam had a singular sweetness and attractiveness of manner, with a love of harmless banter and paradox, and was keenly sympathetic with all his many friends. He won the Second Chancellor's Medal. Through him I became introduced to his father's house, still shadowed by the sudden death of his son Arthur and of a daughter. Mr Hallam was very kind to me, and the friendship of him and of his family† was one of the corner-stones of my life-history....Henry Hallam, like his brother and sister, died suddenly and young, to my poignant grief. His death occurred while I was away in South Africa. I have visited the quiet church at Clevedon, where all the Hallams lie, each memorial stone bearing a briefly pathetic inscription, and kneeling alone in a pew by their side, spent part of a solitary hour in unrestrained tears." (pp. 65–6.)

Another passage I wish to cite bears upon the nature of Time; it should be compared with Galton's view of Time in the *Inquiries into Human Faculty‡*.

"I will mention here a rather weird effect that compiling these 'Memories' has produced on me. By much dwelling upon them they became refurbished and so vivid as to appear as sharp and definite as things of to-day. The consequence has been an occasional obliteration of the sense of Time, and the replacing of it by the idea of a permanent panorama, painted throughout with equal vividness, in which the point to which attention is temporarily directed becomes for that time the Present. The panorama seems to extend unseen behind a veil which hides the Future, but is slowly rolling aside and disclosing it. That part of the panorama which is veiled is supposed to exist as vividly coloured as the rest, though latent. In short, this experience

* Methuen & Co., London, 1908.

† There was another daughter Julia Hallam, who travelled with Emma and Francis Galton: see Vol. I, p. 180, and also pp. 140–1, 171, 191, 205–207, and 238.

‡ See Vol. II, p. 263.

PLATE XXXVII

A reverie, caught " when the spirit was not there."

has given me an occasional feeling that there are no realities corresponding to Past, Present and Future, but that the entire Cosmos is one perpetual Now. Philosophers have often held this creed intellectually, but I suspect that few have felt the possible truth of it so vividly as it has occasionally appeared to my imagination through dwelling on these 'Memories.'" (pp. 277–8.)

In Galton's last chapter, entitled *Race Improvement*, he summarises what he has hoped for and what he has done for Eugenics. He writes:

" Skilful and cautious statistical treatment is needed in most of the many inquiries upon whose results the methods of Eugenics will rest. A full account of the inquiries is necessarily technical and dry, but the results are not, and a 'Eugenics Education Society' has been recently established to popularise those results. At the request of its Committee I have lately joined it as Hon. President, and hope to aid its work so far as the small powers that an advanced age still leaves intact may permit." (p. 321.)

The last paragraphs of the *Memories* reiterate the teaching of 1865*, expressing it, perhaps, more effectively and concisely. It is probably very rare for a man at 86 to gain wide acceptance for a creed which he failed to impress on his contemporaries when at 42 he had the vigour and energy of early manhood. Galton was clearly 40 to 50 years ahead of his own generation. He thus concludes his autobiography:

"I take Eugenics very seriously, feeling that its principles ought to become one of the dominant motives in a civilised nation, much as if they were one of its religious tenets. I have often expressed myself in this sense, and will conclude this book by briefly reiterating my views.

"Individuals appear to me as partial detachments from the infinite ocean of Being, and this world as a stage on which Evolution takes place, principally hitherto by means of Natural Selection, which achieves the good of the whole with scant regard to that of the individual.

"Man is gifted with pity and other kindly feelings; he has also the power of preventing many kinds of suffering. I conceive it to fall well within his province to replace Natural Selection by other processes that are more merciful and not less effective.

"This is precisely the aim of Eugenics. Its first object is to check the birth-rate of the Unfit, instead of allowing them to come into being, though doomed in large numbers to perish prematurely. The second object is the improvement of the race by furthering the productivity of the Fit by early marriages and healthful rearing of their children. Natural Selection rests upon excessive production and wholesale destruction; Eugenics on bringing no more individuals into the world than can be properly cared for, and those only of the best stock." (pp. 322–3.)

" I shall treat," said Galton in his 42nd year, " of man and see what the theory of heredity of variations and the principle of natural selection mean when applied to man†," and his treatment only ended with his life.

7, WELL ROAD, HAMPSTEAD, N.W. *November 5, 1908.*

MY DEAR FRANCIS GALTON, I cannot refrain from sending you a line now that I have finished the *Memories* to thank you for the very kind things you say about my work. I have read the book with great interest and it has been helpful in more ways than you will realise. It was nice to find you also knew and appreciated Croom-Robertson. What a wonderful width of interests you have had, and how delightful that you had not to wedge them in between other things and carry out your work in haste! I spoke to Heron yesterday about work and his appointment and I must look into the original terms of his nomination before discussing it further. I certainly thought it came *ipso facto* to an end in February, but he seems to think it was as in Schuster's case for three years. I do not know that we could get a harder worker at present.

* See Vol. II, pp. 71–78.
† See Vol. II, p. 86.

Let me say exactly how affairs stand. (1) Eugenics Laboratory Memoir No. V has gone to press. It is on the Inheritance of Vision and on the Influence of Environment on Eyesight. It is a heavy bit of work and would have been stronger had we only been able to collect data *ad hoc* of an accurate kind. But I think it definitely shows what ophthalmologists have doubted—the inheritance of the various classes of eyesight, and further that environment, notably school environment, is not the most important factor in shortsightedness. (2) Resemblance of nephew and niece to uncle and aunt—will go to press in the next few weeks. (3) Brainweights of normal and insane. This took a good many weeks' work, but the results are inconclusive. The data were sent by Crichton-Browne, but they lack several needful points, e.g. information as to special type of insanity, and the records filled in at Wakefield from the Asylum Case Books are not accepted by C.-B. I am doubtful whether the results should be published, except to induce some one to start *de novo*. (4) Eugenics Laboratory Publication VI. Occupation of Father and Mother in relation to the Physical Health of school-children. This is based on 20,000 Glasgow returns provided by the Scottish Education Office. It will be ready by Xmas. (5) Eugenics Laboratory Publication VII. Influence of physique (nutrition, tonsils, teeth, glands, etc.) on mental capacity of children. Data for 30,000 London School-children from County Council. This also will be ready by Xmas. (6) *Treasury*: 10 plates are now engraved, or ready for engraver. I hope to have Part I out this month. This represents the last 18 months of work, and I want you to see that the staff have been working really well, but that these heavy bits of research do not come lightly to an end. Our not publishing anything for a year must not be taken as a sign of inanition.

By the by I got a few days ago about 50 folio sheets of pedigree and accounts of the Lushington family! V. L. had asked a nephew to prepare it, but had not written to tell me about his having done so. We have already some 20 pedigrees of distinguished families, with perhaps 200 individuals in each. They will have to go on folding sheets, they are so gigantic.

Nettleship has just found two albino dogs—brother and sister. We are now going to try and discover whether we can create a race of albino dogs from these two. There have only been very vague rumours of such things hitherto. They are from their photographs very beautiful beasts, and I hope not too delicate to survive.

I trust the winter quarters are going to be a success. Let me hear how you go on and what problem you are turning over. Affectionately, K. P.

I have nine biometric research students this term and my new Laboratory is *full*. It is the first time I have had more students than I want on this side. A man came this afternoon for admission wanting to do research work, and I took quite a lordly tone with him and told him to go away for a fortnight and write a paper and I would take him if it was good enough for publication. I have never been able before to pick and choose postgraduate workers—and this man was a Cambridge wrangler!

7, Well Road, Hampstead, N.W. *November* 30, 1908.

My dear Francis Galton, I want to send you a hurried line to say that I hear from all sides that Heron did *exceedingly well* at the Eugenics Education Society the other night. I really think he ought to give a course of lectures on Eugenics next term and that it would do him and the subject good. I feel sure he has a lot in him and only wants to be made to feel more confidence in himself. I shall make an effort to hear Miss Elderton on Wednesday week. Can you send me a line as to the enclosed difference in the Galton pedigree and that of the big Darwin pedigree? Is it possible that James K— M— had issue that died early?

I am sorry to hear about your cold, but I expect that the weather will be more fixed this next month, and that a fixed temperature is what we all need.

I have got Part I of the *Treasury* to Press and I think my talk at the Royal Society of Medicine recently will help it forward. I hope you won't think Part I too medical, but I want if possible to bring the medicals in the first place into line. Now one more point, do you know the P—s or anybody connected with them? There is a very singular inheritance in their family, which they keep screened and I should like to get some clues if it were possible.

Yours always affectionately, Karl Pearson.

I was at Oxford last week, going through mice-work. Mrs Weldon comes up to work for a fortnight in my laboratory and we hope to get clear on some points. The medal for the prize is in hand.

MEADOW COTTAGE, BROCKHAM GREEN, BETCHWORTH, SURREY. *December* 2, 1908.

MY DEAR KARL PEARSON, I write at once and will send the corrected notice of the M— twins as soon as I get it back. The facts had got mixed, and if re-sorted would I believe be right. J. K— M— *had* issue; L— M— (now apparently on his death bed) had *not*. He was *unmarried*.

It gives me great pleasure to hear so favourable an account of Heron's lecture. Mrs Gotto wrote to the same effect as you did. I wrote and thanked *him* for it. It would be a good thing if, as you suggest, Heron could be made to lecture, or hold a class in some form at University College, as he has gifts for success. I do not "know the ropes" well enough to venture to say more than that the idea seems most desirable. Did he show you a long German poem on Eugenics by Sophia Martin, wife of a Professor at Rostock, Mecklenburg? I am told that it is not bad at all, and possibly may be rated as really good poetry.

Enclosed I send a rough idea in outline. It may be a familiar one (and *might* be wrong!), but seems worth sending. Beyond its measurement, there is no fact in correlation that is more interesting than the proportion in which the causes of variation are (1) unknown or neglected, and (2) known; and it is so easy to deduce this from r in the simple cases of linear correlation between normal variables. How far it might be extended to other correlations I have no clear idea, but it seems very improbable that much could be done in that way.

Affectionately yours, FRANCIS GALTON.

7, WELL ROAD, HAMPSTEAD, N.W. *December* 13, 1908.

MY DEAR FRANCIS GALTON, You will have been expecting to hear from me about Miss Elderton's paper, but alas! I could not get to hear it. I have been crippled with lumbago for a week and was perforce absent. I had a very tiring week previously, culminating in a meeting at the Royal Society of Medicine, where I went to ask help for the *Treasury*, but found myself the subject of a very bitter attack from a disciple of Bateson's called N. I am not a ready debater and find it hard to marshal my arguments in reply to a set speech of nearly 70 minutes designed to prove that biometry was sheer rubbish, and that medical men would be fools to give any help to a biometrician. It is on these occasions I miss so much Weldon's ready repartee and light cavalry charges into the foe! I don't know how far I saved defeat; if I did it at all, it was owing to the unmeasured abuse of my opponent. But it put the final touch to the very heavy week and I broke down on Saturday. I have crawled down and round the Laboratory for two or three days, after two days in bed and one on the sofa, but I am back on the sofa again to-day wondering whether the muscles of my back will ever do proper duty again! Heron gave a good account of Miss Elderton's paper, but I wish I could have been there to give her some aid.

On other matters some progress has to be reported. 12 pedigree plates and 4 plates of half tone illustrations for the *Treasury of Human Inheritance* are now ready and some of the text is now set up. We shall have it out by January certainly. Also the first sheets of the long-delayed memoir on the Inheritance of Visual Characters have come in. I shall hope to send you proofs of a Note of mine for *Biometrika* and shall welcome any criticisms. I hope all goes well.

Affectionately, KARL PEARSON.

Nettleship has got two albino bitches satisfactorily crossed by the albino dog. It will be most exciting to see the result of this attempt to create an albino race of dogs. I had an albino hen offered me the other day, but I did not see how to keep it!

MEADOW COTTAGE, BROCKHAM GREEN, BETCHWORTH, SURREY. *December* 14, 1908.

MY DEAR KARL PEARSON, Your bulletin distresses me. Lumbago is so painful and depressing. I have just been reading the biography of Alice Hopkins (daughter of the Cambridge coach). She had sciatica, and spoke of $2\frac{1}{2}$ *feet* of pain. I sympathise much with you, as you may well be assured.

Yesterday I received a pedigree of the M— twins, about whom you sent a paper for verification. Enclosed I send it in a correct form. Also, after reading your Skin Colour of Crosses, I jotted down a recollection of my own which impressed me much. You can make any use of it you like. The paper is very instructive. I have pencilled a few words on p. 3 which seemed wanted.

I am assured from many sides that Miss Elderton did her lecturing *excellently*. Also that Heron did his part as Chairman very well indeed. Miss Biggs was in London Thursday and Friday nights and met several "Eugenicals," full of enthusiasm. I grieve at the rudeness of your Mendelian opponents, which is harmful to progress. Confound them!

Don't hurry one bit, but don't please destroy the little problem I sent you about the proportionate efficacy of known and disregarded causes of variability in two correlated variables. It will keep as long as your convenience requires, and you are over-worked. I have thrown off my chronic cough for three whole days. May it prolong its absence. All goes on well, though of course monotonously.

Miss Biggs was at Miss Parker's wedding and delighted with all she saw.

Ever affectionately yours, FRANCIS GALTON.

7, WELL ROAD, HAMPSTEAD, N.W. *December* 14, 1908.

MY DEAR FRANCIS GALTON, I am sorry to trouble you again so soon and also to write, perhaps, unclearly, but I have only just got through the day and my back is giving me much pain. So please do not give undue weight to any phrase in itself. Miss Elderton came to me to-day and said that she had received an offer of the post of Secretary to a London College from the Principal of that College and that she had been given till Friday to consider her answer. That she had at first made up her mind to refuse, because she much preferred research work to executive work, and had not intended to tell me. But on second thoughts she felt she must ask my advice. Now I know exactly what this means, that home affairs are not too flourishing, and the post at the College means a definite post with steady rise, and a good position if needful for further advance. My impression of her is that she is a remarkably able woman with capacity in more than one direction. The first impulse was to say, but "you *must* stay here, the Eugenics Laboratory will collapse without you," but I felt without knowing your views, that this was hardly fair to her or to you. Now I want your advice before Friday. I cannot think of the Laboratory without Miss Elderton, she is the life and soul of the place, knows the whole of the material, writes all the letters and keeps everything going. I am sure she does not want to go, enjoys the work and is keen on the subject, and would find the secretarial work at a College less to her taste, but it offers an assured future. Now what ought we to do in the matter? I have always considered that you must look upon the Laboratory as on its trial and that if we failed to satisfy you, you must ruthlessly change the system or close the Laboratory as seemed to you best. Am I right therefore in trying to induce her to stay? I have no doubt, she is so valuable that she will always get a post, but suitable posts don't turn up every day and I feel if we advise her to stay we ought to say: If the Laboratory is closed or re-organised we will give you long enough notice to find a new berth. I don't think she would mind for herself, but, as I said before, her contribution to home funds is of some importance. On the other hand, if we go on, she is almost indispensable. It would take years to get any one with the same training, if even they had the same aptitude. Now what shall we do? It is not, I think, a question of money. There was ample when I last saw the accounts, and notwithstanding that there will be a heavy publication expenditure in the next six months (there are four memoirs nearly ready, and there is the *Treasury*) there is quite enough for present purposes and for a future pledging of resources. May I say to Miss Elderton: We will give you a permanent appointment subject to a year's notice, or such shorter period as would seem good to you and fair to her? The problem then is: Ought we in justice to her future to let her go? Or, ought we for the sake of the Laboratory to keep her with a more permanent post and perhaps an increase of salary? May I guarantee her, say £— or £— with a year's notice? I feel the answer cannot rest with me, because it depends to some extent on the future of the Laboratory. I don't want you to keep the Laboratory going for our sakes; we are all keen and ready to go on with the work on the lines which our powers render possible. But whenever you feel that we are not doing what you think best for the acquirement of that knowledge, which I know you have most at heart, then you must simply give me the word and we will bring things to a close. In this matter of Miss Elderton's, by advising her to refuse the College post I might be protracting the life of the Laboratory beyond your wishes, and thus I must consult you on the point. I do not know whether I have put her own wish strongly enough, she wants to stay and would do it at personal sacrifice, but here home calls on her have to be considered. Affectionately, K. P.

MEADOW COTTAGE, BROCKHAM GREEN, BETCHWORTH, SURREY. *December* 15, 1908.

MY DEAR KARL PEARSON,   All you say in favour of Miss Elderton I am *fully* prepared to believe from my much less but still not inadequate knowledge of her. She most certainly ought to be retained if possible, as the far future working of the Laboratory will be much more hopeful if she continues in it.

My feelings about the Laboratory remain the same that they were two years ago when we had so much correspondence and I drew up a Codicil to my will to provide amply for its permanent establishment after my death and to pay for a professorship. I can't undertake to die soon in order to hurry on the endowment, but I have not the slightest desire to do otherwise than continue the present £500 a year so long as I live. I would *increase* it, by say £50, rather than reduce it, if it were clearly advisable to do so. It is worth considering whether Miss Elderton's position in the Laboratory might be altered, by hereafter calling her Secretary, and on the next occasion abolishing the Research Scholarship altogether. It would not do to promote her over Heron, but hereafter when his term terminates it might easily be done. Possibly you may think that the two duties of Secretary and Research Fellow might be worked simultaneously, but if so, it must be clear which of the two is the responsible head, and I do not see my way here. Anyhow on the next vacancy the promotion could easily be made. I am most sorry about the cruel lumbago.   Affectionately yours, FRANCIS GALTON.

7, WELL ROAD, HAMPSTEAD, N.W. *December* 15, 1908.

MY DEAR FRANCIS GALTON,   Your letter gave me *great* pleasure this morning. We do not need more money, and above all things we want you to live to see the work you have set going reach more general acknowledgment. But what, I think, the younger workers, who really have worked hard and toiled forward against a good deal of outside (and even inside *) discouragement need is the knowledge that you really care for their work, and I think your letter really helps in that. You hardly realise how much they think of almost anything *you* do or say! Among the fourteen workers in the Biometric and Eugenics Laboratories at present we have five women and their work is equal at the very least to that of the men. I have to treat them as in every way the equals of the men. They are women who in many cases have taken higher academic honours than the men and who are intellectually their peers. They were a little tried therefore when your name appeared on the Committee of the Anti-suffrage Society! I refer to this merely to show that what you think and do *does* produce effect in the Laboratory, and therefore the knowledge that you really care for their work helps us all round. I think that your approval accordingly counts for a great deal more than you realise. I know Miss Elderton is very keen on the work and wants to devote all her energies to it, but I am sure the feeling that *you* think she is doing good work weighs as much as or more than any opinion of mine. I ventured to tell her that she was indispensable and that there was no immediate fear for the life of the Laboratory. I can trust you to bear this in mind if anything should happen to me.

I have not forgotten your problem, but I wanted to have another talk with Heron over it, before I returned the sheet. Could you not write a note on it for *Biometrika*? It would be quite easy to get a table calculated for you. Bulloch came in to-day with 30 pedigrees of hermaphrodite families. One noteworthy point has come out in collecting this material—a disproportionate number of hermaphrodites, perhaps 25 p.c., are *twins*. This is a very noteworthy point indeed and deserves special investigation. I have heard of hermaphrodites in sheep; were these twins?   Always affectionately, K. P.

Please excuse this handwriting, I am writing on my back.

* P.S. Only last week a lecturer in the College read a paper "On the influence of Heredity on Conduct," which consisted chiefly of abuse of the Eugenics Laboratory work and workers.

MEADOW COTTAGE, BROCKHAM GREEN, BETCHWORTH. *December* 22, 1908.

MY DEAR KARL PEARSON,   This is little more than a sincere Xmas greeting to you and yours. May that cruel lumbago keep its fangs off you. It is sometimes consoling to think of greater suffering than one's own, so imagine the feelings of the Chinaman who, humbly visiting his great superior on whom all his hope of advancement lay, when about to make his kow-tow was suddenly smitten with lumbago!

I have written both to Miss Elderton and to Heron, saying nice things. The latter has sent me the calculated values for the little formula; I have tried unsuccessfully to put my point in a "Note" as clearly as desirable, so that matter must stand over for the present. I will try again later. I see your lectures at the Royal Institution on Albinism are announced. Sir Trevor Lawrence (Pres. of the Horticultural), a great grower of orchids, has his home near here. He has much to say about an albino orchid of his, but I am so weak in botanical nomenclature that I am not at all sure whether I understood rightly what he told me. If you care for more, sufficiently to frame questions, I could easily get answers from him.

Lady Phillimore near Henley on Thames, the wife of the Judge, showed me a breed which she thought unique (as I understood) of white *ducks* in a pool in her grounds.

I must not trouble you with more than to beg you to give my heartiest Xmas greetings to Mrs Pearson and your children. Ever affectionately yours, FRANCIS GALTON.

HAMPSTEAD. Christmas Eve, *December 24, 1908.*

MY DEAR FRANCIS GALTON, I saw Heron yesterday. He has delayed his journey northward in order to work now, and take a few days off at the beginning of next term. The University of St Andrews has asked him to give four lectures, and I thought he had better do it and spread the light there, as Macdonell has done at Aberdeen. He seemed to appreciate your kind letter very much. We arranged a course of Lectures at University College of which I send the rough draft. I think it ought to do well. We tried to get the subject into 5 or 6 but had to give up the idea. Please comment on it. Did you hear whether the white ducks at Henley were true albinos with *red* reflex in their eyes?

The *Treasury of Human Inheritance* progresses, and I think it ought to be out by January. The great difficulty is to get all the material into the same "format." Each man makes his pedigrees, his notes and his bibliography in a different way. But after the first number appears we shall have a more concrete form for future contributors to work by. You take for example "Deaf-Mutism" or "Tuberculosis," and nobody so far has made a bibliography of papers dealing with heredity in these subjects. All sorts of pedigrees have been coming in, and I think when the first few parts are out, we shall have a constant flow. The heredity inquiry is everywhere in the air now.

I hope this cold turn will not be too bitter for you. I am much better, but still tender in the back, and I can't get up or down easily. I must say I like for working purposes a good high temperature. With the best wishes for the New Year for both Miss Biggs and yourself,

I am, Yours affectionately, KARL PEARSON.

Mrs Weldon has been working for more than a fortnight in the Laboratory over the mice skins. It is a big business, but we shall get it through some day.

About your problem in correlated variables, I think you are correct if $A, B, C, D, \ldots\ldots$ which are causes of X are not correlated among themselves, for the reduction of variability σ is then indicated by a standard-deviation

$$= \sigma \sqrt{1 - r_1^2 - r_2^2 - r_3^2 - \ldots}.$$

But if they are correlated, I am less certain about your view. For example for two causes 2 and 3 we have

$$\sigma \sqrt{\frac{1 - r_{12}^2 - r_{23}^2 - r_{31}^2 + 2r_{12}r_{23}r_{31}}{1 - r_{23}^2}},$$

and r_{23} contributes to this reduction by the term $2r_{12}, r_{23}, r_{31}$, as well as by $-r_{23}^2$. Thus I don't feel quite clear about your view when the causes are correlated together.

MEADOW COTTAGE, BROCKHAM GREEN, BETCHWORTH. *December 30, 1908.*

MY DEAR KARL PEARSON, A sharp attack of asthma which has departed this morning as suddenly as it came on sent me to bed with three warm bottles, unfit to do anything but sleep. So excuse delay in answering.

The programme of lectures seems excellently devised, being good in itself and bringing out the subjects of which the lecturers can speak with authority. Hereafter one wishes for lectures on some such subjects as "Effects of small social changes in promoting the birth-rate of capable or of incapable citizens." But only an historian could do that and the Eugenics Laboratory is hardly the place for it. I have no comment other than complete acquiescence with your programme.

Miss Elderton comes here for a week-end on January 30th (I think), so I shall hear many details from her as to what is going on. I am glad that your lumbago, which you have borne so heroically*, is better, though maybe this cold snap of weather has been an enemy to you. Seasonable weather !! Stuff and nonsense—Give me the temperature of an incubator !

I can't think "Germinal Vitality" worth serious consideration. It would require much evidence from horticulturists and breeders to make it at all probable. *His* evidence is *very* lax.

I am very glad that St Andrews has asked Heron and that he will lecture. One of the professors there, Stanley Butler (of physics and mathematics), is a nephew of my Wife's and writes me to-day a letter practically about my book but evidently not forgetful of Eugenics. The white ducks of Lady Phillimore had *not* red eyes so far as I noticed. *Later.* I had intended after getting through arrears of writing, to send a revised statement of my problem, but find myself too tired, so must postpone. Ever affectionately, FRANCIS GALTON.

(15) *Events and Correspondence of* 1909. In the pressure of work upon me during the years 1905–1909 I had scarcely noted the changes taking place in Francis Galton; they were gradual, and so much of the old fire and suggestiveness remained that I did not fully realise how he was failing, though the failure was far more rapid on the physical than the mental side. Re-reading the letters that passed between us in the year 1909, it now seems to me clear that he passed another milestone on the decurrent highway of old age in his 87th year. The only published writings of Galton that I can find for this year are the following :

(i) A brief Introductory Note to the *Treasury of Human Inheritance.* The letters of this year will indicate how keen was Galton's interest in this work. It was designed on a comprehensive scale, and was intended to provide data for the measurement of all phases of human heredity by pedigrees indicating the transmission of ability, mental superiority and defect, physical and pathological characters in stirps. It has now, 1930, reached its third large quarto volume, but the cost of the photographic and pedigree plates and the need of funds to pay contributors have sadly hampered its progress. Occupation could be found in this direction for at least half-a-dozen thoroughly trained workers, but while endowments are always forthcoming for the maintenance of the unsound, there is so far little enthusiasm for building up our knowledge of why the unsound come into existence†. The idea of the *Treasury* was not Galton's, but it met with his full sympathy, and the early costs of publication were defrayed from his grant to the University of London. His prefatory note runs thus :

"The Inheritance of Qualities in Families lies at the basis of the Science of Eugenics, and though much is known about it a much fuller inquiry is urgently needed than has hitherto been possible. Goodness and badness of physique, constitution and abilities are distributed in similar proportions among individuals in successive generations, but the chain-work of hereditary influences through which this is effected has been most inadequately recorded. The facts of Family

* Fortunately Galton was not present to *hear* his future biographer's language !

† Honourable exception must be made of the Committee for Medical Research, which has by its grants enabled the Galton Laboratory to carry forward the section of the *Treasury* dealing with the inheritance of eye-defects.

Inheritance, being unregistered, fall readily into oblivion as generations pass by, and an enormous amount of valuable experience is thereby irrevocably lost. The object of the *Treasury* is to remedy, as far as lies in its power, this deplorable waste of opportunity.

"If the *Treasury* prospers, as is hoped and expected, a vast amount of information will gradually be collected by its means, in a form suitable for analysis, that will enable more exact conclusions to be hereafter drawn and more emphatic advice to be given than is now possible.

"In conclusion I may perhaps be permitted to express my own sincere gratification that the Eugenics Laboratory has already become so well equipped and conditioned as to undertake the publication of this large and important serial." FRANCIS GALTON.

(ii) A "Foreword" to the first issue of the *Eugenics Review*, which appeared in this year, and in which, as the offspring of the Eugenics Education Society, the Hon. President took great interest. Galton stated the aims of the *Review* in the following sentences :

"Its general purpose is, as stated in the Prospectus, to give expression to the Eugenic movement and to place Eugenic thought, where possible, on a strictly scientific basis....

"The EUGENICS REVIEW emphatically disclaims rivalry in any form with the more technical publications issued from time to time from the Eugenics Laboratory of the University of London now located at University College. On the contrary, it proposes to supplement them. There are two sorts of workers in every department of knowledge—those who establish a firm foundation, and those who build upon the foundation so established. The foundation of Eugenics is, in some measure, laid by applying a mathematico-statistical treatment to large collections of facts, and this, like engineering deep down in boggy soil, affords little evidence of its bulk and importance. The superstructure requires for its success the co-operation of many minds of a somewhat different order, filled with imagination and enthusiasm ; it does not require technical knowledge as to the nature of the foundation work. So a navigator, in order to find his position at sea, is dependent on the Tables calculated for him and printed in the Nautical Almanac or elsewhere. But he may safely use these Tables without having acquaintance with the methods by which they were constructed....The field is very wide and varied. To those who carefully·explore it the direct conflict of Eugenics with some of the social customs of the day will be unexpectedly revealed, while its complete harmony with other social customs will be as unexpectedly made clear." *The Eugenics Review*, Vol. i, pp. 1–2, April 1909.

Galton did not foresee that one of the troubles of the remainder of his life would be that the one sort of workers would bitterly attack the other. The members of the Council of the Eugenics Education Society and its journal from the very outset became the harshest critics of the youthful Eugenics Laboratory. No publication of the latter from the day the Society was founded to the present has met with aught but unfavourable reception from that quarter. And this conduct not only rendered much Eugenic investigation still-born, but vexed endlessly the founder not only of the Laboratory but of the Society. Galton did not see that a group of persons of widely diverging views, especially on such topics as sex-problems, several of whom had very highly strung temperaments and little if any real scientific training, might ultimately do small good either to Eugenics as a science or to Eugenics as a creed. As Director of the Eugenics Laboratory I was in a difficult position, and I felt it wise to stand aloof from the Society. I knew that the help I needed and was seeking from the medical profession would hardly be accorded if I were associated with certain then prominent members of the Society. Further I felt morally sure that sooner or later our very different ways of approaching eugenic problems would lead to a divergence of opinion, which would have been harmful inside the Society. My staff consented to give the Society a couple of lectures, but I asked its Council the sole favour of leaving us alone, as we were quite ready to leave them. Not till certain members of its Council, not content with asserting the futility of the actuarial or statistical method of attacking eugenic problems, began to hint that we were wasting Galton's gift to the University did it appear to me necessary to make any reply to such ill-informed criticism*. But there is little doubt that the endeavour to make Eugenics a science in the academic sense—to build up a special technique for it and fix in a vaguely circumscribed field a defined area for cultivation—was much hampered by the action of successive officers † and members of the Eugenics Education Society. It is conceivable that Galton's attempt to appoint for the Laboratory and the Society separate spheres of action as indicated in the " Foreword " just cited did on the whole more harm than good ; nobody likes to be told, however true it may be, that he is incapable—without training—of doing the higher type of work. Be this as it may, Galton in the last two years of his life was—to use a mild word—saddened by the attitude of certain members of the Eugenics Education Society. I recognised myself that the staff of the Laboratory had laboured hard and done good work. I knew that neither they nor myself were biased in one way or the other in such problems as those of the relative effect of inheritance and of environment, of the influence of parental alcoholism on the health and mentality of school-children, of the inheritance or non-inheritance of the tuberculous diathesis, or of mental defect and insanity. We simply desired to reach the truth by applying appropriate scientific methods to such data as were available. The only prejudice permitted in the laboratory was the distrust of all preconceived opinions and the doubt of statements based merely on impressions. Once, however, we had ascertained the conclusions flowing from our data we were not prepared to surrender them because they clashed with the largely sentimental notions of those who had not closely studied these problems. I knew Francis Galton was with us in these points, but I think our opponents were less aware of it, nor to this day have they realised that he was so doubtful of the manner in which the Eugenics Education Society was being conducted, that in December 1910 when he asked my advice, a word from me—not spoken—would have led to his retirement from the Society‡. It is necessary to make these remarks or the letters of 1909–1910 would be unintelligible.

(iii) A preface to W. Palin and Ethel M. Elderton's *Primer of Statistics*. This little book was written to carry out Francis Galton's conception of a series of " object lessons " in elementary statistics as shadowed forth in the

---

* When, many years after Galton's death, we had at last saved enough from the scant publishing funds of the Laboratory to venture on a journal—*Annals of Eugenics*—in which to issue our researches, we were virtually accused by an official of the Society in its *Eugenics Review* of having neglected this duty far too long !

† One President of the Society recently organised a petition of its members to the governing body of another department of the University requesting that they should institute a second professorship of Eugenics !

‡ See my remarks on influencing the judgment of men of genius even when they are old on pp. 408 and 412 of this volume.

Oxford Herbert Spencer Lecture*. For further details of its history the reader must consult Galton's letters of this year. The preface runs:

"In my 'Herbert Spencer' lecture of 1907 before the University of Oxford, I expressed a belief that the elementary ideas on which the modern system of statistics depends, that the quality of the results to which it leads, and that the meaning of the uncouth words used in its description, admitted of much simpler explanation than usual. I sketched out a possible course of lectures to be accompanied with certain simple sortings, with object lessons and with diagrams. Finally, I expressed the hope that some competent teacher would elaborate a course of instruction on these lines. I entertain a strong belief that such a course would be of great service to those who are interested in statistics, but who, from want of mathematical aptitude and special study, are unable to comprehend the results arrived at, even as regards their own subjects. It is, for example, a great hindrance to have no knowledge of what is meant by 'correlation.'

"I learnt with much pleasure that two very competent persons were disposed to undertake the task—namely, Mr W. Palin Elderton, well known as a highly instructed actuary, and his sister, Miss Ethel M. Elderton, who holds the post of Research Scholar in the Eugenics Laboratory of the University of London (now located in University College), and who is a thoroughly experienced worker in the modern methods.

"This primer is the result. It goes forth on its important errand of familiarising educated persons with the most recent developments of the new school of statistics, and, I beg to be allowed to add, with my heartiest good wishes for its success."

September, 1909.

(iv) Galton was much interested in the course of this year in the asserted Deterioration of the British Race, and in the Report of the Commission on that subject. The problem was essentially a statistical one, but the evidence given before the Commission was largely that of witnesses without any statistical sense, who gave merely their opinions and impressions based too often on narrowly local or inadequately transitory observations†. Above all other problems Galton had selected that of the segregation of the mentally defective as a field in which something might be achieved at once. He was roused especially by any appeal to an individual case as confuting a statistical average. Such an appeal drew from him a letter to *The Times* of June 18th in this year:

Sir,—A specious inference was drawn yesterday, in a speech by Lord Halsbury at the luncheon given to Lieutenant Shackleton by the Royal Societies Club. He said (I quote from your report) that: "in view of what Mr Shackleton had gone through it was impossible to believe in the supposed deterioration of the British race." But exceptional performances do not contradict the supposition in question. It is not that deterioration is so general that men of remarkably fine physique have ceased to exist—for they do, thank God—but that the bulk of the community is deteriorating, which it is, judging from results of inquiries into the teeth, hearing, eyesight and malformations of children in Board Schools, and from the apparently continuous increase of insanity and feeble-mindedness. Again the popularity of athletic sports proves little, for it is one thing to acclaim successful athletes, which any mob of weaklings can do,—as at a cricket match,—it is quite another thing to be an athlete oneself.

42, RUTLAND GATE, S.W. *June* 16.   FRANCIS GALTON.

* See p. 317 *et seq.* above. The little book has done extraordinarily well and has passed through several editions.

† I feel bound to quote again here Galton's splendid aphorism of 1879 (see Vol. II, p. 297 above): "General impressions are never to be trusted. Unfortunately when they are of long standing they become fixed rules of life, and assume a prescriptive right not to be questioned. Consequently those who are not accustomed to original inquiry entertain a hatred and a horror of statistics. They cannot endure the idea of submitting their sacred impressions to cold-blooded verification."

This common error, which might be termed statistically—giving a new sense to the Latin phrase—the *ex uno disce omnes* fallacy, seems peculiarly characteristic of ageing statesmen. I can hardly attribute it to mental deterioration with age, but rather to their ever waxing appreciation of the calibre of the minds to which they chiefly appeal*.

In order to break a lance for the segregation of the mentally defective Galton wrote a brief essay entitled "Segregation" which was published in a small book : *The Problem of the Feeble-Minded. An Abstract of the Report of the Royal Commission on the Care and Control of the Feeble-Minded*†. It was planned by the Cambridge Eugenists and had besides Galton's essay an introduction by Sir Edward Fry.

Galton considers that the Royal Commission attacked a eugenic problem of the first order of magnitude with thoroughness and remarkable success, and that the evidence before them emphasised the view that the annual output of mentally defective children admits of being largely diminished in future generations at a slight cost by the policy of segregation. He estimates that slightly under one per cent. of the population belongs to the class which includes the mad, the idiotic and the feeble-minded. On this estimate 300,000 persons might roughly be supposed to fall into the category in England and Wales. Galton says that they fall "little short of a million"; I do not understand how he has reached this result. I believe, however, that if we could form a census of all those who have *at any time in the course of their life* been certified we should reach a higher percentage than Galton's 1 in 118. In addition to those who have once been certified, there are many in the wealthier classes of the community, who are tended without certification in their own homes, as any medical man of large practice can bear witness to. The problem of the insane is not a matter personal only to their relatives, it is a question of the highest national importance. The Commissioners in Lunacy ought to be instructed to keep a *General Register of the Insane*, open to any inquirers under due restrictions, the first of which should be evidence that they belonged to or sought to be connected with the stock inquired about. If *A* and *B* are tending to pass from friendship to a closer relationship, it ought to be possible for them to ascertain—what in many cases is screened from them by their elders—whether one or both come of insane stock. No one can fully appreciate the urgency of a General Register of the Insane, who has not seen close at hand the terrible affliction of a wife, when her husband develops a hitherto screened familial insanity and when she realises that her children may, one or more, be stricken down with it in later years. Until such a register is organised and has

---

* I may be permitted perhaps to cite another illustration of this *ex uno disce omnes* fallacy. In a lecture dealing with mental growth I had cited data to show that on the average the prime of quickness of mind, as judged by mental reactions, occurred about 28 years in males and then slowly but steadily declined. My statement, of course in a distorted form, was duly reported in the press. At a public dinner a few days later, a well-known statesman, ageing but at present active, denied the truth of my statement on the grounds that his own—an exceedingly spry and inventive—mentality had grown brisker with the years!

† Messrs P. S. King & Son, 123 pp. This little book is now unfortunately out of print.

been running for at least a generation, it is not possible to obtain anything but the roughest, and this probably a minimum, estimate of the mental disease in the country, and of the number of persons in whom it is hereditary. Galton, citing the Royal Commissioners, assumes that 66,000 of the feeble-minded are not provided for, and that from the eugenic point of view these form the most dangerous sub-class of the mentally defective. I would venture to suggest that those who have familial insanity of a kind which is not chronic, but permits of return to home and mate, may be equally dangerous*.

Writing of the feeble-minded Galton continues:

"The persons in question are naturally incapable of standing alone. If protected and supervised they may lead harmless, and even useful, lives and do something towards earning their living. But when unprotected and cast upon the world, they go to the bad. They do so, not necessarily through vicious propensities, but from the absence of will-power to resist temptations; and quickly sink into the pauper and criminal classes. The women commonly become prostitutes. The feeble-minded, as distinguished from the idiots, are an exceptionally fecund class, mostly of illegitimate children, and a terrible proportion of their offspring are born mentally deficient. A decorous family life among their children is obviously impossible; the conditions of their nurture prevent it. Some of the issue of the feeble-minded are wholly mad or imbecile and find their way to asylums; others are merely feeble-minded and drift into bad ways as their parents did before them; in others, again, the evil is latent, but may break out in a subsequent generation. So the mischief goes on increasingly, and, judging from the growth of insanity, a considerable part of the population has already become bearers of germs of degeneracy.

* * * * * * * * * * *

"Almost all the evidence printed in the report points unmistakably to segregation for life as the only means of preventing feeble-minded girls from doing great harm to the community. They propagate children freely, as already mentioned, who, whether they be as little, less, or more, mentally endowed than themselves, are in all cases subject to most undesirable conditions of nurture."

Galton then refers to the voluntary homes for feeble-minded girls, and the question of whether they are really happy in them. At that date compulsory detention was not allowed, and accordingly, if a girl was discontented she could leave the home, and one could not really assert that happiness *with* this freedom was a valid ground for believing in an equal happiness when she could not escape. I pointed this out in a letter to Galton (see our p. 373), but he seemed to think that he had evidence for their happiness even under compulsion in the many institutions and labour-colonies where now

"they live happily and feel as if at home, and where they remain for many years. Unfortunately, as yet, no power exists for their compulsory detention. The inmates are taken out, it may be, by their not wise relatives, or they want a change and leave of their own accord. Anyhow, when they quit the shelter of the institution, they usually go to the bad, and after a time very often apply to be again taken in, with an actual or a forthcoming baby.

---

* It is not moral insensibility, but ignorance, which is too often at the root of the evil. I can recall the case of a young man who had as bad a pedigree of familial insanity as can be well met with, and had actually himself been certified. He consulted the Director of the Eugenics Laboratory on his proposed marriage, and confidently believed that if he begot children when he was not insane, he would not place them under the slightest risk of hereditary insanity.

"Feeble-mindedness is of many grades. In a large institution, the inmates, whether men or women, can be graded and be much more easily supervised than in small ones, and be occupied in work, greatly to their own happiness, or in play, according to their several capacities. As regards cost of maintenance, some few of the feeble-minded may wholly or nearly pay for their keep by their work; almost all of them can do something towards the expenses. The cost of maintenance per head, all included, does not necessarily exceed £25 a year. What the average cost of each uncared-for, feeble-minded person may now be can only be guessed, but in work-house and prison maintenance, in thefts, destruction and food, it may be safely reckoned to far exceed that sum."

I cannot believe that the argument from "happiness" is a useful one. How is "happiness" to be defined? Is it what *we* hold to be the state of happiness for another, or the state which that other holds to be for his own happiness? Is it to be the happiness of the moment, or the average happiness over a period of years? The fact that occasionally—probably in conditions of suppressed impulse—certain feeble-minded girls have attacked their matrons or even visiting Commissioners, seems to me to indicate that we must advocate segregation not on the ground of "happiness" for each feeble-minded individual but, as Galton puts it, on the ground that it will be a eugenic victory over ills of long standing, hitherto scarcely noticed, but forming a "very serious and growing danger to our national efficiency."

"Every high form of civilisation brings evils in its train, eating like cancer into the constitution of the people, and surely leading to their gradual deterioration and ultimate ruin, unless they are boldly withstood in good time. The propagation of mental deficiency is one of these evils and the report shows that it is now ripe to be dealt with."

This is the last formal essay which Galton wrote; the urgent plea of national efficiency united for a common end in their last years the two men who had fought for and against Darwin's theory of natural selection (see p. 122 above).

Such are all the public actions with regard to Eugenics that I can credit Galton with in this year*. But his letters show how constantly during the six months of ill-health at the beginning of this year his thoughts turned to Eugenics, his laboratory and popular propagandism.

* Of work not concerned with Eugenics all I can recall is the paper on "Sequestrated Church Property": see Vol. II, pp. 410–11. I may also note that the Eugenics Education Society reprinted in this year in a little volume entitled *Essays in Eugenics*, seven of Galton's lectures and papers, namely the Huxley Lecture, the papers in the Sociological Society's journal, the Herbert Spencer Lecture and the paper on Local Associations. It is a very serviceable little volume for propaganda work. In the preface Galton says that the progress of Eugenics in the last few years is such that its practice is not a mere Utopian vision; the influential power behind Eugenics is Public Opinion, and it is amply strong enough for the purpose when fully aroused.

"It is above all things needful for the successful progress of Eugenics that its advocates should move discreetly and claim no more efficacy on its behalf than the future will confirm; otherwise a reaction will be invited. A great deal of investigation is still needed to show the limit of practical Eugenics, yet enough has been already determined to justify large efforts to instruct the public in an authoritative way, as to the results hitherto obtained by sound reasoning applied to the undoubted facts of social experience."

Galton to the end inculcated safe, but slow and continuous advance, educating, but never hustling Public Opinion.

Correspondence of 1909.

7, WELL ROAD, HAMPSTEAD, N.W. *Jan.* 3, 1909.

MY DEAR FRANCIS GALTON, I had intended to write a line sending every best wish of the New Year to Miss Biggs and yourself, but letters don't get written when they are of the type which one writes for pleasure! I have just got the last bit of copy of the next issue of *Biometrika* to press. It will be rather a fighting number I fear; but it runs from opsonins to cocks' combs and tadpoles' blood-corpuscles and so ought to be of general interest. I enclose the first forty pages of the memoir on Vision to be published in the Eugenics Laboratory series. I fear you will find it dull reading, but it means nearly a year's work on Miss Barrington's part. She insists on my adding my name to hers, but she has done the real "grind" of it, and I am only responsible for the direction of the work and more or less putting it into form. I think it opens up almost virgin soil, and ought to attract some notice. Your suggestion as to apparently small causes affecting the fertility of populations would no doubt form an excellent topic for an essay, but it would demand a long period of careful collection and investigation. One instance I can cite for you. Before the passing of the Factory Acts children were a valuable commodity in Bradford. Their parents are said to have nurtured them well and taken great care of them; they strove to have a good supply, because they were a source of revenue. Since the passing of the Acts, a child hardly contributes to the family income before he starts his own establishment. He does not repay his cost of nurture. The result has been that mechanical checks to conception and abortion are rife in Bradford; there is a general decline in the birth-rate, and the child is looked upon as a burden.

I am planning a great change in my work of the next half year which may possibly come off. I have asked for a holiday for six months, on the grounds: (i) that I have been feeling inert and below par, and (ii) that the arrears of research work are so great, that nothing can get finished. I have not had a real holiday since my marriage tour, eighteen years ago, and, I think, the College may be willing to permit it now, for the grant for the Biometric Laboratory from the Drapers' Company expires this year, and it is desirable to have a good show of completed research work with a view to its renewal. If my request be granted I shall have no elementary teaching of any kind. I shall be a "half-timer" at College, spending my afternoons with the research students and with them only, while I devote my mornings to polishing off arrears of work, so that in the summer I can take a complete holiday. I hope this will be a restorative. But as one gets older the tissues don't seem to respond to rest in the old way. I now seem to understand better why so many men give up research work for attending committees at 50. Even retire and sell their working tools like Huxley!

Still I am going to look upon my six months as a restorative to pristine vigour—until at least I have shown it won't work. Affectionately, KARL PEARSON.

MEADOW COTTAGE, BROCKHAM GREEN, BETCHWORTH. *Jan.* 8, 1909.

MY DEAR KARL PEARSON, The probability of your receiving a half-year's remission will be grateful news to all your friends, for you are by now dangerously over-worked, I feel sure. What a solid piece of work that is on the Inheritance of Vision. It bears on its face the mark of a full year's labour. You did not ask me to return the paper so I do not. I shall be eager for a postcard to tell me that the half-year's relaxation is fixed. Pray send one when it is so. Yesterday, Charles Galton Darwin, the mathematical son of Sir George, lunched and spent some hours with me. His fate will be decided in June, just about the date of the Darwin Centenary. If he gets the Senior Wranglership—the last that can be got—it will be delightful. He is very bright and capable looking, but too modest to be bothered with questions about his chances, which, as I hear from many quarters, are good but not certain.

Your instance of the Bradford birth-rate being affected through an Act of Parliament is very instructive. From time to time other more or less similar instances will be noted, especially as to the effect of changes in purely social usages.

I am just living on, capable of very little useful work, but very comfortable, and having everything to be grateful for that can reasonably be expected now that the winter fogs are ceasing. Ever affectionately yours, FRANCIS GALTON.

HAMPSTEAD. *January* 9, 1909.

MY DEAR FRANCIS GALTON,   Yes, I have got my leave and for six months I am to be a half-timer! In July and August I shall try to take an absolute holiday. I shall devote all my mornings to clearing off arrears and afternoons to the higher lectures and research folk. If I keep away from my students in the mornings I shall probably be able to work off arrears. The difficulty will be if my substitutes fail to get a grip on the students or break down with 'flu, which is always rampant this term.

There have been several friendly notices of the forthcoming *Treasury*. We have difficulty in settling whether to bind the plates into the parts, put them loose or make a pocket to hold them. What do you think? Some users will want to compare all the plates dealing with the same characteristic, and of course these may be Plates III, XX and LXI.

I am sorry to hear you feel less active, but does not the winter always teach us that our ancestors back in some grade or other hibernated? My first lecture at the Royal Institution is to preach the doctrine that the white man may be descended from a manlike dark-skinned ancestor—say a *Pithecus satanas*—but not the negro from a manlike *white* ancestor.

Affectionately, K. P.

I suppose you will be at the Cambridge Darwin Commemoration? In some respects I should like to have gone, but it is too much a glorification of what is not Darwinism to please me*!

MEADOW COTTAGE, BROCKHAM GREEN, BETCHWORTH. *January* 12, 1909.

MY DEAR KARL PEARSON,   I am so glad that the half-time arrangement is settled. You ask suggestions about the plates for the *Treasury*. A number of loose pages is objectionable, but could you not have them all in a paper wrapper which should itself go into a pocket in the cover? Could you conveniently instruct the proper person to send me a copy *here*?

Good luck to your *Pithecus satanas* lecture! I see that Taylor is about to give a course of lectures at Hampstead which, judging from his prospectus, may contain new and interesting points.

I have been bothered by letters in the *Times* about my share in Identification by Finger-Prints and have sent a reply. This morning I had a note "The Editor of *The Times* hopes to publish at an early date the communication kindly sent him," so I suppose that other letters have been received and that he will dispose of them all at once. I don't by any means acknowledge the justice of what my adversaries say.

This change of weather suits me well and I have got out a good deal both yesterday and to-day. Ever affectionately, FRANCIS GALTON.

* Neither Francis Galton nor I attended the celebration, nor did we find ourselves able to contribute to the memorial volume.

The time, perhaps, has arrived when it is permitted to record an anecdote of what occurred at a meeting of the organising committee of which Adam Sedgwick and Bateson were members. It was suggested that I should be asked to contribute a paper to the memorial volume. Bateson said that, if I were asked he would have nothing more to do with the Committee or the volume. Sedgwick said: "Are you the Pope?" and the incident ended in laughter, with the compromise that I should be asked to write on a definite topic—which did not permit of my breaking a lance for the threatened stronghold of Darwinism. A few lines from a letter of Thiselton-Dyer to Galton dated March 29, 1909, may, perhaps, be quoted here:

"You have done exactly the same thing that Darwin did. He gave a working hypothesis of the production of a species. You have succeeded [with the theory of Regression], where Herbert Spencer and everyone else have conspicuously failed, in showing how structures are got rid of when they become superfluous. But incidentally you have enormously enhanced the inevitableness and potency of Natural Selection. I have long wanted to say all this, so pray forgive this long letter. There is a sorry reaction against Natural Selection at the moment. The Darwin family seem to me to have practically thrown over their father's theory. But I verily believe that you have set it firmly on its feet again."

I do not think tho "Darwin family" ever consciously threw over their father's theory, but they certainly did not perceive how the novelties they fostered tended to a "sorry reaction."

THE GALTON EUGENICS LABORATORY, UNIVERSITY COLLEGE. *January* 23, 1909.

MY DEAR FRANCIS GALTON, I have been approached by the Provost to know whether you would not take the chair at the first Eugenics Lecture on February 23. I need hardly say that it would give me very great pleasure if it were possible for you to do so. *But* I would on no account urge this on you, if it would in any way involve a risk in the travelling up to Town and the possibility of damp or cold here. You must give this its due weight, and I shall fully understand your decision.

Of other points there is little to record. I hope shortly to send you the remaining proofs of Miss Barrington's memoir and Part I of the *Treasury*, but the Press moves but slowly. I gave my first lecture at the Royal Institution on Albinism in Man last Tuesday, my chief point being that the manlike ancestor of man had a darkly pigmented skin. The evidence shows that no blacks are ever thrown from a white, but copper and white with yellow or red hair come as variants from jet black negroes. I think I have got a real point, but the audience while ready to accept a pithecoid ancestry were not prepared for a negroid! My second lecture is on Tuesday.

Yesterday I heard Wallace on Darwinism. The Royal Institution was packed to the roof. Wallace was quite audible, but not very original and lacking in the vivacity needful to keep his audience alive. But it really was worth hearing him. He gave examples of 20 individuals to show variation and to *prove* that two organs of the same individual were *not* correlated!! But he made a strong attack on progress by mutation, and used one very good argument, namely that from mimicry, which I had not heard before. He said that a beetle exactly like a wasp could have reached the wasp condition by gradual approximation to open wings and colouration, but it could by no conceivable jump suddenly become a wasp in all external appearance. I think the argument was valid and a strong one.

A white man married an albino negress, the offspring were two *mulattoes*. What can we make of that as far as Mendelism is concerned? It seems to me a very curious state of affairs, and it means that the black colour was latent in the albino negress and blended in her offspring.

Ever yours affectionately, KARL PEARSON.

I thought the enclosed might show you that correlation is going to be a real tool in Astronomy.

MEADOW COTTAGE, BROCKHAM GREEN, BETCHWORTH. *January* 25, 1909.

MY DEAR KARL PEARSON, You know well how willingly I would have done what the Provost suggests, but my infirmities put it wholly out of the question. It would be dangerous for me to attempt the task. Moreover, as a much less important matter, my deafness shuts me out from presiding. Please convey all this to the Provost and beg him to excuse me.

I was eager to hear more about your Royal Institution lecture, of which the only printed account I had seen, viz. in the *Graphic*, was worthless. But Mrs Gotto, who has just been here on a week-end visit, assured me that she had heard it was excellent.

The mimicry argument, on which Wallace laid stress, is in the air. Butler I think started it, and a man, of whose wife I know something, Professor Walker, a cytologist of Liverpool, sent me the MS. of a forthcoming book of his; he lays great stress on it.

The white man marrying an albino and producing two mulattoes, is paralleled by Sir Trevor Lawrence's experience with albino orchids: they rarely if ever produce albino offspring. They have colour *latent* in them.

Thank you for the astronomical paper. I have as yet only glanced through it, but am delighted to find that two different samples of stars give the same result.

I have very recently lost a dear brother-in-law, Arthur Butler; also the widow of my cousin Sir Douglas Galton. The elder of her two daughters lives near here and I heard of her mother's gradual sinking, from senile gangrene of a foot, which I suppose poisons the system with septic matter.

Thank—what? or whom?—the days are lengthening, and hope is in the air! I trust your comparative rest is acting as favourably as you could wish. In a little more than a month I shall be moving on, possibly to the New Forest for another month before venturing to London.

Ever affectionately yours, FRANCIS GALTON.

I wonder if you happened to see a column of mine based on Schuster's work, about Sequestrated Church Property? The *Daily Mail* ordered a Reporter to interview me, who learning at Rutland Gate that I was here, asked for instructions and was ordered to Dorking by the Manager. He had to walk 2 miles each way on a bitter evening and all for a paragraph. What hustlers Editors are!

7, WELL ROAD, HAMPSTEAD, N.W. *February 4, 1909.*

MY DEAR FRANCIS GALTON,   I wrote to Hartog, the University Registrar, asking him to get Heron reappointed for a third final year, and Miss Elderton's scholarship also extended and raised to £—. I told him that you had been consulted on the point, and that you generally approved. It might be well for you to write a line to him to show that we have talked over the proposal. I have had a good deal of worry and delay over the *Treasury of Human Inheritance.* It is a gigantic task. I think the disease pedigrees alone run to thousands, mostly in out of the way journals and dissertations, not accessible in England. But I hope the experience of this first part will make the others easier, and get the contributors running smoothly in definite grooves. I had simply no idea of the amount of material that really exists nor of what this work may do to bring it to a focus!

Another point has been troubling me which I want to write to you about. Mrs Gotto has asked for Miss Elderton's and Heron's lectures for publication. I hope she will not think me churlish in feeling compelled to refuse. This refusal arises from more than one cause. Miss Elderton gave material and some results of work which is not yet finished, and which it is our duty to finish and publish in a form rather more academic than the publications of the Society. Heron not only gave work which he hopes shortly to publish in the Galton Laboratory *Memoirs,* but I gave him free run of my diagrams, some of which relate to work in progress, and of which it would not do to anticipate the publication. Neither had at the time thought of publication but only of interesting the Society in *work in progress.* I think you will see that it is not churlish, but practically desirable not to anticipate full publication.

Affectionately, K. PEARSON.

MEADOW COTTAGE, BROCKHAM GREEN, BETCHWORTH. *February 6, 1909.*

MY DEAR KARL PEARSON,   I have written to Hartog about Heron and Miss Elderton adding that I suppose he will hardly think it necessary to summon a Committee, but that if he does I am too infirm to attend it.

The *Treasury* will give great trouble to you, but you will, I hope, be able to divert a yet larger part of the office work to it. (I wonder if among the thousands of disease pedigrees you have included the important one given by Bedford Price, pp. 110–111 of the *Report on the Feeble-Minded...,* Vol. 1?) It is as you say, a gigantic work, especially at the outset. How I wish I could help; but I cannot, my working powers are now so small.

It will never do to allow the Eugenics Education Society to anticipate and utilise the Eugenics Laboratory publications. I will write to Mrs Gotto about it. I have written a brief seud-off to their forthcoming Review, in which I emphatically insist upon the difference between the work of the two classes of publication, that they are *supplementary,* and in no sense rivals. The Laboratory gives the foundation, the Society the superstructure.

We leave here towards the end of the month; as at present arranged for Lyndhurst in the New Forest, but I will write further.

I have got drawn into a publication about the Feeble-Minded, in which there are to be two collaborators, one being Sir E. Fry; if that falls through I retract also. In the meantime I have got all 8 vols. of the Report—a mighty mass of letterpress. Would it be acceptable and useful to the Library of the Eugenics Lab. if I sent it there when done with?

I hope your "half-time" gives a sensible amount of relief.

Ever affectionately yours, FRANCIS GALTON.

HAMPSTEAD. *February 7, 1909.*

MY DEAR FRANCIS GALTON,   Thank you most heartily for your very sympathetic letter. I agree so wholly with what you say—there is need for the purely scientific research, and for propaganda. I feel that the former demands two essentials: we have got to convince not only

47—2

London University but the other universities (i) that Eugenics is a Science and that our research work is of the highest type and as reliable and sober as any piece of physiological or chemical work, (ii) that we are running no hobby and have no end in view but the truth. If these things can be carried out we shall have founded a science to which statesmen and social reformers can appeal for marshalled facts. If our youthful efforts were mixed up in any way with the work of Havelock Ellis, Slaughter or Saleeby, we should kill all chance of founding Eugenics as an academic discipline. Please don't think I am narrow, or that I do not admit that these men have done or may do good work. All I say is that I could not get the help we are getting from the medical profession, from pathologists or physiologists, if we were supposed to be specially linked up with these names. Rightly or wrongly it would kill Eugenics as an academic study. All I want is to stand apart doing our scientific work, not in any way hostile to the Eugenics Education Society, giving it any facts we can or an occasional lecture, but not being specially linked to it in any manner. For this reason I am rather sorry that D. has gone on to its Council, because it makes a link, which I think it is better for Laboratory and Society not to forge—it will hamper the freedom of both. My policy, however, with my young people is to show them my own standpoint, but in no way to control their action. Unofficially and privately I shall always be ready to aid the Society.   Yours affectionately, KARL PEARSON.

I think we have a copy of the Feeble-Minded Report, but it is needless to say that we shall hail your gift if we have not. I know that Miss Barrington has been at work on the pedigrees in it.

I am in a state of most irrepressible excitement! I believe I am on the track of a far-reaching clue, namely the effect of presence or absence of *internal* pigment, especially that of the brain centres in mammals. I think it is going to explain why deaf-mutism, imbecility and albinism occur in the same stocks. Don't reveal my secrets! But I believe the ordinary albino *has* internal pigment; the imbecile lacks at one or more brain centres internal, but he does not lack external pigment, and the deaf-mute lacks pigment in the membrane of the perilymph chamber, the "retina" so to speak of the ear. The imbecile deaf-mute albino, *who spins like a waltzing mouse*, lacks pigment everywhere. The waltzing mouse is a partial albino. The partial albino cat with blue eye and white coat is a deaf-mute. The wall-eyed horse tends to "spin." The perishing of the internal pigments of the brain leads to senile insanity. Most of us lose only our external pigment with age. Everything fits in and it ought to give a grand connected theory of degeneracy. Of course it may all prove a dream! On Friday night there was an autopsy on an albino and much may turn on it, if the internal pigments are shown to be there. Mott is examining the pigmented centres in the brains of imbeciles and deaf-mutes for me. Of course my hypotheses may all collapse, but so far it seems to be the first connected theory of why imbecility, albinism and deaf-mutism run in the same stocks. I expected that imbeciles might "spin," and I find from inquiry that the "spinning imbecile" is a known type. There is an American at Colney Hatch, Mott tells me, who continually spins like a whirling Dervish. I shall put that down, if my theory works out as akin to Hamilton's prediction of the "conical points" of the wave surface! Don't laugh at me too heartily!

7, WELL ROAD, HAMPSTEAD, N.W. *February* 10, 1909.

MY DEAR FRANCIS GALTON, Your *Foreword* is most kindly and all we could possibly want. I saw Mrs Gotto to-day and tried to explain to her that our position was one of sympathy but independent action. We must not make ourselves in any way intimately associated with propagandism. The medical men are coming in and giving us splendid material for the *Treasury*, often confidential and personal histories. But Saleeby and others on the Eugenics Education Society's Council are red rags to the medical bull, and if it were thought we were linked up with them we should be left severely alone. I think it a very great thing to have won even partial confidence from a portion of the medical world, and if we can keep it and extend it, we shall have really done a great stroke in forwarding the scientific side of Eugenics. I have mentioned all this because I believe Mrs Gotto thinks me unreasonable, but we should only hamper each other's movements, and to make Eugenics an academic study and get the medical world to aid us will be one definite piece of work done on one side of the movement, and as you say this sort of work can be a foundation for the other.   Yours affectionately, K. PEARSON.

MEADOW COTTAGE, BROCKHAM GREEN, BETCHWORTH. *February* 20, 1909.

MY DEAR KARL PEARSON,   Can you give me a line of guidance as to the value of Miss Mary Dendy's data on Feeble-Minded Children, of which she sent you copies? (She fears they were not satisfactory.) My reason is that I have corresponded with her and she estimates that, for every F.M. her institution takes in, two F.M.'s are prevented from coming into existence. I asked for the grounds of this estimate and she writes offering to send masses of original or of copied data, which I do not want. My object in writing to you is merely to learn in a *general way* whether her grasp of statistics seems to you to be fairly good, or otherwise? I was much struck with the goodness of her evidence.

All this arises out of a forthcoming little book from Cambridge, about which the Horace Darwins and the Whethams are keen. Its purpose is to give a short account of the contents of the Blue Book. They have persuaded Sir Edward Fry to write a short (and excellent) preface, and me to write a short paper also, which I have done, calling it "Segregation." The weakest points in this are want of good evidence for the great *average fecundity* of the F.M. women, and for the *happiness* of the segregates in labour-colonies, etc. It was as to the former of these that I wrote to Miss M. Dendy, and I have suggested that she might be asked to contribute also as to the latter. I am sorry to bore you with all this rigmarole.

We leave here on this day week, Saturday 26, for the Crown Hotel, Lyndhurst, where I have taken rooms for a week certain, with power of staying on. How lovely this weather is!

Affectionately yours, FRANCIS GALTON.

UNIVERSITY COLLEGE, LONDON, W.C. *February* 21, 1909.

MY DEAR FRANCIS GALTON,   We have in the Laboratory eight or nine MS. volumes covering the records of nearly 1000 feeble-minded children provided for us by Miss Dendy but copied at our expense. We have only partially analysed these, and we did not go steadily at them because we had our doubts as to whether in the cases of relatives no entry meant in all cases that the relatives were sound, or that it was not really known whether they were sound or not. An examination of our data for Birmingham and Manchester showed such very different percentages of alcoholism and insanity in the F.M. stocks, that it did not seem feasible to advance farther without more certainty of the method of examination and record. Miss Dendy was *most* kind, and, of all the people working at the feeble-minded that I have come across, the most business-like in her record and her talk. *But* in a long personal interview which Miss Elderton and I had with her, we did not feel confident that the categories "sound" and "nothing known" had been really kept apart. In few cases had the inquirer gone beyond the mother and investigated the weight to be given to her answers. We could not press the point further, because Miss Dendy rather resented our cross-examination as a charge on her own credibility. I did not feel that her data were untrustworthy, but I did not feel confident enough about the point mentioned to undertake heavy work on them, while we had better material unreduced. I believe that as far as the size of fraternity of feeble-minded goes, the results are *quite* trustworthy and I have used them, but I should use them with regard to heredity somewhat cautiously.

The next point we come to is exceedingly difficult. You may take it as certain that the feeble-minded stocks are very prolific. But the feeble-minded girl or woman is not generally selected as a wife. She is seduced and often bears illegitimate child after child in one or other workhouse. You will find a good deal of evidence for this in the Report. Often she becomes a prostitute and loses her power of bearing children. I do not think that in any of these degenerate cases, the actual degenerates are so socially dangerous as the degenerate-bearing stocks, which are generally most fertile. A great many epileptics are, however, married and appear to bear largely feeble-minded, albinotic and insane, as well as epileptic children. I should certainly think Miss Dendy was correct, however, in saying that to segregate a feeble-minded girl is to save society from one or two feeble-minded, or more accurately degenerate, children. I have heard from more than one woman who works among the feeble-minded, that at certain ages and times they cannot be allowed out for five minutes without offering themselves to the first man they meet. You have in their cases the imperial passion unrestrained. Does not this answer your second question? Given such a dominant impulse, and prevent its fulfilment by segregation, how can the segregated be "happy"? It will be like a caged and foodless animal with plenty outside its cage. You can hold that the restraint is better than the freedom in its ultimate bearings on the happiness of the individual. But you will not get these purely animal and uncontrolled natures to regard it from this standpoint. The justification for the segregation must be, I fear, not their "happiness," as judged by themselves, but the profit to society at large. The madman might be "happier" seated in the market-place with a paper crown on his head, but we wisely segregate him for our own and not his happiness, even if he is quite harmless. I think this is the only line we can take with regard to the feeble-minded.

Mrs Hume Pinsent, one of the Feeble-Minded Commissioners (address Lordswood, Harborne, Nr Birmingham), has much evidence on the need for segregating the feeble-minded. She is a sister of Mr Justice Parker and a friend of mine. I feel sure, if you were to write on this point as a *definite* point, i.e. whether feeble-minded girls were responsible on an average for one or two degenerate children, she would reply to you with her wide experience. At the same time we were not able to do more with her inheritance data than I fear we can with Miss Dendy's, there was a lack of the requisite information, and especially of a distinction between "known to be sound" and "no information."

I am rather anxious about the success of our first lecture on Tuesday. I do not in my MS., alas! seem to say effectively what I want to say. Always affectionately, KARL PEARSON.

MEADOW COTTAGE (on Saturday I shall be at the Crown Hotel, Lyndhurst).
February 24, 1909.

MY DEAR KARL PEARSON, G. K. Chesterton's paragraph is too grotesquely absurd to be worth noticing. His name and paragraph might however be kept in mind for a future "Dunciad," in which specimens of current nonsense might be quoted. I am most desirous to hear about your lecture and the audience. Ever affectionately, FRANCIS GALTON.

My only surviving brother* died yesterday; the result of an accident, practically.

7, WELL ROAD, HAMPSTEAD, N.W. *February* 24, 1909.

MY DEAR FRANCIS GALTON, I am so sorry to hear of your loss, which I know will mean another rending of the ties with the past. I appreciated your feelings towards the various members of your family so much better after reading the "Memories," and have very vividly in mind what you say of this brother.

There is another point I have been thinking about very much and I want Miss Biggs to second my endeavours. I wish so strongly you would have a bust made by a first-class man. Pictures are excellent but only one person can possess them, whereas a good bust means fairly easy economic multiplication in plaster. My thoughts have been turned to it recently because of Hope Pinker's present to me of Weldon's bust, and I have just purchased a cast of Montford's Darwin to match it in the Biometric Laboratory.

About the lecture there is I fear little to be said. The audience, about 57, was very attentive and I think quite earnest. But I made a mistake, I *read* and did not speak, so I only got about half through my material. My points were non-inheritance of acquired characters and so little permanent effect of nurture, small direct effect of nurture compared with nature, old maintenance of standard by relative death-rate, and our need to replace that by a selective birth-rate; but I did not properly get to the latter. I shall probably give a second lecture next time, and postpone Heron's for a week. I cannot say that I was satisfied with either my material or my treatment of it, but I am not "in fettle" just now.

Affectionately yours, KARL PEARSON.

CROWN HOTEL, LYNDHURST. *March* 1, 1909.

MY DEAR KARL PEARSON, There is much to read and write about, but I am hardly up to much just now. We motored here on (snowy) Saturday and I reached my limit of resistance; consequently I was tucked up in bed all yesterday, with happily perfect results, so that except for weakness I am quite at my normal again. These quarters are singularly comfortable and I expect to stay here for a month. What an issue of Eugenic work this past week! I was re-reading your Oxford lecture (*Mem.* in Table X, correct Female to Male in the 2nd or 3rd line) and had

---

* Erasmus Galton, aged 94.

received Miss Barrington's (and your) paper when I left. It will be read as soon as pressing arrears are worked off. The *Western Morning News* of Friday has just reached me. What an unusually sensible and forcible article! It shows that Eugenics is being taken seriously at last.

You may recollect that I told you of a clever *Punch* cartoon of the Lord and the Bull, which I failed to find again. Miss Burnand, half-sister of the caricaturist, has made it out for me, at last. It was by Du Maurier, and is in *Punch*, March 20, 1880. When you are in easy reach of a collection of *Punch's* volumes do look at it. It ought really to be utilised somehow, possibly by the Eugenics Education Society. I will suggest it to them. About the bust you suggest—is not a bust rather a "White Elephant"? Eva Biggs and I will talk it over. A small thing that could stand on a chimney piece with other things, seems better.

Thank you for your kind words about my brother, whose cremation took place on Friday, very quietly by his express wish and no mourning to be adopted by his relatives. So I do not write on black-edged paper. It is strange that a living human being should so quickly be reduced to four handfuls of ash, and that scattered over the soil of a garden. The whole thing has rather upset me, as is but natural. Affectionately yours, FRANCIS GALTON.

7, WELL ROAD, HAMPSTEAD. March 7, 1909.

MY DEAR FRANCIS GALTON, I am so glad that at any rate you will consider the possibility of a bust. It has, from my small experience, seemed so sad that this sort of thing should be badly done after a man is dead, when it can be effectively done only when he is alive. Besides this, we have the (to me) all important point that a bust is capable of good reproduction at a moderate cost, if the initial cost be great indeed.

Punch, March 20, 1880, is within 10 feet of me, but we can't open the cupboard because the wall has settled and jammed the doors, and *Punch* from 1850 onwards is at present inaccessible, until a carpenter is forthcoming. I brought it from my Father's, but the weight was too much for the wall, and the result is as above! We will put up a copy in the Laboratory.

Now I want to ask you if you remember closely any more Darwins. Horace Darwin has a wen near the right ala of his nose. I had often noticed it and thought nothing of it. Yesterday on my bust of Darwin I noticed that there was a projection in the same position and see it is is a wen! Then I find it also on the 1881 portrait of Darwin. Do you know if it occurs in George, Frank or Leonard, or any of the older generation? It is extraordinary how blind one is at times. I knew Horace's wen quite well and never realised it as a marked inheritance, until my eye caught it on the Montford bust, and I had verified it on Charles Darwin's portraits!

As to cremation, both my parents were by their special desire cremated. It seemed to me so far less repulsive than the ordinary earth burial, but the preliminary ordeal was very galling. The law assumes that you have probably poisoned your relative and proceeds to a system of cross-questioning attendants and nurses which may be most painful in the hands of an unsympathetic local officer or magistrate. Affectionately, K. P.

CROWN HOTEL, LYNDHURST. March 8, 1909.

MY DEAR KARL PEARSON, The chill caught motoring sent me fairly to bed, starvation, and doctor (a good one), but I am now in the drawing room again and convalescent. Otherwise I should have written to say how gratified I had been at the account sent me by Heron of the success of your last lecture. That 10°/ₒ of the population are producers of half the next generation shows the possibility of promoting the well-being of the nation by concentrating attention on a comparatively few families. I rejoiced too at the slashing conclusions of the memoir on Hereditary Vision.

About the bust, it seems that the sculptor brother of Charles Furse lives within easy railway distance of here. Eva Biggs is making various inquiries before fixing anything.

Methuen has sent me a substantial cheque (£66) on account of my *Memories* up to the end of December, and the sale has proceeded since then and is proceeding—which will go towards paying for the bust. There are other funds also, similarly available. I have not been out of the house since I arrived here 9 days ago, but am convalescent now and look forward to seeing soon something of the pretty neighbourhood.

Heron wrote that you looked very *fit* at the time of your lecture. The overpowering weight of *Punch* is amusing. Ever affectionately, FRANCIS GALTON.

7, WELL ROAD, HAMPSTEAD, N.W. *March 9, 1909.*

MY DEAR FRANCIS GALTON, I am extremely sorry you should have had this invalid time after your change, and rejoice that you are now able to come down again. The past fortnight has been very trying for one and all of us, and it is sad to find so many of one's young folk laid up. I have had two assistants down and many students—influenza as usual.

I heard Heron to-day and, though he was very nervous, he did quite well and held his audience—one of about 45 to 50. He has a good delivery and will, I feel sure, become a first-class lecturer. He did not grow monotonous, and he had plenty of material and resource. Perhaps he might reiterate his points, as he makes them, a little more; it always helps a general audience to be told beforehand what is about to be proved, and to be told afterwards that such and such a point has been proved. But this omission is general with young lecturers, who do not know the density of the average human, and it is capable of easy correction. I must not write more now. Affectionately yours, KARL PEARSON.

Hartog wrote to me about the Report for the year on the Eugenics Laboratory, saying that it ought if possible to be in this month. I have sent him some account of our work, suggesting that it should go to the Committee first. I have—I fear rather hurriedly—detailed what has been done, what is in the doing, and what possibly might be done as to staff, etc. I think when you come back to town, perhaps in May or so, it might be well to have a meeting of the Committee and discuss the future work, and if the Laboratory goes on, what line it should take.

CROWN HOTEL, LYNDHURST. *March 10, 1909.*

MY DEAR KARL PEARSON, I hope to be fit in May (latish) to take part in the proposed meeting of the Committee to discuss future work of the Eugenics Laboratory. I feel that its work depends so largely on yourself, that I shrink from suggesting anything. You, not I, know what is feasible, and I bear in mind that you want, and may ask for and get, a complete holiday for a year or so. Whatever under the circumstances commends itself to you as the proper course to lay down, I am practically certain to agree to, but I should make a hash if I endeavoured to do so myself. How far can Heron and Miss Elderton stand alone? With your support and supervision they do their work admirably, but without it I should fear errors in planning. There are so very few besides yourself competent to supervise, and you may begin to feel the onus of doing so too great for continuance. Tell me, please, exactly what you think about this. I am very glad that Heron's lecture pleased you, and that you think so highly of his powers and promise. Miss Elderton (with her brother) has just concluded and sent me a typed copy of the elementary book that I proposed should be written (in my Oxford lecture). I have seen some parts of it already, and must go through it to-morrow. Excuse more now.

Ever affectionately yours, FRANCIS GALTON.

7, WELL ROAD, HAMPSTEAD, N.W. *March 18, 1909.*

MY DEAR FRANCIS GALTON, This is a line to say that Heron's second lecture went quite nicely. He discussed physical inheritance in man and dealt with attacks on your eye-colour data from the side of Davenport and Hurst, who assert that two true blue eyed parents always have blue eyed children. His audience was quite good and there were some new faces. My report on the Eugenics Laboratory was drawn up rather hastily, because Hartog said it must be in by February. Your kind letter as to the future of the Laboratory shall be replied to with a suggestion or two during the Easter vacation, so that you may have time to consider matters before the meeting in May.

Now as to the Eldertons' booklet. I have not yet seen it; it is something which they have done quite off their own bats, and I am very curious to read it. I think it would probably be quite an addition to our smaller format series and help that on, but I am not sure whether it would get the same sort of circulation that a well-known publisher would procure for it, as we spend very little on advertisement. We must consider that point. There is a stupidly hostile article on Eugenics in the *Nation*. I have got my *Punch* cupboard forced, and we are much pleased with the "Bull and the Earl." I think we must get an enlargement in sepia made for the Galton Laboratory, so don't give it away as a crest for X.

Did you get a copy of the *Treasury*, Parts I and II, last week? I should like to know if you would care for other copies, and also that you are not very disappointed with it.

Yours affectionately, KARL PEARSON.

CROWN HOTEL, LYNDHURST. *March* 18, 1909.

MY DEAR KARL PEARSON, Excuse pencil. The demon *lumbago* has planted beak and claws into my loins and sent me helplessly to bed. I have a good doctor and a skilful man-nurse, besides my niece and her maid. Also this hotel is most comfortable, so there is no cause for murmuring.

Hartog will fix some day for the Committee in the latter half of May that will fit in with other work. He is much pleased, as well he should be, with the work done under your supervision at the Eugenics Laboratory. What an immense amount of information, closely packed, there is in the *Treasury*! I congratulate you heartily upon it. Excuse more for I write under difficulty. I am curious to learn how you will arrange about the Laboratory during your holiday absence, which I trust will be both enjoyable and healthful. It is good news that Heron lectures so well and is so promising.

The "Bull and the Earl"—I suggested it for a vignette, if arrangement could be made, in the new *Review*. M. Crackanthorpe rose to the idea, but I don't know what will come of it. I wonder whether you noted a *most* Eugenic undertaking in last week's *Spectator*, signed by (Lady) Constance Grosvenor. I sent it—just in time—to M. Crackanthorpe for the *Review*. It is worth reading and digesting.

Ever affectionately but crippled now, FRANCIS GALTON.

HAMPSTEAD. *March* 20, 1909.

MY DEAR FRANCIS GALTON, I am indeed sorry to hear of the lumbago and I know by recent experience, how trying it is. The only point I know in its favour is that it goes as quickly and mysteriously as it comes! And I trust this will be your case. Curiously enough only a post or two earlier I heard from my chief craniological worker—Dr Crewdson Benington—that the fiend had seized him. I don't think you need anticipate that I shall be a long time away, probably I shall only try to get a complete summer holiday. I have got Dr Goring and two assistants coming on May 1st for a year to reduce the measurements made on the criminals in H.M. Prisons. It will be a gigantic piece of work as there are about 40 characters, physical, mental and moral, in more than 3000 criminals, and it ought to be the first real piece of criminal anthropometry effectively reduced and discussed. The Treasury are paying for the assistants and granting Dr Goring a year to do his work in. This will keep me fairly closely at it, and it ought to throw light on many Eugenics questions. It is the first "semi-official" recognition of our statistical laboratory.

I am rather anxious to see what support the *Treasury* meets with. We have made a bigger venture than anything since *Biometrika* was started, and I don't know whether the medicals will rise to the occasion. Let me have a card to say how you get on.

Affectionately, KARL PEARSON.

CROWN HOTEL, LYNDHURST. *March* 22, 1909.

MY DEAR KARL PEARSON, The lumbago, after one week in bed, *is* "mysteriously disappearing." Allah be praised!!

I write now about the Eldertons' little elementary book, for the cost of publication of which I am responsible. It never occurred to me before, but the Eugenics Education Society are just the people to publish it. It is exactly in their way. They *have* published several essays and are about to republish mine (I received a letter this morning), and as the writing of the Eldertons' book was due to the suggestion in my Oxford lecture, it would come into their series of publications with aptness.

Thank you for all you tell me. We shall be turned out of this hotel about the end of March by a crowd of hunting men who make a practice of coming to the New Forest in April. Whether I then return to town, or stay out longer, depends much on the caprices of the mysterious lumbago. It is wonderful what capable and well educated young doctors one finds now in such small places as this. Ever affectionately yours, FRANCIS GALTON.

I am writing to Miss Elderton.

7, Well Road, Hampstead, N.W. *March* 25, 1909.

My dear Francis Galton, I have not yet seen the Eldertons' MS. but I suppose I shall eventually. I shall be quite ready to publish it as a Laboratory publication if that seems desirable to those concerned. I am not at all sure, however, that it would not be well to try it with a good publisher first of all. It would save the expense of publication and get a reasonable amount of notice from the Press and advertisement. I think Mr Elderton is a little frightened of the idea of the Eugenics Education Society. The Society, *I* think, is doing good work, but there are some names associated with it, that some people (probably without basis) fight shy of, and of course he has to be rather cautious that he does not link himself with anything that would affect his Office. Please remember I have no authority for this view, but it is a possible inference from what I found was the feeling about the Eugenics Education Society publication and I thought it might be worth suggesting as possible. You will no doubt consider the matter all round. As I said, I am ready to do anything in regard to publication.

You will be amused to see that you and I are " Moralstatistiker," whatever that may mean.

We go away to Great Missenden on Friday, April 2nd. I hope the lumbago is now quite mastered. Affectionately, Karl Pearson.

I have sent in to the Royal a criticism of Darbishire's recent paper " An Experimental Investigation of the Influence of Ancestry," which I should like to have your views on when in type. I have also found out that while Mendelism gives—judged by *somatic* characters—a correlation of $\frac{1}{3}$, if we correlate *gametic* characters, this rises to $\frac{1}{2}$, so that it is the assertion that the hybrid shows the dominant character which makes the difference.

Forest Park Hotel, Brockenhurst, Hants. *April* 4, 1909.

My dear Karl Pearson, I put off writing until we were safely established here, where we arrived yesterday and where two of my nearest relatives are coming to us for a few days. My five weeks in Lyndhurst have been those of an invalid, but I already feel the good from a change of air and am less pessimistic. You will be, I know, at Missenden, but your letter gave no more exact address, so fearing miscarriage I send mine round by Hampstead. You have duly impressed the *Medical Gazette* with the Eugenic microbe, I am glad to see. The proof sheets of the Education Society's *Review* have at last reached me. It seems on the whole creditable, but more definite work by them is needful, and will come in time. I am curious to learn what you think of the Eldertons' attempt. I hope you are all well placed at Missenden. This is a very nice and cheerful hotel. I have nothing to say that you would care to hear, having been in a sick chamber so long, and dependent on a man-nurse—a very good and quiet man, by the way. May invalidism long keep away from you and yours, is my hearty wish!

Ever affectionately yours, Francis Galton.

Ickenham, Great Missenden, Bucks. *April* 6, 1909.

My dear Francis Galton, I was very glad to get your note this morning and hear that you had got into comfortable new quarters. I feel sure this springlike weather will do us all good. I ran eight miles on my cycle to-day—a great achievement for me now-a-days—and I do not feel the worse for it. Sedley Taylor of Trinity came in with a friend of ours to tea, and I had some Eugenics talk with him. One point he told me—namely that they had a portrait of you at Trinity—was news to me and good news. It is much that at Cambridge anything but Batesonism should be recognised. I have been hoping to get on with the Albino paper, but much old work has stopped the way. I had first to finish the two Mendel papers for the R.S. and then to write a rejoinder to the attack in the R.A. Society's *Monthly Notices*. Yesterday I got the Eldertons' book and read it through last night. I think it on the whole very good. One has got to remember that they have not 30 years of experience in lecturing behind them. There are many things I should myself have illustrated more copiously with diagrams and models, and I think the chapter on the probable error wants further illustration. This I have suggested to them. But take it all in all I think they have done a difficult thing creditably—better than I in the least anticipated. The right thing would have been in the old days one of Macmillan's Science primers, but I don't know whether they still issue them or anything like them. You will perhaps remember Clerk Maxwell's on "Matter and Motion" and Huxley's "Introductory

Primer"? This would be exactly the right "format" and express the extent and aim of the work. I think it should pay its own way and a good publisher would take it, especially if *you* wrote a few introductory lines.

I sent the last plates of the next part of the *Treasury* back yesterday, and the text is all ready but for a section—bibliography and introduction—on inheritance of the insane diathesis, which I can't get out of the man who has promised it.

I shall look forward with much curiosity to the *Eugenics Review*. It seems to me that at present there is so very much spade work to be done, and that we are apt to go astray if we merely discuss without the necessary groundwork of facts. I suggested to the Society that it should set about definite pieces of work in the way of collecting material, but I doubt if many of its members have the true scientific instinct. Mrs Gotto said she had two academic persons who wanted to do eugenics research work and asked for suggestions, and Heron naturally said "Why not send them to the Laboratory, we can always find work for them to do, and are endowed to do it." But Mrs Gotto seemed to think that they ought to work under the guidance of the Society. She also asked for copies of our schedules concerning family characters and diseases. These I was perfectly willing to provide her with, on condition that when filled in they were returned to the Laboratory. This did not, however, seem to be satisfactory to her. I think it is not possible for us to provide schedules, which are issued for definite pieces of work in progress in the Laboratory, to be used independently by other investigators. This has actually been done in America and Scotland, persons borrowing our schedules on the excuse that they were going to return them to us, and then using them to collect facts for themselves! It does not seem quite playing the game! One enthusiastic American got 100 of my schedules, which he said he would return to me. He used them for his own purposes and never a one did I see again! My own view is that our work lies in different fields and is supplementary, but I fear the Eugenics Education Society will not accept this view, and does not fully grasp that we can be quite sympathetic, but must do our own work in the narrower field of statistical research.

I have a nice letter from Lady Welby, but asking a question rather beyond me—why a grouping of three is more frequent than other groupings. I fear she would think me flippant if I suggested she should make a frequency curve of the odds at the principal race-meetings during the year, to see if there is a basis for her statement! Have you noticed the effect of political feeling between England and Germany? Formerly I could always get a civil answer if I wrote to a German librarian or scientific man asking a question. Now I rarely get any answer at all! Affectionately, KARL PEARSON.

FOREST PARK HOTEL, BROCKENHURST, HANTS. *April* 8, 1909.

(I stay here until April 17 and then move on.)

MY DEAR KARL PEARSON, You tell me much of interest. You will, of course, gently snub Mrs Gotto, if she goes too far in her zeal. I have expressed as emphatically as I can, in the "Foreword" to the forthcoming Eugenics Education Society's *Review* (due a week hence), my view of the distinctive character of the work of her Society. It can only popularise, and work upon foundations laboriously laid elsewhere.

I am glad you approve on the whole of the Eldertons' primer.

I have written to Miss Elderton by this post, suggesting that the consideration of *where* to publish should be deferred till after this number of the *Review* has appeared, which will indicate its probable future *status*, and may advertise the books it is intended by them to issue, among which, if their proposed programme is carried out, the primer might suitably be included and get advertised where likely purchasers would see it.

I am very glad that you feel your strong physical powers returning. About the portrait in the College Hall of Trinity, Cambridge; shortly after they elected me to an Hon. Fellowship, two College Dons saw Furse's portrait of me at my house and suggested that I should offer a copy of it to the College. This was done, and I must say that it is an effective addition to their collection, both because the picture is a good one and because its background is somewhat light coloured and shows up very well against the dark oak panelling.

I congratulate you on the forwardness of the next part of the *Treasury*.

Lady Welby is irrepressible in her inquiries. She was with us at Lyndhurst for more than a week, full of mystical *triads*, etc. and much else. Socially she is very charming and good.

What painful evidence you give of the modern tone of German feelings towards us innocents! Wishing you all possible Easter pleasure and success.

Ever affectionately, FRANCIS GALTON.

42, RUTLAND GATE, S.W. *April* 22, 1909.

MY DEAR KARL PEARSON, It has been a grief to me, that my doctor who has just been, will not allow me to go to the Royal Society this afternoon to hear your paper. We came up yesterday, a day earlier than intended, and stay in town till next Wednesday! I write this to Hampstead not being sure of your present address. Affectionately yours, FRANCIS GALTON.

It did me good to read your solid writing in *Biometrika* about the mulattoes, etc.

ICKENHAM, GREAT MISSENDEN, BUCKS. *April* 22, 1909.

MY DEAR FRANCIS GALTON, ...I have to thank you for a copy of the *Eugenics Review.* I heartily approve of the *Foreword* and your clear statement of the position of affairs. I think the review will stir people up and lead them to think about this all important matter. A good deal of the text is a little "thin," and some statements a bit misleading. For example, the penny-a-liner report of what Mr Gilbey said at the Police Court regarding the *deaf-mute* woman ought not to have been inserted without verification. The case, of which I have the full pedigree, is worse than appeared at the Police Court, but Mr Gilbey did not make the absurd statements attributed to him. He has been working so loyally to help the Eugenics Laboratory collect deaf-mute pedigrees, that I am sorry to see this stupid misstatement of one of the $\frac{1}{2}d$. papers reprinted, and hope he will not suppose us in any way responsible. I think the Journal will be a success and do good work the nearer it approaches the standard set by the *Archiv für Rassenbiologie*, which I see noticed in their pages. But, of course, it would be hypercritical to expect that standard at first.

I am returning to town on Monday, and send this to 42, Rutland Gate, as I do not know your present address. I trust that you are well, and in a warm district. I have been renewing many sad memories of W. F. R. W.'s and my early excursions and his pond dredgings on this side of the Chilterns in the first autumn of his Oxford life. Affectionately, K. P.

42, RUTLAND GATE, S.W. *April* 23, 1909.

MY DEAR KARL PEARSON, You will probably get this about the same time as one I wrote yesterday. The Doctor was quite right; when the time of meeting approached I felt quite unfit. My circulation is playing tricks.

R. wrote me a long letter, not a word in it either about yourself or about *Biometrika*, but appealing for pecuniary help. Knowing that there must be some long story in the background I did not answer it, the more so as I do not see my way to do what he asks.

The *Eugenics Review* is rather feeble, but may mend and I think will do so. I wonder if you noticed Crackanthorpe's blunder about improving the sight of hawks by breeding! I have just come from a hawking district and am assured that hawks never breed in confinement, but are caught wild when young and trained afterwards, as elephants are. Thank you for what you tell me about Mr Gilbey. I shall see Mrs Gotto here this afternoon and will tell her. You speak of W. F. R. W.—how memories crowd upon us, unexpectedly. I do not leave here until Wednesday, if then, for I suspect the Doctor may be averse to the exertion of the change. Coming here tired me a good deal. I am most curious to read your paper of yesterday.

Ever affectionately, FRANCIS GALTON.

ICKENHAM, GREAT MISSENDEN, BUCKS. *April* 24, 1909.

MY DEAR FRANCIS GALTON, I enclose two notices which you might miss. You shall have copies of my R.S. papers as soon as I have them. I have only seen a single proof of each: they do not appear to send round copies as of old to the author. I hope to have in a few days my lecture on the "Groundwork of Eugenics" for you. I hope you won't judge it too severely, as it had to be done in the pressure of term time and I was in much haste. Still, I think it fills a gap in Eugenic literature. If you are in Town may I come and see you, to-day, Saturday, week? I won't stay long, but I should like to see you after the winter absence, if I may, and you are not feeling too overdone. Affectionately, K. P.

42, RUTLAND GATE, S.W. *April* 26, 1909.

MY DEAR KARL PEARSON,   Yes, *do* come on Saturday. Tell me beforehand at what hour I may expect you? The doctor has forbidden my going into the country again just now, at which I feel much relieved, not feeling up to moving again so soon. All you say of what you have published and are about to publish of course interests me greatly.

Affectionately yours, FRANCIS GALTON.

I return the "cuttings" with many thanks.

7, WELL ROAD, HAMPSTEAD, N.W. *May* 2, 1909.

MY DEAR FRANCIS GALTON,   I owe you a word of apology and Miss Biggs also. It was not till I got to the Station that I realised how late it was. You, I fear, must have found our talk very trying, but the time went so quickly that I was quite unconscious of how stupidly I was tiring you. You must put it down to your own power of not wearying others and forgive me; but if I come again I will keep my watch out! My heart was very full at seeing you so fixed to your chair, but ten minutes' talk showed me that you were really as active as ever and that consoled me. The wonderful part of life is that the problems are so manifold and as long as we retain our mental curiosity, there is no cessation to our activity or to the pleasure of life. I have felt this even in moments of physical disablement.

I have been thinking over the difficulty I saw was in your mind about the future, I hope the very distant future, of the Eugenics Laboratory. You must remember that at present the training in statistics does not lead to paid positions. It is beginning to, but the posts available are few and the best men who want to get on in life won't enter this field. But if your Foundation ever becomes a reality, there will be something for a strong man to look forward to, and this will act itself as an inducement. Also the time is coming when governmental and municipal work will demand men of the kind we are training. We are only a little bit (not very much) ahead of public needs. My strong view is that in a very few years there will be plenty of good men in this field. Now might it not be well to give the University a few years' grace, if the authorities thought fit to use it, before appointing a professor, after the endowment becomes actual? This would suffice to bring men into the field and save the University from the need of making an *immediate* appointment, if the right man were not forthcoming at once. Lecturers could be appointed for a year or two, and the Library extended and developed. For example, a period of five years fixed, in which the University would have time to look round, and until a professor was appointed 50°/₀ of the endowment might be used for continuing the Eugenics work by lectureships, etc., and 50°/₀ go towards a permanent endowment for publication and library funds. I believe that this period might never be used at all, or only some of it, but it would save the University from a compulsory appointment if the right man were not ripe for the work at the first opportunity. I feel so strongly that you have in this matter just met a great future need, that one would deprecate any first appointment which would not be an all-round success, or of appointing someone who would not be willing to make use of all the material and the connections which the Laboratory has now established. Given a few years' grace and the man will be forthcoming. In the last four or five years I have had at least two or three really strong men pass through my hands, but I could not frankly say: "Stick to statistics and throw up medicine or biology because there is some day a prize to be had." I feel sure, however, with a future, such men will naturally turn to Eugenics work. Only this last winter one of my American students said: "I wish I could go in for Eugenics, but my bread and butter lies in doing botanical work. I know that definite posts are there available." And that was precisely the case with Raymond Pearl, who has now got the control of an Agricultural State Breeding Station—he was far keener on man than on pigs and poultry, but the public yet has not realised that it needs breeding also! Well, if you will only make up your mind to stay with us a few more years this will right·itself! There must be sooner rather than later a government statistical bureau, and this will demand trained statisticians. Once we have a flow of such men who mean to make statistics their profession in life, there will be ample material to select from. At present the biometrician is the man who by calling is medical, botanical or zoological, and he dare not devote all his enthusiasm and energy to our work. The powers that be are against him in this country.

Now I shall weary you as much with my letter as with my talk. The only further matter I have in hand would be this. There is some talk of Heron going to St Andrews as lecturer on statistics. I would rather see him in a government appointment, e.g. Scottish Education Office or something of that sort. But he knows the ropes so well now that it would be desirable to keep him—if this does not come off this year—until he is appointed. I should therefore suggest for the next year, i.e. from February, 1910, this sort of monetary arrangement:

Fellow:	£200
First Assistant:	£120
Second Assistant:	£ 80
Draughtsman:	£ 45
Petty Cash:	£ 25
	£470

This would leave only £30 for publication, but, I think, there is still a balance which may be added to this and leave us enough to get through the year's publication. The draughtsman's or rather "draughtswoman's" appointment is the new feature. I have been paying Miss Ryley 10/- for each pedigree plate for the *Treasury* and we ought to get about 60 to 70 ready on the material we have now collected in the year. The market price would be hardly less than 15/- or 20/- according to the amount of work and we should have a full control of her time and energy. The work is very beautifully done, as I think you will see from the engraved sheets, and I should think it worth while to retain her for at least a year. If the above meets with your approval I will suggest it for sanction to the Committee and the University. If you approve and still feel not able to attend a meeting in May, I could bring it forward and we might see what the Committee think. Affectionately, KARL PEARSON.

42, RUTLAND GATE, S.W. *May* 4, 1909.

MY DEAR KARL PEARSON, Your visit was a treat, and did not tire me. When you come next, which I hope will be soon, don't look at your watch at all! My leg gets better.

I quite agree with your views in all their detail about the future of the Eugenics Laboratory, but delayed writing until I could look at the copy of my Will to see whether or no the circumstances you have in view would make delay in filling up the post impossible. I think not, but have written for a copy of the clause in question, for you to see. If a codicil be thought advisable, it can be supplied. It might be desirable to add a phrase after the words referring to the Establishment of the Professorship "within five(?) years after my decease," but I will think further about it, and will write when I send you the copy, in about three days. It is hard to steer between too much rigidity and too much slackness.

About your test for acuteness of eyesight, why not simply wind the test card to and fro, until just readable, when viewed through an "isoscope*", which the subject adjusts to his focus. The absolute value of the distances for a normal eye in terms of the just-perceptible difference, to be determined once for all by the operator. The personal index, due to + or − magnification by the cornea, etc., to be determined by the distance between the lenses of the isoscope, when vision is clear, at some definite distance or distances. *How* to calculate it is another matter.

Ever affectionately yours, FRANCIS GALTON.

42, RUTLAND GATE, S.W. *May* 6, 1909.

MY DEAR KARL PEARSON, Here is a copy of the paragraph in my Will concerning the future Professorship, which *might* be advantageously relaxed a little by a codicil. The codicil might state that the Professorship may be unfilled if the University so desire for a period not exceeding five years, its duties being carried on in the meantime in such way as the University may determine. Is that what you desire? Is there any other change or addition desirable? Very sorry to trouble you about this, but it is important.

Ever affectionately, FRANCIS GALTON.

I have £500 all ready for next year, to be paid to the University.

* See our Vol. II, p. 332.

7, WELL ROAD, HAMPSTEAD, N.W. *May* 7, 1909.

MY DEAR FRANCIS GALTON,   Reading the section you have sent me it seems so thoroughly good that it appears almost a pity to modify it. At the same time there might be a difficulty supposing at any time—at starting or after a successful career—there was no immediately available man for the post. I have made inquiries and find that in France a University is not forced by the existence of an endowment to fill a chair unless there be some person suitable to hold it. Thus at the Paris " École des Langues orientales vivantes," it is a recognised principle that under circumstances of this kind a professorship may be held in suspense until a suitable man be found, the teaching being meanwhile conducted by a lecturer who has no claims to fill the chair unless he should prove himself suitable in the course of his work. I think this is a very good provision. I should, however, be inclined to limit the vacancy to a period not exceeding five (or even four) years. That is time enough to test a man or two, and for the University to look round. The lawyers would probably know quite easily how to word a codicil, if they consider the present section does not permit of the power required. What is needed is power to hold the professorship in suspense for any period not exceeding four or five years if some person suitable to hold it is not, in the opinion of the Senate, immediately available, and to carry on the work by lecturer or lecturers until a suitable man be found.

Your talk led me to thinking over possible men. It is rather difficult to do so, because the man who might be good now might be too old, or changed in aspirations and form of work, long before any appointment will be made. Besides this, I cannot plan a future contingent on the death of a friend, who is one of the few sympathisers with all the work in hand and without whom it would seem all stale and unprofitable. But I want you to remember that there are still men like Palin Elderton and Raymond Pearl, who are thoroughly keen, full of the vigour of youth and strenuous to any extent, and that there is no reason to fear they will not have successors.

Now to another point which may amuse and interest you. Some account of the Francis Galton Laboratory has got into the Chicago papers, I do not know how or by whom it may have been written. An American writes and says that he is a friend of the wealthiest man in B——, who is immensely interested in Eugenics, and has been experimenting on horses to measure the effect of sympathy on conception ! All this seemed rather vague ! But he continues that his friend wants to know if we want help in any way at the Laboratory ! Of course I am going to write to him, and though nothing may come of it, I shall give ample details. Looking at the future, what I should suggest would be : (i) endowment for publication and education-work, (ii) the formation of a complete library of Eugenics, (iii) if he is inclined for a big thing, the building of an institute in which the future work of the Galton Professor and Laboratory can be carried on. Of course, nothing will probably come of it. These wealthy men are strange in their ways, and change their minds frequently. But if he seems to be keen and likely to do something, I shall, perhaps, ask you to write to him. Of course, one began to dream golden dreams of a Eugenics Institute, a hive of workers, under the control of your future professor—something like the Institut Solvay at Brussels—but I must not be hopeful on such a very slender basis ! Affectionately, KARL PEARSON.

I return the Extract. I have just been looking at Bateson's book. I take the place of Weldon as the butt for his contempt. There is not the least recognition of the fact that almost every one of the dogmatic statements made by the author a few years back have now been quietly dropped !

42, RUTLAND GATE, S.W. *May* 12, 1909.

MY DEAR KARL PEARSON,   Here is the draught codicil suggested by my lawyers, to whom besides writing on my own behalf, I enclosed your last letter to myself, as helping to make the position clear to them. About the "library," the addition of that word in the codicil as marked in pencil at the side of Clause 10, seems sufficient, but would you prefer more detail ? I should mention that the endowment will be considerable. Allowing 10 °/₀ for legacy duties, I reckon that its value will exceed fifty thousand pounds or, say, an income of £1500. But Chancellors of the Exchequer may make larger inroads than hitherto upon bequests. Excuse bad writing. It is upon my knees. Ever affectionately yours, FRANCIS GALTON.

7, WELL ROAD, HAMPSTEAD, N.W. *May* 13, 1909.

MY DEAR FRANCIS GALTON, I think the codicil will achieve what you want. The University should provide and I have no doubt will provide rooms, etc. for your Professor, so that I think there is no need to provide for buildings of any kind. Looked at with the experience of the last few years before me, it seems to me that what your man will need is (i) a fairly ample expenditure on books and journals; there is an immense amount of literature coming out now which wants collecting—-reports, journals, isolated monographs, etc., and (ii) an adequate publishing fund. If at any time the professorship were for a year or two vacant no better use could be made of any surplus, after paying for a trial lecturer, than *investing* it as a library or publication fund.

I think I told you a man wrote to me about a year ago and asked what he should do with £1000 he wished to leave in his will to Eugenics. I told him to leave it to the University of London for a fund for popular lectures and publications in Eugenics. I don't know whether he carried this out. To the American, who has recently written and asked what the Laboratory needs most, I have answered: Library and Publication funds. Now these may very likely not come off and then a free hand to the University in respect of *investing* unspent income for special funds of this kind would be of much value. A strong man (and your foundation will bring a strong man) wants freedom of this kind enormously. I know even in my own smaller way what the Drapers' Grant has been! I had not to think about how to get funds for a special bit of work, but had, within its limits, power to go and do a thing or get an instrument made without worrying over how it was to be paid for! It would have been an immense boon had I had it between 30 and 40, when one was in the prime of one's working powers, and had only a small private income.

You will see I have marked in pencil three points in the codicil. I think the word "extension" would indicate that you did not wish the work necessarily not to develop and grow during any interregnum. I have inserted the words "in the University," but these are only suggested on the assumption that you have them in view. If not, they would of course not be right. If not inserted, there might be claims on the University for aid in many ways—where you may have given aid. I think, perhaps, the word "initiated" might be introduced as helping to cover any or all work of the Eugenics Laboratory, which it might be considered worth preserving under altered conditions.

Miss Elderton gave quite a good lecture. She speaks with great clearness and is perfectly self-possessed. Her audience was about 45, and quite a good one in quality. Her lecture on Tuesday ought to be an interesting one as it will be the first attempt to give a quantitative comparison of Nature and Nurture.

I have got rather a heavy cold—the result of a chill on Sunday—-but I have not yet been kept to the house. I did not go to the Royal last night, but we sent some exhibits from the Laboratory, which you would have been the one person to appreciate fully, and you, also, would not be there! Affectionately, KARL PEARSON.

You say nothing about yourself, but I hope the leg if not yet down is better.

42, RUTLAND GATE, S.W. *May* 18, 1909.

MY DEAR KARL PEARSON, Best congratulations on the Eugenics pamphlet. It is so *massive* as well as popular, and nothing could be better than the letterpress. The diagram of the square box with peas, intended to be clustered thickly towards one corner, does not, however, tell its own tale. In the other diagram some dotted lines are wanting in parts. It is a real starting point for popular Eugenics literature of a high class.

I have received this morning from my lawyers the codicil amended in accordance with your suggestions. They have done it uncommonly well. Affectionately yours, FRANCIS GALTON.

7, WELL ROAD, HAMPSTEAD, N.W. *May* 18, 1909.

MY DEAR FRANCIS GALTON, Thank you very heartily for your kind letter as to the pamphlet. I hope it may do some good. The printing has been rather hurried because we wanted it out before the lectures came to an end. The plates were all made by simply taking photographs of the actual diagrams, models, etc. used in the lecture itself, to save the time and labour of redrawing or sketching. Thus they are rather crude and faulty. But this process, which I had never ventured on before, has the merit of great rapidity. It did not, however, show up properly the red amid the yellow peas in the box diagram.

I am glad you have got the clause worded satisfactorily. I think it will be a safeguard but I hope it will be unnecessary and that the right man will be found when the time comes.

Miss Elderton gave a very good lecture to-day—I think *quite the best* of the series—but for some reason her audience was rather smaller. It was a pity as the material was very good and she lectured fluently for over an hour. Affectionately, KARL PEARSON.

THE GALTON EUGENICS LABORATORY, UNIVERSITY COLLEGE. *June 9, 1909.*

MY DEAR FRANCIS GALTON, I think the enclosed letters may interest and amuse you. To Letter I, I replied that we need aid most for (a) a publishing fund and (b) a library, and that if Mr H.'s views went beyond these lesser matters he could build an Institute for the Galton Laboratory! Letter II follows! Just think of those "old and crooked" mares forwarded to England and arriving in Gower Street! We should have to tether them in the quadrangle! I have written again endeavouring in a friendly way to show that the highest results are not to be obtained by any experiments on "old and crooked" mares! But some Americans are very weird, and a better man than myself might have made something out of Mr H. "the wealthiest man in B——"! Affectionately, K. P.

42, RUTLAND GATE, S.W. *June 11, 1909.*

MY DEAR KARL PEARSON, What fools this world contains!—even in the U.S. A man with the persuasiveness and moral standard of a dealer in horses or in works of art, might possibly succeed in diverting the coin to more promising Eugenic purposes than the effect of sympathy on the conception of mares, but *quere.*

I heard from Heron, that M— shied at the idea of publishing the little book by Miss Elderton and her brother. Of course I am prepared to contribute towards the cost of the publication, if on those terms only a good publisher would accept it.

Ever affectionately, FRANCIS GALTON.

7, WELL ROAD, HAMPSTEAD, N.W. *June 15, 1909.*

MY DEAR FRANCIS GALTON, You will be glad to know that Messrs Adam & Charles Black have accepted the Eldertons' book at their own cost giving to the authors a 10 °/₀ royalty on copies sold. I think these are as good terms as we could expect.

There is to be a meeting of the Galton Laboratory Committee to see the Report and if approved to forward it to the University Senate next week. I don't think it means more than sanctioning the Report Hartog sent to you. He wrote to me that there will be a vacancy on the Committee and asked me unofficially if I could suggest any member of the Senate to go on. He sent me the enclosed list and asked for suggestions. But I know nobody with a Eugenic bent. Do you? They all seem to me folk rather lacking in imagination. Gregory Foster would do quite well, but I don't know that he has any special knowledge in this direction.

I had two very different men to see the Laboratory yesterday. Dr Woodward, the President of the Carnegie Institute of Washington and Dr Chau-tao-Chen, First Secretary of the Chinese Finance Board. I suppose a sort of permanent Secretary of State for the Chinese Treasury. I got him to promise me some Chinese Skulls*!

Yours affectionately, KARL PEARSON.

I am so sorry about Charles Galton Darwin! Still he has done quite well and will no doubt get his fellowship. Will you write a brief introduction to the Eldertons' booklet? What shall we call it—A Primer of Biometry or of Statistics or what?

42, RUTLAND GATE, S.W. *June 16, 1909.*

MY DEAR KARL PEARSON, Of the names in the Senate suitable for our Committee, A. C. Headlam, Principal of King's College, seems one of the most suitable. He has given help in various ways—lecture room, etc.—to the Eugenics Education Society. Roscoe is very

---

* He thought the skulls of decapitated criminals might be available, but even these never reached me.

sympathetic, I think, and very businesslike, as you know. Sir Philip Magnus is also worth considering, besides Gregory Foster. I am prepared to pay in the £500 for next year's cost of the Laboratory as soon as the Committee has met and done what it ought to do!

You have done unexpectedly well about the Eldertons' little book. If, as you suggest, it is called a Primer, it ought to be of Biometry *and Eugenics**. The two latter words are important.

I shall be very happy to write a few words of introduction, quoting from my lecture at Oxford (the Indian Anarchist's Foundation), on the need of such a book.

With you, I am very sorry at C. G. Darwin's ill fortune; but I take it, he knows quite enough maths. to make them his effective *servant* in future work, and I hope he will do so.

How amusing about the Chinaman! You will not I suppose extract pecuniary help through Dr Woodward. Ever affectionately yours, FRANCIS GALTON.

Please excuse bad writing. I am placed, on account of swelled legs, in an uncomfortable position.

Private. A letter came to me the day before yesterday from the Premier to the effect that I was to be knighted on the "approaching" King's birthday (i.e. on Nov. 9). A precious bad *knight* I should make now, with all my infirmities. Even seven years ago it required some engineering to get me on the back of an Egyptian *donkey*! and I have worsened steadily since†.

THE GALTON EUGENICS LABORATORY, UNIVERSITY COLLEGE. *June* 17, 1909.

MY DEAR FRANCIS GALTON, I am so pleased that among all the humbug of this world—and science is no more free from it than politics—the work you have done should be officially recognised. My Chinaman and Dr Woodward were only a trifle previous in their use of "Sir Galton" last Monday. My memory of poetry is very misty, but has not Wordsworth a poem "Who is the perfect Knight"? Certainly I don't think it was the man who could mount his steed best.

Will you let me have your views on the Galton Laboratory Report, if you are unable to be present? Particularly as to how far you would wish us to proceed to the election of a new fellow in or before next February, or are content with the staff remaining at present as it is. Also what you think of appointing Miss Ryley at £45 a year to do the pedigree plates, provided we do not exceed our funds. Of course, if you feel able to come to the meeting, I need not trouble you to write. If there are any views you would like to have expressed, please send them to Hartog or to me that they may be read to the meeting. The Senate met yesterday and through an oversight I fear on my part in telling you *to write to College*, I did not get your letter till dinner time. Meanwhile the Senate had put Cyril Jackson, Chairman of the L.C.C. Education Committee, on the Galton Committee. I think this is really a good appointment. The L.C.C. Committee has lent us 10,000 schedules of London children and if we can get really into close touch with that body, we shall have the finest material accessible anywhere. Affectionately, KARL PEARSON.

If you are not at the Committee on June 25, may I come in after the meeting and talk over its doings with you? It would be about 4 or 4.30 to 5 o'clock.

7, WELL ROAD, HAMPSTEAD, N.W. *June* 26, 1909.

MY DEAR FRANCIS GALTON, There was a point the Principal of the University asked me, and which I forgot to mention to you yesterday. He said "Could you tell me whether Sir Francis Galton would object to the University seeking for further funds to increase the activity and possibilities of the Eugenics Laboratory?" I said that I would sound you on the matter but that I thought I knew your answer would be: That anything that helps forward the cause of Eugenics had your approval. I said to him that as far as I personally was concerned the points I should emphasise strongly would be: (i) that if any further aid should come to the

---

* (Or rather, "of Biometric and Eugenic Calculations." This would be the long title. F. G.) The little book was finally called: *A Primer of Statistics.*

† Another friend said to Galton: "Why they ought to have made you a K.C.B. years ago!" and he replied with a twinkle in his eye—it was on his morning "trundle"—"Well, I am a sort of K.C.B.—I am a Knight of the Chair of Bath."

Laboratory, it must remain, as long as I had any share in the work, the *Galton* Laboratory; (ii) that the provision should take directions which would give greater scope to the future Galton Professor and not in any way anticipate that foundation, e.g. establishment of a permanent publication fund, annual grant for purchase of books, and increase of staff or accommodation for workers. In fact I told him what I have told the Americans, that "any contributions would be gratefully received," but that this work was going so far as it lay in my power to be associated with your name. I don't know that the University will do much for us—they did not succeed in doing much when I set about building the observatories—but there is no reason why they should not try. The fact that they did ask for money to help forward the work would to that extent be a sign that "Eugenics" had been finally accepted as a part of a University's work. Will you let me have your views, possibly in a form I could unofficially communicate to the Principal of the University? Affectionately, KARL PEARSON.

7, WELL ROAD, HAMPSTEAD, N.W. *June* 28, 1909.

MY DEAR FRANCIS GALTON, Your letter was *admirable*, and I think will be helpful. The "accident" about the slip was that I forgot to put in a note about it! I want a few words to express the idea that since Darwin we no longer look upon a "race" or "nation" as fixed in type or character, but as always in a state of change towards the better or worse, and that the statesman who realises this and works for the future will be the one whom history, which is ever written "in the future," will commend. I have expressed myself very clumsily, but I want some words to this effect to sum up my lecture on the *Problem of Practical Eugenics*, and I thought perhaps you could give the paragraph a more apt phrasing.

Yours affectionately, KARL PEARSON.

42, RUTLAND GATE, S.W. *June* 28, 1909.

MY DEAR KARL PEARSON, It is grateful to me to hear from you that the University of London is so favourably disposed towards Eugenics as to consider the propriety of seeking aid to increase its utility. Such aid would be acceptable in the direction that has commended itself to you, namely to the establishment of a publication fund, an annual grant for purchase of books, and for an increase of workers together with accommodation for them. Would you kindly convey these views, together with any others of your own, unofficially to the Principal? I may add that I have just sent to Hartog the promised cheque of £500 for the maintenance of the Laboratory next year. Affectionately yours, FRANCIS GALTON.

42, RUTLAND GATE, S.W. *June* 29, 1909.

MY DEAR KARL PEARSON, I wrote to Hartog to say that you and I were quite at one in respect to modifications in the *status* of the personnel of the Eugenics Laboratory (I forget the words I used, but they were to that effect) and explained to him that I was much too infirm to attend the Committee.

As regards the last paragraph in your Report, as we have got (thanks to you) satisfactory workers at the Eugenics Laboratory one cannot do better than give more permanence to their positions than at present. If the funds allow, by all means include Miss Ryley on the Staff. In fixing the future titles and emoluments of Heron, Miss Elderton and Miss Barrington, if you can get in a word to absolve us from granting pensions on retirement, it might be well. I have known much grievance created on the part of those who had "expectations" in other Societies and Offices. For my part, I think it will be much more satisfactory to rearrange the titles of our officials, as you propose. I wonder what those titles will be, out of "Secretary," "Librarian," "Editor," "Computer," and so forth. Miss Elderton should if possible have a title to herself, and not "Assistant...." All this is merely suggestion. You know the ropes so far better than I do, that I am sure to acquiesce in whatever proposal you may make as to these not unimportant details. Ever affectionately, FRANCIS GALTON.

NYSTUEN, PAYABLES, WOODCOTE, NEAR READING. *July* 4, 1909.

MY DEAR FRANCIS GALTON, In the hustle of getting away, I could not write to thank you for many things. I have adopted most of your re-wordings for my sentence which are clearly "betterments." I am very glad the Eugenics Laboratory arrangements are settled for another year from February. I feel that with the present staff we can do good work. The Laboratory might achieve more with another personnel, but the new blood would want training afresh, and I fear facing anything extra at present. I feel very deeply the kindness with which you have fallen in with all my suggestions, and made it possible for me to work the Laboratory with a minimum of additional labour. As you know, I never can find the right words to express what I think or feel, but you will try and interpret the spirit under them.

What is producing the most unfortunate effect at the present time is the recent sneering attack of Bateson *. It is difficult to determine whether it is better to spend energy on replying to such criticisms or to leave them unregarded and go on with my own work. In the latter case the unthinking public assumes them to be valid and that no reply is possible. In the former, one wastes the energy that should be spent on permanent work on attacking a man whose whole position changes from year to year. For example, he used to assert that Albinism in man was a Mendelian unit character; now that with six years' work I have got some data and facts as to albinism, Bateson, knowing this, finds it "a case to which Mendelian rules do not apply." Controversy in such a case is impossible, it becomes wrangling.

The enclosed may interest you, if Heron has not sent it to you. We are here in our "wooden house" with much pleasure as to our environment and hopes for restfulness. The rooms give one a sense of space and light, and the furnishing is graceful and comfortable. The "stoeps" are pleasant and the green fields run up to us on every side. Our landlady, to judge by her books, must have a wide range of taste in French, German and scientific literature. I find Huxley's scientific (not popular) essays and Frank Balfour alongside George Meredith and Zola! She has been round the world (to judge from her photographs) and has qualified as a medical practitioner!

I hope your summer resort will be an equal success. If there be any chance of your being within motor distance, please let me know, for I might get to you, if the fatigue were too great for you to get here. Always affectionately, KARL PEARSON.

42, RUTLAND GATE, S.W. August 7, 1909.

MY DEAR KARL PEARSON, On Tuesday we go for two months to Fox Holm, Cobham, Surrey, and really there is now good hope of a belated summer. I am very wishful to know how you all are. For my part, I creak on, not unhappily but little usefully. However, I have small jobs on hand which interest me. One of these I shall want before long to consult you about. It is due to the proposal of Ploetz's society to give some sort of diploma to those who rank eugenically in the uppermost quarter of the population. I have long considered how some such scheme could be practically worked out, and am putting my ideas on paper. Ploetz (I strongly suspect on your initiative †) has asked me to accept the Hon. *Vice-Presidentship* of the Society. They have only five *Hon. Members*, among whom are Haeckel and Weismann. It is a great honour.

You may like to hear that, overpersuaded by you and by Miss Biggs, I have had my bust modelled by Sir George Frampton, R.A. Friends quite approve of it. It is to be cast in bronze and will be ready before Xmas. The various operations are tedious but are now in the hands of specialists. How carefully good artists work! It was a delight to watch his touch. The model was finished two days ago. I have got a very nice house. Fox Holm is between Byfleet and Cobham. It is just South of St George's Hill; 3 miles from Weybridge (via Byfleet), 2 miles from Cobham. I should indeed be grateful if you could come over to us some day. I am far too infirm to get about, without much care, or would find my way to you. We motor down on Tuesday. Possibly I may find motoring less fatiguing than hitherto. Two hours of it a few days ago was as much as I could bear. Ever affectionately yours, FRANCIS GALTON.

PAYABLES, WOODCOTE, NEAR READING. August 8, 1909.

MY DEAR FRANCIS GALTON, I was very glad to get your letter this morning, for I was beginning to fear that you might be ill, as I had no news direct or indirect. We have had several "ups" and not a few "downs," since the beginning of July. *Item*, My boy has got into College at Winchester, which we hardly expected and he is carrying off his honours nicely and is thoroughly enjoying his holiday. *Item*, I went to Oxford to give away the prizes at his school,

---

* See above, p. 288, and compare pp. 406–408.
† I knew nothing about the matter. K. P.

and found myself at the appointed time landed at Uffington in sight of the White Horse, having forgotten to change at Newbury, and only got to Oxford when the ceremony was all over! *Item*, I had to go up to see my Doctor; but he has made me feel distinctly better, granting me a sound heart, lungs and arteries, but a crippled digestive machinery. *Item*, I have nearly got a number of *Biometrika* ready, and am really getting forward with the albinos and other work. You will receive shortly two more Eugenics Laboratory Lectures and Part III of the *Treasury*. *Item*, I have had some unpleasant American experiences with the man who wanted to help the Laboratory. In my second letter I simply said that we could not approve his horse-breeding experiments, and that I regretted we could not send a man out to America to explain our projects, but that he could hear any particulars he wanted of the work of the Laboratory from Professor R. Pearl. That, I thought, was the end of the matter. However it appears he wrote to Pearl and was so pleased with his account, that according to his Father Confessor, the Baptist Minister, he determined to hand over his fortune to the Laboratory! Considering that I had written pretty frankly that I thought his ideas were folly, this was a sign of wisdom on his part! Now come his relatives on the scene and they, according to the Baptist Minister, have been writing the would-be Eugenist letters *in my name* to prove that I am insane! Really the Americans are a wonderful people and full of resource. I don't mind the two letters I have written being treated as public property; they concern only the purpose of the Laboratory and the foolish character of the American's breeding schemes, but it is a bit rough to have forgeries put out in one's name even in a foreign country. The whole thing, however, has its humorous side.

I think Ploetz is a sound man, and keen on Eugenics. I should not, however, allow his "International" Society to absorb yours as a branch, which he may suggest. I fancy he is working in the first place to accumulate material with regard to families.

I am so heartily glad about the bust and so grateful to Miss Biggs for seeing it through. I knew it ought to have been done, because it is idle to disguise the fact that there will be a need for it, and it is so feeble to get recognition only of this fact, when it is too late to get a true portrait. I am sure you will enjoy this weather, if your new quarters are at all airy. We have been taking two meals a day in the open air. I shall look you up on the map and certainly come over if cycle and train will work in. I have got three albino puppies born since we came down, so that now we have ten albino dogs. It is strange to see how motherhood has converted our fearful, shy little Pekinese into a furious little vixen. She sprang at a huge English sheepdog the other day and drove it right out of the croft, and she promptly nipped my fingers when I touched one of her pups. They have not opened their eyes yet, but to judge from their coats they are all albinos. I propose next to try a cross with a pug, the offspring should not be albinos, and then if we cross them, we might get a race of something approaching albino pugs. These Pekinese albinos are not as graceful as the normal Pekinese, and are very inert. Always yours affectionately, KARL PEARSON.

Fox Holm, Cobham, Surrey. *August* 11, 1909.

MY DEAR KARL PEARSON, This is mainly to report arrival at this pretty, small house, with lawn, gardens and acres of wood-land. Best congratulations on your boy's successful entrance into Winchester.

What strange people the Americans are! Don't get dragged into a law-suit there!!

About the bust, I was over careful about praising the work, which I did not see after Frampton's final handling in his studio. Miss Biggs did, and is enthusiastic about it, both as a likeness and as a work of Art. So that is well.

I hope the puppies' eyes are now open and that they are as red as you could desire. It is excellent news that your Doctor passes you as quite sound. Alas, I am *not*, and no better in essentials. It will be delightful to see the two additional lectures, and Part III of the *Treasury*.

Ever affectionately, FRANCIS GALTON.

PAYABLES, WOODCOTE, NEAR READING. *September* 4, 1909.

MY DEAR FRANCIS GALTON, I hope all is well with you, and that your quarters have fulfilled your hopes of them, and not proved too cold during this sunless month. I meant to write to you before, but I have been rather depressed and somewhat over-worked. I have been suffering from teeth troubles, not exactly toothache, which would be settled by one or two losses, but a general sort of neuralgia in the jaw, which passes from one tooth to a second and hardly allows itself to get fixed. Then I have set myself too big a task with this albinism monograph. I cannot get it done, and have spent most of my vacation over the geographical chapter. I am still on the African section although Asia and Australasia have gone to press. It is reaching too large proportions already, and the Piebalds, Heredity and Statistics chapters are yet undone. I find on measuring up the map that Weybridge is beyond my cycling powers. I should like to induce a motoring neighbour to carry me across, but he has not given me a chance yet of leading him to an offer! He keeps us, however, alive with a flow of guests. Among the last were Professor Turner and his wife from Oxford. My bairns have learnt to cycle and I have gone short runs with them along the old Peppard lanes, but they will soon outride my distance. We had a very pleasant day in Winchester to see Egon's new surroundings. I was immensely struck with the beauty of the College and hope his life there will be a happy one. The environment of a great school like this ought to excite the boys to be and to do.

I have heard no more of the Americans, so I trust they will leave me in peace. The three puppies are getting about now, but I don't think I shall be able to keep four albino dogs, and must seek a home for them. I hope the Laboratory publications reached you safely. I enclose two notices. I believe some of the daily papers also had notices. Miss Elderton's Lecture on Nature and Nurture ought to be out this week. I am printing a paper by Dr Goring on the " Inheritance of Phthisical and Insane Tendencies based on criminal Observation," which I think is very good. My neighbour here is a great pig breeder, but he *will* not take any interest in actual measurements for heredity, only in the prize and show work.

Ever yours affectionately, KARL PEARSON.

Fox Holm, Cobham, Surrey. *September* 6, 1909.

My dear Karl Pearson, I had delayed writing, hoping vainly that you might discover a way of getting here comfortably, some day. I am truly sorry you feel unrested. You will discover, as all your elders have discovered, how strict our bodily limitations are. We are each of us machines, each of his individual horse-power which we cannot strain safely by tying down safety valves, or the like.

Let me offer a tribute of admiration to your lecture, which I have read and re-read and look upon as a masterpiece. As for the *Treasury* it speaks for itself of the immense care in compilation.

I wonder whether you could conveniently turn some of the Laboratory folk on to a simple, but, I think, important inquiry, for which the collection of family histories affords ample material. It is, how many relations, *on the average of that collection*, has each person in the following degrees :

(1) Grandfather's (Paternal) brothers,   (2) ditto (Maternal),
(3)   „   „   sisters,   (4)   „   „
(5) Father's brothers,   (6) Father's sisters,
(7) Mother's brothers,   (8) Mother's sisters,

and *quere* the sons and daughters of 5, 6, 7 and 8.

The above *8*, plus the 4 grandparents and 2 parents = 14 in all, form a large body of individuals and it is well worth while, in the frequent absence of exact knowledge of their number, to appraise the *average significance* of heredity in such and such a degree. If you think this feasible, I will draw up a more careful scheme, excluding half-brothers and the like.

I am so glad about your boy at Winchester and about your other "bairns" on their cycles. You would be amused to see the mechanical appliances that Gifi and my man-nurse use to prise me into a Victoria—I am so helpless!

I hope the puppies prove to be thorough albinos.

X. has written an uncommonly good paper in the *Sociological Review*, of which I received an offprint this morning, on the obstacles to Eugenics. It is the best piece of writing that I have seen of his. Perhaps the American will come down unexpectedly with a big gift after all!

So you have a great pig establishment close by. My heart rather leans to pigs, but I wish they did not smell. Ever affectionately, FRANCIS GALTON.

PLATE XXXVIII

Francis Galton, aged 87, on the stoep at Fox Holm, Cobham, in 1909, with the faithful Gifi
and the Albino puppy Wee Ling.

PAYABLES, WOODCOTE, NEAR READING. *September* 10, 1909.

MY DEAR FRANCIS GALTON, What you want about the average number of relatives is of importance and shall be done, but it will need one or two points considering first. In the first place, the younger generations are not always complete and it may not always be easy to ascertain whether this is so or not. I think it would very much diminish the available pedigrees if one had to be certain on this point. In the next place, there has been such a great change in the past thirty years, the modern *complete* families are 1, 2, 3 or 4, but 30 or 40 years ago they were anything up to 6 or 11. There is also another point, do you mean to include *all* born, or only those living to a definite age? A generation ago, perhaps, ¼ died in infancy and childhood, even in the professional classes; now perhaps only ⅛. You will see that this may, without some agreement as to treatment, introduce difficulties. I am not at all sure that the best way would not be to work at the Quaker family histories or the older Herald's Visitations. But we shall always have to remember that the problem reduces to the size of the family in a certain definite class, and this is modified by custom, by period and by the infantile and child death-rates. Could we not reach your point by discovering the average size of family and the sex ratio in each grade? I enclose a rough copy of Miss Elderton's Lecture. It ought to be out to-morrow.

Here is a rough postcard my boy has made of the albino Pekinese Spaniels. They are very jolly little beasts—and quite of the harmless lap-dog order. Would Miss Biggs like Wee Ling? He will want to have a little training, but I don't think he would give much trouble. If at any time he became a nuisance I daresay I could find another home, but I should like to know where he was, if he had to be united in holy matrimony at any time with one of his cousins or half-sisters!

The pigs of our neighbour, who has some 300 acres, are very lordly and go with attendants, one pig, one man, for their daily exercise. Yours always affectionately, KARL PEARSON.

I have heard no more of the Americans! Why cannot Cook and Peary behave like Darwin and Wallace?

FOX HOLM, COBHAM, SURREY. *September* 11, 1909.

DEAR PROFESSOR PEARSON, The photo of Wee Ling is most attractive and I should of all things enjoy to bring him up—but this alas is prevented by the "cruel uncle"! Possibly your powers of persuasion might move him. Since you induced him to sit for a bust, you might prevail over this matter too, won't you try? and I will bring up the pup in the way he should go, having had much experience with dogs in my life.

MY DEAR KARL PEARSON, The foregoing appeal from Eva Biggs has melted away my antagonism to dogs. Yes! send Wee Ling and much care shall be lavished on him*.

Ever affectionately, FRANCIS GALTON.

Hurray! E. B.

FOX HOLM, COBHAM, SURREY. *September* 12, 1909.

MY DEAR KARL PEARSON, I answered about Wee Ling yesterday, in a hurry, to save the Sunday post. This refers to the other part of your kind letter. It had been my intention to write about some of the points you raise, all of which are important.

Respice finem. My object is to procure the desired data from one or more well defined and homogeneous groups, defined by convenient limits as to date and minimum age of children; this latter has to be regarded : say 20, or other early marriageable age. The dates are a more serious matter. You know better than anybody, the times over which childbirth has continued normal in any particular group. The Quakers, as you suggest, would serve well. So eminently would the Jews, if returns exist.

Have you ever, by the way, inquired about what I understand to be an immense storehouse of family facts, viz. the *printed* pedigrees, taken under affidavit, of the families of *intestates*, whose property comes into Chancery? I have no lawyer at hand to consult afresh.

---

* I cycled over from Woodcote to Cobham taking Wee Ling in the basket on my handlebar. Plate XXXVI was a result of this visit, and Plate XXXVIII shows Wee Ling in good company shortly afterwards.

My authority was the late Vaughan Hawkins, whose account was graphic and most interesting. But this was half a century ago, and the procedure may have changed since, and the old Records be inaccessible; but it is worth inquiry into. He said that the difference between the sizes of family of rich and poor was most conspicuous, the limit of eight in the former corresponding, if I recollect aright, to sixteen in the latter.

To return to the point after this episode. If you have time to think out a *moderate* inquiry of this sort, and see your way to set some clerk to work on it, I should be very glad.

Thanks for Miss Elderton's lecture. How well and clearly much of it is written. It would tax the power of a consummate literary genius to make statistical reservations easy to grasp.

My friend Lt.-Col. Melville, the army physician, was delighted with Heron's lectures, which he attended. He contemplates sending some of his best students for statistical instruction at the Biometric Laboratory, if they can be taken in. So he tells me!

Affectionately yours, FRANCIS GALTON.

PAYABLES, WOODCOTE, NEAR READING. *September* 13, 1909.

MY DEAR FRANCIS GALTON, You must not have the dog to be a nuisance. It was a mere idle suggestion on my part, as they really are rather nice as dogs go and almost unique. If you have it, and it does not fit in, then we will find another home for it.

I have got, I think, the person to put on to do the work on the relatives—a new recruit coming in October—I will look up the Chancery data on my return. I suppose copies will be preserved at the Record Office, but I will inquire.

It is very hard to make people understand, that one has no aim but to get at the facts in this "Nature and Nurture" business. When we came to the problem, I expected to find the two factors about equipollent, but the insignificant character of "Nurture," as compared with heredity, soon became transparent. Even now when we find a fairly high (e.g. 0·2) correlation between environment and physique, it is very doubtful whether it is not a secondary effect of heredity, the feebler parents having a worse environment, because their wages are less. But the view that "Nature" is the fundamental factor is stirring up, as I feared it would, a whole hornets' nest.

Affectionately yours, KARL PEARSON.

7, WELL ROAD, HAMPSTEAD, N.W. *October* 18, 1909.

MY DEAR FRANCIS GALTON, Miss Elderton asks me to answer your card, because she is not quite sure as to one or two points. It depends to some extent on two matters: (1) How the midparent is defined:

Midparent Deviation = $\frac{1}{2}$ {Father Deviation + (Mother Deviation increased in ratio of father's variability to mother's variability)}.

This is theoretically the best definition and agrees with your original one provided

$$\frac{\text{Father's variability}}{\text{Mother's variability}} = \frac{\text{Father's mean value}}{\text{Mother's mean value}}.$$

This equality is very nearly true for many human characters, but not quite for all.

(2) The existence or absence of assortative mating between father and mother. Let us call the correlation between father and mother ρ. This correlation coefficient is rarely over ·2 and lies between ·1 and ·3. Assuming this, we have:

$$\text{Midparental } correlation = \sqrt{\frac{2}{1+\rho}} \text{ (Mean of parental correlations)},$$

and again:

$$\text{Ratio of Mean Filial to Midparental Deviations} = \frac{1}{1+\rho} \text{ (Mean of parental correlations)}.$$

Now there is no sensible difference between the parental correlations that we have been so far able to discover. Hence if r = parental correlation,

$$\text{Ratio of Mean Filial to Mid-parental Deviations} = \frac{r}{1+\rho}.$$

Now r is very close to ·5—it varies from about ·46 to ·52 for the best series in man. Weldon's results for mice *not yet published* give almost the same values. *But he has so selected his pairs of mice that ρ runs up to ·8!* For man ρ may be safely put ·2. Thus the ratio you want is $= \dfrac{·5}{1·2} = about ·4$.

The ratio of mean filial deviation to parental deviation, i.e. for a single parent, is ·5 but of course the prediction in this case is subject to a larger probable error; these errors in the two cases being about in the ratio of $\sqrt{·75}$ to $\sqrt{·60}$, the latter corresponding to the midparental estimate.

I hope this will not be too complex, and that I have given what you want. Pray write again if there be any further point I could make clearer.

I had a letter from the Principal of the University saying that the University was drawing up a list of their needs and asking me to say what the Galton Eugenics Laboratory needed. It was a somewhat difficult question to answer since if the University is in the way of getting money, there is no reason why the Laboratory should not have a considerable share. I suggested that £100 a year for books, £200 for publications, and £500 to pay a man to give the bulk of his time to supervision, could be easily assimilated! If we get $\frac{1}{4}$ of all this from the University we may be happy, but it really is a sign of the times that they ask us if they can aid. We are very full this session. In the Biometric and Eugenics Laboratories together we have I think 16 research workers, and practically no vacant tables.

I shall shortly send you the average numbers of certain classes of relatives—aunts and uncles. I fear we cannot work cousins because the records are too incomplete.

Has Wee Ling behaved himself, or has he become a nuisance? Don't hesitate to return him if he has become a difficulty.

Affectionately yours, KARL PEARSON.

THE RECTORY, HASLEMERE. *October* 25, 1909.

MY DEAR KARL PEARSON, You can with difficulty understand how incompetent I am to do mental work. I have blundered much in putting the enclosed into shape, desiring to avoid needless complexity, and now if the suggestion (B) be adopted the problem becomes apparently simple enough. Still I dare not trust myself to do it. I only want a rude approximation, but want one very much.

Nettleship lunched with us on Saturday and inspected Wee Ling's eyes. The puppy is a joyful little beast with a now tightly curled tail and is a friend with all the servants. But he has a horrid temper, and bites with his little sharp teeth and swears in Chinese dog-language, a quite different language to that of English pups. He had a sharp lesson from the cat, in social usages; for trying to oust her from her chair, he received a wipe from her claws across his little pink nose. No real harm done, but it must have hurt.

We are well placed and the air of Haslemere suits me perfectly, but I do very little. Sir Archibald Geikie tells me of scientific events. He was delighted with Birmingham and remarked that among the men selected for degrees were two brothers (Haldane), one brother and sister and brother-in-law (Balfour, Mrs Sidgwick and Lord Rayleigh).

I asked Nettleship about you, whether he thought you were not working too hard. He evidently thought so, but added that you were like a racehorse, difficult to keep quiet. And here am I bothering you about a problem! How I wish you could be relieved from routine work. I wonder if you will come down to see your friends hereabouts?

My niece is happy, after $2\frac{1}{2}$ weeks out of the allotted 4 weeks in bed, for rest-cure. She *hopes* to get abroad to S. France in early winter, leaving me in charge of another niece (Mrs Lethbridge). I am fortunately well-nieced; three are at the moment hereabouts, two in this house and one hard by. Ever affectionately yours, FRANCIS GALTON.

This letter contained the following problem of which a solution was sent to Francis Galton as a New Year's Greeting, 1910, and was published in *Biometrika*, Vol. x, pp. 258–275.

Problem.

October 25, 1909.

An array H is made of husbands arranged in estimated order of civic worth (see remarks below). Gauss's Law is supposed to apply throughout. Let the standard deviation of H which does not need measurement be unity. Cut off a segment G from the upper end of H, including $1/n$th of the whole of H ($1/n$ is here wanted only for the two values ·02 and ·04, to which the corresponding deviates in Sheppard's Table, *Biometrika*, Vol. v, p. 4, are 2·0537 and 1·7507). Make an array F in order of civic worth of all the male adult children of G as calculated from the formula for parental Heredity. It will be a skew array. Let the mean (or better the median) of all the values in F be f, and let the position of that value in the array H be $1/w$ of its length from the upper end. Required : the ratio of w to n for the two values of $1/n$ mentioned above, and consequently that of the deviates at those class-places (from Sheppard's Table).

Remarks.

A. It seems impossible to obtain a satisfactory numerical value of civic worth, but it is not more difficult to classify it by judgment, than it is to select recipients of honours, members of Council, etc., out of many eligible persons. Therefore the method here adopted is to compare class-places and to derive the corresponding deviates from Sheppard's Table.

B. Some law of fertility must be assumed that shall give limits to the possible error from ignorance of the true one. Perhaps the assumptions (i) that infertility so balances deviates that the F values are much the same as those of the children of parents at $1/n$th of the array from the upper end, and (ii) that they are the same as those at $1/2n$th of the same, might be adequate.

7, WELL ROAD, HAMPSTEAD, N.W. *October* 26, 1909.

MY DEAR FRANCIS GALTON, I have only just got back from Newcastle, so you must excuse a hurried note. I was down by 5 o'clock yesterday and back by 1 o'clock to-day. But I always feel heartened by lecturing to the north country folk. They came between four and five hundred strong, had $1\frac{1}{2}$ hours' lecture, and nearly 100 standing all the time and so keen and interested.

One point I can tell you at once, the *average* civic worth of your array of offspring would be to the average civic worth of your array of fathers (both measured from their respective means) in the ratio $r\sigma_1$ to σ_2, where r is the correlation coefficient, σ_1 and σ_2 the standard deviations of fathers and sons respectively. This would be true for linear regression quite independently of Gauss, wherever you cut off your array of fathers, and quite independently of any law of fertility, if $r =$ correlation of father and son, and σ_1 and σ_2 their standard deviations. But r will not be the r for a stable population, and σ_1 will not $= \sigma_2$ as for a stable population, if you make fertility a function of the inherited character. Their values will then turn on the law of fertility and this may upset the whole story if it alters very markedly the value of r. I will, however, look into the point and let you know. I expect this is the kernel, however, of what you really want, i.e. the alteration in variability σ_2 of sons and the new value of r. Still the other value may interest you— i.e. that regression does not apply only to the group of offspring of one parental value, but the whole population of offspring due to any series of parents has a mean regreding on the mean value of any section of the parental population, precisely in the same way as mean of array of offspring regredes on a single parental value.

I am extremely sorry to hear about Miss Biggs, and hope the trouble may not be of long duration. You will miss her very much during the winter.

I don't understand Wee Ling's temper; his brother and sister are very frisky, but angelic in temper. I gave you Wee Ling because we had decided he was the most intelligent*. I hope you got the letter about the midparent. Yours always affectionately, KARL PEARSON.

Have you seen the Whethams' book? Or Riddle's paper on Pigment and Mendelism?

7, WELL ROAD, HAMPSTEAD, N.W. *October* 31, 1909.

MY DEAR FRANCIS GALTON, I think I have got out a general theory of the problem you suggested on the following lines: Given a differential fertility, what changes will it make (1) in the mean and variability of the offspring and (2) how will it change the coefficient of heredity in the population. From these results I can at once deal with your special problem of a certain percentage of the population having a desirable character but lessened fertility. The chief difficulty is the form of the law of fertility. Now the distribution of the size of families in any population is not Gaussian, it rises steeply and falls slowly, thus:

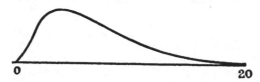

It certainly does not closely approach any mere straight line relation. It would therefore seem reasonable to suppose such a curve to give the fertility distribution with any character. It seemed to me better than taking a straight line to see where we arrive by supposing the fertility is somewhere a maximum and drops in Gaussian fashion on either side of the modal value. To take the cases in which the fertility is greatest with the worst values of the character we have only to place the fertility curve much to one side, e.g.

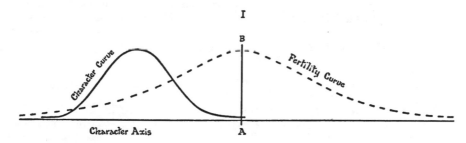

To the right of the line *AB* no individuals occur and accordingly the fact that there would be fertility, if individuals occurred, is of no importance. We can also take the case when a very small part of the population is fertile, thus:

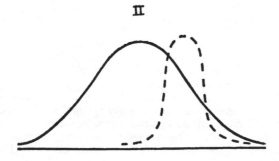

* See our Vol. II, p. 76, as to breeding for intelligence in dogs. Unfortunately Wee Ling, while markedly intelligent, and long a dear friend of the biographer's family, turned out to be incapable of reproducing his kind!

and by pushing the curve to the extreme left we obtain practically an increasing fertility instead of a decreasing fertility with the character.

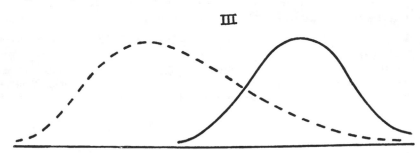

I and III correspond fairly closely to your suggested linear relation, and, I think, the algebra is easier on this assumption. The formulae are fairly complex, as we have the three constants fixing the fertility curve at our choice, but I have got out the distribution of the offspring generation in terms of these constants. As usual the formulae are perfectly idle until we turn them into numbers. The only point to be noted is that the character of the population takes in a very few generations, as we might anticipate, the value corresponding to that of maximum fertility.

I will get illustrative numerical cases worked out, but they will take some calculation and this may delay matters some time. Unfortunately Mrs Weldon has had to give up the mice work although incomplete, and my computer, Miss Bell, who succeeded Dr Lee, is pressing on with that. For Weldon's sake this work ought to have been published long ago. I think in the course of three weeks the fertility numbers can be ready. Heron and Miss Elderton are both struggling to finish their big memoirs, and Miss Barrington and Miss Ryley are respectively at work on pedigrees of locomotor ataxy and on cataract pedigree plates. We have never had such a crowd in the laboratories and the "lecture series" has popularised the work and brought a number of keen but theoretically (statistically) rather weak students, who want to do Eugenics work, but have not a preliminary training.

Now as to your other problem. I find that from the family schedules we can get no trustworthy record of the average number of cousins, because selection has been made of cousins, but the aunts, uncles, nieces and nephews seem to work all right. The cousins must be taken from *full* pedigrees and I am uncertain at present whether our material will suffice. Here are the results for uncles, etc.

On paternal side: Each uncle has 3·23 married brothers and sisters who provide him on an average with 6·4 nephews and nieces apiece.

Hence each uncle has 20·7 nephews and nieces.

Each paternal aunt has 3·23 married brothers and sisters with average families of 6·2, or there are 20·0 nephews and nieces.

On maternal side: Each uncle has 3·33 married brothers and sisters with 6·1 offspring, or there are 20·3 nephews and nieces.

On maternal side: Each aunt has 3·33 married brothers and sisters with 6·0 offspring, or there are 20·0 nephews and nieces.

Roughly therefore each family consists in our data of 6 offspring, 3 male and 3 female, of whom *more than* 3·3 marry and produce *more than* 20 offspring. I say more than 3·3 marry, because while an aunt or uncle has 20 nieces and nephews, she or he may or may not themselves be married, and in either case they would still be aunts or uncles. Roughly I make out that a man has between 40 and 50 first cousins of both sides together. I have 19 paternal and 14 maternal first cousins, but I am below the average as only two of my mother's brothers and sisters had children and several of my father's had very small families. However we will try to get some real data.

There is another point about which I want to write you a few lines. For some time past there has been a series of attacks, some signed, some unsigned, by X. upon the Eugenics Laboratory. I do not know X., nor do I want to know him. I have never spoken to him, nor have I ever directly or indirectly criticised his *books*, although I think they have done great harm to the cause of Eugenics in the minds of many who would otherwise have been sympathetic. But the recent tone he has taken amounts to an accusation that the Laboratory has been and is wasting the money provided by you. I feel that the time has come when it is necessary for me to reply to the sort of charges X. scatters. It will be unpleasant work because brushing off mud is always unpleasant, but if you leave thrown mud long enough on the best of coats, you are ultimately mistaken for a vagabond. I have delayed writing to you on this point, because I felt sure you would not like any controversy between two supporters of Eugenics. I have rigidly left X. alone on this very account, hoping that he would have the good sense to treat us in the same manner; he has not done so, and my patience is practically exhausted. He has anticipated that I should not reply to him, because of his connection with the Eugenics Education Society and my connection with the Galton Laboratory. It seems to me, however, that the time has come when some step must be taken. If you feel, *as I do*, that any attack on a member of the Council of the Eugenics Education Society is incompatible with my official relationship to your Eugenics Laboratory, I will resign officially as from the end of this year. This will allow of completing Part IV of the *Treasury*, for which we are pledged to subscribers for this year. I will give every aid—no less than at present—for the forthcoming year to Heron and Miss Elderton so that they can finish the work they have in hand, and there will be time to think of the future during the year. This will relieve you of any anxiety for the continuity of matters, but I shall give the aid, not as director but as a personal friend of the young people here. At the same time I shall feel perfectly free to reply to the criticisms which X. has been making of their work and of the expenditure of the Laboratory, which they are not able to make themselves.

I had hoped that the Eugenics Education Society would do its own work and leave us to do ours, but some members of the Council think otherwise, and as they choose to throw down the gauntlet, I must take it up, though I do so very reluctantly, and particularly because I feel it can but pain you. Still, I think you will, if you imagine yourself trying to work the Laboratory in my place, admit that you could not pass by charges of what really amounts to wasting the founder's money. Please remember that I am and shall always be,

Yours very affectionately, KARL PEARSON.

THE RECTORY, HASLEMERE. *November* 2, 1909.

MY DEAR KARL PEARSON, It is painful news to me about X., whose articles I have not seen. Of course, if he attacks your work directly or otherwise, the right to reply rests with you and I do not see that the closeness of his connection with the Eugenics Education Society need deter you. It is of course bad for the progress of Eugenics when two workers in it disagree, and gives an opening to ill wishers to say nasty things, but all that must be faced. I do not know whether my name has been used in the attacks, but I authorise you to say, if it helps your argument, that the conduct of the Eugenics Laboratory under your control meets with my complete confidence and satisfaction. Thanks many for the number of relations in near degrees. The paternal and maternal figures strongly corroborate one another. I hope it will eventually work out thus:

A man (or woman) has ( ) brothers ( ) sisters
( ) father's brothers ( ) father's sisters
( ) mother's brothers ( ) mother's sisters

and so on for first cousins, so far as the data permit.

Thanks also, very many, for the heredity problem. What I want is the mean deviation of the offspring of parental couples in whom the father deviates (by Gaussian calculations) not less than d from the average. It would thus take the form of class-places. Thus, if the fathers are all selected men, in the ratio of the best out of 50 (or out of 100) of the general adult male population, what would be the class-place of the *mean* value of their offspring?

Enclosed is a photo for you of Wee Ling held by my man-nurse, Charman. He strengthens and grows weekly, and is petted though he gnaws perpetually.

Ever affectionately yours, FRANCIS GALTON.

BIOMETRIC LABORATORY, UNIVERSITY COLLEGE. *November* 8, 1909.

MY DEAR FRANCIS GALTON, Thank you very much for the photograph of Wee Ling; he is clearly progressing. Thank you also very much for your letter. I enclose a sample of X.'s type of attack. The article is worth reading to show the hopeless character of this man's work. There is not a single appeal to demonstrable facts, to statistical data, in the whole paper. It is simply rhetorical, wholly indefinite in result and meaning. But any reader of the obscure paragraph on p. 9 must, whatever else he makes out of it, come to the conclusion that we have wasted the resources of the laboratory in a "sterile logomachy" and that we have made no attempt to trace the origin of alcoholism or measure its influence on the offspring and the individual. The essential fact is that we are the only people who have really endeavoured to measure the relation of alcoholism in parents to the mental and physical condition of the children, and that only in this Laboratory is the relation of alcoholism to crime and insanity actually known and its statistical correlations to environment and class have here alone been worked out. I believe that we only have seen what relation alcoholism has to feeblemindedness. The rest is "impression," "opinion," rhetoric and fustian like that exhibited by X. I think if you carefully read the paragraph—and it is only one among many which have emanated from the same quarter—you will see that we cannot continue to leave such charges unreplied to. I hate this sort of controversial work, but sometimes it must be undertaken, if only to prevent the truth from being swamped. I feel very strongly about this, and must write to you exactly what I feel. But if this criticism of an active member of the Eugenics Society seems to you undesirable, I will do it from outside the Laboratory altogether. Yours affectionately, KARL PEARSON.

THE RECTORY, HASLEMERE. *November* 9, 1909.

MY DEAR KARL PEARSON, I have read and re-read the marked passages pp. 9–10 in the *British Journal of Inebriety*. They seem to me more suited for a bantering reply, than for the fire of heavy guns. I mean, for a paragraph in the sense of "What does X. really want? He seems to object to statistical inquiry showing the extent to which feeblemindedness is transmissible. But that is a fact that statesmen must take into account and of which it is of primary importance that the information should be trustworthy. He thinks it a serious matter that Eugenists are not acquainted with physiology and pathology, but that is certainly not true of many contributors to the Eugenics Laboratory and other Biometric Publications. He wants inquiry into the origin of defects; by all means let it be attempted by those who are capable and see their way to fruitful inquiry. But that is a special line of research with which the Eugenics Laboratory is not occupied. Lastly, what is meant by the sonorous phrase 'sterile logomachy'?"

I have scribbled the above just as I should do in a first draft, to ease my mind and get my thoughts in presentable order. Don't think more of it than that. The great point is not needlessly to *embitter* any controversy, but to show that the opponent is ignorant and presumptuous. I feel sure you can do this.

I have writing now near my elbow a very good lady assistant, Miss Augusta Jones, who tells me that her sister is now working at your laboratory.

My niece left her bed yesterday, much better for her month's rest-cure, but will require I fear somewhat prolonged care. She goes to Rutland Gate for the week-end, to be doctored and set up with winter clothing. Ever affectionately yours, FRANCIS GALTON.

I may be amusingly embarrassed in relation to X., because he has undertaken to boil down for Harmsworth's forthcoming big serial publication four of my books, and I have assented, the publisher assenting also. I have not seen any advertisement of this $\frac{1}{3}$ to $\frac{1}{2}$ million issue but a favourable allusion to it in *Public Opinion*.

7, WELL ROAD, HAMPSTEAD, N.W. *November* 11, 1909.

MY DEAR FRANCIS GALTON, I ought to have written to you to thank you for your letter and now for your extract from the *Cambridge Review*, but I have been very busy and just about fit for the sofa when I get home at night. You, seeing things from a reposeful distance, can judge more wisely than I, but I feel very strongly the general harm that all exaggeration and rhetoric does to a good cause and I am sorry that your books are to be taken in hand by this prophet of the age. He can no more understand the *Natural Inheritance* or the *Hereditary*

Genius than he can grasp the principia of statistics. Why are there so many baneful journalists of the type which seems to delight in strangling all genuine scientific spirit? This may sound harsh, but I fear it is none the less true. I wish your books were in other hands. They will survive any treatment our friend may deal out to them, but the men you want to interest will be repelled, not knowing how far the rhetoric and froth lies in the account or in the original. You see how strong I recognise our friendship to be, that I venture to write thus! For myself I feel your advice is a wise view, and I shall endeavour to bear it in mind in making some reply to the criticisms which have been made.

We have been a good deal troubled with Wee Ling's sister. She has had a bad bronchial attack, and we have had a Vet. almost daily, but hope she is on the safe side now.

Yours always sincerely, KARL PEARSON.

7, WELL ROAD, HAMPSTEAD, N.W. *November* 24, 1909.

MY DEAR FRANCIS GALTON,   Here are more of our friend X.'s productions! *Six* years ago he wrote a letter to the *Daily Chronicle* stating that if tuberculous persons kept their bedroom windows open, they might intermarry without danger to their children. That was just when we had first reached the perfectly definite conclusion that the tuberculous diathesis or constitution is inherited. I wrote—this was before the days of Eugenics and I had never heard of the man before—that it seemed to me criminal to suggest that the tuberculous might freely marry without danger to their children. Since then I have not criticised him nor referred to his opinions. It may be that my words have rankled in his mind and produce effusions of the type I again send you. I don't want to worry you with these matters, but X. writes as the accredited representative of the Eugenics Education Society and his words and actions are damaging the whole movement in the minds of those who are worth convincing.

Yours always affectionately, KARL PEARSON.

THE RECTORY, HASLEMERE. *November* 25, 1909.

MY DEAR KARL PEARSON,   It so happened this morning that while I was writing to Miss Elderton, the post brought Press cuttings, including this of the *Pall Mall* (Nov. 23). I wrote to her about it, quoting a sentence from a recent speech of the Poet Laureate: "Do not resent criticism and never answer it," which seems to contain much of value.

X.'s article contains 3 detached pin-pricks or goads: (1) Mendel, (2) children of inebriates, (3) children of consumptives. I can't see that (1) has anything to do with the present question and might be put off by a sentence. (2) and (3) attack the appropriateness of the statistics used, and might perhaps be usefully answered, not as a reply to this particular attack but in a *brief memoir* dealing only with the point in question. Newspaper controversy will lead to rejoinders and re-rejoinders and will hardly convince in the end. Every pronouncement admits of an opposing argument. Think of dear old Euclid, whom we once thought infallible, and of Paley, whom generations of the ablest men of their day considered proof against attack! I sincerely hope you may see your way to do what I have ventured to suggest.

I was glad to see in the newspapers that you have given help to the "Child-Society" (of which I otherwise know nothing) in framing their "questionnaire."

All goes on smoothly here. E. B. has been to town for a few days' doctoring and has returned very well in spirits. Ever affectionately yours, FRANCIS GALTON.

I send my little volume of reprinted lectures. The little book by the Eldertons will surely do some good.

THE RECTORY, HASLEMERE. *December* 3, 1909.

MY DEAR KARL PEARSON,   I was *very* glad to see in some Press cuttings received this morning, that you were about to test the effects of environment on Jewish children. This seems to me a far better response to what X. has written, more dignified, than a controversial argument.

Ever affectionately yours, FRANCIS GALTON.

The puppy progresses; so do we all. The puppy grows very like a portrait I once had of Confucius.

BIOMETRIC LABORATORY, UNIVERSITY COLLEGE, GOWER STREET, W.C.
December 3, 1909.

MY DEAR FRANCIS GALTON, Most hearty thanks for your volume of collected papers. It will be most useful to have them in one book to refer to. I only regret you did not republish one, which I think very good. It appeared in, I believe, the *National Review*, Admiral Maxse's journal*? If I recollect rightly it first defined "Genetics," and was earlier than all now republished.

I have been very busy, mostly with matters leaving no permanent trail, and have had no time for controversy. The work on children, as to the influence of environment, is not undertaken in reply to X.; it is part of work long planned and intended to verify conclusions already reached. For the general conclusion that environment has not an influence one-quarter of heredity, we have overwhelming evidence now.

I fear we shall not see eye to eye with regard to X. It is not criticism that does any harm, it is repeated misstatement by a man who is ignorant, which does, not us, but the whole progress of Eugenics harm when it comes from, and apparently with the sanction of a Society established for distributing Eugenic truth. I won't write any more about it, or bother you any further in the matter. The evil is done, and can't be undone, but it seemed to me that I ought to tell you at least once how strongly I feel about it; and I am glad to think how strongly your friendship reacts to the strain. In one respect X.'s abuse does the Eugenics Laboratory a service—the medical profession—which with rare exceptions finds X. impossible—will not be rendered less sympathetic to the work here by a knowledge that our Eugenics differs both in method and results from his.

We have had a good deal of anxiety with our puppies; I expect we cannot give them the space and exercise they require, and I fear we must part with them, although it "will tear our hearts." Yours always affectionately, KARL PEARSON.

(16) *Events and Correspondence of* 1910. We now reach the last year of Galton's life. It was, perhaps, more active than the previous year, 1909; but the signs of physical failure became more marked; his handwriting was now and then for the first time difficult to read, but his mind remained ever suggestive, and to me personally his help and sympathy were ungrudgingly poured out until the very end. His simple nature rejoiced in the honours conferred on him. If the long-delayed knighthood was a pleasure to him, the conferment on him in this year by the Royal Society of the Copley medal† was a still greater delight. In particular, when I saw him soon after the award he was full of appreciation for the words of the President on St Andrew's Day:

"But it was not only in geography and meteorology that Sir Francis Galton manifested his versatile energies. He was much interested likewise in biological studies, especially in regard to questions of relationship and heredity. So far back as 1871 he began what has proved to be a voluminous and important series of contributions to these subjects. From his first paper, 'Experiments in Pangenesis,' down to his last volume on 'Eugenics,' his successive papers have shown a continuous development of ideas and conclusions. He was led from his early ethnological inquiries into the mental peculiarities of different races to discuss the problems of

* See above pp. 88–93.
† Galton received numerous letters of congratulation, which he thoroughly enjoyed. Professor H. H. Turner, writing to express his pleasure at the award, added that no doubt the recipient would find an amusing problem in their classification. It has fallen to another to do so. There is only one that breathed the love and affection of a long friendship. "Another gold medal! How glad I am! I feel a golden glow too, a much bigger glow than I expect that you are feeling"—so Lady Pelly expressed herself. Only two letters ventured to address the medallist as "My dear Galton." They were from his old mid-Victorian friends Lord Avebury and Sir Henry Roscoe. For the remainder he was "Dear Sir Francis." Galton had outlived the friends of his youth and most of those of his prime!

hereditary genius from the fundamental postulate that a man's natural abilities are derived by inheritance under exactly the same limitations as are the form and physical features of the whole organic world. To obtain further data for the discussion of this subject he carried out the elaborate statistical inquiries embodied in his *English Men of Science.* Confident in the results of these researches, he proceeded after the manner of 'the surveyor of a new country who endeavours to fix, in the first instance, as truly as he can, the position of several cardinal points.' His results in this quest were given in his *Inquiries into Human Faculty and its Development* published in 1883. A further contribution was made by him in 1889, when his work on *Natural Inheritance* appeared. His subsequent papers and essays on 'Eugenics' have still further stimulated inquiry into a subject of such deep and transcendent importance in all efforts to improve the physical and mental condition of the human race. It has seemed to the Council fitting that a man who has devoted his life with unwearied enthusiasm to the improvement of many departments of natural knowledge, whose career has been distinguished by the singleness and breadth of its aims and by the generosity with which he has sought to further them, should receive from the Royal Society its highest award in the Copley Medal." *Nature,* Dec. 1, 1910, Vol. LXXXV, p. 143.

I have extracted only a portion of the summary by the President of Galton's life-work, but it will suffice to indicate that before the end of his life the highest English scientific body sealed with its approval his labours.

We have to note of his actual writings two slender papers in the *Eugenics Review* and one or two letters to the newspapers. Beyond these we have his personal letters to friends. Let us first consider the two papers in the *Review.* The earlier is entitled : "Eugenic Qualities of Primary Importance*." Galton states that his few lines are offered as "a contribution to the art of justly appraising the eugenic values of different qualities†." Galton considers that certain broad qualities are needful in order to bring out the full value of special faculties. We can ascertain what these broad qualities are by considering what are the differences between prosperous and decadent communities.

"I have studied the causes of civic prosperity in various directions and from many points of view, and the conclusion at which I have arrived is emphatic, namely, that chief among those causes is a large capacity for labour—mental, bodily, or both—combined with eagerness for work. The course of evolution in animals shows that this view is correct in general." (p. 75.)

Galton then cites birds and mammals as replacing the more sluggish reptiles. Mammals, he says, are so constituted as to require work ; when they cannot exert themselves they become restless and unhealthy. Prosperous communities are conspicuously strenuous, decadent communities conspicuously slack.

Galton admits that circumstances may raise the tone of a community ; a cause seizing the popular feeling may arouse a potentially capable nation from apathy, but it would do so still more if the community had inborn "strenuousness," a simpler word would be "grit." To make his argument complete Galton ought to have demonstrated that "grit" is a hereditary character. I have little doubt that it is so, but I know of no investigation on the point. According to him this strenuous quality is built up of a sound body and sane mind enlightened with intelligence above the average and

---

* Pp. 74–76 of Vol. I.

† The reader will note that Galton in July, 1910, used the word "qualities" and not "faculties": see our p. 225, above.

combined with a natural capacity and zeal for work. It would thus appear that strenuousness is compounded of three or more simpler factors, and it is needful to suppose that these are individually either linked or highly correlated. My own investigations of school-children demonstrated that Health, Ability and Athletic Power were certainly inherited. Hence a compound of these would be so without doubt. Further, I found that these three factors were themselves intercorrelated, if not so highly as each separately was correlated in brothers*; thus it appears probable that strenuousness is an inheritable quality. Galton contrasts the strenuous and slack communities in apt sentences worth preserving :

"A prosperous community is distinguished by the alertness of its members, by their busy occupations, by their taking pleasure in their work, by their doing it thoroughly, and by an honest pride in their community as a whole. The members of a decaying community are, for the most part, languid and indolent; their very gestures are dawdling and slouching, the opposite of smart. They shirk work when they can do so, and scamp what they undertake. A prosperous community is remarkable for the variety of the solid interests in which some or other of its members are eagerly engaged, but the questions that agitate a decadent community are for the most part of a frivolous order. Prosperous communities are also notable for enjoyment of life, for though their members must work hard in order to procure the necessary luxuries of an advanced civilisation, they are endowed with so large a store of energy that, when their daily toil is over, enough of it remains unexpended to allow them to pursue their special hobbies during the remainder of the day. In a decadent community, the men tire easily and soon sink into drudgery; there is consequently much languor among them and little enjoyment of life." (pp. 74–5.)

Some of the critics of Eugenics have said that men of genius, who are so valuable to a nation, are often epileptic, crippled or semi-insane; this is the old fallacy of pointing to isolated exceptions, which prove the rule, when once we have demonstrated how few such exceptions are. There is no link of Nature which binds intellects of exceptional ability to sickness of body or mind; and if such a bond existed our first object as Eugenists would be to rupture the chain and breed men noble in body as well as in mind. Shall we permit epileptics to breed that another Napoleon may be given to mankind? He may come soon enough without that! Shall we refuse to segregate morons†, because Byron was a poet? Dante and Goethe were not morons and were greater than he! Shall we cease to discourage the mating of the tuberculous, because Keats was consumptive? The outrooting of phthisis is of more importance to a strenuous nation than even the possession of the man who wrote *Hyperion*! Surely Galton has reached a truth when he tells us that to work for the strengthening of our nation is higher philanthropy than we are wont to meet with in the current and very restricted meaning of that word. The practice of Eugenics is something greater than the practice of charity. Let us, he says in conclusion, interest ourselves

"in such families of civic worth as we come across, especially in those that are large, making friends both with the parents and the children and showing ourselves disposed to help to a reasonable degree, as opportunity may offer, whenever help is really needful." (p. 76.)

* See "On the Relationship of Health to the Psychical and Physical Characters in School Children," *Drapers' Company Research Memoirs, Studies in National Deterioration*, No. IV, Cambridge University Press.

† The accepted technical term for a mentally high grade abnormal person.

Those who preached salvation for men through good works, never thought of adding that the object of such charity must not be an enemy of society. To give to a beggar increased the grace at the disposal of believers, even if the mendicant's poverty and sores were the product of his own licence. Then came those who taught that charity must be organised and due inquiry made as to the character and needs of the recipient. This destroyed the spontaneity of charity, the desire to do at once and easily a good work, and reap immediately that feeling of grace acquired which has descended traditionally from the older faith. Lastly, we have Galton's view of philanthropy, propounding as it does a third view of charity: seek the family of civic worth, the individuals of eugenic stock and confine your help, " whenever help is really needful," to these alone. Our statesmen

"should regard such families as an eager horticulturalist regards beds of seedlings of some rare variety of plant, but with an enthusiasm of a far nobler and more patriotic kind. For since it has been shown elsewhere that about 10 per cent. of the individuals born in one generation provide half the next generation, large families that are also eugenic may prove of primary importance to the nation and become its most valuable asset." (p. 76.)

Thousands of pounds are willed every year to charities, not infrequently without knowledge of, or inquiry into the social value of the institutions benefited; it is the old seeking for grace by good works regardless of the recipient. Yet not even mere hundreds of pounds are left by testators, as by Galton, to increase our knowledge of what really makes for national efficiency, or to put into practical use the knowledge so acquired. Year by year the property and endowment of charities, and the number of those living upon them, some good, many worthless, few really under national control, increase to an alarming extent.

Let us turn to the historical source of the Reformation and remember what happened when unthinking belief in "good works" poured into the lap of the Church endowments and estates for the support of masses of men, who did little to increase the efficiency of the nation; in Galton's sense of the words, many monastic bodies were decadent communities—indolent, slouching, conspicuously slack. The danger to-day appears to come from a different side, but the false principle which is at work is the same, and we can study the analogy with profit.

Galton's second paper is entitled: "Note on the Effects of small and persistent Influences*." Our author was always urging that small but repeated influences will like drops of water ultimately wear away the hardest rock. He preached it to his too impatient followers, who with less insight into the workings of Nature, and into the religious and social evolution of mankind, largely failed to be impressed by it; some were eager for immediate eugenic legislation, when Galton would have had them give repeated if almost impalpable shoves at the right instant to the swing of public opinion. It was in the persistent action of small influences that Galton trusted for a revolutionary change in public opinion with regard to Eugenics. He refers as an analogous illustration to cases in which travellers are deflected from their

* *Eugenics Review*, Vol. I, pp. 148–9.

proposed course, and return to the point from which they started, thus really walking in a circuit or making a complete revolution. He cites the actual experience of a young friend who in a walk of $2\frac{1}{2}$ miles actually came back unconsciously to her starting point. It was a problem to delight Galton—he reckons it all out and concludes that in $7\frac{1}{4}$ paces she turned on the average through half a degree—which is roughly about 4′ of angle in a single pace, an amount quite inappreciable by ordinary observation*. Then we have the long experience of a sagacious old man:

"So it is with public opinion. It may be slow to deflect, but if deflected gently and continuously in the same direction by reasonable advocacy, it may be ultimately turned quite round by that agency alone. ...For although, if watched for a short time only, public opinion appears to be stable, few things are more unstable in the long run." (p. 149.)

Thus Galton would have his disciples turn public opinion in favour of Eugenics. And bearing his caution in mind, his biographer thinks always of the Odenwald, and gives Mrs Grundy a mild O. B. cannon—a friendly but persistent shove in the ribs at every third pace.

It was a grave misfortune that in this, the final year of his life, Galton should have been drawn into three controversies which sadly interfered with the last piece of work he had in hand.

The first attack came from a member of the Eugenics Education Society against the investigations which the Eugenics Laboratory had been making on the influences of order of birth on health and longevity. Now let us suppose those researches were wholly erroneous—which I do not admit—then the fitting way to criticise them was to show that they were statistically in error. Instead of that we were treated to an outpouring of turgid rhetoric—"The biometricians—so called, one fancies, because they measure everything but life"—"Things like that are trifles in biometrics, where anything may happen. The point I wish to make is that statements about the first-born can mean nothing, and investigations can discover nothing, until we abandon this preposterous worship of Number as Number—in which our Neo-Pythagoreans remind one of nothing more than the superstitions of the seventh son of a seventh son"—"I am not concerned here to defend the House of Lords, nor primogeniture in any of its forms, but I am concerned to protest against the tendency to identify the divine cause of eugenics or race-culture with these mathematical divinations," and so on†.

---

* I may be permitted, perhaps, to cite my memories of a similar case in relation to Oscar Browning, a well known character of my Cambridge days. While I was in Heidelberg studying after my Cambridge career, O. B. came there for a week-end in a very hot July, and we arranged on the Sunday to take a walk in the Odenwald, lunching about one o clock at X. (I forget its actual name). The day was so hot that we determined to leave the road and walk in the same direction through the forest itself. Now those who knew O. B. will remember that when walking with him he cannoned against you at every third pace. As a result of this slight but persistent series of impacts, we emerged at 12.30 on to the high road again, moving in the right direction—not at X., but a few yards behind where we had first entered the forest. We returned to Heidelberg for lunch.

† Dr C. W. Saleeby in the *Pall Mall Gazette*, May 10, 1910, but there was much of the same character elsewhere also.

These denunciations were called forth by letters to *The Times*, March 21 and 31, from Francis Galton and myself in regard to the reform of the House of Lords. Galton drew attention to the fact that a distinction must be drawn between the principles of primogeniture and of heredity. The latter does not involve the former, and whereas a strong stirp may show an adequate number of scions of marked ability, it does not follow that we shall catch able legislators by sending eldest son after eldest son to the House of Lords. My thesis was that the Upper House has been too often recruited by mere plutocrats, by political failures, or by the sons of men who have not taken the pains necessary to found or preserve an able stock—the mother of the eldest son may have been the sister of a Cecil, or a chorus-girl. The House of Lords wants more, rather than less of the hereditary principle. As Galton put it : "There seems to be a regrettable amount of ignorance among our legislators of the facts and statistical methods upon which Eugenics is based."

In *The Times* of May 21 appeared a summary of the memoir on the Children of Alcoholic Parents, issued by the Eugenics Laboratory, and on the whole a favourable leader upon it. This led to an endless controversy, and somewhat violent statements* on the part of those *The Times* termed "the enthusiastic advocates of what they are pleased to call temperance." It is not my intention here to renew old controversies but to account for the feelings that were raised in Galton's mind with regard to the Eugenics Education Society in the last year of his life. One of Galton's chief missions in life had been to develop statistical theory, to obtain scientific measures of variation and correlation and thus to ascertain whether differences between classes were or were not significant. The development of his methods applied to living forms, including man, had been termed "biometry," and solely by means of such biometric or actuarial methods is it possible to answer many social and medical problems. "General impressions are never to be trusted. Unfortunately when they are of long standing they become fixed rules of life, and assume a prescriptive right not to be questioned. Consequently those who are not accustomed to original inquiry entertain a hatred and a horror of statistics †." Rightly or wrongly the ideas conveyed in the above sentences formed Galton's method and his conception of scientific research; to contemn them was to set at naught Galton himself.

Our statistics were good for the purpose we had in view and there was more than one series; from them came indubitably for the relative health of children of school age the result expressed in the words "the balance turns as often in favour of the alcoholic as of the non-alcoholic parentage"—in short we were unable to state that by the time children reached the school age, those of the alcoholic were less healthy than those of the temperate.

The Chairman of the Eugenics Education Society, Mr Montague Crackanthorpe, wrote at once to *The Times* to state that the result was "contrary to general experience"—but not a single datum did he bring forward. "General

* It was confidently asserted that the staff of the Laboratory were "in the pay of the brewers"!

† Galton: see our Vol. ii, p. 297.

experience" was another term for Galton's "general impressions" which had assumed a prescriptive right not to be questioned. The Chairman administered one blow after another to the Honorary President of the Society ! The latter had asserted that "probability is the basis of Eugenics"; the former thought he knew of a better method, though there is no evidence that he ever described it, still less attempted to apply it.

"To those, however, who are familiar with the methods of eugenic...research the Report [that of the Eugenics Laboratory] causes no surprise at all. It simply confirms their belief that, serviceable as biometry is in its proper sphere, it has its limitations, and that a complex problem such as that of the relation of parental alcoholism to offspring is quite beyond its ken....

"First the biometrical method is based on the 'law of averages' which again is based on the 'theory of probabilities,' which again is based on mathematical calculations of a highly abstract order. From this it follows that in this particular problem, biometric research supplies no practical guide to the individual....

"I agree that some of the new technical phraseology used by the biometricians is at first rather repelling—notably, their coefficient of correlation...."

But this, we are told, is not so bad as their probable error, which they had to borrow ready-made from the astronomers.

"Further: the biometrical method deals only with patent and not at all with latent characteristics or qualities. Herein it differs markedly from Mendelism...."

And so the Chairman of Galton's Society wandered on, talking of matters he did not understand and of a memoir—as he admitted afterwards—he had not at the time read*. Heredity was not discussed in the memoir, and accordingly the reference to Mendelism was meaningless. What the Chairman of the Eugenics Education Society imagined would be the effect of his letter I cannot say; that it moved Galton so that a word would have led him to resign his honorary presidency of the Society I do know. As for the members of the Eugenics Laboratory their irritation was far greater at the attack made on Galton's scientific creed than at the idle criticism of their own work. Galton himself wrote the following letter published in *The Times* of June 3rd:

ALCOHOLISM AND OFFSPRING.

TO THE EDITOR OF THE TIMES.

SIR, Mr Crackanthorpe's letter under the above heading casts doubt on the value of biometric conclusions because they are "based on the 'law of averages,' which again is based on the 'theory of probabilities,' which again is based on mathematical calculations of a highly abstract order." So far as I can understand this account it seems to me inaccurate, but I have no idea of what is meant by "law of averages." Allow me to give my own version of biometric methods—i.e. that they are primarily based on observations, after they have been marshalled in order of their magnitudes—the little figures, say, coming first and the larger ones last—by drawing diagrams, and by countings. This much suffices to give a correct idea of the distribution of any given set of variables; it is also sufficient to give a fair idea of the closeness of correlation, or of kinship, between any two sets of variables. Here exact correspondence counts as 1, no correspondence at all as 0, and intermediate degrees are counted by intermediate decimal fractions. However, in usual biometric computations, where large numbers of figures are

* He sent round the very morning his letter appeared in *The Times* to 42, Rutland Gate to borrow the memoir "as he thought he ought to see it."

discussed, the greatest possible precision has to be reached, and the measure of the accuracy só determined has to be ascertained; then elaborate mathematical methods must be employed, which cannot be briefly described except in highly technical terms.

I do not at all agree that "the relation of parental alcoholism to offspring is quite beyond the ken" of biometric methods. The memoir that is criticised discusses that relation in regard to offspring in their early life. The simple question, divested of all connotation, whether or no adult offspring suffer, and in what degree, seems to me perfectly within the ken of biometry. But the interpretation of the results so obtained is quite another consideration.

Francis Galton.

My admiration for Galton was never higher than when I read this letter. He had a right to be indignant, but he very quietly expressed his complete dissent from the views of the Chairman of his Society.

The controversy concerning the memoirs on alcoholism of the Eugenics Laboratory continued almost to the end of 1910. There was in the Temperance Press a good deal of the usual type of biased criticism; it was even boldly asserted that the memoirs had been published in opposition to the wish of Sir Francis Galton, and the manifest antagonism of the Eugenics Education Society to these memoirs needed some public statement; there were those who thought that the Laboratory had some relation to the Society, or even that the former was in rebellion against the latter, its supposed creator! The point had been reached when the paths of Society and Laboratory must diverge, a point I had foreseen, but had not expected to meet with quite so early on the journey. Galton was indeed in a difficult position : on the one hand there was a small group of workers endeavouring to the best of their ability to apply his own methods to reach safe conclusions with regard to important social problems; on the other hand he had called into existence a very miscellaneous group of persons—held together by a faith which had not yet its "confession"—many of whom had little scientific training and still less capacity for judging statistical work; a few were cranks, and some of these were rendered septic by their own verbosity.

This body Galton felt to be needful as a force to spread Eugenic ideas. He was only slowly learning that a "confession" is requisite to hold together the members of a sect, and that without this there will be just as many creeds taught as there are individual propagandists. To this miscellaneous crowd Galton's name was merely a symbol or flag; they had never studied his scientific methods, nor did they know the stress he had laid on various results deduced from them*. To them Eugenics was a matter of sentiment and of "general impressions," and they were not prepared to submit their sacred opinions to any numerical test, nor were they "sufficiently masters of themselves to discard contemptuously whatever may be found untrue" (see our Vol. II, p. 297). Not yet had Galton given up hope that

---

* One member doubted whether psychical characters were inherited at the same rate as physical; another whether "nature" was markedly more influential than "nurture," although he did not know what Galton understood by "nurture"; a third muttered "lies, damned lies and statistics," regardless of the truth that the trouble is not that figures lie, but that liars figure. In short, all that Galton held certain, and therefore held most dear, was called in question by members of the very society he had brought into being.

his propagandist Society might in the end prove useful; only when I saw him on Dec. 28–29 of this year was his judgment inclining him to resignation. I refused, although begged, to turn it either way. Galton expressed to me his grave doubts as to whether the Society was not doing more harm than good and whether it was not desirable to resign his presidency. I turned the conversation to other matters, believing that no attempt should be made by the relatives and friends of men of genius to control their decisions even when they are very old. You may aid them in their work, but must not attempt to mould their opinions. Their opinions may seem to us everyday folk unwise, but we have only first sight for the past, the present or the future—they have second sight, the prescience which in itself constitutes genius.

Professor Marshall, Sir Victor Horsley, Mr M. Crackanthorpe, Mr J. M. Keynes and Dr Saleeby joined hands in an attack on the Eugenics Laboratory memoirs. The latter in particular ventured in the *British Journal of Inebriety* to hint that Galton himself was not in sympathy with the work of the Laboratory. The latter wrote to me as follows:

The Court, Grayshott, Haslemere. *Oct.* 27, 1910.

My dear Karl Pearson, Saleeby is obnoxious to the cause. I send a copy of the enclosed by this post to the *British Journal of Inebriety* and another to Saleeby with a few curt but civil lines. I shall be rejoiced to hear from you. All goes well here. In great haste.

Ever affectionately, Francis Galton.

To the Editor of the *British Journal of Inebriety*. *Remarks by Dr Saleeby*. My attention has been directed to an article by Dr Saleeby in the last number of your *Journal*, at your request. I suppose that you will feel so far responsible for its contents as to print in your next issue my disclaimer of a prominent part of it.

The article implies that an antagonism exists between the views of the Eugenics Education Society and those of the Directorship of the Eugenics Laboratory of the University of London. That an antagonism exists between at least one member of the Society, namely Dr Saleeby, and the Laboratory is absolutely shown in this article. But I have no reason to suppose that the opinion of the Society at large, as held by its Council, is antagonistic*. If it were, I could not occupy the post I now hold of its honorary presidency, because so far from depreciating the work of the Laboratory, I hold it to be thoroughly scientific and most valuable, and I rejoice that I was its founder. Francis Galton.

It may not be amiss to state here that *all* the memoirs issued by the Laboratory were first read by Galton in proof and many as well in manuscript. He had never made a condition that he should see them, he left us complete freedom in every respect, but they were sent because even to the last his suggestions and criticisms were invaluable. To *The Times* a few days later, Nov. 3rd, Galton wrote thus:

THE EUGENICS LABORATORY AND THE EUGENICS EDUCATION SOCIETY.

Sir, It is frequently implied, especially by lecturers and writers of articles on alcoholism, and the belief appears to be widely entertained, that the Eugenics Laboratory of the University of London and the Eugenics Education Society are connected. Sometimes it seems to be thought that the laboratory is partly under the control of the society, or, on the other hand, that the two are more or less antagonistic. Permit me, as the founder of the one and the

---

* Galton chose to overlook at the moment the action of the Chairman of its Council!

honorary president of the other, to say that there is no other connection between them. Their spheres of action are different, and ought to be mutually helpful. The laboratory investigates without bias, and with the help of highly-trained experts, large collections of such *data* as may throw light on some of the many problems of eugenics. The business of the society is to popularize results that have been laboriously reached elsewhere and to arouse enthusiasm in the public. It is active in doing so. I wish to take this opportunity of saying that I wholly approve of the fairness and scientific thoroughness of the laboratory work under the direction of Professor Karl Pearson.

It is unfortunate that much of the criticism on the work of the laboratory is by those who write under a strong bias. That on the effect of alcoholic parentage upon offspring is an instance of this. I have neither need nor wish to say more about this question, because I understand that a discussion of these criticisms will appear in a second edition of the Memoir in question, which is now at the press*. Also that a new Memoir on extreme alcoholism in adults will appear in a few weeks. FRANCIS GALTON.

Enough has been said to indicate that Galton strongly sympathised with the staff of his Laboratory under the criticism poured out on it, much of which was written by those " under a strong bias." It worried him greatly because the attack originated in a group which had been labelled "Eugenists" by Galton himself, and was largely directed against the employment of methods, which he himself had devised.

Heredity and Tradition. The boundary line in the case of mankind between tradition—that is, the handing down of acquired experience in the form of knowledge, habits and institutions—and heredity—that is, the physical transfer to offspring of germinal matter which controls the development of their qualities or of their descendants' qualities—is not a very easily defined one. It does not admit of obvious experimentation in the case of man. Certain languages, for example, have nasal, guttural and even vowel sounds, which are difficult of acquirement by members of races which have not spoken those languages for generations. Is there a physical heredity of the organs of speech which carries with it differences of vocalisation in the different races of man? The song of birds is specific; do they acquire their individuality of song by heredity solely, or by tradition? The cry of the baboon can express at least pleasure, fear, rage and love-thirst; it is the same with the dog. We know too little of the development of language in the earliest stages of mankind to fix a definite boundary to the hereditary and the traditional. There are many other such instances which may be cited. Generally we must admit that it has been too customary to attribute to traditional knowledge in man what in other animals we term hereditary

---

* Replies were made to our critics not only in the public press, but in the following publications of the Laboratory:

A Second Study of the Influence of Parental Alcoholism on the Physique and Intelligence of the Offspring. By Karl Pearson and Ethel M. Elderton.

The Influence of Parental Alcoholism on the Physique and Ability of the Offspring. A Reply to the Cambridge Economists. By Karl Pearson.

An Attempt to correct the Misstatements made by Sir Victor Horsley, F.R.S., F.R.C.S., and Mary D. Sturge, M.D., in their Criticisms of the Memoir: "A First Study of the Influence of Parental Alcoholism, etc." By Karl Pearson.

All published by the Cambridge University Press.

"instinct*." Ray Lankester, in a letter to *The Times* (May 30th, 1910), used the following words:

"There is no reason to suppose that any structural condition of the brain corresponding to knowledge or belief can be handed on from generation to generation by organic continuity—that is to say by the reproductive particles—whatever fancies and suggestions of a contrary tendency may have been indulged in by those who prefer mere speculation to scientific method."

Besides this passage much else in Lankester's letter on Heredity and Tradition was scored by Galton in his copy which I possess. He was moved to write as follows:

HEREDITY AND TRADITION.

TO THE EDITOR OF THE TIMES.

Sir, In your issue of May 30 Sir E. Ray Lankester maintains it to be almost unthinkable that "definite belief, or what we call specific knowledge," could be transmitted organically from one generation to another, and that very much of what is commonly ascribed to organic inheritance is really acquired through education. The question, in short, refers to the parts played respectively by Nature and by Nurture. I am not sure of the exact meaning to be attached to the terms "specific knowledge" and "definite belief," as applied to other animals than man, but it seems to me that a hen-reared duckling shows a specific and definite belief that water is suitable for swimming by taking to it, notwithstanding the cries and gestures of its foster-parent.

Similarly that the terror of monkeys in a menagerie at the sight of a snake, or that of an artificially incubated chicken at the cry of a hawk, or, again, the impulse that seizes on the neuter females of a hive to massacre their brothers, whether the hive be reared from a single queen or otherwise, all rank as specific and definite impulses. Very many other illustrative cases could be adduced that will occur to most readers.

Sir E. Ray Lankester quotes Speech as part of the great tradition of man. It is so, no doubt, in its developed form, but not in its elementary condition of mere cries expressive of elementary wants. Each kind of animal has its peculiar cry. I have long since instanced the cuckoo, which, though nurtured in the nests of birds that chirp and twitter, utters its familiar note as soon as it is grown up.

Much more is inherited than educability—namely, the propensity to act in the same way under similar circumstances which characterises all animals of the same race, whether they have been reared from eggs and had no maternal teaching, or otherwise. Fowls reared in incubators, fish in fish farms, dragon-flies, moths bred for silk or for show, each species behaves after its kind in well-known ways, whether the individuals have been taught or left wholly to themselves.

To some persons it seems almost profane to place the so-called material and non-material matters upon the same plane of thought, but the march of science is fast obliterating the distinction between the two, for it is now generally agreed that matter is a microcosm of innumerable and, it may be, immaterial motes, and that the apparent vacancy of space is a plenum of ether, that vibrates throughout like a solid. Francis Galton.

Ray Lankester's reply was, I venture to think, by no means a strong one. He introduced the word "human" and stated that we had no right to consider that animals were, when exhibiting a particular behaviour, i.e. when following animal instincts, in a state of mind which corresponds to that of human knowledge or human belief. But the whole problem of the boundary line between heredity and tradition is whether, and where, we have the right to draw

---

* I have heard that certain primitive races, after defaecating, throw earth over their excrement. No doubt this is attributed either to fear of magic being wrought on themselves or to a nascent *knowledge* of sanitary welfare. But many dogs promptly cast with their hindpaws—very ineffectually under domestication—sand or earth over their faeces; this is of course attributed to *instinct*.

a distinction between instinct and traditional conduct. We have not at present any liberty to assume that animals are not conscious of and do not think about their so-called "instinctive" actions. The last paragraph of Galton's letter was merely a reminder that material and non-material are at present undefined terms, and that such terms as "corporeal" and "extra-corporeal" as used by Lankester are very vague. Further the statement—in our present ignorance—that it is barely possible to imagine a mechanism by which the reproductive germ-cells could carry from one generation to another the extremely complicated and precise structural conditions which are the material correlatives of what we call "a definite belief" or of what we call "specific knowledge," can be met by asking how it is possible to imagine a similar mechanism by which the chaffinches born last year are guided to build this year a nest of the most perfect workmanship

> "...that seems to be
> A portion of the sheltering tree."

Assimilated to its environment, is such a nest the product of hereditary knowledge* or hereditary instinct, and whichever it may be, is the material mechanism which can produce this any easier to imagine than one which might carry a "definite belief," the belief, for example, that the development of the herd instinct, which we call patriotism, is essential to national welfare?

(17) *Francis Galton's Utopia.* I have described, if briefly, the controversies of the last year of Galton's life; they undoubtedly hindered his other work†. But his active mind was still busy with the idea of spreading, even more widely than his Eugenics Education Society could achieve, his creed for the regeneration of mankind. Thinking over the problem of books that have had lasting influence on mankind his thoughts turned to those ideal polities, Plato's *Republic*, More's *Utopia*, Harrington's *Oceana*, and Butler's *Erewhon*. Why should he not exercise a similar influence on generations to come by writing his own *Utopia*, a story of a land where the nation was eugenically organised‡? A modern Gulliver should start his travels again and seek a bride in Eugenia. Only a fragment of this *Utopia*, which was termed "Kantsaywhere," has reached me, it deals with "The Eugenic College of Kantsaywhere." The book purports to be "Extracts from the Journal of the late Professor I. Donoghue§, revised and edited in accordance with his request by Sir Francis Galton, F.R.S." On my last visit to Galton on Dec. 28–29, 1910, I was told with an air of some mystery by his niece that he was writing a "novel," that he probably would not mention it to me, but that if he did, I must persuade him not to publish it, because the

---

* The young birds certainly never watched their parents building the nest. Nor has anyone to my knowledge ever seen, or at least reported that he has seen, a young chaffinch or a young swallow studying the architecture of the parental home with a view to his or her own future needs!

† Besides the two Eugenics papers (see pp. 401–404 above), he only published the paper in *Nature* on "Numeralised Profiles," see our Vol. II, pp. 326–328.

‡ "Let us then give reins to our fancy and imagine a Utopia—or a Laputa if you will." Galton in 1864: see Vol. II, p. 78. The idea was not originated in 1910.

§ "I don't know you"!

love-episodes were too absurdly unreal. It is perhaps needless, in view of what has been said above*, to say that I should have given no such advice. Galton was failing in physique but not in mind, when I talked with him less than three weeks before his death; and to recommend him to destroy what he had thrown time and energy into creating would have seemed to me criminal. If Swift had died before the issue of *Gulliver's Travels*, or Samuel Butler before the publication of *Erewhon*, their relatives might possibly have destroyed with equal justification those apparently foolish stories. I do not assert that Galton had a literary imagination comparable with that of Swift or of Butler, but I feel strongly that we small fry have no right to judge the salmon to be foolish or even mad, when he leaps six feet out of our pool up a ladder we cannot ascend, and which to us appears to lead into an arid world. We must remember that Galton had set before himself in the last years of his life a definite plan of eugenics propagandism. He wanted to appeal to men of science through his foundation of a Eugenics Laboratory; he had definitely approached separate groups like the Anthropologists in his Huxley Lecture and the Sociologists in his lecture before their Society and in his subsequent essays; he had appealed to the academic world in his Herbert Spencer Lecture at Oxford, and to the world that reads popular quarterlies in his Eugenics Education Society. But there are strata of the community which cannot be caught by even these processes. For these he consented to be interviewed, and for the still less reachable section who read novels and only look at the picture pages of newspapers, he wrote what they needed, a tale, his "Kantsaywhere." His scheme for proselytism was a comprehensive one, but I think Galton knew his public better than most men.

An Ibsen or a Meredith with far more imaginative power would, if they had taught Galton's creed, have struck above the level of those for whom Galton intended his tale. Its actual composition was started in May or June of 1910, when Galton had returned to Rutland Gate from Haslemere. It received many modifications in characters, names and actions during the following six months. In December he was sufficiently satisfied with it to submit it to a publisher, but the publisher would have none of it! Galton— as I realised, once he began to send me papers for criticism—was so modest that a moderately adverse judgment on a single point might lead him to discard many months of work; one learnt to mix praise with every suggestion for amendment. Almost anyone's adverse judgment, even that of a publisher or his reader who must assess solely by the likelihood of profit, was enough to shake Galton's confidence in his own work. To his niece, Mrs Lethbridge, he wrote on Dec. 28 :

You and Guy more especially must have had a wretched time of floods and tempests. We on the high ground feel like Noah on Ararat....

The glorious frosty sunshine of this morning picks me up. I have been "throaty" and obliged to rest a good deal. Karl Pearson comes this afternoon for one night. I am saving my voice for him. "Kantsaywhere" must be smothered or be suspended. It has been an amusement and it has cleared my thoughts to write it. So now let it go to "Wont-say-where." My very best New Year wishes to all of you and best love. Ever affectionately, FRANCIS GALTON.

* See above, p. 408.

Might not that which had cleared Galton's thoughts, in time, have cleared ours also ?

The following letter will explain how the fragment of " Kantsaywhere " came into my possession. It is from one of Galton's nieces to whom the task of destruction had been committed by his executors, and written to one of these ; it was forwarded to me by the latter with the fragments.

I was just thinking of writing to you about "Kantsaywhere" when your letter came. When I began the work of execution, my heart misgave me so much that I thought I would begin by merely "Bowdlerizing" it, and then see. So I destroyed *all* the story, all poor Miss Augusta, the Nonnyson anecdotes, and in fact everything not to the point—but there were a good many pages that I felt myself incapable of judging. So I am returning the mutilated copy, hoping (if you and Eva* could agree on the point) that Professor Karl Pearson might see it. Unfortunately Eva is not well enough just now to be consulted, so we must wait. Mutilated as it is, poor "Kantsaywhere" can never be published, and it is as safe from *that* as if it were destroyed altogether, but I think what remains might interest Prof. Pearson, and possibly, though I doubt it, be useful. Besides if *something* survived, I should not feel quite so much like a murderess! The duplicate copy is destroyed altogether....Anyhow it seems to me that if any one has "Kantsaywhere," it should be Professor Pearson or one of the Darwins. But this is only *my* view, and I don't want to urge it. You and Eva will be better judges than I.

[Dated: March 27, 1911.]

No doubt those who took upon themselves to pass judgment on Galton's last work were fully conscious of the responsibility they shouldered. But the fealty of a biographer is of a different kind ; his duty is to give a *full* account of his subject ; if there were weaknesses, they were compensated by strengths ; if he is called upon to describe the actions of his subject when young, he must equally describe those of his old age. Whatever the duty of a literary executor† may be in determining whether the issue of an unpublished paper will tend to increase or lessen the reputation of the testator, this duty does not fall to the biographer; he must give an account of all that has come within his cognizance and which he thinks can illustrate the character and opinions of his subject. He must not emphasise strength by omitting to notice what some may consider weakness. Nor is the temptation to omit repeating to the reader what I know of " Kantsaywhere " at all overwhelming. I think it may help " to clear the thoughts " of all of us regarding what a society organised eugenically should strive to achieve. As for the story itself it was a mere driving band to carry the force of Galton's ideas into the working parts of minds differing widely from his own. The thread of the story, as far as I have been able to ascertain it from one who had read it before destruction and from the fragments in my possession, ran as follows : A professor of vital statistics after certain adventures reaches Kantsaywhere. He is a man of some parts and meets with a young lady of that country, who is about to take her Honours Examination at the Eugenics College. The professor is much interested in the customs of Kantsaywhere as well as in the young lady, and determines if possible to obtain for himself as high a Eugenics degree

---

* Lucy Evelyne Biggs, Galton's great-niece and companion.

† I do not think that Galton by his will appointed any literary executor, so that his papers, published and unpublished, would appear to have become the property of the residuary legatee, the University of London.

as Miss Allfancy. The hero of the tale is the " I " of the following extracts. Apparently a colony of Kantsaywhere had been founded and its government entrusted to a Council. My fragment opens with the statement that a Mr Neverwas had died :

"leaving all his property in the hands of trustees for the use of the Council of Kantsaywhere and their successors. He desired in his testament that the income should be employed in improving the stock of the place, especially of its human breed. The methods of doing so in force at the time of his death were to be continued with such future changes directed to the same end as experience might suggest.

"The College was to grant diplomas for heritable gifts, physical and mental, to encourage the early marriages of highly diplomaed parents by the offer of appropriate awards of various social and material advantages to relieve the cost of nurturing their children, to keep a minute register of results, and to discuss those results from time to time. He laid down the principle with much emphasis, that none of the income of his property was to be spent on the support of the naturally feeble. It was intended, on the contrary, to help those who were strong by nature to multiply and to be well-nourished. The practice of charity in the ordinary sense of protecting the feeble, however commendable in itself, was to be left to such other agencies as might be formed independently of the College and not disapproved by it.

"The 200 inhabitants of 1820 have now become 10,000, partly through natural increase, which is equal to the full rate of the present [1910] population of Russia, where in every decade, 100 becomes 140. At this rate in 90 years the 200 have become 1000. Immigration accounts for the rest.

"The Trustees of the College are the sole proprietors of almost all the territory of Kantsaywhere, and they exercise a corresponding influence over the whole population. Their moral ascendancy is paramount. The families of the College and those of the Town are connected by numerous inter-marriages and common interests, so that the relation between them is more like that between the Fellows of a College and the undergraduates, than between the Gown and Town of an English University. In short, Kantsaywhere may be looked upon as an active little community, containing a highly-respected and wealthy guild. So much for the early History of Kantsaywhere."

Our hero remarks that on his arrival in this strange colony he found himself more " keenly looked over " than ever in his previous experience.

"It is the way of Kantsaywhere, for everybody is classed by everybody else according to their estimate or knowledge of his person and faculties.

"Let me explain at once that what they are concerned with in one another are the natural, and therefore the only heritable characteristics. We have heard much in political talk of the 'prairie value' of land, that is to say, of its value when uncultivated, neither fenced nor drained, ploughed nor planted, only to be reached over the waste, and having neither houses nor farm buildings. Applying this idea to man, as if he were land, it is the prairie value of him that the Kantsaywhere people seek to ascertain. His 'brute value' would be a proper expression if employed in the original sense of that word, but 'brute' has acquired so many disagreeable connotations that if used here it would be misunderstood.

"I learnt that I was only just in time to undergo the first of my two examinations. It was merely a 'Pass' one, but a necessary preliminary for admission to the 'Honours' examination, in which the more successful candidates are classed in order of merit. The Honours examination of girls for the year was just over, and the lists were to be published that very night. The eldest daughter of the house, Miss Augusta Allfancy, was a candidate and all the family were keenly anxious to learn the result, for it would have an important effect on her after-life. It seemed tacitly agreed that nothing should be said on this matter until the results arrived, so they were only too happy to have their thoughts diverted to English topics and to my own affairs.

"In Kantsaywhere they think much more of the race than of the individual, and on my expressing a faint surprise, the family argued to the following effect: 'Suppose a person to be one of the two parents of four children. He or she contributes a half share to each, which is much the same as a whole share to two*. This process may continue indefinitely in a growing population like their own, so his or her influence on the race may increase in geometric proportion as the generations go on. A person is therefore more important as a probable progenitor of many others more or less like to him in constitution than as a mere individual.' I learnt that the object of the first examination was to give a Pass certificate for 'Genetic' qualities. By 'genetic' is meant all that is transmissible by heredity, whether it be of ancestral origin or a personal sport or mutation. The refusal to grant a Pass certificate is equivalent to an assertion that the person is unfit to have any offspring at all. By a second-class certificate that permission is granted, but with reservations, of which more will be said later.

"In reply to my expression of diffidence as regards my own success, I was emphatically reassured by my late scrutineers as to my *personal* capabilities, which Tom was pleased to rate at '30 at least,'—a term which will be explained later. But what my *ancestral* claims might be valued at, was another matter. They assured me that my sponsor, Mr Allfancy, had already submitted an outline of them to the examiners, in as favourable terms as the information warranted, and that he was quite satisfied with them for pass purposes, but was sure that they were insufficiently authenticated to receive adequate credence from the examiners for honours. Consequently far fewer marks might be awarded me for my ancestry than I probably deserved. They all expressed surprise at foreigners knowing so little with exactness about their grandparents and other ancestors, saying, that everyone in Kantsaywhere knew their own as well almost as if they had been their playmates and comrades, and that they all possessed an abundance of well authenticated facts about them†....

"I was told on inquiry that those who were placed high in the list, as Miss Augusta was, were justified in expecting numerous advantages on their marriage, that as many of them as there were vacancies in the College—there were ten in the present year—were elected Probationers, and therefore future recipients of those advantages if their husbands were adequately diplomaed, but not otherwise. What the girls most thought of, as Tom afterwards told me, was a marriage between two probationers whose joint marks exceeded 200 and who had at least two stars, of which more will be said later. It gave the right of having the marriage conducted with special ceremony‡, and of its being known and recorded as a 'College marriage.' The offspring of such marriages are reckoned foster children of the College during their childhood, and they and their 'College parents' are helped in many important ways. But Tom added that his sister, in order to obtain one, must marry a man with at least 107 marks and one star, and that very few of such unmarried men are available. I took full notes of what Tom told me of the advantages attached to a College wedding, and to others which were a little short of having a 'joint 200 marks and two stars,' but I must get them verified before putting the results into my Journal."

We now reach Chapter V of the work, entitled: *Pass and Honours Examinations*. I have reproduced above all that remains of the first four chapters of the work; the bulk of the extracts given are certainly from Chapter IV, but some possibly from Chapter III. I do not know even the title headings of the first four chapters of the story. On March 21st Tom Allfancy takes the stranger to the Examination Hall for the Pass Examination, where, he tells us:

"I went through physical tests, which I need not describe particularly, as they were similar to those which all Englishmen undergo before admission into the Army, Navy, Indian Civil Service§, etc. But the examination was more strict and minute and in the medical part it was

---

* A "share" in this sentence must be taken of course to comprise all that an individual's germ-plasm involves, not merely his apparent characteristics.

† Galton was undoubtedly thinking here of his books the *Record of Family Faculties* (1884) and the *Life-History Album* (1884); see our Vol. II, pp. 362–370.

‡ This idea, as well as others in "Kantsaywhere," closely resembles that of Galton's first paper on Eugenics, that of 1864; see our Vol. II, p. 78.

§ See our pp. 231–2 above.

such as a very careful Insurance Office* might be expected to require. I was much questioned about the papers that Mr Allfancy had sent in, as regards my personal knowledge of the authorities for the facts there set forth. They then smilingly gave me a first-class P. G.—Passed in Genetics—degree, and I had to imprint my fingers in their Register, for future identification if necessary †. So I returned to my host with one small portion of a load of anxiety taken off my mind.

"I heard a little now, but must inform myself more particularly hereafter, as to the fate of those who failed to pass. A Bureau was charged with looking after the unclassed parents and their offspring, and much was done to make the lot of the unclassed as pleasant as might be, so long as they propagated no children. If they did do so kindness was changed into *sharp severity*.

"Labour Colonies are established where the very inferior are segregated under conditions that are not onerous, except that they must work hard and live in celibacy. It is difficult to describe the indignation and even the horror felt in Kantsaywhere, at acts that may spoil the goodness of their stock, of which they have become extremely proud and jealous. They look confidently forward to a coming time when Kantsaywhere shall have evolved a superior race of men. As it is the people who are born there and emigrate nearly always excel most of their competitors on equal terms, and return in after life with sufficient means to end their days in tranquillity near their beloved College.

"In the evening I found the Allfancy party much saddened by ill news to the effect that one of their dearest friends, who had made a 'College' wedding with much éclat a few years previously, had given birth to a deformed child. I had expected to hear from Mrs Allfancy some severe remark on the subject, but was mistaken. She was most sympathetic with the family and the child. The College was responsible, she said, for its existence: the marriage of its parents had its highest approval; it was brought into the world in accordance with the rules they advocate. The misfortune was due to some overlooked cause, which might or might not be of a kind that would hereafter be understood and could be provided against. No blame whatever attaches to the parents who should be whole-heartedly condoled with. The child should be in no way discouraged on account of its natural defect, except as regards absolute prohibition hereafter to marry."

Our hero now enters for the Honours Examination, and the description of the anthropometric tests and even the place of examination remind us at every turn of Galton's South Kensington Anthropometric Laboratory: see Vol. II, pp. 257–262 and 370 *et seq.* The reader who has followed the course of Galton's labours in Vol. II will recognise how in his Utopia he draws together all the threads of his apparently disconnected efforts to unite them into a strong eugenic strand. The following is Professor I. Donoghue's account of his experiences on March 25th:

"This was the first of the four days to be occupied in the annual examination of about 80 candidates for Honours, one quarter of them on each day. The examination consists of four divisions. The first is mainly anthropometric, the second is aesthetic and literary, the third is medical, and the fourth is ancestral. Many examiners are employed and a staff of skilled clerks in addition. The examination is conducted in batches, each batch being assigned a particular hour for beginning, and for being thenceforward submitted to the four sets of Examiners successively.

"My batch had to present itself at 12 noon. At that hour I handed in my Pass Certificate to an official, who sat in the Hall, by the entrance to a long enclosure of lattice-work ‡, through which everything was easily seen from the outside. The enclosure contained a row of narrow

* See our pp. 243 ftn., 268 above and Chapter XVII.

† See our pp. 154 and 159 above.

‡ The whole passage is a description of Galton's First South Kensington Laboratory; even the lattice-work—the beginning of which is seen in Plate L, p. 371, of our Vol. II—was in use there.

tables ranged down its middle, on which most of the measuring instruments were placed, the heavier ones standing on the ground between them. Those instruments were duplicated that required a longer time for their use than the rest. A passage ran between each side of the tables and the walls of the enclosure. Five attendants, each having one candidate in charge, were engaged all day long in making a tour of the tables in succession. The candidate emerges and is dismissed at an exit door, which is separated from the entrance by a low gate, over which the official can lean while he sits.

"Immediately after entering the enclosure, my attendant made me sign my name and impress my blackened fingers on a blank Schedule. It contained numerous spaces with printed headings, which the attendant filled in with pencil as he went on. He took me round the enclosure, testing me in turn by every instrument and recording the results. They referred to stature, both standing and sitting, span of arms, weight, breathing capacity, strength of arm as when pulling a bow, power of grip, swiftness of blow, reaction time, discrimination (blindfold) between weights, normality of eye, acuity of vision, colour sense, acuteness of hearing, discrimination of notes, sensitivity of taste and of touch, and a few other faculties. Lastly the states of my teeth, which are particularly good, and of my mouth, were inspected. The entries to my schedule now and later on were, as I heard, to be examined and checked by clerks whose business it was to translate the Measures into Marks, according to a definite system. For related faculties, Weight and Strength in combination, a sheet of paper ruled in squares was prepared, in which a series of successive weights was written down its side and a series of strengths along its top. In the square where the line of the one was crossed by the column of the other the appropriate mark was written. This was copied out by the clerk for the use of the Examiners. But more will be said later on of their Measures and Marks.

"I was next taken to another part of the Hall and submitted to an examination for aesthetics and literature. I was given both prose and poetry to read aloud before the Examiners, a copy of these extracts having been handed to me to peruse beforehand. Then some simple singing was asked for. After this, a few athletic poses were gone through as well as some marching past, and the Examiners noted their opinions on my Schedule. Then I was allowed an hour to write four short essays on given subjects. This was the only literary test.

"I should say that they lay much stress on the aesthetic side of things at Kantsaywhere. 'Grace and Thoroughness' is a motto carved over one of the houses for girls in the College, and I have seen it repeated more than once in embroidery and the like. A loutish boy and an awkward girl hardly exist in the place. They are a merry and high-spirited people, for whose superfluous energy song is a favourite outlet. Besides, they find singing classes to be one of the best ways of bridging over the differences of social rank. Musical speech and clear but refined pronunciation are thought highly of; so is literary expression, and this examination is intended to test all these. The 'arry and 'arriet class is wholly unknown in Kantsaywhere.

"I was then medically examined in a private room, very strictly indeed, and much was asked about my early ailments and former state of health. Here again I need not go into details, for they can be easily imagined in a general way, even by a layman. It is wonderful how adroit the skilled medical examiners become in their task. Nothing seemed to escape their sharp observation, whether of old scars or any internal abnormality. My few defects were unimportant; I thought my vaccination marks had become invisible but they were quickly noted and minutely examined. The principles on which marks are to be awarded are fully laid down in printed directions.

"Lastly came the consideration of my ancestry. The papers communicated by Mr Allfancy were produced and again looked into and criticised, but much more minutely than before, and the value of the authorities for the facts stated in them was keenly discussed. I lay under a difficulty here. The official records made at Kantsaywhere are so minutely kept, that the requirements of the examiners have grown to be extremely rigorous as regards the evidences of ancestral gifts and maladies. All immigrants are more or less suspected. Besides this, such evidences as would require little confirmation in England, owing to public knowledge of the characters of their high authorities, may, and do, require more confirmation here than can easily be collected at home. I deeply resented my own ill-luck in this matter. The examiners told me only what I was fully prepared to hear, but expressed at the same time much regret that they were unable to give as many marks for my Ancestral Efficiency as I possibly, or even probably, deserved. In fact, I only got 5 marks for my ancestry.

"This concluded all that I had to undergo. I had spent about one hour under anthropometric tests, and from half-an-hour to one hour under each of the other three, besides the hour in essay-writing, or about four hours in all, exclusive of intervals. Candidates were undergoing examinations in different parts of the Hall at the same time, but not necessarily in the same order. The Medical Room was wholly separated from the rest. The Examination Hall was in full use during 6 hours, so with duplicated examiners, more than 20 candidates could be wholly and easily examined in a single day. Four such days dealt with all the 80 candidates. The clerks were simultaneously employed, each in copying and in reducing entries and adding up figures, which after being checked by other clerks were submitted to the chief examiners. Those gentlemen had also acted as overseers and taken some part in the examinations.

"The maximum number of positive marks that could be gained by each candidate is four times 30 or 120. A star (*) might also be gained in each subject. The marks were totalled, and about half of these totals usually range between + 45 and + 70. None of the candidates were given negative marks, those who would otherwise have received them having been weeded out by the Pass Examination. The names and marks of those who gained 70 marks and upwards are published in the newspaper, together with such brief notes as each case might call for. This part of the publication is official and wholly under the editorship of the Registrar. I learnt that supplementary marks might be, and often were, accorded for especially good service to the community subsequent to the examination. They had to be proposed by the Board of Examiners, and the grounds for the proposal had to be set forth in their Annual Report. This was submitted to the final approval of the General Meeting, which was almost always given as a matter of course. These Supplementary Marks are supposed to attest that the natural capacity of the person who receives them really exceeds that which was expressed by the number of marks he had received at the original examination.

"I do not know much in detail about the examination for girls. It is carried out by women examiners who had taken medical degrees elsewhere, and is, I was assured, as thorough as that which I had myself undergone, and was considered to be as trustworthy.

"There is a bifurcation of the Examinations both for girls and boys, part of each of them being intended for the more cultured class and part for the hard workers, whether on farms or in town. I need not go into particulars.

"I inquired minutely whether they were unable to devise some test for endurance or staying power, which seemed to me one of the most important of those they had to consider. It seemed that they had not as yet succeeded in eliminating the effect of practice. Neither were they enabled to examine into character directly as a separate subject, partly because it was not fully developed at the usual age of examination, and partly because of the extreme difficulty at that age of estimating it justly, the teachers and the comrades of a girl or boy often making sad mistakes of judgment.

"I was assured that no doubt was felt as to the trustworthiness of the marks given by the examiners, as a general rule, subject rarely to exceptions such as might be expected. The sons of College Marriages were unmistakably superior in bodily and mental gifts to those of the ordinary folk of Kantsaywhere, and these again compare very favourably with those of neighbouring colonies. Besides this, numerous results are published in which comparisons are made between the children of high-diplomaed parents and of those who are less highly graded. All concur in showing the general superiority of the former, just as much but not more than would be expected of the offspring of various qualities of any domestic animal. A general conviction of this truth forms the firm basis of the customs and ideals of Kantsaywhere.

"CHAPTER VI. *The Calendar of Kantsaywhere.* I returned to my host's house, where I was congratulated on having gone through my ordeal. I felt sure of success in the anthropometric part because I was something of an athlete, having rowed in a University race. I was also good in other respects, being reputed by good judges to be so prompt and sure a shot, that I have been urged, in all seriousness, to go to Monte Carlo and compete there for the valuable pigeon-shooting prizes. I knew I was all right medically, and thought I might do fairly in aesthetics. I, however, saw clearly that I was not even yet received with perfect freedom, except by Tom; the others evidently waited to learn how I should be placed, before letting themselves go, so to speak. They did not as yet invite me to accompany them to the houses of their friends, so I had much spare time, and thought the best way of occupying it until the lists were out, was to stay indoors and to make a careful study of the Calendar of Kantsaywhere College. I saw little of Miss Augusta at this time, as she was invited to a succession of parties. The first four were official invitations given to ensure that each girl probationer should be made acquainted with an equal number of male probationers three or more years older in standing. The male probationers are divided more or less at random into two groups A and B, the females into F and G, then the four official invitations are to A and F, A and G, B and F, and B and G. They have an amusing old-fashioned method of grouping and re-grouping the guests at these entertainments, in order that each girl should have a full half hour of conversation with each young man. It approached merry-making and banished diffidence. It seems however that marriages between two newly made probationers are not particularly approved. It is thought best that the girls should marry young, say about 22 years of age, which admits of more than 4 generations being produced in each century. As for the men, they have to establish themselves in some occupation before they can support a wife, which cannot usually be done till nearly the age of thirty. Consequently many social gatherings are arranged to bring together young girl probationers and older unmarried men, also of the rank of probationers. Persons may fall in love in Kantsaywhere as they do in England, on grounds more or less unaccountable to others, but it is felt here that the best girls and the best men should have frequent opportunities of becoming friends and the earliest chance of falling in love with one another.

"I was surprised to learn from the Calendar of the large extent of the College possessions in farms, houses, hostels, and funds, which were used to encourage early marriages among the most highly diplomaed; I also perceived that the Collegiates must look upon themselves, as they did, as a great family community, out of which about one half of the members of each new generation were obliged to seek their living elsewhere, just as it usually happens in English families now. The Calendar contained the names of all who, since the date of the preceding edition, had either received marks exceeding + 70 or any special award. The record in the Calendar of their doings was minute. It corresponded in length to the paragraphs of Burke's or Debrett's Peerages, but differed totally from them by containing anthropological facts, and little else. It was a mine of information for inquirers into heredity, yet it was described as being only a brief abstract of what was preserved in MSS. in the records of the Registrar.

"Tom had hinted to me that he thought his sister was slightly chagrined at her marks falling short of one half of those required for the great honour of a College wedding. The number of names of the men amongst whom she must marry, in order to secure one, was very small, and could easily be found from the Calendar. I looked for them and found only twelve, some or all of whom might be already engaged.

"The large property of the College consisted, first, of the original endowment, of which the income was now retained in England and had been accumulating during recent years to form an Emergency Fund. Secondly, of the fee simple of the district and of all the houses, etc., that had been erected on it since the beginning of the Settlement. Thirdly, of gifts and bequests from former Collegiates, in gratitude for their rearing and in payment for its cost. Fourthly, the annual Eugenic Rates from Kantsaywhere. The inhabitants submitted as cheerfully to as heavy a rate in support of the College, as we do for the support of our Fleet, namely three quarters of £1 per head of the population. We in England, numbering some 45 millions, contribute about 35 millions of pounds annually to the maintenance of our Navy. Here, the 10,000 inhabitants contribute £7,500 to the College, and could easily be persuaded to contribute more, if it were really needed. In very round numbers one half of the income from the last two sources, from gifts and from rates, goes to the Examining, Inspecting and Registering Departments, which together form the soul of the place. The other half goes to collegiates who really need help to enable them to give proper nurture to their large families. This is done very judiciously on the joint recommendations made to the Committee of Awards, by a Board of District Visitors in conjunction with the District Inspector. The Chief Medical Inspector is one of three High Officials, the Rector and the Registrar being the other two. These are elected by the Senate at its Annual General Meetings for a term of three years, and are re-eligible. The Senate consists of all resident Collegiates of either sex, who had gained at least 70 marks, or who are parents of children whose average marks exceed 70 and whose total marks exceed 200, and is the supreme Authority, but in quiet times, the above-mentioned three High Officials, together with a Council, annually elected at the General Meeting, manage matters very much in their own way. This constitution works very well on the whole, though with occasional jars, much as those which occur in our leading Scientific Societies at home.

"An important Committee of this Council is charged with the care of those who fail to pass the Poll examination in Eugenics. Such persons are undesirable as individuals, and dangerous to the community, owing to the practical certainty that they will propagate their kind if unchecked. They are subjected to surveillance and annoyance if they refuse to emigrate. Considerable facilities are afforded to tempt them to go, and agents of the College who are settled in the nearer towns to which they are most likely to drift, are prepared to take charge of them on their arrival. Their passage out is paid, small sums are granted to them at first, on the condition of their never returning to Kantsaywhere. They must renounce in writing all its privileges before being allowed the cost of deportation. Not a few of these persons do well enough especially when the principal reason of their rejection is some hereditary taint, and not personal feebleness. As regards the insane and mentally defective, suitable places for their life segregation are maintained in Kantsaywhere. With so small and eugenic a population, the cases are few and easily dealt with.

"The Regulations printed in the Calendar confirmed the view I had already formed, that the propagation of children by the Unfit is looked upon by the inhabitants of Kantsaywhere as a crime to the State. The people are not misled by the specious argument that there is no certainty whether the anticipations of their unfitness will be verified in any particular case and the individual risk may be faced. They look on the community as a whole and know the results of unfit marriages with *statistical* certainty, which differs little from *absolute* certainty whenever large numbers are concerned. For instance, they say 1000 unfit couples will assuredly produce a number of children that can be specified within narrow limits, of each grade of unfitness, though they cannot foretell whether these children will be the offspring of A, B, C or X. This same statistical certainty forms a large part of the foundation of laws and penalties in every part of the world. There are many grades of expected unfitness, ranging from that of the offspring of the idiots, the insane and the feeble-minded, at the lower end of the scale of civic worth, to whom the propagation of offspring is peremptorily forbidden, whether it be by forcible segregation or other strong measures, up to the moderate unfitness expected in the offspring of parents who rank only a little below the average in eugenic worth. The methods of penalizing, taken in the order of their severity, are social disapprobation, fine, excommunication as by boycott, deportation, and life-long segregation. The degree of restriction varies from the limitation of the offspring of unfit parents to a small number, up to its total prohibition. They say that limitation of families is now a recognised institution among most of the cultured and many of the artisan and labouring classes in Europe and America, and there is no reason why a sentence demanding it for the protection of the nation should not be passed, and the infraction of that sentence punished as a criminal act. As regards fines, if the defaulter cannot pay them, he is treated with severity as a bankrupt debtor to the State, being placed in a Labour Colony with hard work and hard fare until it is considered that he has purged his debt. With so small a population as the 10,000 of Kantsaywhere, and with the general high level of breed of its inhabitants, the cases of marked unfitness are not sufficiently numerous to require formal classification in different asylums. They can be more or less individually dealt with by the Board of Penalties.

"The difficulty must again be discussed here, relating to the introduction of unfit immigrants. Municipal laws have been enacted, that are quite as severe as those in America and elsewhere, to exclude impecunious immigrants, but they are enacted here for the purpose of excluding the immigration of the constitutionally unfit into Kantsaywhere. Ships, as already mentioned, are only allowed to disembark their passengers subject to the fulfilment of certain accepted conditions. If unfulfilled, the ship-owners are obliged to convey them back to whence they came. Registered medical men are established at the principal ports from which immigrants arrive, whose certificate that a person has passed the ordinary test for fitness in body and mind is accepted. It exempts them from the somewhat more severe and tedious examination of which I have already spoken, which is conducted in a building attached to the Custom House and must be successfully gone through before they are allowed to disembark even for a short residence. They are required later on to pass the Poll examination which allows them to become citizens of Kantsaywhere.

"The grades of unfitness on the part of those who are married are determined by the number of their joint marks. Immigrant parents both of whom have received positive marks at the Poll examination may keep their children with them, but not otherwise.

"The restriction placed by public sentiment and, in extreme cases, by penalty, on the number of offspring that a couple may propagate in Kantsaywhere, is based on that of their joint marks. If these exceed + 20 the restriction is nil and large families are encouraged. If between + 10 and + 20 they are restricted by public sentiment to about three children. If over 0 and under + 10 they are restricted to two children. If between 0 and − 10 they are restricted by law as well as by sentiment to one child. If below − 10 offspring are wholly prohibited to them. The above concessions were established as compromises, after balancing conflicting claims. It was necessary to take into account the need of the parents, the advantages of family life and the well-being of the children, as well as that of the race."

Chapter VII is entitled *Measures and Marks*. It commences with the following words :

"A paragraph in the Calendar headed "Measures and Marks" greatly interested me in connection with my previous statistical studies. These enabled me to understand easily the methods used in Kantsaywhere, which must seem puzzling and fanciful to others to whom they are wholly new. Such persons will I fear skip this chapter."

Next follows a description of Galton's process of ranking by size. Then the quartiles are defined and we are told that half of their difference is taken equal to 10 Q-Vars*, while half of their sum is accepted as the middlemost value (median) of the series.

"Each measure is translated into the middlemost value of the series *plus* or *minus* so many Q-Vars. The quickly increasing variety of larger values than 30 Q-Vars and the fear of untrustworthiness in applying them have led the examiners in Kantsaywhere to limit their measures to a maximum of 30 Q-Vars, in each of the four principal divisions of the Examination. If the candidate obviously deserves still higher marks, they add a star (*) with accompanying explanation. Tom's exclamation that I was 'at least 30 in personal qualities' was thus explained.

"Measurement by Q-Vars, or indeed by any kind of Var, in the case of all 'Normal' variables†, has the further advantage of affording means whereby class-places may be converted into measurements, or vice versa, notwithstanding that they run at very different rates.

"...It is reasonably inferred that such faculties as cannot yet be directly measured, but which can be classified by judgment, will also obey the 'normal' law. The suitability of candidates for a particular post, or the goodness of essays written by different candidates, are cases in point. Whenever the objects in a 'normal' group of values can be classified, their class-places can be converted into Q-Vars.

Conversion of Q-Vars into Centesimal Class-Places.

Q-Vars	− 30	− 20	− 10	0	+ 10	+ 20	+ 30
Class-places (Centesimal)	2	9	25	50	75	91	98

"Measures made in Q-Vars are converted into marks by multiplying them by a factor appropriate to the importance of the faculty measured."

Thus Galton says if the civic worth of one faculty be ½ that of a second, the marks of the first will be multiplied by 0·5, before combining the two.

* The "Var" is thus the tenth part of the "probable error" = ·06745 × standard deviation. Thus 30 Q-Vars equal about 2·0235 times the standard deviation, and roughly about 2 °/₀ of the population exceeds this.

† "Normal" variation is described in simple terms and attributed (erroneously) to "the great mathematician Gauss"; it is stated to be "with a useful degree of precision" the rule of distribution in the case of most anthropological measurements.

He does not explain how the proper weighting is to be reached. It is this system of scoring that the clerks used, when, working in pairs independently for the sake of checking, they reduced the marks for the Examiners in the eugenic tests (see p. 418 above). It will be seen that Galton is here reproducing the ideas and methods of his paper of 1889 on "Marks for Bodily Efficiency" and his preference for the use of percentiles (see pp. 387–390 of our Vol. II). In that second volume I have said on p. 401 that Galton, when 85 years old, broke a last lance for the use of the ogive curve, the median and quartiles. I had not at that date examined the fragment of "Kantsaywhere." It will be clear from the above *résumé* of its Chapter VII, that within two months of his death, in his 89th year, Galton again illustrated in popular language the advantages of his method of ranking.

We now reach the last chapter that has been preserved of Galton's Utopia. Here he largely drops his Eugenic State and gives expression to his own ideas on male and female beauty, on immortality and on funeral services, bringing in of course composite photography : see our Vol. II, pp. 283–298. I reproduce the whole of this chapter.

"CHAPTER. VIII. *Marks gained by me—Society of the Place.* The lists came out on March 30. I got 17 marks less than Miss Allfancy, i.e. 77, but under the circumstances it was a very fair performance and I at once noticed the change in the reception given to me. It was distinctly more genial and intimate than before, and I was begged to accompany my host's family to half a score of different places to which they were invited. The loss of marks I had sustained owing to an "English" ignorance of my ancestry, became generally known and allowed for. I will describe in a few words my general impressions of Kantsaywhere society, to which I was now freely introduced. I had carefully guarded myself against exaggerated expectations of what might have been achieved by selective breeding at this place. It is but a small community and though of a high general level, the highest variations from that level cannot be expected to exceed those of an enormously larger population whose level is somewhat, and even considerably, lower. There are nearly 50,000,000 inhabitants in the British Isles and only 10,000 in Kantsaywhere; that is, they are 5000 times fewer. Again, however far gone a population may be in its decadence, it will retain enough organisation to bring forward its best specimens when there is a demand for them. I was greatly impressed by the tone and manner at the social gatherings that I attended, which were at first those of the more cultured class. The guests were gay without frivolity, friendly without gush, and intelligent without brilliancy; they were eminently a wholesome set of young people, with whom one could pass one's life, not only in serenity but with satisfaction and even a large share of keen pleasure. The physique of the girls reminded me of that of the "Hours" in the engraving of the famous picture of 'Aurora' by Guido in Rome. It is a favourite picture of mine and I recall it clearly. The girls have the same massive forms, short of heaviness, and seem promising mothers of a noble race. The simple way of gathering the hair in a small knot at the back of the head, shown in the dancing 'Hours,' is the fashion at Kantsaywhere. So is the general effect of their dresses, only they are here more decorously buttoned or fastened, than are the fly-away garments in the picture. As for the men they are well built, practised both in military drill and in athletics, very courteous, but with a resolute look that suggests fighting qualities of a high order. Both sexes are true to themselves, the women being thoroughly feminine, and I may add, mammalian, and the men being as thoroughly virile. No petty gossip or scandal is to be heard in their conversation, but a great deal is said about family histories and the prospects of the coming generation. These subjects occupy almost as much of their talk as athletic topics do at a public school, or as the performance of horses in racing circles. And it was genuine interest too; for they looked upon themselves, as I have mentioned more than once, with obvious pride as a chosen race for the purpose of furthering humanity, and were as suspicious and guarded against unknown outsiders as a Jew against a Gentile, or

PLATE XXXIX

Guido Reni's Picture of Apollo and the Hours preceded by Aurora, from the Casino of the Palazzo Rospigliosi, Rome.
Reproduced by kind permission of *The Architect* (March 14, 1885).

a Greek against a Barbarian. This gave a prevalent, and not a disagreeable mannerism. It suggested a constant sense of *noblesse oblige*, far removed from that disagreeable but not uncommon "Oxford tone" which implies that the speaker is a superior person to his listener. I think the selection of Kantsaywhere College folk may be rated as about equivalent to at least the best quarter of that of the population of Kantsaywhere town, which itself has a high level. The Collegiate average must be fully equal to the best twelfth of an English population. Now 1 in 12 is that of the foreman of a jury, and, unquestionably, the foremen play their parts, as a rule, very respectably. We are accustomed to appreciate bodies of picked men in many ranks of life and know well how superior they are. The crew of an Arctic research vessel are said to be a magnificent set of men : so are the Sappers and Miners. At a somewhat lower, but yet conspicuous degree of selection, stand the persons attached to those great and well-managed estates and firms, whose service is so popular that they have always more candidates to choose from than there are vacancies to fill.

....

"Nothing struck me more than the photographic workshops, for besides their immediate interest, a religious parallel was drawn from them which will be described farther on. There is a great demand in Kantsaywhere for composite portraits of families. The material for making these is abundant and excellent, as it has long since become the fashion, now grown into an obligatory custom, for everyone to be photographed at reasonable intervals, both in full face and in profile, under similar and standard conditions of light, in addition to whatever more artistic representation may be desired. I am a bit of a photographer myself, and was delighted at the punctilious and exact way by which composite photographs were made. There was no unacknowledged faking but the work was strictly truthful throughout the whole process. The object is, I need hardly say, to superimpose the images of many different portraits, all of the same size, aspect and shading, in succession for a short time, upon the same photographic plate. The scale of the portraits and their emplacement require much precision. Here the various reductions and adjustments are leisurely made for each portrait and in a separate frame. When the photography begins, the frames are dropped in succession into their exact place, guided by pins and resting on a horizontal board below a fixed vertical camera*.

....

"I saw several beautiful composites in the Studio, of men and women, respectively. Every family desires at least four family composites, one of the Grand-parental series, including Great Uncles and Aunts on both sides, another of the Parental series, including Father and Mother, Uncles and Aunts, and yet another of Self, Brothers and Sisters. Lastly, one made from the four grandparents and the two parents, allowing one half of the exposure time to each grand-parent that was allowed to either parent. A peculiar interest lies in the close analogy between composite portraits and their religious imagery, as will be seen from what is now about to be said.

....

"Their creed, or rather, I should say, their superstition—for it has not yet crystallised into a dogmatic creed, is that living beings, and pre-eminently mankind, are the only executive agents of whom we have any certain knowledge. They look upon life at large, as probably a huge organisation in which every separate living thing plays an unconscious part, much as the separate cells do in a living person. Whether the following views were self-born or partly borrowed I do not know, but the people of Kantsaywhere have the strong belief that the spirits of all the beings who have ever lived are round about, and regard all their actions. They watch the doings of men with eagerness, grieving when their actions are harmful to humanity, and rejoicing when they are helpful. It is a kind of grandiose personification of what we call conscience into a variety of composite portraits. I expect that many visionaries among them— for there are visionaries in all races—actually see with more or less distinctness the beseeching or the furious figures of these imaginary spirits, whether as individuals or as composites. There seems to be some confusion between the family, the racial, and the universal clouds of spirit-watchers. They are supposed to co-exist separately and yet may merge into one or many different wholes. There is also much difference of opinion as to the power of these spirits, some think them only sympathetic, others assign the faculty to them of inspiring ideas in men, others

* See our p. 215 above.

again accredit them with occasional physical powers. Everyone here feels that they themselves will, after their life is over, join the spirit legion, and they look forward with eager hope that their descendants will then do what will be agreeable and not hateful to them. I have heard some who likened life to the narrow crest of the line of breakers of a never-resting and infinite ocean, eating slowly and everlastingly into the opposing shore of an infinite and inert continent. But that metaphor does not help me much, beyond picturing what, in their view, is the smallness in amount of actual life with the much larger amount of elements of potential life. It is quite possible that if their confused ideas were worked out by theologians, who in a general way firmly believed in them, and who were able to define on valid grounds the extent of influences that the spirit world exerts over the living world, a very respectable creed might be deduced. Their superstition certainly succeeds, even as it is, in giving a unity of endeavour and a seriousness of action to the whole population. They have no fear of death. Their funerals are not dismal functions as with us, but are made into occasions for short appreciative speeches dwelling lovingly on the life-work of the deceased.

....

"The houses near the town are practically villas, for the use of town dwellers, each with a small garden for flowers, vegetables and fruit. The extent of garden and agricultural land is about twenty square miles. There are about 500 holdings in all, of a rough general average of 40 acres each. About one half of these are let at a low rent, especially to highly diplomaed parents. Though every married couple has perfect freedom in choosing his residence here, or in emigrating elsewhere, the attractions offered to those who settle in the country are so large and many that the pick of the Collegiates occupy farms or villas. A country life is considered to be so highly conducive to the health and size of families that a large part of the wealth of Kantsaywhere is gladly allotted to its encouragement. It is a great convenience to the Registrar to have so large a part of his charge located close at hand and for his inspectors to have means of easily verifying doubtful statements by conversation with neighbours. Nearly every household undertakes some unpaid office connected with administration and there is abundance of local pride and patriotism in doing this work well. With a less gifted people these customs would hardly answer, but here it is otherwise.

"The character of the farming of Kantsaywhere is in many respects such as is described as ruling in Denmark, but for the most part it must bear a closer resemblance to...."

Here my fragment breaks off, the remainder having been removed, so that it is not possible to say what agricultural system Galton thought superior to that of Denmark.

Galton himself wrote very little of "Kantsaywhere down; he dictated it to his Secretary, and was much diverted by his own characters. On one occasion he had to be reminded that he had already killed a personage, whom he badly needed later, and accordingly, much to his amusement, the slaughter had to be revised.

It is needful here to recall a point which Galton as an anthropologist strongly insisted on. He held that any form of superstition held by a tribe or nation as a whole—even the worst type of fetishism—was a source of strength to the believing group. A religion might be false, but anthropologically it was better than no religion[*].

Galton was a firm agnostic, that is to say while fully recognising the infinite mystery behind life, and indeed behind the physical cosmos as well, he did not think that man could fill the void either by his own reasoning or by revelation of a transcendental kind. Nevertheless he believed that every nation required its peculiar "superstition," and he devised in the above paragraph a curious one for the inhabitants of Kantsaywhere. It appears to

[*] See pp. 88–89 above.

centre in what I have termed the "Generant" of the stirp*, the composite individual who represents the entire ancestry of any person. Galton thinks of this in connection with his composite photography, and then introduces these Generants as an improved version of the Chinese worship of ancestors. They were to act as conscience to the new generation, in a land where each citizen studied and was proud of his forebears. That Galton himself thought of this spirit world as more than a valuable "superstition" I very much doubt.

(18) *Further Letters of* 1910, *concerning Eugenics, etc.*

THE RECTORY, HASLEMERE. *January* 1, 1910.

MY DEAR KARL PEARSON, What a noble New Year's greeting you send me! I prize it among the highest of honours, for it will be a landmark in the path of progress of Eugenics. How I admire the forcible and confident beats of your mathematical wings! Certainly, as you have phrased it, to Francis Galton, not "Sir," which under the circumstances sounds like tinsel. I rejoice in your work all the more, as it covers and includes much that I dearly wanted to see done, but had not strength or capacity to do. *Biometrika* is just the most suitable form of publication, too†.

You must kindly tell me soon about my contribution to the Eugenics Laboratory for 1911, about which Hartog will wish to know. I am quite prepared to go on as before if you see your way to its continuance, either in the present or in some modified form, consequent on the possibility of Heron wishing to follow an independent line, or more especially to your own desire to be freed from the care of its oversight (I hope not).

Give please my warmest wishes for the New Year to all your party. Wee Ling prospers and grows, and is a favourite. He enjoys a dry bone to chew. What an inexpensive and wholesome Lord Mayor's banquet might be provided if the Guests were supplied each with a plate and a dry bone, and nothing else!

The half sheet of a letter that was mistakenly sent to you, has since been identified.

Ever affectionately and gratefully yours, FRANCIS GALTON.

7, WELL ROAD, HAMPSTEAD, N.W. *January* 9, 1910.

MY DEAR FRANCIS GALTON, I hope you will not have thought me ungrateful in not replying to your very kind letter before. But I have been very, very busy. Fundamentally trying to get another chapter of the piebalds—i.e. one on the albinotic skin and dealing with pied folk and leucoderma and modern and ancient theories of pigment changes—to press. It is practically finished to-day—my last day of holiday. I have also revised in proof 80 pages on albinism in the negro; got Miss Elderton's paper on parental alcoholism finally passed and to press; and written 20 pages of suggestions to Heron for his big memoir. In addition I have read 10 papers for *Biometrika* and had to refuse four, which is always unpleasant for it makes foes. I have another half-dozen papers which want writing up and will again be postponed. I don't know whether I told you that last September old Dr Crewdson Benington died. He had been working in the Laboratory for two years, nothing finished, and a wheelbarrowful of manuscripts on skull measurements have come from his friends—"to be edited and finished." He was a curious old fellow—really able in many ways and affectionate, but difficult. He had divorced his wife, and his life was a failure, but he just settled down in the Biometric Laboratory and worked like a lad of 20, and I think we more or less kept him on the tracks. He came four years ago and then disappeared to the upper reaches of the Amazon, but Biometry brought him back again! The last two years he worked away without a break—and then last long vacation he was all alone in London and there was nobody to look after him. Poor old fellow, I always feel that if I had had time to write him weekly letters, he would still have been measuring skulls!

---

* See our pp. 20–21, 29.
† See our pp. 392–397 above.

We have got the scheme for recording eyesight and home-environment of Jewish children started. But we want more volunteer social workers. We propose doing 800 boys and 800 girls, but seeing and scheduling 500 or 600 homes will be a heavy task for one worker, Miss Rosenheim. I wish we could find a couple more Jewish ladies. Then I have been interviewing a Prison Commissioner to try and get an extension of the time for those working at the Criminal Statistics.

I have recently found some rather good Eugenics materials—on the lives of girls committed to Industrial Schools and afterwards followed for perhaps 6 to 10 years. Also very good material on *physically* defective children in Liverpool which I hope to get access to—they keep a fairly full account of the parents and what becomes of the children. Did you see Major Darwin's address? It seemed to me quite well put and likely to do good.

I have put in the "New Year's Greeting" a paragraph suggesting the sort of statistics that are needed, and hope to let you have a revise as soon as the diagrams are engraved. It is very good of you to say so much about the paper. My regret is that I am so slow in fulfilling requests and suggestions.

Now as to the Laboratory. Of course I want very much to see it go on and develop. The time is ripe for this sort of work and if we make it a success, it will be taken up at other places and then a real knowledge of what makes for true national greatness will be reached. Heron will certainly, I think, leave this year. His big paper is completed. He has grown a good deal, and has been very sympathetic and helpful. He has been helping much more in the laboratories and did quite good teaching work with the six workers we had in the Laboratory last term. I have got to try and find him a berth, but I think it ought to be possible. I expect a medical officer will come for training this term and there are sure to be one or two others. Miss Elderton also has done a good deal of teaching work last term. Miss Rosenheim, who is to do the Jewish homes, was in her charge; Miss Barrows, who has gone out to Jamaica to investigate the characteristics of half-breeds, and Miss Jones were both more or less in her hands. I hope Miss Jones will come back and take up the Industrial School girls—it would be a good bit of work. We want badly these trained social workers to go out and work for the Laboratory. But we have made a beginning. I think the work has been very good all round this last term and we really had not a vacant seat some days in the week. Unfortunately Miss Ryley got rheumatic fever, and this checked the *Treasury* work till I found Miss Jones was very excellent at drawing pedigrees.

Harelip is done, the section on Cataract nearly done, and Haemophilia and Dwarfism practically complete and ready for engraving. Miss Barrington has prepared nearly 100 pedigrees of ataxy and atrophy and muscular failures. So that the *Treasury* has material ready for at least another year. I think really there has been a great deal of very thorough and honest work done and that is why I lament outside criticism (!) of the kind that has appeared from inside the Eugenics fold. I hope you were not disgusted with the *Standard* notice. The interviewer cut out a long bit in the middle of the article and stuck the two halves together crudely! I hate being interviewed, but he said Hartog wished it when I refused the first time.

I think we ought to consider the right man to succeed Heron, if you settle to go forward. You might see when you come back to Town one or two of the men working in the Biometric Laboratory now. L—, who is a clear-headed Cambridge mathematician, has had an engineering training, but has been two years doing statistical work. I consider him very good. And M—, who has been also two years, is a very strenuous person and distinctly able, but he has not specialised as yet on man, and is rather rougher in manner.

I am glad to hear about Wee Ling. We miss our puppies very much and I fear we shall not have, as we had hoped, another litter this January. Even the matrimony of dogs is not always a success!

No more now. You will hardly read through all this. Always affectionately yours, K. P.

THE RECTORY, HASLEMERE. *February* 26, 1910.

MY DEAR KARL PEARSON, The sale of the Eugenics Laboratory memoirs and papers is very gratifying and encouraging. You will doubtless receive suggestions as to the kind of change that would make the *Treasury* more sought after. X.'s article does not impress me, because he has made no proper study of the "positive" aids to Eugenics. He sent me a programme of a recent lecture in which the "positive" influences were hardly alluded to. I wrote, and pointed that out to him, but it was too late. Removal of influences that obstruct fertility *is* positive Eugenics. No one has yet studied the conditions under which a population has made sudden advances, and there are many such cases—after pestilences—in colonies—etc. The way in which the Dutchmen of the Cape multiplied was remarkable. Hopefulness seems a powerful aid; despondence is a powerful check.

Have you seen Whetham's singularly clear and powerful lecture, delivered at Trinity College, Cambridge, of which he is a tutor? If you have not, I would send you my copy to read.

Bernard Shaw is about to give a lecture to the Eugenics Education Society. It is to be hoped that he will be under self-control and not be too extravagant.

Wee Ling now weighs 16½ lbs., and though usually the reverse of aggressive, flew at a bigger puppy than himself of a commoner breed, to the loudly-expressed disapprobation of its owners, who were taking him for a walk. All goes on as usual with us.

Affectionately yours, FRANCIS GALTON.

THE RECTORY, HASLEMERE. *March* 6, 1910.

MY DEAR KARL PEARSON, We hope to be back for good at 42, Rutland Gate, on March 21st (Easter being March 27). It would be only too delightful if you could come and see me during that week. Select your own date and I will make my plans suit. Possibly you might be persuaded to spend a quiet night with us? Will you? Eva and I had doubts as to where the pretty card with the quotation from Meredith came from. We *suspected* it was Mrs Pearson, now that I *know*, please thank her from me, gratefully.

A letter from Heron about the antagonism of leading members of the Eugenics Education Society makes me unhappy. A quotation from a paper by Dr Slaughter justifies his contention fully. The passage seems to me inappropriate, untrue and in the worst taste. I have written to Mrs Gotto, who sees much of him, to point out this privately to him and otherwise to help in the cause of harmony. I don't like, just yet, to take a stronger course. How unwise many people are!

Like you, I in my small way have been a little plagued by retarded printing. An article of mine, long in type, will I expect really appear in this week's *Nature*. The profiles in it may amuse you. Ever affectionately yours, FRANCIS GALTON.

THE RECTORY, HASLEMERE. *March* 10, 1910.

MY DEAR KARL PEARSON, Heron's paper on Environment and Intelligence is indeed a credit to the Laboratory. How greatly he has improved, under your eye and help, since he first came. There is a weight and fulness in his writing now, that can hardly, I think, be further improved. I do not write to him myself, simply as a matter of discipline. It is better that praise should come from you or through you.

I have done all I can, within reasonable limits, to put a stop to the vagaries of members of the Council of the Eugenics Education Society, in which I am warmly seconded by Crackanthorpe and Mrs Gotto. Bernard Shaw* has been another difficulty but I trust that matters will now improve. *If they had the men*, the Society might do really good work in emphasising such points as those brought forward by Heron, whether on the incompleteness of the present school statistics, or, as in a former paper, on the registration of the insane.

Ever affectionately yours, FRANCIS GALTON.

42, RUTLAND GATE, S.W. *March* 25, 1910.

MY DEAR KARL PEARSON, We are safe back and I have now thrown off some bad effects of the little journey. Do come soon. Wee Ling is grown, of course, and Miss Biggs is very fond of him. But London is a bad place for pet dogs and Wee Ling cannot be trusted loose, as he runs wildly after stray dogs to play with them. We think of finding a home for him during the summer and have two possibilities in view. According to the first plan, he would be taken by my niece, Violet Galton, to Warwickshire on Tuesday next and be left with my nephew, Edward Wheler, who is knowing about dogs, acting as Judge in some great shows and sending his retrievers to win prizes at others. So if you could come here Saturday, Sunday or Monday you would see the little creature before he goes. Fix your own time. I have no engagements and am always rejoiced to see you. Ever affectionately yours, FRANCIS GALTON.

* Anecdotes of the famous have always a peculiar flavour. The Eugenics Education Society had asked Bernard Shaw to give a lecture, and some members of its Council had been somewhat in doubt about the matter. All Galton's contribution to the quandary was: "I don't mind good jokes, but Bernard Shaw makes such bad ones."

42, RUTLAND GATE, S.W. *April 6, 1910. Dictated.*

MY DEAR KARL PEARSON, I have been so knocked about with cough, that I am still unfit for almost anything, with ever so much in arrear. I did not even wish you a happy return for your birthday, nor have I been as eager as I should be to hear about Wee Ling. It is joyful news that he gets on so well with you all*. I feel a sort of apology is due for having wandered so far from my regular track as to write the article in *Nature*, but I wished to finish off such bits of unfinished work as I could hope to achieve. As you say, life does not seem long enough for all the possibilities of interesting work. I feel like Tennyson's Ulysses:

"Life piled on life
Were all too little, and of one to me
Little remains."

So please count my apparent vagaries as merely an attempt to get some things now in disorder into shipshape form before I die.

If I find, as I expect to do, that Miss Jones can be trusted with the work you suggest and which is precisely one of the things I had in view, I will set her steadily at it.

Ever affectionately yours, FRANCIS GALTON.

P.S. I am so sorry I forgot to send a card about Uncle Frank's health, but I have been in bed with fever three days. E. B.

42, RUTLAND GATE, S.W. *May 7, 1910.*

MY DEAR KARL PEARSON, The King's death will throw much out of gear, and may prove a disaster to our country!

The account you give of an apparent wish of the University of London authorities to dissever Eugenics, so far as locality is concerned, from Biometry, seems to me most unfortunate for the former. It may be logical to unite it with Sociology, but practically it would almost give a death-blow to its scientific status. Whenever you think I could intervene with advantage, pray tell me. I had seen the *New Age* and was vexed at Dr S.'s remarks. The allusion by the Editor to myself and to my letter may be quite correct, but I cannot properly recall the circumstances, owing to having had recently to reply to 2, 3 or more letters in the sense "I sympathise with your object, but am too infirm to give any active help."

The Council of the Eugenics Education Society have, I learn, extruded Dr S. by not putting his name on the candidate list. As I am told, certain members of the Council strongly objected to serving longer with him, and Mrs G. undertook to tell him so, which she did, doubtless with all practicable tact, but I have reason to know his feelings are much wounded. He however spoke very nicely to Mrs G.

Like yourself I missed the notice of Mr Justice Parker's lecture in the *Times*, and I am sorry.

Crackanthorpe's address, as it appeared in the *Times* summary, was weak, but a little better in the original, which I read. Ever affectionately yours, FRANCIS GALTON.

42, RUTLAND GATE, S.W. *May 11, 1910.*

MY DEAR KARL PEARSON, Saleeby's letter to the *Pall Mall* lies near the frontier between "do-nothing" and "do-something." I wish that somebody, other than our two selves, could be posted up to reply to him. I am still in favour of "doing-nothing" ourselves. Whatever either of us might write, would be responded to and the issue be perplexed. Perhaps some opportunity may arise before long of pronouncing emphatically on all such inapt criticisms as those indulged in by C. W. S. and showing their futility. Ever affectionately yours, FRANCIS GALTON.

42, RUTLAND GATE, S.W. *June 9, 1910.*

MY DEAR KARL PEARSON, Your letter is a useful reminder, but you must not accuse yourself of having forgotten to tell me any part of its contents. Enclosed I send the pestilent copy of the *New Age*. Do not trouble to return it. I am very glad to have made it now clear to them that I will not take any part in it.

Now, will you pardon me if I ask for a few minutes of your time, to look over the little memoir herewith enclosed. I may be a fool, but I think the simple results to be both new and important. Are they so, or merely rubbish, or anything between?

Ever affectionately, FRANCIS GALTON.

* He had rejoined us at Hampstead.

42, RUTLAND GATE, S.W. *June* 10, 1910.

MY DEAR KARL PEARSON, You are over-good to have taken such pains about my little problem. It shall now lie in a state of suspended animation. I must think well over the doubt you point out, whether the m_μ, etc. might be taken as of *equally probable* occurrence. At present, the difficulty does not strike me as it should. On all the other points I am fairly well prepared to give justification and explanation. Once again very many thanks.

The *Times* of to-day contains no rejoinder by Crackanthorpe to your paper, but he sent in the morning for a copy of your memoir, which I lent him.

Affectionately yours, FRANCIS GALTON.

42, RUTLAND GATE, S.W. *June* 23, 1910.

MY DEAR KARL PEARSON, Hartog came here yesterday and gave a most satisfying account of the friendly disposition of the University towards the Laboratory. I especially asked him whether there was anything in the supposition that it was proposed to transfer it to Sociology. It is quite unfounded, so he assured me. Then he went on about the enlargement of the accommodation, by buying the next house and bridging across, at a cost of £1500. After he left, I thought that perhaps (if you thought it pressing) I might hasten this if I offered £750 on condition of the University supplying the rest. So I wrote privately to him to that effect, asking if he and the Principal thought it *probable* the money could be raised, adding that I made the offer subject to your approval, for though I had heard of the scheme from you I did not know exactly how far you approved of it. Hartog replies in a letter, just received, that I should consult you at once. So I do, hereby. I feel much less disposed to offer this money unconditionally than under the proviso that the University should meet it by an equal contribution.

I shall think of you all—including dogs—to-morrow evening.

Ever affectionately yours, FRANCIS GALTON.

42, RUTLAND GATE, S.W. *June* 27, 1910.

MY DEAR KARL PEARSON, I can assure you that I acted *proprio motu* and that Hartog had not given the slightest hint, direct or indirect, on the matter. He came, partly to explain a misapprehension which, for some reason, he thought I was under, that the ultimate direction of the affairs of the Laboratory was circuitous. On the contrary, he assured me it was direct through the small Committee and thence to the Senate. Partly it was a personal visit, and I naturally asked many questions. Afterwards, turning over what he had told me, I wrote the letter. I have now written again to him wholly exonerating him from the suspicion of having, in any way, suggested that I should give more at present, adding that seeing, now, that matters were more complicated than I supposed, I would withdraw the offer. But that it might be repeated, probably in an altered form, if it seemed likely to draw an equal or larger contribution from or through the agency of the University, to match it. Both my letters to Hartog were "private." So glad to hear the Soirée was a success. Affectionately, FRANCIS GALTON.

42, RUTLAND GATE, S.W. *July* 4, 1910.

MY DEAR KARL PEARSON, Thanks for both of your sendings, (1) the cutting from the *Medical Times*, which I return, (2) for the letter—how on earth it ever reached you is a mystery—from Eva Biggs' servant (and more than servant), who is now married and settled in New Zealand. Yesterday I got together to tea, Miss Elderton, Crackanthorpe and Ploetz. C. made himself very agreeable in a long tête-à-tête with Miss E., but I fear was insufficiently penitent to receive full forgiveness.

About my little problem, I was appalled, on re-reading what I sent you, at its crudeness. I was ill when it was dictated and I find that an important sentence must have been omitted. Moreover it is deplorably wrong in one part. Please banish it from your memory and allow me shortly to send you a revised version. I have had two baddish days, mostly in bed, but am better and was fit for yesterday's tea. Eva is out all to-day, at and about Haslemere, looking at houses for the autumn there, of which we have had some offers.

I trust that Yorkshire retains its attractiveness to you all. We have here thunderstorms and most un-Julylike weather. I have not ventured out of doors for more than a week.

Ever affectionately yours, FRANCIS GALTON.

42, RUTLAND GATE, S.W. *August* 4, 1910.

MY DEAR KARL PEARSON, It is pleasant to hear that you are thriving in Yorkshire. I am still in London, not going to Grayshott until August 16. We have had much of very unenjoyable weather, but the last 3 days have been pleasant. Asthma has plagued me, but I stave off the worst bouts now, by smoking a cigarette of *bhang* (Indian hemp-hashish). It is curious to perceive the spreading of the narcotic effect over the lungs and everywhere.

Q. and his elder brother have just had tea here. He is simply a *beautiful* youth, of the very best Jewish type—simple and very intelligent. He thinks that there is a mine of information bearing on Eugenics that could easily be worked in Manchester, and said that he would like to write to you about it. I encouraged him to do so. So you will understand. I heard from him about his Russian and mystical Grandfather and the Kabbala (? spelling).—A good spiritualistic story is told of him.

So Marshall is at you again now, and with reinforcements about to come on the scene! Anyhow he is a worthy antagonist.

What pleasure and health you must have given Miss Elderton.

Ever most affectionately, FRANCIS GALTON.

Kindest remembrances to you all.

THE COURT, GRAYSHOTT, HASLEMERE. *August* 18, 1910.

MY DEAR KARL PEARSON, At last I am most happily settled. Your letter reached me in London just before motoring here. I had to spend that afternoon and all yesterday in bed, but am now up and eager, having got over a horrid asthma! It is pleasant to hear of your excellent weather and of much else. You know of course of the treatment bestowed on a *big* dog for sheep-chasing, viz. coupling him to an old ram, but Wee Ling's life would soon be pounded out of him in that way.

It is too bad of Victor Horsley. Of course Crackanthorpe's letter justifies him, but I feel myself to be incidentally referred to. If ever I know of any such *direct* reference, I will certainly disavow it.

You must be glad at feeling in sight of the end of Albinism—yet it suggests something more in respect to Melanism. I wonder whether the singular blackness in the R. family has been traced to a negro ancestor? I mean the present Lord R. and most of his sisters. His father also was very dark.

You will like Q., I am sure, when you know him personally. He is as modest as he is capable.

The tuberculous inquiry will not, I imagine, cause so great an outcry as the alcoholic. You have accustomed people to suspect the truth of current beliefs. I wonder what Sir Donald MacAlister thinks of all this? He is very favourably disposed towards Eugenics and is, as you know, a vigorous mathematician.

Try and excuse this bad writing. It is performed on a board, while sitting in a wheel chair, and with a scratchy pen, brought to me. Very best wishes to you all.

Ever affectionately yours, FRANCIS GALTON.

In October of this year the attacks on the work of the Eugenics Laboratory were in full progress and Galton wrote the letters to *The Times* and the *British Journal of Inebriety* cited on pp. 408–9 above. He was peculiarly moved by the half-hints made by certain writers to the press that he was out of sympathy with the work of the Eugenics Laboratory. All my letters to him directed to Haslemere in the last year of his life together with most of the letters he received during the same period appear to have been destroyed after his death, probably when Grayshott House was restored to its owners. Thus the correspondence for this last year must appear one-sided.

THE COURT, GRAYSHOTT, HASLEMERE. *October* 30, 1910.

MY DEAR KARL PEARSON,   Will the enclosed draft of a letter to the *Times* fulfil what you think desirable? Pray make suggestions freely. I have heard from X. in a long "private" letter replying to what I sent him. He writes nicely but impenitently*. He is about to give numerous lectures. Very asthmatically, but affectionately yours, FRANCIS GALTON.

X. is no longer even a member of the Eugenics Education Society.

48, GROSVENOR STREET. *November* 8, 1910.

MY DEAR GALTON,   I must write a line, as one of your oldest friends†, to congratulate you on the great honour of the Copley Medal.

I hope you have been keeping well.   Yours very sincerely, AVEBURY.

GRAYSHOTT HOUSE, HASLEMERE. *November* 13, 1910.

This will henceforth be my address.

MY DEAR KARL PEARSON,   You must indeed have been "rushed" as you say. The Press cuttings reached me of the letters of you and the antagonists, whom it seems to me you bowl over easily.

Thanks about the Royal Society. I shall not, could not, attend however much I wished it, and had thought of asking you, if Sir George Darwin failed, to receive the medal on my behalf. But he *will*, anyhow, be there. So I have asked him to do so.

People die so fast that I can find only five other living Englishmen, with *Copley* after their names, in the Royal Society list of Fellows; they are—Sir Joseph Hooker, Lord Lister, Lord Rayleigh, Sir William Crookes, Alfred R. Wallace. How *age* counts!

Thank the Staff for me for their joint telegram of congratulations. There is no news here that you would care for. What a political turmoil is at hand!

Ever affectionately yours, FRANCIS GALTON.

GRAYSHOTT HOUSE, HASLEMERE. *December* 6, 1910.

MY DEAR KARL PEARSON,   Who is Mr Snow? You seem to have found a worker after your own heart. I wish that you or he could throw more light on the paradox that cousins are no more unlike than uncles and nephews. It would seem a reasonable deduction that cousins to the nth degree are as much alike as first cousins. Then, again, statistics make out (unless I am quite wrong) that husbands and wives are as much alike as first cousins. I wish you could clear my puzzled mind. Also one wants to know more precisely about the compound effect of hereditary influences. What is that of bi-parental—of the same kind—as compared with uni-parental? What is that of all four grandparental + bi-parental? and so forth. The whole lot together cannot exceed 1·0‡.

I congratulate you on the last number of *Biometrika*.

How are you all? Your Winchester son will soon be with you. All goes on quietly here, but I am not allowed out of doors in such weather as we have recently had. In fact, I have been imprisoned now for 14 days and begin to crave for open air.

Sir Archibald Geikie comes not infrequently over the 5 hilly miles that separate his house from mine, and tells me scientific news.

If you care to rear a breed of dogs *who eat woollen cloth*, there is one in this house that does so. He began by nibbling off and swallowing the lappet of my man-nurse's coat, who had been caressing him, and subsequently found his way into the butler's pantry at night, and ran away with a beautiful new pair of trousers of mine, dragged them to his kennel and gnawed out a piece bigger than the palm of my hand and ate it. It has strained my Xmas feelings to pardon him!

* Galton's singular gentleness of disposition rarely allowed him to give expression to some of his deeper feelings about the proceedings of certain of his rasher self-styled followers. One incident, however, has been preserved: a letter came at mealtime; it went flying across the dinner table with the exclamation, "*My* disciple indeed!"

† Lord Avebury was 76 years old, twelve years younger than Galton, but they had been associated in many projects.

‡ It seems to me now in the light of experimental determinations that it *can*, and that this is the source of progressive evolution when small groups are isolated or there is intensive in-breeding.

It is wonderful how skilfully my tailor has patched the hole. The "fine-drawing" of the edges of the patch are invisible without scrutiny, such as no stranger would venture to make*!

I do hope all is well with you. Send me a line, even on a postcard.

Ever affectionately yours, FRANCIS GALTON.

GRAYSHOTT HOUSE, HASLEMERE. *December* 14, 1910.

MY DEAR KARL PEARSON, We are delighted that you can come. You will be most welcome as early as is convenient to you on the 28th, and as late as you care to stay on the 29th.

The Report of the Committee to the Senate, which I return, gives solid grounds for its application for a further grant to Eugenics and I am glad it was written by Hartog†, as it shows that the opinions of the Chief Executive Officer are strongly in its favour.

Few things would gratify me more than that you should be relieved from the drudgery of teaching engineering students, etc., and be kept free for Biometry and Eugenics. I return the Report, which I cordially approve, wishing, in vain, that I was familiar with the hidden springs by which the Senate of the London University is moved and was able to give indirect influence towards its acceptance.

Snow has kindly sent me an off-print of his Memoir.

Poor Tong‡! Ever affectionately yours, FRANCIS GALTON.

(19) *The Last Scenes.*

Galton had fretted his one hour upon Life's stage; the panorama, to use his own simile, had reached its final turn on the roller. This was the last letter I received from Francis Galton. On December the 28th and 29th I was with him at Grayshott. The weather was favourable and we sat out in the sunshine, Galton warmly wrapped up, talking about the work of the Eugenics Laboratory, the shortcomings of some members of the Council of the Eugenics Education Society, which were much troubling him, and again about the grave reaction against Darwinian evolution. One thing I remember very well, Galton's intense pleasure about the Copley Medal (I had not seen him since the award) and the numerous friendly congratulations he had received, even from some who had long passed from his circle. At dinner the conversation took a lighter tone. We had two recent converts to the Catholic Church, and we gravely considered why the Devil devotes so much more attention to Catholic than to Protestant countries and individuals. "You don't stick a knife into Professor Y. or Dr X. as I should probably try to do in your place," interjected one ardent convert. "That is because I have not your security for absolution," I urged and added: "Is your main thesis correct, did not the Devil disturb Martin Luther when he wanted to get on with his own work? I fear other minor devils cause me also to waste good ink." Galton took his full part in the talk. He seemed to me physically frail, but mentally active, and I saw no greater cause for anxiety than at

---

* The following letter from Galton's tailors may serve to give colouring to the incident:

10, CLIFFORD STREET, BOND STREET, W. *December* 14, 1910.

SIR FRANCIS GALTON, SIR, We have received the pair of Trousers and are carefully repairing the holes torn by the dog, which we are pleased to learn has been placed in eternal exile. They will be forwarded to you as quickly as possible. We remain, Sir, Your obedient servants, STULZ, BINNIE & Co.

† Galton was under a misunderstanding; it would be the Report on the Galton Laboratory based upon material provided by its Director.

‡ An albino bitch I had been obliged to send to a painless death owing to the development of an incurable disease. She was the mother of Wee Ling.

PLATE XL

Francis Galton, aged 88, from a sketch made by Frank Carter, twelve days before Galton's death.

PLATE XLI

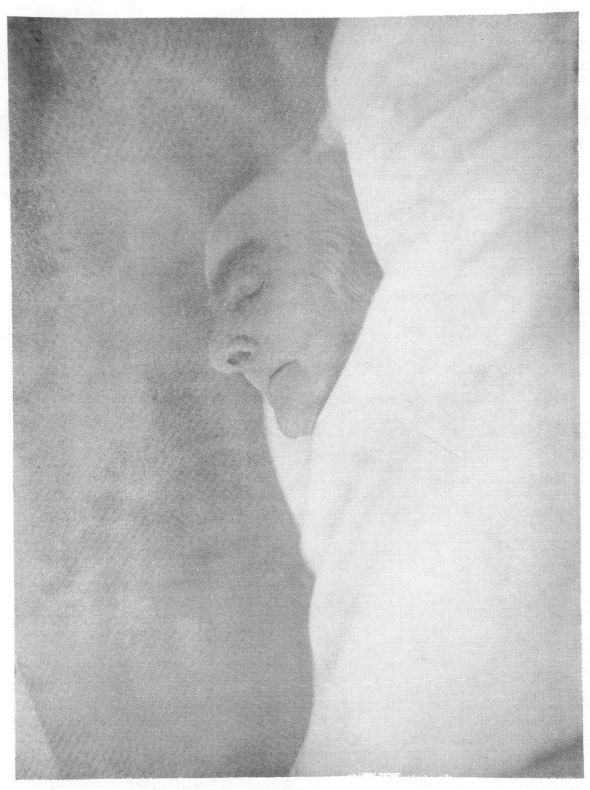

Francis Galton, January 17th, 1911, from a photograph taken after death.

any period in the past five years. Indeed I had been more anxious in 1909 than I was in 1910. There was no thought in my mind that I should not see him again, and that in another three weeks I should be standing at his grave-side.

GRAYSHOTT HOUSE, HASLEMERE. *December* 31, 1910.

DEAR PROFESSOR PEARSON, Uncle Frank has given me your letter. I don't believe the devil leaves you Protestants and Agnostics alone, but he doesn't torture you as he does Catholic communities....Who but the devil prevented you from doing what I asked, namely persuading Uncle Frank off that worrying Eugenics Education Society. You and your pupils do not let your names appear among that tiresome crowd, so why should Uncle Frank's name be put at the head of them?

The Doctor has been and keeps Uncle Frank in bed all day to rest, but this is the rule now once a week. He did so enjoy your visit. I wish you came more often as it cheers him up.

Ever sincerely yours, L. E. B.

GRAYSHOTT HOUSE, HASLEMERE. *January* 2, 1911.

DEAR PROFESSOR PEARSON, Uncle Frank is splendid again, and had a certain Dr Lyon Smith to tea and talk an hour yesterday. I honestly believe your visit did him good, but the cold tried him on Saturday [December 31st].

I daresay you are right about the E.E.S., but thought you might say in a quiet way some time that you were sorry he left them such a free hand. I don't like him, at the end of his life, being mixed up with such a set and who knows that some day he may not be made answerable for their actions, for after all he invented Eugenics.

I was joking about the devil, and shouldn't dream of taking notice of anything Luther either said or did! But its a great pity you folk blind yourselves to the existence of devils, and regard their tricks as a twist in the brain or something hereditary! A happy New Year to you from L. E. B.

GRAYSHOTT HOUSE, HASLEMERE. *January* 15, 1911.

DEAR PROFESSOR PEARSON, Uncle Frank is one degree better to-day but still in danger. He is not the least worried about the Laboratory affairs. I only told him the teetotallers were attacking you, and that a good leading article in the *Times* had snubbed them. He was much interested. He is quite easy in his mind and very clear when he speaks, but too weak to speak more than a word or two. My cousin Edward Wheler, a very dear nephew of his, is here—we never leave him a minute. Will write again. Very sincerely, L. E. B.

GRAYSHOTT HOUSE, HASLEMERE, SURREY. *January* 18, 1911.

MY DEAR LADY PELLY, I have the saddest news for you—dear Uncle Frank died last night—he had a sharp attack of bronchitis and died of heart failure, not having the strength to fight against it—he suffered much discomfort but very little pain, and just at the last he was very peaceful.

Edward Wheler and I were with him and, just before sinking, he looked and smiled as one of us spoke, or Gifi came and looked at him. Up to the last few hours he was bright and keen, and if strong enough to articulate would quote some poetry or make an amusing reply. He looked so sweet when dead I could hardly believe he had gone. He is to be buried at Claverdon on Saturday next (January 21st), the home of his father, a few miles from Warwick. He was truly fond of you. He was ill just a week. With much love and great sorrow at having to give you such sad news. Yours affectionately, L. E. BIGGS.

On January 18, three days after his niece's letter to me, a telegram reached me that Francis Galton had passed quietly away on the previous night. Early in the month he had caught a bronchial cold from one of his attendants, and his strength was inadequate to carry him through the attack. I personally had lost the master in whose footsteps I had trod since I met his *Natural Inheritance* in 1889, and the man with whom my friendship had grown closer and closer year by year, even to his death. Beyond near friends, then few in number, and his younger relatives, in two generations of descent, to whom his sweet disposition rather than his scientific talents had endeared him, there was an outside world which hardly realised that, with Francis Galton, the last of the great Victorian scientists had passed away. The friend of Darwin, of Wallace, of Hooker, of Tyndall*, of Huxley and of Herbert Spencer† had survived them all, and closed their era with the creation of a new science. It is too early yet to say whether the truths that it may give to the world are destined to form the religion of progressive states, the creed of a new type of mankind; or whether those who understand are still too few to impress upon the inert masses that by studying and then applying biological laws to his own species man may step over the corpses of his failures into a hard-won kingdom. There was little in the obituary notices that showed a real understanding of what Galton had achieved, still less any recognition of the possibilities of his work for the future of mankind.

Even the memorial tablet in the church at Claverdon Leys, prepared as

IN MEMORY OF

SIR FRANCIS GALTON. F.R.S.

BORN 16TH FEBRUARY 1822
DIED 17TH JANUARY 1911

YOUNGEST SON OF SAMUEL TERTIUS GALTON

AND FRANCES ANN VIOLETTA HIS WIFE

DAUGHTER OF ERASMUS DARWIN. F.R.S.

MANY BRANCHES OF SCIENCE OWE MUCH TO HIS LABOURS,
BUT THE DOMINANT IDEA OF HIS LIFE'S WORK
WAS TO MEASURE THE INFLUENCE OF HEREDITY
ON THE MENTAL AND PHYSICAL ATTRIBUTES OF MANKIND.

it was under the care of loving friends, lacks any reference to the crowning achievement of Galton's life. I would add to its last line the words: "in order that a true knowledge of natural inheritance might enable man to lift himself to a loftier level."

There was a unity underlying all Galton's varied work—even to that last creation *Kantsaywhere*—which only reveals itself when, after much inquiry and retrospection, we view it as a whole and with a spirit trained to his modes of

---

* MY DEAR GALTON, Herbert Spencer and myself are both low in health, would you come with us to-morrow (Saturday) to the Isle of Wight and help the sea air to restore us?

Yours ever, JOHN TYNDALL.

ROYAL INSTITUTION (dated " Friday " only, but in the '70's).

† DEAR GALTON, This day week, the 16th at 7, a few friends will dine with me *here*— Frederic Harrison and Morley of the *Fortnightly* among others. Will you give me the pleasure of your company as one of the number? Sincerely yours, HERBERT SPENCER.

ATHENAEUM CLUB, 9 *March*, 1878.

PLATE XLII

The Church at Claverdon, with the iron railings surrounding the vault where Galton's body lies.

thought. Twenty years of almost continuous reflection on Galton's labours have enabled me to see, using his own words, the whole as a "permanent panorama, painted throughout with equal colours," and to grasp better how great diversity of production may nevertheless be consistent with a marvellous unity in the main aim of a life. The skilful steersman may set the vessel's head to many points but never leaves out of mind his final port. The aimless wayfarer may voyage over strange seas and seek many havens, but without a steadfast purpose in life will never reach a firm anchorage. From 1864 to 1911 Galton achieved in many fields, yet in 1864 he had realised his life-aim—to study racial mass-changes in man with the view of controlling the evolution of man, as man controls that of many living forms. Has the time come for man to put the bridle on himself? To tame by science the nescient waywardness which lays waste his stock? Galton believed the hour had struck, and his fame in the future will largely depend on the accuracy of that judgment. If he was right, it will give him in the history of human civilisation a place equal to that of Darwin in science, but there will be a second place for him, to which Darwin has no claim; Galton taught a new morality, an unwonted doctrine of altruism—like all new creeds, difficult to accept and easy to pour scorn on : "Help the strong rather than the weak; aid the man of to-morrow rather than the man of to-day; let knowledge and foresight control the blind emotions and impetuous instincts wherewith Nature, red-clawed, drives man, mindless and stupefied, down her own evolutionary paths." "Awake, my people," was Galton's cry, like that of a religious prophet of the olden time. He was an agnostic, in that he saw the weakness of the creeds so far proclaimed by man, suffice, as they may, for many less deeply-probing minds; yet, as his niece* said to the biographer, he was a *religious* agnostic; the term seems to me an apt one. Galton believed in a recondite purpose in the Universe, which we men cannot unriddle, and he urged his fellows with religious earnestness to take up the burden of their task and further develop their species in fitness to its environment. Increased vigour of mind and body appeared to him the aim of the power which we seem to discern working obscurely, and as if with difficulty, behind the apparently blind forces of Nature.

Such thoughts were hardly present with most of us as we stood around the open grave in Claverdon church-yard†. What we felt deeply was the personal loss of that gentle, affectionate and modest nature, generous in thought and in practice, here bestowing an idea and there a helpful hand; rarely saying a harsh word, and often moderating the acerbity of others; taking life

---

* Mrs Millicent Galton Lethbridge.

† Francis Galton's remains were placed in the family vault constructed in the church-yard by his father. He lies by his parents, Samuel Tertius and Violetta (Darwin) Galton. The vault is surrounded by the iron railings seen in our Plate XLII. The stone contains the name and the dates of birth and death. Galton no doubt expressed a wish to lie there, and a simpler village church-yard, more remote and peaceful, could scarcely be found. Yet cremation, as in the case of Herbert Spencer, or of Galton's own brother, Erasmus Galton, would have seemed to his biographer a more fitting end for what must one day perish. It is with pain that I think even to-day of Francis Galton's mortal remains coffined in a vault.

earnestly, but with a saving sense of humour; he would have been of earth's elect even if he had never achieved high rank in science. It was the loss of that ever-flowing spring of understanding human sympathy that we felt most bitterly. His teaching days were already over, and his clearly stated creed would remain with us, if he himself had passed away; but the gracious friendship and the long-continued series of affectionate letters were for ever broken. There was no one left who would have the same keen and enlightened interest in all forms of biometric work, nor indeed anyone to whom a Report on the work of Galton's own Laboratory would be in future of capital importance.

From many talks with Francis Galton about the future of his Laboratory, I knew he desired the whole time and energy of a relatively young and strong man of science to carry it successfully through its infancy. It did not occur to me to think of myself as the first director of the Laboratory to be created by his testament, for I should have been wholly unwilling to give up the superintendence of the Biometric Laboratory I had founded and confine my work to Eugenics research. It was because in 1909 after much discussion we could not hit upon the really suitable man for the first Galton Professor, that Francis Galton added the codicil to his will allowing the University to delay for a few years the appointment to the chair. I only learnt after his death the clause relating to myself which, after showing the codicil to me, he had added to it, granting me the liberty, if the University were willing to elect me to the professorship, of continuing my Biometric Laboratory. He had realised I should not desert it, even to be freed from elementary teaching. This was for me a last token of affection, and the creation of an obligation which I have sought in the past nineteen years to repay to the extent of my powers. May this book in part bear witness thereof.

Fortunatus ego, cui in vestigiis ejus, tametsi graviter claudicans, spatiari conceditur !

APPENDIX I.

The Codicil to the Will of Sir Francis Galton.

The Will to which this is a Codicil was made on October 20, 1908, and the Codicil is dated May 25, 1909. In both Will and Codicil the word "faculties" replaces the "qualities" of the University Committee's definition* There is a further clerical blunder in clause (4) (*a*); the word "Professional" is a *lapsus calami* for "Professorial." The former adjective has no sense in its present situation, and I know the latter to be what Galton intended, for I drafted at his request in 1906 a statement of the duties of the proposed professor, and the adjective "professorial" was introduced in order to distinguish teaching of an academic character from that communicated in public lectures.

I DEVISE AND BEQUEATH all the residue of my estate and effects both real and personal unto the University of London for the establishment and endowment of a Professorship at the said University to be known as "The Galton Professorship of Eugenics" with a laboratory or office and library attached thereto AND I DECLARE that the duty of the Professor who for the time being shall hold the said Professorship shall be to pursue the study and further the knowledge of National Eugenics that is of the agencies under social control that may improve or impair the racial faculties of future generations physically and mentally AND for this purpose I DESIRE that the University shall out of the income of the above endowment provide the salaries of the Professor and of such assistants as the Senate may think necessary and that the Professor shall do the following acts and things namely:

(1) Collect materials bearing on Eugenics

(2) Discuss such materials and draw conclusions

(3) Form a Central Office to provide information under appropriate restrictions to private individuals and to public authorities concerning the laws of inheritance in man and to urge the conclusions as to social conduct which follow from such laws

(4) Extend the knowledge of Eugenics by all or any of the following means namely:

   (*a*) Professional instruction

   (*b*) Occasional publications

   (*c*) Occasional public lectures

   (*d*) Experimental or observational work which may throw light on Eugenic problems.

He shall also submit from time to time reports of the work done to the Authorities of the said University.

AND I DECLARE that the receipt of the Principal for the time being of the said University shall be a sufficient discharge for any moneys payable to the said University under this my Will and shall effectually exonerate my Executors from seeing to the application thereof AND I ALSO DECLARE that the said University shall be at liberty to apply either the capital or income of the said moneys for any of the purposes aforesaid but it is my hope that the University will see fit to preserve the capital thereof wholly or almost wholly intact not encroaching materially upon it for cost of building fittings or library Also that the University will supply the laboratory or office at such place as its Senate shall from time to time determine but preferably in the first instance in proximity to the Biometric Laboratory I state these hopes on the chance of their having a moral effect upon the future decisions of the Senate of the University but they are not intended to have any legally binding effect whatever upon the freedom of their action AND I HEREBY DECLARE that it shall be lawful for the Senate of the said University if they shall think fit so to do to postpone the election of the first or any subsequent Professor of Eugenics for a period of not exceeding four years from the date of my death or from the date of the occurrence of any vacancy in the office as the case may be AND I DESIRE that in the meantime and until the appointment of the first Professor the Senate shall out of and by means of the income of my residuary estate make such arrangements

* See p. 225 above.

as may be necessary to ensure the continuance without interruption and the extension of the work in connection with Eugenics initiated by me and now carried on on my behalf at University College and that during any subsequent vacancy in the Professorship the Senate shall out of and by means of the said income make such arrangements as may be necessary to ensure the continuance without interruption of the work being carried on for the time being at the Eugenics Laboratory of the said University AND I HEREBY DECLARE it to be my wish but I do not impose it as an obligation that on the appointment of the first Professor the post shall be offered to Professor Karl Pearson and on such conditions as will give him liberty to continue his Biometric Laboratory now established at University College AND in all other respects I confirm my said Will IN WITNESS whereof I have hereunto set my hand this twenty fifth day of May One thousand nine hundred and nine,

FRANCIS GALTON.

APPENDIX II.

Scheme by Sir Francis Galton for a Eugenics Discussion Committee.

I have several times in the course of this work pointed out Galton's belief in Committees. In particular I have noted that in 1905 he set up an Advisory Committee for his Eugenics Record Office. The Minute Book of this Committee indicates how little could be achieved in this manner. It really only hampered the Research Fellow (see p. 233 above). When the Eugenics Record Office was reorganised as the Eugenics Laboratory, there was again an "Advisory Committee," but it was not to be and never was summoned; it consisted of experts in various fields, who were individually consulted when our work led us in the direction of one or another branch of science. These experts were of much service, and we were very grateful for their aid; but there were no periodic meetings designed to discuss what the future work of the Laboratory ought to be; or any excuse for much talk by those who were ignorant of the difficulty of collecting data, or what it was possible to deduce from them when obtained.

Advisory Meetings at the Eugenics Record Office, 88, Gower Street, W.C.

Mr Galton would be glad to utilise the room of the Eugenics Record Office, after office hours, for the occasional meeting of a few invited persons who seriously desire to promote *Inductive Research* in matters connected with Eugenics.

In Mr Galton's absence, Mr Schuster would act as host.

Under these conditions, Mr Schuster would arrange the day of each meeting, in conference with Mr Branford and Dr Slaughter, and the hour of opening and closing it.

He would similarly arrange as to the persons to whom invitations should be sent on each occasion, bearing in mind that exigencies of space make it inconvenient for more than eight persons to be present at the same time.

He would also draw up the Agenda, a copy of which will accompany each invitation.

The meetings will be somewhat informal, but its members may proceed to elect a chairman for the evening if any two of those present desire it. Whenever the votes including that of the chairman are equal, the host shall have a second and casting vote.

Minutes of each meeting shall be kept by Mr Schuster, who shall cause them to be typed in duplicate, one copy to be retained by the Office.

The Secretary, Miss Elderton, will do all necessary typewriting and posting.

The primary purpose of the meeting will be to propose and thoroughly discuss suitable subjects for eugenic research, including time, cost, the persons who might undertake them, and the value of the expected results. Definite proposals of this kind should take precedence in the Agenda.

Other topics connected with Eugenics might afterwards be discussed, preference being given to those that bear on the future work of the Office.

(signed) FRANCIS GALTON, *October*, 1905.

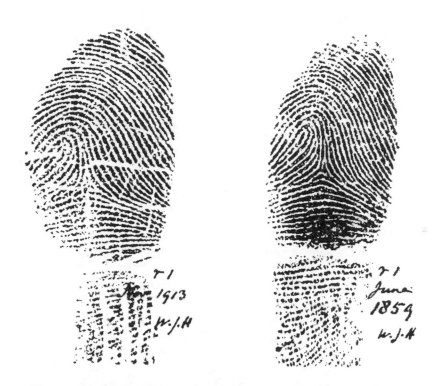

Finger-prints of Sir William J. Herschel's right forefinger at 54 years'
interval, the longest known proof of persistence. The 1913 print
shows the creases which develop with old age. Cf. p. 142 above.

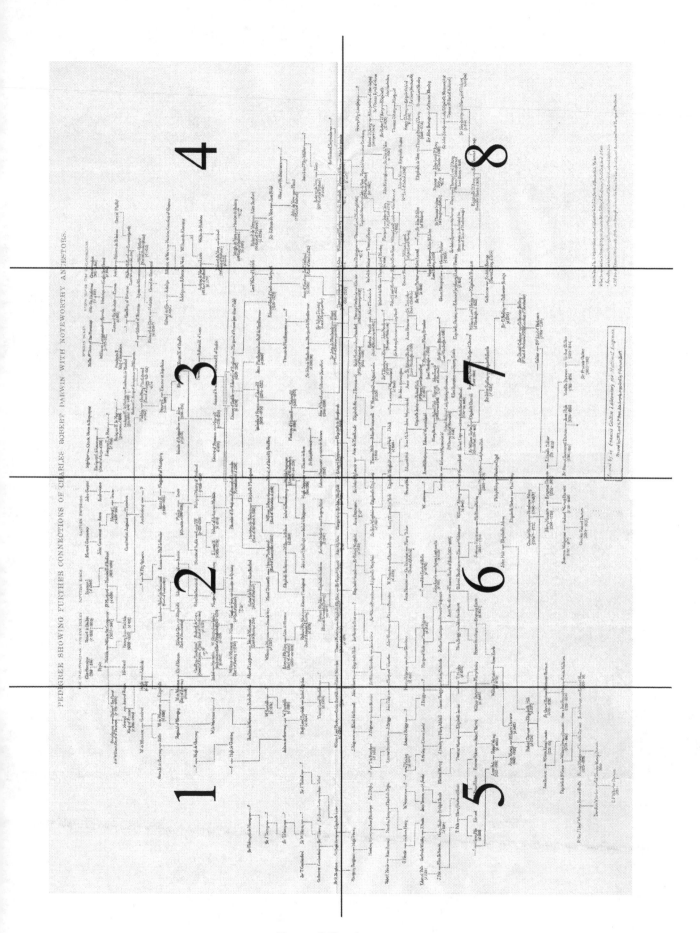

PEDIGREE SHOWING FURTHER CONNECTIONS OF CHARLES ROBERT DARWIN WITH NOTEWORTHY ANCESTORS.

Key to following pages

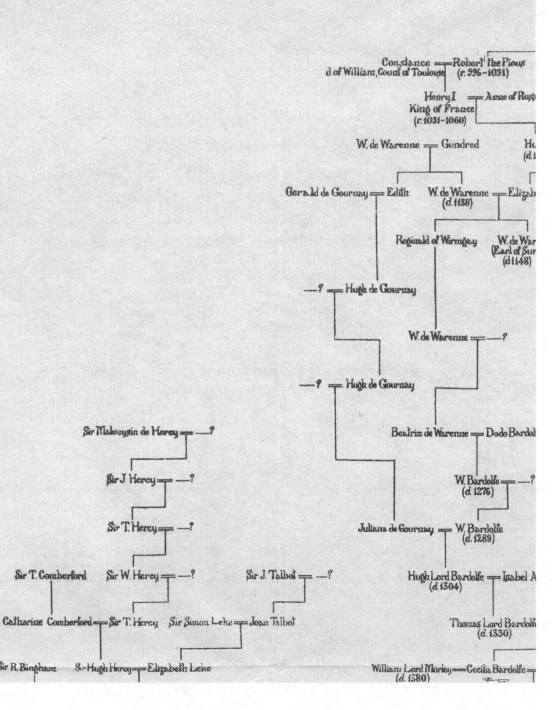

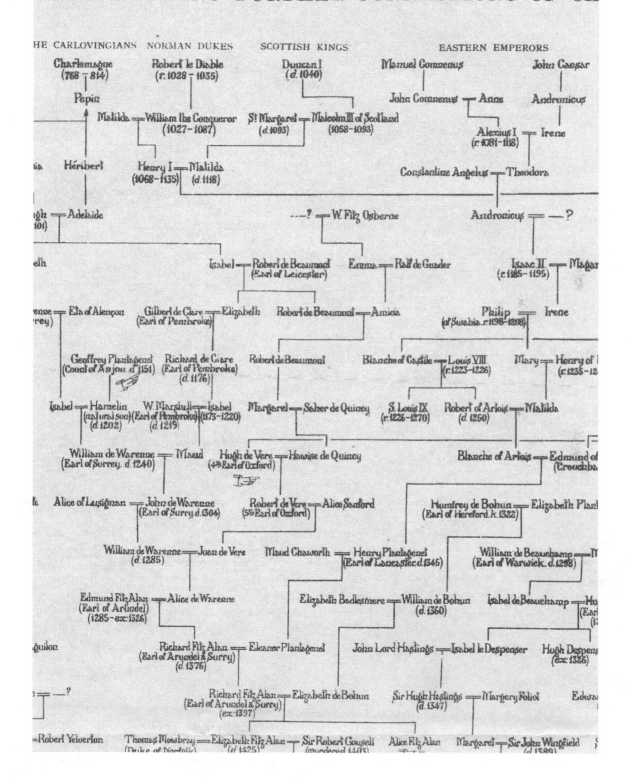

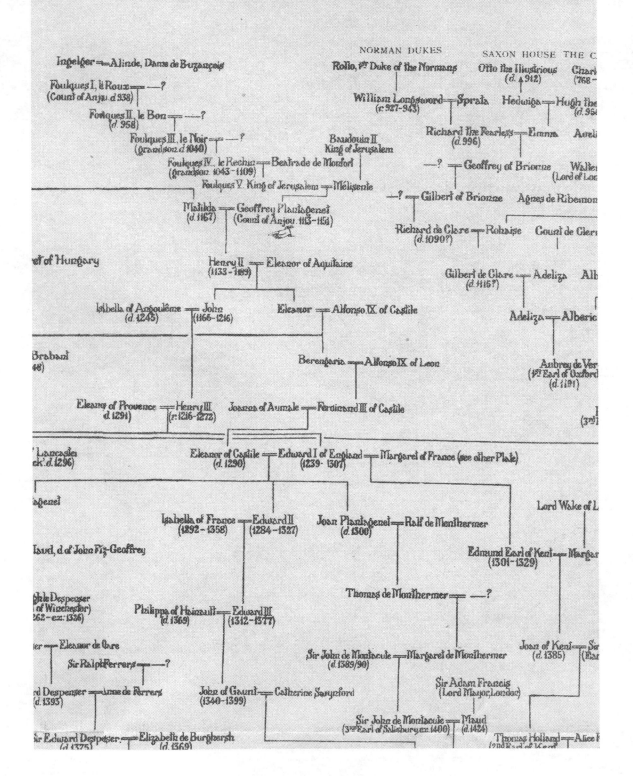

ARLOVINGIANS

emagne
814)

Great
6)

ne ——— Osborn de Bolobec   Gerald Flaitel

Giffard ——— Ermengarde
gueuille c.1060

——— Walter Giffard
(Earl of Buckingham)
(d. 1102)

oni

orie de Vere ——— Beatrix, Countess of Ghisnes

de Vere   William de Abrincis

—— Lucia   Walter de Bolebec

Robert de Vere ——— Isabel
Earl of Oxford)
(d. 1221)

Hugh de Vere ——— Hawise de Quincy
(4th Earl of Oxford)
(d. 1263)

iddell   Robert de Vere ——— Alice Saxford
(5th Earl of Oxford)
(d. 1296)

ot   Sir Alfonso de Vere ——— Jane Foliot

Giles, Lord Badlesmere ——— ?

John de Vere ——— Maud
(7th Earl of Oxford)

John Lord Fitz-Walter ——— ?

T. Holland
d of Kent d. 1360)   Aubrey de Vere ——— Alice
(10th Earl of Oxford)
(d. 1400)

Sir Richard Serjeaulx ——— ?

tz Alan   William Lord Morley ——— Cecilia Bardolfe   Richard de Vere ——— Alice Serjeaulx

4

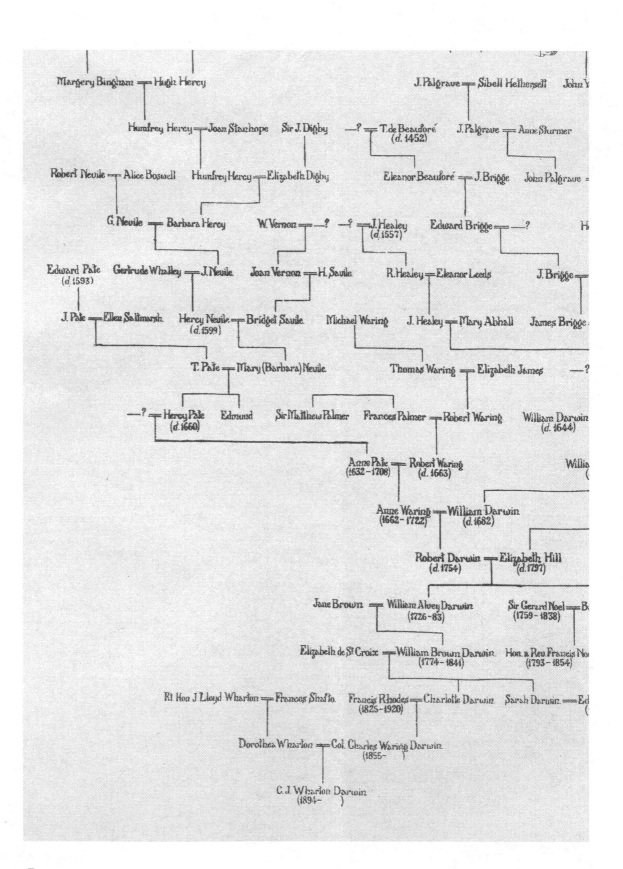

5

Yelverton — Elizabeth Rede   Sir Oliver le Gros — ?   Elizabeth Gousell — Sir Robert Wingfield (d 1431)   Anne Despenser   Sir John F

Sir William Yelverton — Joan le Gros   Sir William Brandon — Elizabeth Wingfield   Sir John Wingfield (d 1481) — Elizabeth Fitz Le

Margaret Yelverton   John Glemham — Eleanor Brandon   W. Brandon (d.1485) — Eleanor le Bruyn   Henry VII — El: of York   Elizabe

Henry Palgrave (d 1517) — Anne Glemham   Anne Browne — Ch: Brandon (Duke of Suffolk) — Mary Tudor   Francis Hal

? — Margaret Rede — Clement Palgrave (d. 1583)   ? — John Earle of Salle (d. 1570)   W. Leake — ?

Mary Stuteville   Arthur Fountayne — Frances Palgrave   John Earle (d 1611) — Agnes Locksmith (d 1579)   Joan Le

T. Healey (d. 1652)   Mary Brigge — John Fountayne   Anne Fountaine — Thomas Earle of Salle (1563-1605)   Gabriel Rawlinson — Eleanor Hutchinson   William Alvey (d. 1649) — Fran

Mary Healey   Frances Fountayne — Erasmus Earle (d.1667)   Susanna Rawlinson (1644-1724) — Matthew Alvey   William 5th L (1609-1

Darwin (d.1675) — Anne Earle

John Hill — Elizabeth Alvey   Philip Foley —

Elizabeth Turton — Paul Fo

Charles Howard (1706/7- 1771) — Penelope Foley (1708-1748/9)

Baroness Barham   Mary Howard (1740-1770) — Erasmus Da (1731-1802

— Cecilia Methuen   Susanna Wedgwood (1765- 1817) — Robert Waring Darwin (1766-1848)

ward Noel (1825-99)   Charles Robert Darwin (1809-1882)

6

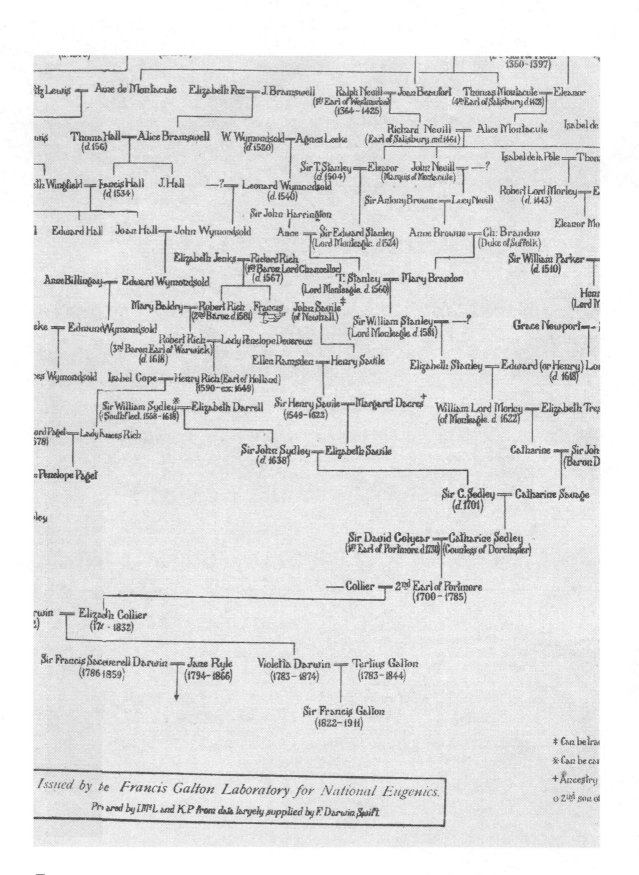

7

(d. 1380) (11th Earl of Oxford) (d. 1417)

Anne Despenser = Thomas Lord Morley (d. 1418)

Elizabeth Howard = John de Vere Robert de Vere = Joan Courtenay
                    (12th Earl of Oxford)
                    (exc. 1461)

Henry Fitz-Langley = ?

... Molins = Thomas Morley

Robert D'Arcy = Alice (widow of John Ingoe)
(Lawyer's Clerk)
                 Sir Thomas Tyrell of Heron

... Lord Morley (d. 1436)

Margaret = John de Vere
           s.p. (13th Earl d. 1513)

Alice Ki Irington = Sir John de Vere (d. 1450)

Sir Robert D'Arcy = Elizabeth
(fl. 1420)
                    John Harleston

...izabeth de Roos William Lord Lovell = Alice Deincourt (d. 1455)

John de Vere = Elizabeth Trussel
(15th Earl of Oxford d. 1540)

Thomas D'Arcy = Margaret

...ley = William Lovell (styled Lord Morley)

Roger D'Arcy = Elizabeth, (d of Sir Henry Wentworth)
(d. 1508)

Alice Lovell   ? = Sir John St John (of Bletsoe)

Elizabeth de Vere = Thomas Baron D'Arcy (1506-1558)

Thomas Lord Stanley

...y Parker = Alice St John
...orley d. 1455)

Sir John Savage = Catharine Stanley

...Sir Henry Parker (d. 1553)

Sir Thomas Kytson (of Hengrave, Suffolk)

Frances = John Lord D'Arcy
Rich   (of Chiche d. 1580)

Sir John Savage = Lady Elizabeth Manners, (d of Thomas 1st Earl of Rutland)

Sir John Spencer = Katherine   Mary = Thomas Lord D'Arcy
...d Morley   (from whom is descended the   (Viscount Colchester and
                      present Duke of Marlborough)   Earl Rivers. d. 1639)

Sir John Savage = Mary, (d of Richard Allington)
(d. 1611)

...ham

Elizabeth D'Arcy = Sir Thomas Savage
(Countess Rivers. d. 1650)

... Savage
...rcy. d. 1654)

...ed back 11 generations without distinction to Sir John Savile of Savile Hall, Yorks.

...rried back through 3 generations to John Selley of Scadbury in Southfleet. Kent. d. 1500

...can be traced through Dacres of Cheshunt to Dacres of Westmoreland, without distinction

...? Thomas Wentworth, 1st Baron, through whom in the female line there is direct descent from John Nevill Marquis of Montacute

8

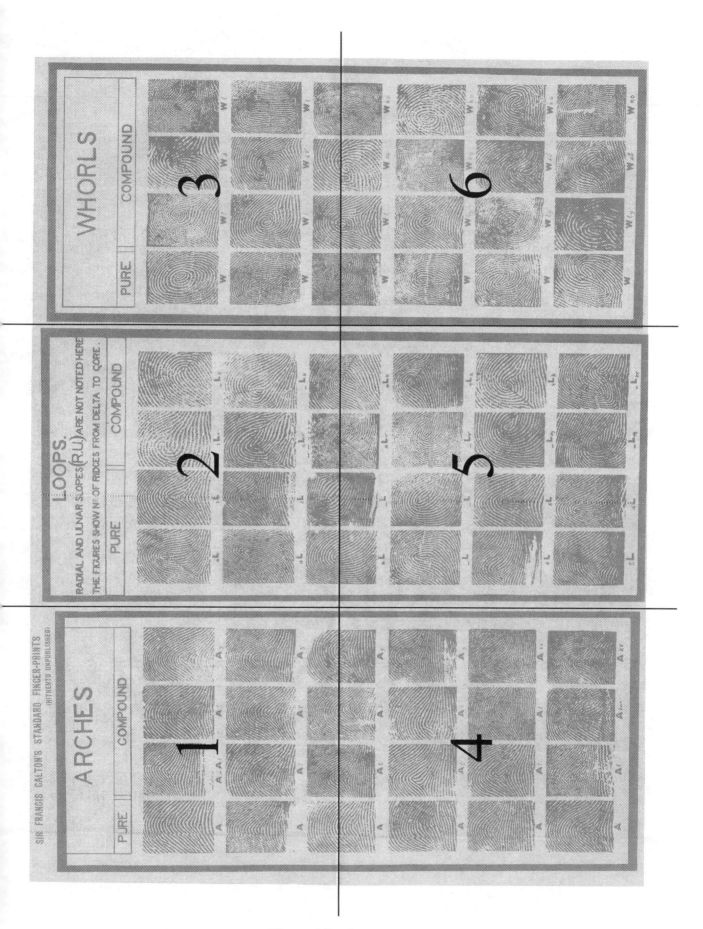

Key to following pages.

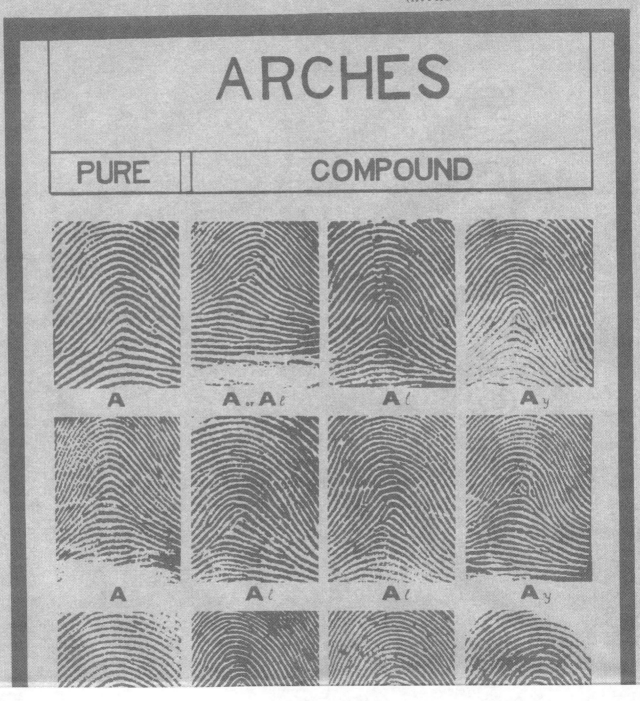

LOOPS.

RADIAL AND ULNAR SLOPES (R.U.) ARE NOT NOTED HERE
THE FIGURES SHOW Nº OF RIDGES FROM DELTA TO CORE.

PURE	COMPOUND

LOOPS.

RADIAL AND ULNAR SLOPES (R.U.) ARE NOT NOTED HERE

THE FIGURES SHOW Nº OF RIDGES FROM DELTA TO CORE.

PURE	COMPOUND

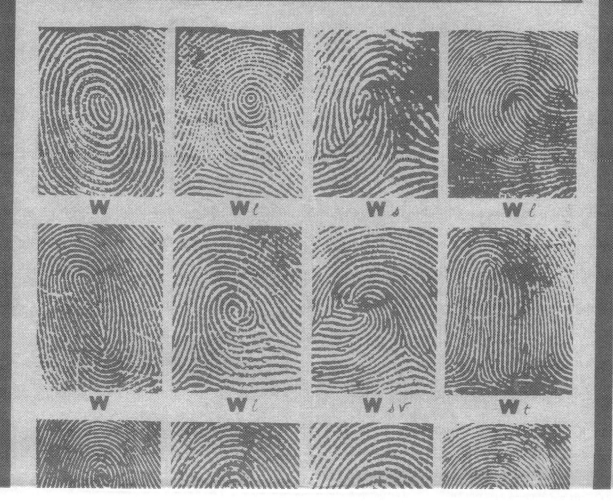

WHORLS

PURE	COMPOUND

W   W *l*   W *s*   W *l*

W   W *l*   W *sv*   W *t*

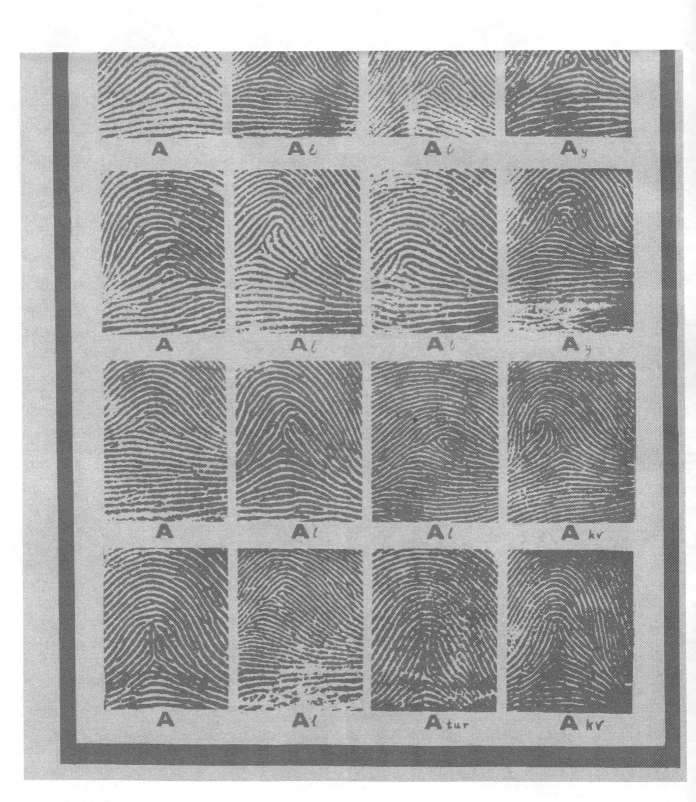

4

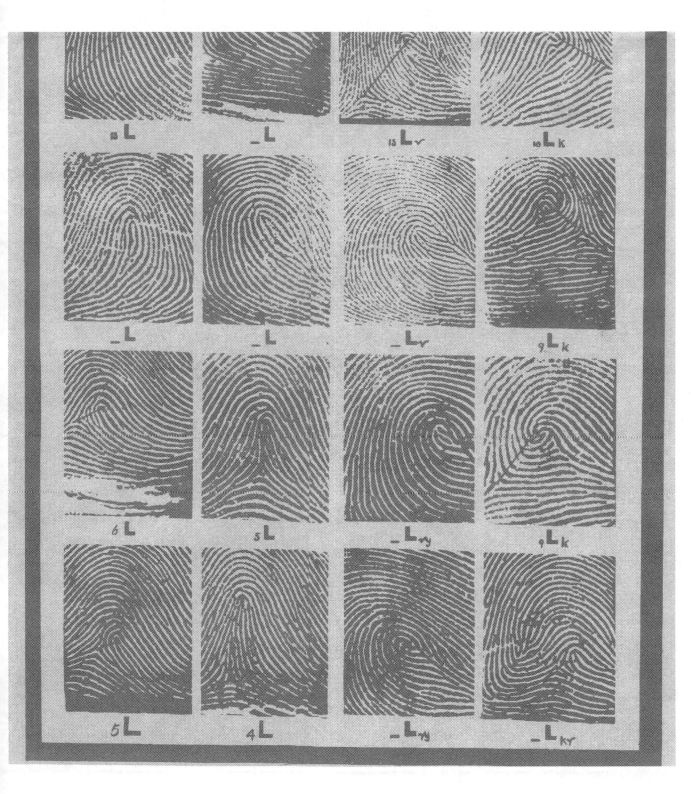

5

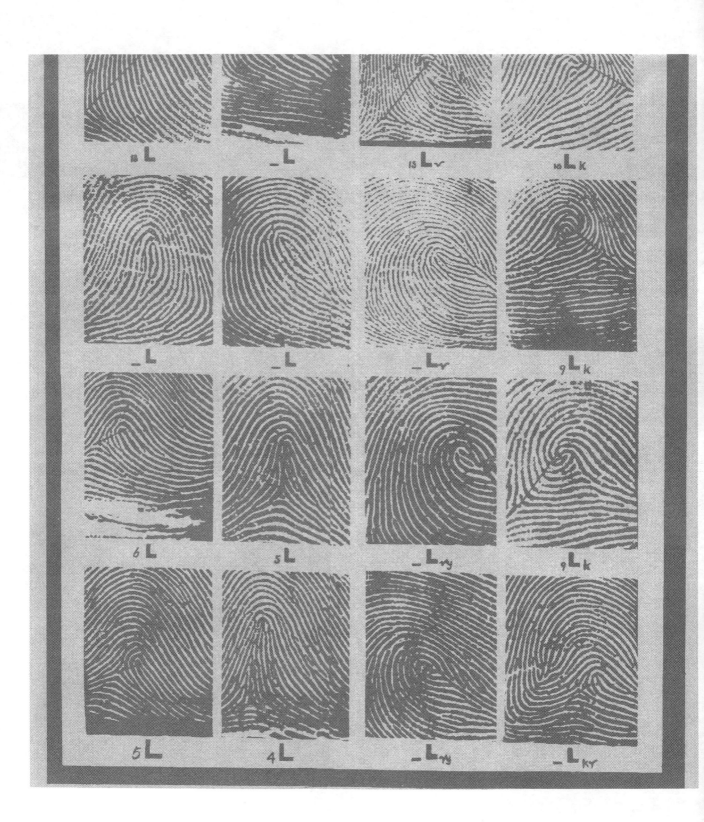

5

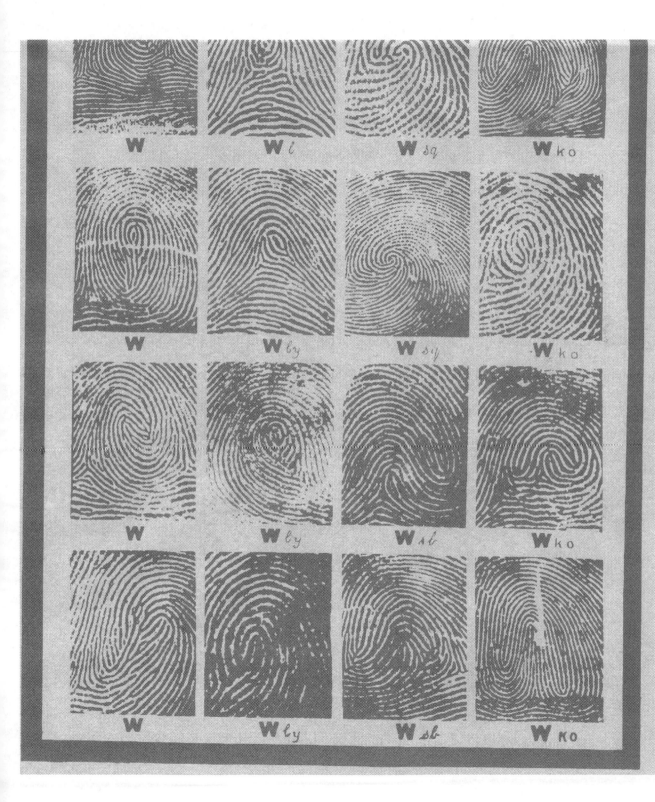

Printed in the United States
By Bookmasters